Quantum Mech

Scientific Perspectives on D

Inaugural Volume – 1988
Physics, Philosophy, and Theology:
A Common Quest for Understanding
Edited by Robert John Russell, William R. Stoeger, S.J., and George V. Coyne, S.J.

Published by the Vatican Observatory in commemoration of the 300th anniversary of the publication of Isaac Newton's *Philosophia Naturalis Principia Mathematica*

A Series on "Scientific Perspectives on Divine Action"
Robert John Russell, General Editor

First Volume – 1993 (Revised Edition 1996)
Quantum Cosmology and the Laws of Nature:
Scientific Perspectives on Divine Action
Edited by Robert John Russell, Nancey Murphy, and C.J. Isham

Second Volume – 1995
Chaos and Complexity:
Scientific Perspectives on Divine Action
Edited by Robert John Russell, Nancey Murphy, and Arthur Peacocke

Third Volume – 1998
Evolutionary and Molecular Biology:
Scientific Perspectives on Divine Action
Edited by Robert John Russell, William R. Stoeger, S.J., and Francisco J. Ayala

Fourth Volume – 1999
Neuroscience and the Person:
Scientific Perspectives on Divine Action
Edited by Robert John Russell, Nancey Murphy, Theo C. Meyering, and Michael A. Arbib

Fifth Volume – 2001
Quantum Mechanics:
Scientific Perspectives on Divine Action
Edited by Robert John Russell, Philip Clayton, Kirk Wegter-McNelly, and John Polkinghorne

Jointly published by the Vatican Observatory and the Center for Theology and the Natural Sciences

Supported in part by a grant from the Wayne and Gladys Valley Foundation

Quantum Mechanics

Scientific Perspectives on Divine Action

Robert John Russell
Philip Clayton
Kirk Wegter-McNelly
John Polkinghorne

Editors

Vatican Observatory
Publications
Vatican City State

Center for Theology and
the Natural Sciences
Berkeley, California

2001

Robert John Russell (General Editor) is Professor of Theology and Science in Residence at the Graduate Theological Union, and Founder and Director of the Center for Theology and the Natural Sciences, in Berkeley, California, USA.

Philip Clayton is Professor and Chair of the Philosophy Department at the California State University (Sonoma), USA.

Kirk Wegter-McNelly is a doctoral candidate at the Graduate Theological Union in Berkeley, California, USA.

John Polkinghorne is Past President and now Fellow, Queens' College, Cambridge, and Canon Theologian of Liverpool, England.

Jointly published by the Vatican Observatory and
the Center for Theology and the Natural Sciences

Distributed (except in Italy and the Vatican City State) by
The University of Notre Dame Press
Notre Dame, Indiana 46556
U.S.A.

Distributed in Italy and the Vatican City State by
Libreria Editrice Vaticana
V-00120 Citta del Vaticano
Vatican City State

ISBN 0-268-03978-X (pbk.)

Cover – A quantum consideration of identical particles, such as electrons, leads to the famous Pauli *exclusion principle*, according to which two or more bound electrons can never be in the same total quantum state, i.e., never in the same space *and* the same spin state. As a consequence of the exclusion principle, the total wavefunction for two electrons in quantum states α and β must be antisymmetrical (i.e., the sign of the wavefunction is inverted when the particle labels, 1 and 2, are exchanged) so that ψ is zero if α equals β: $\psi_{\text{total}} = 1/\sqrt{2}[\psi_\alpha(1)\psi_\beta(2) - \psi_\beta(1)\psi_\alpha(2)]$. The figure on the cover depicts the space wavefunction along the x-axis for two electrons in an infinite square well potential of width a, with the first electron (x_1) in the ground state and the second electron (x_2) in the first excited state (e.g., a crude model of a Helium atom). Because the *total* wavefunction must be expressed using a minus sign as above, the *space* wavefunction must also have a minus sign if the electrons' spins are parallel: $\psi_{\text{space}} = 1/\sqrt{2}[\psi_a(x_1)\psi_b(x_2) - \psi_b(x_1)\psi_a(x_2)]$. This leads to a quantum phenomenon that has no classical analogue called the *exchange force*, causing electrons with parallel spins to move as if under a repellant force (i.e., the space wavefunction approaches zero near $x_1 = x_2$, thus the probability of finding the electrons close to one another vanishes). In the case of anti-parallel spins, the "force" becomes attractive. For details see Robert Eisberg and Robert Resnick, *Quantum Physics of Atoms, Molecules, Solids, Nuclei, and Particles* (New York: Wiley and Sons, 1974), 334–46.

ACKNOWLEDGMENTS

The editors wish to express their gratitude to the Vatican Observatory and the Center for Theology and the Natural Sciences for co-sponsoring this research. Particular appreciation goes to George Coyne and William Stoeger, whose leadership and vision made this series of conferences possible.

Editing for this volume began with an initial circulation of papers for critical responses and a pre-conference at the Imperial College at Wye of the University of London (September, 1999) before the main conference at the Vatican Observatory in Castel Gandolfo, Italy (July, 2000), and continued with interactions between editors and authors after the main conference. The editors express their gratitude to all the participants for their numerous written responses to pre-conference drafts and for their enthusiastic discussions during the conferences.

Special thanks go George Coyne for organizing the main conference and for overseeing printing and distribution, and to Chris Isham for organizing the preconference. The editors wish to express their appreciation to Kevin Koczela for his help in producing the cover figure. On behalf of all the participants, the editors also thank staff members at CTNS, the Vatican Observatory, and the Imperial College at Wye.

The editors thank Abner Shimony and the publisher for permission to reprint the article, "The Reality of the Quantum World," *Scientific American*, January 1988.

CONTENTS

CONTENTS

INTRODUCTION

Robert John Russell

1 Background to the Volume

In July, 2000, twenty-three scholars, with cross-disciplinary expertise in physics, astronomy, cosmology, mathematics, philosophy of religion, philosophy of science, history of science, philosophical and systematic theology, and theological ethics, met for a week-long conference at the Vatican Observatory (VO) located in Castel Gandolfo, Italy. The purpose of the conference was to explore the implications of quantum mechanics for philosophical and systematic theology with a particular focus on divine action, and to explore the philosophical and theological elements in quantum physics. The resulting essays form the contents of this volume.

The conference is the last in a series of five such research conferences planned for the decade of the 1990s. The overarching goal of these conferences has been to engage theology, philosophy, and natural science in a process of constructive and creative mutual interaction. The conferences have been jointly sponsored by the Observatory and the Center for Theology and the Natural Sciences (CTNS) located in Berkeley, California. The Observatory is housed within the Papal Palace in the picturesque town of Castel Gandolfo, overlooking Lake Albano thirty miles southeast of Rome. Since 1935 it has been the site of basic research in both observational and theoretical astronomy. It is also here that Pope John Paul II resides during the summer months. CTNS is an affiliate of the Graduate Theological Union in Berkeley, a consortium of nine Roman Catholic and Protestant seminaries located just north of the campus of the University of California, Berkeley. CTNS sponsors and conducts a variety of local and international research, teaching, and public service programs in the interdisciplinary field of theology and science.

This series of conferences grew out of the initiative of Pope John Paul II, who commissioned the Observatory to hold an initial major international conference at the Observatory in September, 1987. The resulting publication, *Physics, Philosophy and Theology* (*PPT*), includes eighteen essays as well as a message by the Pope on the relations between the church and the scientific communities.[1] Based on this work, George Coyne, Director of the Observatory, initiated this series of conferences. A long-term steering committee, including Coyne, Bill Stoeger, Nancey Murphy, and myself, set the overall agenda: to expand upon the research begun in 1987 by moving further into areas in the physical and biological sciences as well as into new areas such as the neuro- and cognitive sciences. Coyne then invited CTNS to co-sponsor this decade of research conferences and publications with the Observatory. The first conference focused on quantum cosmology and the origin of the laws of nature. It built on the initial exploration of cosmology in *PPT* and included such issues as $t = 0$ and the Anthropic Principle.[2] Next came an examination of the

---

[1] Pope John Paul II, "Message of His Holiness Pope John Paul II," in *Physics, Philosophy, and Theology: A Common Quest for Understanding*, ed. Robert J. Russell, William R. Stoeger, S.J. and George V. Coyne, S.J. (Vatican City State: Vatican Observatory Publications, 1988), 15-17.

[2] Robert J. Russell, Nancey C. Murphy, and Chris J. Isham, eds., *Quantum Cosmology and the Laws of Nature: Scientific Perspectives on Divine Action*, 2d ed. (Vatican City State;

sciences of chaos and complexity, followed by evolutionary and molecular biology, both of which greatly expanded the scope of research presented in *PPT*.[3] The fourth conference focused on the neurosciences, engaging mostly new scientific topics in the series.[4] Now, with the fifth conference and volume, we return to one of the central themes of *PPT*: quantum mechanics. The first and third conferences were held at Castel Gandolfo in 1991 and 1996; the second was held in Berkeley in 1993. For the fourth conference we gathered at Pasierbiec, Poland, at the invitation of Michael Heller and the Pontifical Academy of Theology in Cracow. We returned to beloved Castel Gandolfo for the quantum mechanics conference in 2000.

2 Guiding Theme of the Series of Conferences: Scientific Perspectives on Divine Action

A major issue in the way research is carried out in the field of theology and science regards the role science ought to play. Too often science tends to set the agenda for the research with little if any initiative taken by theology. From the beginning it was the clear intention of the steering committee that our research expand beyond this format to insure a two-way interaction between scientific and theological research programs. In order to achieve this goal we decided on a two-fold strategy. First, we searched for an overarching theological topic to thematize the entire series of conferences. The topic of divine action, or God's action in and interaction with the world, was quickly singled out as a promising candidate. Clearly it permeates the discussions of theology and science in both philosophical and systematic contexts, and it allows a variety of particular theological and philosophical issues to be pursued under a general umbrella.

Since the topic of God's action in the world was chosen as the guiding theological theme for the conferences, a brief introduction to the topic was provided in the volume on chaos and complexity. The introduction included a working typology of terms used in the ongoing discussions of divine action by scholars at this series of conferences. Since many of these terms also appear in the present volume, we will reprint the key section here for convenient reference, including figure 1.

2.1 Working Typology of Theological Positions on Divine Action in Light of Science[5]

The following typology (see figure 1) presents a correlation of various theological views with the types of claims their proponents tend to make about divine action, including new developments in light of science. It is meant primarily as a framework to guide the reader and is presented here as work in progress. Its heuristic nature should be underscored, as well as the fact that, like any typology, it represents an

---

Berkeley, Calif.: Vatican Observatory; Center for Theology and the Natural Sciences, 1996).

[3] Robert J. Russell, Nancey C. Murphy, and Arthur Peacocke, eds., *Chaos and Complexity: Scientific Perspectives on Divine Action* (Vatican City State; Berkeley, Calif.: Vatican Observatory; Center for Theology and the Natural Sciences, 1995); Robert J. Russell, William R. Stoeger, S.J., and Francisco J. Ayala, eds., *Evolutionary and Molecular Biology: Scientific Perspectives on Divine Action* (Vatican City State; Berkeley, Calif.: Vatican Observatory; Center for Theology and the Natural Sciences, 1998).

[4] Robert J. Russell, Nancey C. Murphy, Theo C. Meyering, and Michael A. Arbib, eds., *Neuroscience and the Person: Scientific Perspectives on Divine Action* (Vatican City State; Berkeley, Calif.: Vatican Observatory; Center for Theology and the Natural Sciences, 1999).

[5] Adapted from "Introduction," in *Chaos and Complexity*, sec. 3.4, pp. 9–13.

abstraction from the extant literature; moreover, authors may take differing positions at various places in their writings.[6]

The upper portion of the table includes five positions frequently found in historical and contemporary theology regarding divine action. It serves as a backdrop to the current discussions in theology and science, as found in the lower portion of the table. For completeness, the table begins with the possibility that no events are God's acts; this is obviously the atheistic position that denies God's very existence. The second position is that God acts only "in the beginning,"[7] creating the universe and the laws of nature. After this, the universe runs entirely on its own according to these laws. This is the deist option. If science could show that there was no beginning, the deist view would be seriously undermined.

According to a third view, it is the existence of the universe at each moment of time, and not primarily its beginning, that requires an explanation in terms of God's action. Thus, God acts not only in the beginning to create the universe; God acts uniformly in all events throughout the history of the universe to sustain its existence moment by moment. This means that without God's continuous action the universe would simply cease to exist, since matter does not contain its own sufficient principle of existence. Whether the laws of nature prescribe or merely describe the regularities of natural processes (i.e., whether causal efficacy is contained in the laws governing nature or in nature itself), God is required as the ground of natural (secondary) causality and the primary cause of every event in the universe. Still, on this view, no events are *special* acts of God in any sense. We call this view "garden-variety theism," although pantheism could also fall into this category.

A fourth view is held by those who agree that God acts both in the beginning and uniformly in all events. However, they also claim that certain events in nature and history can be said "subjectively" to be special acts of God: they are *seen as* revealing something special about the character or intentions of God even though God does not really act differently in an *objective* way in these events. We call this the "liberal" view. Recently, a much more dynamic conception of the universe has come to prevail over the older, static cosmology. God's action in sustaining the universe has come to be seen as God's "continuous creation" of the universe. Those who view the universe in these more dynamic terms and speak about continuous creation are often eager to attribute special significance to what appears to be the occasional appearance of genuine novelty, even if in the end they admit that all events are in fact uniformly caused by God through the unfolding realizations of the potentialities of nature represented by the laws of nature.

Finally, it is possible to hold that there are events that in some *objective* sense are special acts of God. When we call these events "special" we do so as an acknowledgment of what God is doing in a special way to bring them about. Such divine action has most often taken the form of *intervention*: God performs such acts by intervening in or suspending the laws of nature. We will call this the "traditional" or conservative view. Note that the traditional view *includes* the theist's claim that the existence of the universe—both its continued sustaining in being and its beginning—is a product of God's action.

---

[6] Since its inception in 1991, the typology has been developed further through discussion with a number of scholars including Nancey Murphy and Thomas Tracy.

[7] Assuming there was one, such as Big Bang cosmology supports! For a detailed discussion of the scientific, philosophical, and theological complexity of the assumption, see *Quantum Cosmology and the Laws of Nature*.

Figure 1. Working Typology of Divine Action

	Creation	Uniform Divine Action		Objectively Special Divine Action		
		Sustenance	Subjectively Special	Interventionist	Noninterventionist	
					Apparent with religious presuppositions	Apparent without religious presuppositions
Atheism						
Deism	X					
'Theism'	X	X				
'Liberal'	X	X	X			
'Traditional'	X	X	X	X		
Theology and Science:						
(1) Top-Down or Whole-Part (e.g., mind/brain)	X	X	X		X	?
(2) Bottom-Up (e.g., quantum indeterminacy)	X	X	X		X	?
(3) Lateral Amplification (e.g., chaos)	X	X	X		X	?
(4) Primary/Secondary	X	X	X		?	?

Much of the *current* discussion in the field of theology and science regarding divine action now turns on the question of whether there are *objectively* special divine acts that are *neither* interventions nor suspensions of the laws of nature. We will call these "noninterventionist" views. Among noninterventionists, it is still an open question whether these events are objectively perceptible as special (i.e., whether they would be seen as God's acts by anyone present regardless of their prior religious presuppositions, or whether it is precisely these presuppositions which allow observers to recognize that God has objectively acted in a special way in the event in question).

Among those who opt for objectively special divine action, some authors distinguish between direct and indirect divine acts, or equivalently between basic and nonbasic divine acts. The argument here is that if God acts indirectly through secondary causes to achieve an objectively special act, there has to be at least one direct or basic act somewhere that initiates the chain of events, the outcome of which we call God's indirect, objectively special act. This is a logical point, and not a theological point: if an agent is to do something indirectly, at some point the agent has to do something directly which eventuates in the indirect act. This issue is related to the problem of the "causal joint."

We should note that it is possible to affirm objectively special divine acts without deciding whether these acts are themselves direct acts or are the indirect result of other, perhaps hidden, direct acts of God. On the one hand, the special event in question may be seen as the direct act of God. On the other hand, most of the current arguments tend to see the events we call "special" as indirect acts of God (i.e., as the product of secondary causes stemming from direct acts of God located elsewhere). But if this is the case, then the question in turn becomes: where is this "elsewhere," this real domain of God's direct or basic action?

In surveying the continuing conversations with scholars in theology and science, it has become clear that there are four distinct *noninterventionist* approaches to this question, though combinations of them are also viable: i) "top-down" or "whole-part" causality; ii) "bottom-up" causality; iii) what we might call "lateral" causality, and iv) "primary/secondary" causality. First of all, then, some describe divine action in terms of "top-down" or "whole-part" causality. Here, a localized, special event in the world is viewed as the indirect result of God acting directly in one of two ways: either in a top-down way from a higher level in nature (using such analogies as "mind/brain"), or in a whole-part way starting either at the physical boundaries or environment of the system (an analogy here is the formation of vortices in a liquid heated in a container), or, ultimately, at the boundary of the universe as a whole. The second, or "bottom-up," approach views a special event in the macroscopic world as the indirect result of a direct act of God at the quantum mechanical level, amplified by a stream of secondary causes linked in a bottom-up way. This view presupposes that quantum uncertainty can be given an indeterministic ontological interpretation, while recognizing that other interpretations are also possible.

Some authors in theology and science have pointed to a third noninterventionist option. They stress the "supple" and "subtle" nature of chaotic and complex systems in physics, meteorology, biology, and so on (the analogy here is the "butterfly effect"). Since this view is entirely restricted to the classical level, we can call this view "lateral" causality. Perhaps, then, chaos and complexity, or at least new holistic laws of nature, might lead to new insights into divine action. Like quantum mechanics, this approach requires that chaos theory (or such new holistic laws as may be found) can be given an indeterministic interpretation.

Fourth, some authors are committed to accounts of divine action that work strictly with the distinction between primary and secondary causality. These authors see no need to speak in terms of objectively special divine acts, with the possible exception of miracles.

3 Why Quantum Mechanics and Divine Action?

Among the many facets of the Western intellectual landscape inherited from the rise of modern science in the sixteenth and seventeenth centuries, the "received view" is of nature as a machine. According to classical physics, the universe is governed by a set of deterministic equations formulated by Newton and building on the work of Galileo, Kepler, and others. By the eighteenth century, scientists were convinced that these equations would allow the exact prediction of the future state of the world given precise knowledge of its present state and all relevant forces. As Pierre Simon Marquis de Laplace put it succinctly, to one of sufficient intelligence, "nothing would be uncertain and the future, as the past, would be present to its eyes."[8] In such a causally closed world, there would be little need or possibility for God to act in the special ways and on those particular occasions as the biblical witness had abundantly recounted—unless God intervened in natural processes.

Now with the rise of quantum mechanics early in the twentieth century, a fundamental rethinking is taking place regarding our view of nature. During the period of 1900–1930, a variety of new theoretical approaches were produced in light of the astonishing experimental data arising from atomic physics, including the quantization of energy and angular momentum, the Pauli exclusion principle, and the Heisenberg uncertainty principle, and such bold theoretical formulations as the Bohr model of the atom, wave mechanics, matrix mechanics, and (nonrelativistic) quantum mechanics. In the process, fundamental physical concepts inherited from classical mechanics underwent a radical revision, including our notion of the state of a system, its motion in space and time, the causes of its motion, its composition in terms of constituent parts, the relation between the observer of the system and the system, and the relation between parts of a system widely separated in space and time.

By the end of this period, quantum mechanics had emerged in the form that is still used today, and it has been stunningly successful in predicting and explaining phenomena over a wide range of scales from the subatomic to the macroscopic. Nevertheless, the philosophical issues raised in attempting to give quantum mechanics a satisfactory interpretation have been debated from its inception to the present. Can the wave function, the statistical data resulting from measurement, and the properties of separated parts of a system, be given a realist interpretation as classical science had routinely allowed, even if it should be a highly revised form of realism, or must we settle for a positivist or an instrumentalist interpretation—or even an anti-realist interpretation? If we work within the Copenhagen school, as developed by Niels Bohr, Werner Heisenberg, and other early physicists, does wave-particle duality, or more generally, quantum complementarity, signal the end of a "particle and causal trajectory" view of elementary processes? Does the Heisenberg uncertainty principle point to ontological indeterminism and thus pose a challenge to the closed causal world of classical physics, or is it merely an epistemic feature of quantum statistics, leaving classical causality relatively untouched? And why do determinate results emerge when a measurement is taken on a quantum system if

[8] Pierre Simon Marquis de Laplace, *A Philosophical Essay on Probabilities*, 6th ed., trans. F.W. Truscott and F.L. Emory (New York: Dover, 1961), 4.

such systems take on indefinite states between measurements? More generally, if quantum mechanics is the correct fundamental theory, in what ways is classical mechanics its "limiting case"? And can a historical perspective on the move from formalism to ontology shed light on the problems for such a move here?

Alternatively, given all the issues raised by quantum mechanics, could it be an incorrect theory after all, or is it a correct but incomplete theory, as Einstein believed. And if so, should we modify it somehow? Perhaps we should reinterpret quantum mechanics as offering a deterministic account of physical processes, following David Bohm who replaced the ontology of the Copenhagen school by a "pilot-wave" and "quantum potential" approach. Or perhaps we should view quantum processes from Hugh Everett's "many-worlds" perspective (and its more current and radical "many-minds" formulations), where all quantum states are realized in an infinitely bifurcating world. But there are still other approaches to the question of how determinate physical states arise during measurement out of indeterminate quantum processes. With John von Neumann we might invoke consciousness as essential to this process or, following Werner Heisenberg, we might recall Aristotle's distinction between potential and actual reality—or we could construct a "consistent-histories" interpretation of quantum mechanics as Robert Griffiths suggests in order to account for the discrete results of measurement. Even more radically, we might through out classical Boolean logic and, with David Finkelstein, search for a new "quantum logic" to account for the strange features of the quantum world. What implications will these interpretations have for our understanding of the world as a whole and our relation to it? The complexity of such questions has deepened enormously as a result of the mathematical discovery and experimental testing of crucial theorems discovered by John S. Bell in the 1960s. For example, do the delicate (nonlocal) correlations between distantly separated parts of a system as described by Bell's theorem suggest the presence of superluminal processes in violation of special relativity, or do they suggest instead that such systems are ontologically nonseparable, challenging our classical understanding of ontology? And how does the classical world come about from the quantum world if, in the final analysis, there is only one world?

Some of these issues have been taken up occasionally by scholars in various fields. Wave-particle complementarity has been compared with classical christology and with personal and impersonal models of God. The holistic character of quantum systems has been used in arguments against epistemic and ontological reductionism. The role of observation in bringing about discrete states has been seen as suggesting Berkeleyan idealism and the ontological status of consciousness. Quantum physics has been used as a source of metaphor in scores of writings on world religions. Perhaps most germane to the goals of our conference, the challenge to causal determinism and the promise of genuine openness in nature for discussions of free will and divine action have often been acknowledged by scientists and philosophers.

Still, Christian theologians throughout the twentieth century have, by and large, continued to work with a view of nature dominated by classical mechanism. Clearly such a view makes the enactment of free will problematic, at least from an incompatibilist point of view. More to the point here, a causally closed view of nature confronts a theology of special divine action with a forced option between two alternatives: noninterventionist, subjectively special divine action and interventionist, objectively special divine action. The first view, favored by liberal theologians, argues that what we claim to be special divine action is our subjective attribution of "special" to an otherwise thoroughly ordinary and routine event in nature. Here we gratuitously interpret certain events as if they were due to God's special action although in fact they are not. In actuality, God's action in these events is no different than it is in any other event; all events are objectively ordinary. In the second option, we correctly interpret certain events as objectively special in themselves, since God really did act in a special way in bringing them about—but God did so by intervening in the natural order, either by violating or suspending the ordinary laws of nature.

Now, however, quantum mechanics seems to many scholars to point to a possible *tertium quid*: objectively special noninterventionist divine action. Such an option promises to combine the best of both liberal (noninterventionist) and conservative (objectively special) options into a single, new approach to divine action. In addition, quantum mechanics, as noted above, holds out the promise of a radically new ontological perspective on nature at the most fundamental level. At the same time, tremendous interpretive challenges await those who wish to engage in serious theological reflection on quantum mechanics. It is to this engagement that we now turn in the following survey.

4 Summary of the Volume

4.1 Part I: Scientific and Historical Context

The volume opens with introductory essays on the scientific and philosophical issues involved in quantum mechanics and their historical precedents. **Abner Shimony** begins by describing two essential concepts in quantum mechanics. The first is the quantum state or wavefunction, which specifies all the quantities of a physical system "to the extent that it is possible to do so." This caveat is crucial since, according to the Heisenberg uncertainty principle, not all such quantities have simultaneously definite values. The wavefunction does, however, give the probability of each possible outcome of every experiment that can be performed on the system. The second is the superposition principle, according to which new quantum states can be formed by superposing any two allowable states of the system. From these two basic ideas Shimony delineates three crucial features which distinguish quantum physics and our ordinary experience of the world: objective indefiniteness, objective chance, and objective probability. Thus quantum quantities before measurement are objectively indefinite, their definite but unpredictable value after measurement implies objective chance, and the probability of finding that value after measurement is objective. Shimony then describes in detail a fourth counterintuitive property: non-locality. Here two particles which once formed a single system and have been widely separated show an uncanny correlation between their properties, challenging the relativistic concept of locality (i.e., that effects cannot propagate faster than light).

How are we to handle these remarkable features? One way would be to reject the premise that the wavefunction gives a complete specification of the quantum state. Instead, there might be as-yet unknown, or "hidden," variables at work that explain these strange features. In 1964, however, John S. Bell proved that the predictions of local hidden-variables models are incompatible with the predictions of quantum mechanics. Crucial experiments, such as those by Clauser and later by Aspect (and proposed in part by Shimony), vindicated quantum mechanics at the expense of local hidden-variables theories. Does this mean that quantum mechanics involves unacceptable kinds of nonlocal action-at-a-distance? Not according to Shimony, who points out that quantum correlations between separated parts of a system do not allow one to send information faster than light. Thus Shimony suggests that we think in terms of "passion-at-a-distance" instead of instantaneous action-at-a-distance.

Shimony then describes other experiments which reveal further elements of quantum strangeness. Delayed-choice experiments underscore the difficulties in interpreting how and when quantum properties become definite in the experimental context. Schrödinger-cat type experiments raise the possibility of quantum indefiniteness in the macroscopic world. Here something like an "irreversible act of amplification" is involved, but for Shimony, we may need to discover new physical principles if a full account is to be achieved. Finally neutron interferometry and the Aharonov-Bohm effect underscore additional highly nonclassical features of the quantum world. In sum, these highly nonclassical features of quantum systems raise profound philosophical issues for our understanding of the physical world.

The purpose of **Raymond Chiao**'s essay is to show, by a careful discussion of specific experiments, that the world possesses at least three kinds of nonlocal action-at-a-distance. Chiao first defines action-at-a-distance in general as a correlation between effects, events, or conditions separated by a spacelike interval. (If a light signal cannot be sent between two events, their separation in space and time is called a "space-like interval.") Quantum nonlocality, in particular, is a form of action-at-a-distance which has no classical explanation. But does quantum non-locality violate special relativity? Not according to Chiao, for two reasons: quantum nonlocality never reverses the order of cause and effect and in any case, and quantum events cannot be used for signaling because of the fundamentally probabilistic, uncontrollable nature of quantum events.

Chiao then interprets all three kinds of quantum nonlocalities as resulting from the superposition principle (i.e., quantum interference) in which the sum of allowable states is also an allowable state. In the first and third examples, namely the Aharonov-Bohm effect and the tunnel effect, nonlocality arises out of single-particle interference. In the second case, the Einstein-Podolsky-Rosen effect, non-locality involves two-particle interference, i.e., an entangled state.

1. In the Aharonov-Bohm experiment, a beam of electrons is split, one beam passing through the hole of a superconducting torus, the other around the torus. After being rejoined, the beam displays a single-particle interference pattern whose phase shift depends on the magnetic flux contained by the torus. Chiao sees this kind of quantum non-locality as topological in nature: it is the global topology of the split beams that lead to the local, interference, effect. The phenomenon arises from the local gauge invariance of the electromagnetic interaction and it can be used to explain the Lorentz force.

2. In the EPR experiment with two particles, quantum nonlocality arises from the "nonfactorizability" of the quantum states: since the two-particle state of the system is the superposition of the products of the two states of the individual particles, the superposition cannot be factored mathematically into separate states for each particle. Being so entangled, the state of the system, when measured, depends on the states of both particles in the system regardless of the distance separating them. In the 1960s, John Bell showed that EPR results violate the philosophical assumptions Einstein and his colleagues made in defending "local realism." Bell then proposed that properties of particles, such as position, momentum, spin, etc., do not exist until they are observed, reminiscent of Berkeley's idealism. Chiao presses this point further, claiming that the nonfactorizability of the entangled states implies the "non-separability" of the quantum world.

Chiao then describes the EPR experiment performed in his lab, in which pairs of photons are prepared in an entangled state by spontaneous parametric down-conversion in a nonlinear crystal. Chiao used Franson's modification where

Mach-Zehnder interferometers replaced Fabry-Perot and Michelson interferometers in the detection process. This leads to a modified Bell inequality in which the twin photons possess neither definite energy (color) nor a definite time of emission prior to their detection. Moreover, nonlocality is further demonstrated by the fact that a change in the path length of one of the interferometer arms changes the behavior of the photon which passes along the other, unchanged, interferometer arm.

3. The third kind of non-locality occurs in quantum tunneling where certain kinds of superluminal velocities are possible during tunneling. Here two photons are emitted simultaneously and their arrival times at equal distances are measured. If a barrier is inserted into one of the paths, the difference in the time of arrival constitutes a precise definition of the tunneling time. But will the photon traversing the tunneling path arrive before or after the photon following along the free path? In theory, a superluminal result is possible, in which the tunneling photon arrives first. Chiao shows why this result does not violate relativity: relativity allows for superluminal group velocities and only forbids superluminal front velocities. Moreover, such superluminal effects are governed by the uncertainty principle, and thus cannot constitute a controllable signal.

He then describes in detail the resulting experiment using a Hong-Ou-Mandel interferometer. Chiao's results help decide between three conflicting theories about how to define tunneling time, and they showed that the tunneling process is indeed superluminal in the allowable sense. According to Chiao, such superluminal tunneling implies a third kind of non-locality: an observer moving past the barrier at close to the speed of light would infer that the particle exists simultaneously at both the entrance and the exit faces of the barrier!

Chiao concludes with some additional philosophical and theological reflections in light of these results and his Christian faith. He supports a "neo-Berkeleyan" point of view in which the free choices of observers lead to nonlocal correlations of properties of quantum systems in time as well as in space, giving Berkeley's dictum, *esse est percipi*, temporal as well as spatial significance. Theologically he uses this generalized Berkeleyan point of view to depict God as the Observer of the universe. Here God creates the universe as a whole (*ex nihilo*) and every event in time (*creatio continua*). The quantum nonseparability of the universe is suggestive of the New Testament's view of the unity of creation. In the process Chiao discusses such ideas as the quantum entanglement of all events in the universe given their common origin in the Big Bang, and he responds to the challenge of the quantum Zeno paradox.

Michael Berry's essay addresses the problematic relation between the presence of chaos in classical mechanics and its absence in quantum mechanics. If classical mechanics is the limit of quantum mechanics when Planck's constant h can be ignored, why does a system appear nonchaotic according to quantum mechanics and yet chaotic when we set $h = 0$? Moreover, if all systems obey quantum mechanics, including macroscopic ones like the moon, why do they evolve chaotically? Berry's approach is to locate this problem within a larger one: namely the mathematical reduction of one theory to another. His claim is that many of the problems associated with reduction arise because of *singular limits*, which both obstruct the smooth reduction of theories and point to rich "borderland physics" between theories. The limit $h \to 0$ is one such singular limit, and this fact sheds light on the problem of reduction in several ways. First of all, nonclassical phenomena will emerge as $h \to 0$. Secondly, the limit of long times ($t \to \infty$), which are required for chaos to emerge in classical mechanics, and the limit $h \to 0$, do not commute, creating further difficulties.

To illustrate the role of singularities in the semiclassical limit, Berry first considers a simple example: two incident beams of coherent light. Quantum mechanics predicts interference fringes, and these fringes persist as $h \to 0$ due to the singularity in the quantum treatment. But in the geometrical-optics form of classical physics (where the wave-like nature of light is ignored) there are no fringes, only the simple addition of two light sources. To regain the correspondence principle between classical and quantum mechanics we must first average over phase-scrambling effects due to the influence of the physical environment in a process called "decoherence."

A second, more complex, example illustrates the relation between these singularities and chaos. Berry describes the chaotic rotational motion of Hyperion, a satellite of Saturn. Regarded as a quantum object, Hyperion's chaotic behavior should rapidly be suppressed. Remarkably, however, the suppression is itself suppressed due to decoherence: even the "kicks" from photons from the sun on Hyperion are enough to induce decoherence. This means that, while it is true that chaos magnifies any uncertainty, in the quantum case the magnification would wind up suppressing chaos if this suppression were not itself suppressed by decoherence induced by interactions with the environment.

Finally, Berry turns to emergent semiclassical phenomena. These phenomena do not involve chaos, and unlike more familiar examples of macroscopic quantum phenomena such as superfluidity, their detection requires magnification. His first example is the focusing of a family of light trajectories, such as rainbows or light patterns in a swimming pool. These patterns, or caustics, are singularities in geometrical optics. But upon microscopic examination, caustics dissolve into intricate interference patterns which catastrophe theory describes as emergent semiclassical phenomena called diffraction catastrophes. His second example is spectral universality: if we consider quantum systems whose classical mechanical treatment is chaotic, we find that the statistics of the spectra of all such systems is the same. Spectral universality is nonclassical, because it is a property of discrete energy levels, and it is semiclassically emergent because the number of levels increases in the classical limit, $h \to 0$. Berry's conclusion is that, as we generalize to a deeper theory, the singularities of the old theory are dissolved and replaced by new ones.

According to **Ernan McMullin**, there are several crucial challenges to the relations between quantum mechanics and theology. First, the discussion of both quantum mechanics and theology often relies on a realist interpretation, but some of the most energetic critics of realism are philosophers of quantum mechanics. Second, in dealing with mechanics in general, and with quantum mechanics in particular, the move from the mathematical formalism to its ontological interpretation is highly problematic. The fact that quantum formalism yields two different ontologies (those of Bohr and Bohm) leads to further problems: how can this be so, and how is one to decide between them? Finally, one is faced with the "troubling strangeness" of quantum ontology.

In this essay, McMullin limits his concern to the history of the relation between formalism and ontology in astronomy in order to show that similar difficulties arose throughout this history. First he clarifies terms. Mathematical formalism tells us nothing about the world until it is interpreted in terms of measurable quantities, thus becoming a physical formalism such as quantum mechanics. Both the Copenhagen interpretation and Bohm's interpretation are then second level interpretations between the physical formalism and ontology, and it is here that issues arise which divide scientific realists and nonrealists.

With this in place, McMullin turns to the early history of astronomy and finds similar difficulties in moving from formalism to ontology. He starts with Greek astronomy, where the celestial regularities invited explanation in terms of an underlying physical structure. Aristotle, for example, regarded the mathematical formalism of concentric spheres as implying a causal explanation based on the ontology of a single interlocking system of spheres. But Apollonius and Hipparchus showed that a complex, mathematical model based on the eccentric and the epicycle could explain the data equally well, and this posed something similar to the problem we find in quantum physics: how can two radically different ontologies explain the same phenomena? Perhaps there is no genuine connection between formalism and ontology; if there is, the criterion of "saving the appearances" is clearly not enough in itself to reveal it.

McMullin next examines the Copernican revolution from geocentrism to heliocentrism. Although both formalisms could equally well save the appearances, Copernicus's system offered important advantages. It eliminated unwanted elements of Ptolemy's formalism and it explained phenomena that were mere coincidences in the Ptolemaic system. In addition, Copernicus could specify the order of the planets outwards from the sun, their distances from the Earth, their periods of revolution, and their retrograde motion. All of this disclosed such a degree of harmony that, for Copernicus, it proved the reality of the earth's motion and pointed to God as its creator. McMullin closes this section by challenging Kuhn's assessments of the epistemic merits of Copernicus's system.

Kepler, we are next told, strongly supported Copernicus's realist arguments for the motion of the earth, both for religious reasons and because Copernicus offered such a convincing explanation of the phenomena. His extensive analysis of Tycho Brahe's observations of Mars led Kepler to explain its orbit as elliptical and to suggest a physics that could make such an orbit possible. Kepler's "theory of gravity" was analogous to magnetism and gave the Sun causal primacy in determining planetary motion. But it was only with Newton that we at last have a case "sufficient to warrant reasonable belief" in the heliocentric system. What then of ontology? Certainly Newton's system employed terms like 'force' and 'attraction', but with his rejection of action-at-a-distance, the challenge of finding an ontology to match the formalism remained. As McMullin points out, this situation is obviously similar to the current problems in interpreting quantum mechanics.

McMullin closes by drawing a philosophical moral from this history of mechanical systems: moving from a valid formalism to an underlying ontology has always been a "contentious matter. Mechanical agency has all along proved to be an uncommonly elusive quarry." On the other hand, this history has forced us to expand our intuitive notions of agency and, given the challenges raised by quantum mechanics, it seems clear that further expansion is necessary. He also offers us a theological moral from his historical account: if our explanation of motion at its most basic level is more complicated and mysterious than earlier generations imagined, how much more complex and mysterious must divine agency be, and how hesitant ought we to be in discussing it.

4.2 Part II: Philosophical Interpretations of Quantum Mechanics

In Part II, five essays offer in-depth analyses of various philosophical issues involved in interpreting quantum mechanics. In the process, these essays open diverse avenues of discussion for the themes of human and divine action.

As a prelude to the problem of divine action and quantum physics, **William R. Stoeger** explores the epistemological and ontological implications of quantum physics. Clearly a discussion of divine action in nature requires our confidence that scientific theories actually represent processes and features of the world that are, in some ways, independent of how we know them. Using these theories, we may then be able to constrain our description of divine action in important ways. But to gain confidence in our theories, we must first sift through their interpretations, assessing them in terms of their adequacy and fruitfulness; only those that survive this process will warrant further consideration. What then constitutes a "canon of adequacy" for our assessment?

Stoeger responds first by noting that there are several levels of interpretation involved: i) basic interpretation at the level of the physical theory itself (e.g., the probabilistic interpretation of the square of the wavefunction); ii) consistency and coherence both within the theory, and iii) with other physical theories (e.g., with special relativity); and iv) epistemological and ontological interpretations by which quantum theory may give us knowledge of the underlying reality. Clearly levels (i)–(iii) constrain level (iv) without determining it completely. Using levels (iii) and (iv), Stoeger argues that we can exclude both hidden-variable and other strongly deterministic interpretations of quantum theory. Next, he proposes a "principle of parsimony" in which we minimize our assumptions about what reality is like, allowing the results of quantum physics to "speak for themselves" even if the result is counterintuitive and puzzling. That our interactions with the quantum level are "recalcitrant and resistant" suggests that we are dealing with aspects of the world independent of our measurement of it. We may not have any direct knowledge of the underlying states which produce the phenomena we measure, but our experimental and theoretical knowledge place significant constraints on the properties these underlying entities can have. This assumption is warranted since the models we construct successfully predict and explain other phenomena. Stoeger relies on Ernan McMullin's emphasis on retroduction to support these points. Still, he acknowledges that there may be many significant features of quantum reality that completely escape our detection. Some of these may never be knowable even in principle, while "reality for us" may have features that are not functions of the actual underlying features of the world. Stoeger's essential metaphysical presuppositions, then, are the principle of sufficient reason (what we observe in some way points to an underlying cause) and the principle of relationality (the reality with which we interact is a part of a network of relations with processes and objects at other levels of the world).

Stoeger then briefly describes several key features of quantum physics, including nonseparability, quantization, objective uncertainty, complementarity, objective chance, correspondence, entanglement, measurement, and decoherence. Returning to his criteria for choosing an interpretation, Stoeger notes that "the family of Copenhagen-like interpretations" (including the consistent-histories approach) involves most of these features and is "by far the most satisfactory" interpretation, compared with hidden-variable and many-worlds interpretations. Thus our indirect knowledge of reality is "weakly objective": an independent reality exists and is manifest to us through our interactions with it, but we cannot assess our knowledge of it from these observations. Regarding the question of the epistemic and ontological status of the laws of nature, Stoeger sees these laws as but imperfect and incomplete descriptions of those that obtain in nature. Moreover, these laws are descriptive, not prescriptive.

Several implications for divine action follow from this. God's universal creative action is realized behind the "veil" of natural laws, and it appears in the form of these laws. Isolated cases may seem to violate these laws, and God's action may occur at the level of consciousness and personal relationships. Special divine action may involve top-down influences on matter and thus transcend science. Divine action, as acts of love and care, may be taken to be interventions only if we assume that the laws hold absolutely under all circumstances. The key problem, particularly compared with that of human agency, is our lack of understanding of how an immaterial God can act on the material world.

James T. Cushing sees the question of determinism versus indeterminism as "*the* fundamental issue" regarding the possibilities for particular divine action, and thus the importance of quantum mechanics. His central point is that "considerations of empirical adequacy and logical consistency alone" do not force one to chose the indeterministic view of quantum mechanics as found in the Copenhagen interpretation—Bohm offers an empirically valid deterministic alternative.

Cushing begins by defining a theory as a formalism (i.e., a set of equations and a set of calculational rules for making predictions that can be tested) and an interpretation (i.e., what the theory tells us about the underlying structure of the world). Quantum mechanics presents us with one formalism but two different interpretations, and thus two different theories. The Copenhagen version affirms the completeness of the state-vector description, the principle of complementarity, nonclassical probability, and a prohibition against "any possible alternative causal description in a spacetime background." Physical processes are thus both indeterministic and nonlocal. Bohmian mechanics is an objective (realist) and deterministic account in which the positions of the particles of the system function as "hidden variables" and must be included in a complete state description. As in the Copenhagen interpretation, the Schrödinger equation governs the evolution of the wavefunction, but an additional "guidance condition" governs the evolution of the particles' positions. With the inclusion of a quantum-equilibrium statistical distribution, Bohm's theory is empirically identical with standard quantum mechanics. Its ontology depicts particles following definite trajectories that are completely deterministic and observer-independent. The ontology, however, is nonlocal: instantaneous, long-range influences are included. Still Bohmian nonlocality is "benign," since the "no-signaling" theory of quantum mechanics prohibits sending messages faster than light.

In Bohm's theory, the quantum potential U conveys the influence of the environment on the particle, while U is determined, in turn, by the wavefunction. This means that the measurement process is "an act of discovery—there is *no* quantum-mechanical measurement problem." All observations are, ultimately, position measurements, a feature which reflects our own existence in coordinate space. The classical limit corresponds to U being negligible and is thus more precise than $h \to 0$. From an empirical perspective, Bohm's theory is not only completely equivalent with standard quantum mechanics, but it also captures Bohr's concept of quantum holism and his principle of complementarity. As Cushing puts it, observed values in Bohm's theory are "contextual."

Bell's theorem shows that our world cannot be both objectively real and local. Cushing suggests that locality is the real problem, but reminds us that Bohm offers a *non*local, deterministic hidden-variables theory. In order to discuss Bohmian ontology, Cushing points to "relational holism" since it seems to offer a better conceptual framework than one which distinguishes between separability and locality. It also suggests a world of temporal becoming since it includes a preferred frame for instantaneous action. Still this world is one in which everything, including the future, is determined. Such a world is reminiscent of Newton's idea of space as the divine *sensorium*. It certainly poses a challenge to our ideas of free will and divine action—as does the problem of evil.

In short, then, the choice to accept the Copenhagen view and reject that of Bohm is not a forced move based on logic or empirical adequacy; it is made on other grounds. Similarly, one might chose an indeterministic view of quantum mechanics for theological reasons, but one should not claim that quantum mechanics provides independent, scientifically arguable grounds for such a choice.

The over-arching aim of **Jeremy Butterfield**'s essay is to discuss the strange ontologies that emerge from various interpretations of quantum mechanics. He begins with various proposed solutions to the measurement problem and then focuses on the Everettian interpretation developed by Simon Saunders and David Wallace. At the very outset, however, Butterfield stresses the highly problematic character of quantum indeterminism: it only appears in some interpretations of quantum theory and it involves a highly nonclassical ontology. He acknowledges the enormous empirical success of quantum theory but notes that considerable problems arise in reconciling it with special and general relativity, and he argues strongly against reductionism. He then provides a brief summary of the formalism of quantum theory, including a discussion of pure states, mixed states, and the meaning of probability in quantum theory.

Butterfield begins his discussion of the measurement problem with the orthodox view: during the process of measurement both the quantum system and the measurement apparatus go to a definite state (the "collapse of the wavepacket" for both system and apparatus). Yet according to the linearity of the Schrödinger equation, there is no such "collapse"—the indefiniteness of the quantum system should be transmitted to the measurement apparatus, leaving the position of the "pointer" indefinite. How then can we justify ascribing a definite state to the pointer? According to Butterfield, recent work on decoherence suggests that the continual interaction of the pointer with its environment brings the pointer "very close" to a definite state by leaving it in a narrow mixture of definite states. Still, a complete solution to the measurement problem requires more than what decoherence can provide.

We thus face two choices. The first choice is either to abandon orthodox quantum theory's law of temporal evolution (the Schrödinger equation) and to propose new equations so that the collapse of the wavepacket is a physical process (Butterfield calls this choice *Dynamics*), or to assign values to additional quantities not given by orthodox quantum theory (*Extra Values*). The second choice is either to secure definite values for familiar quantities like the position of macroscopic objects (*DefMac*) or to be satisfied if the macrorealm only appears definite (*DefApp*). Combining these two choices yields four broad strategies for solving the measurement problem. An example of the *Dynamics/DefApp* strategy is that of Wigner and Stapp, where mind or consciousness produces the collapse of the wavepacket, while the *ExtraValues/DefMac* strategy of de Broglie and Bohm ascribes definite values to the position of the quantum system and introduces a "pilot-wave" to guide the system.

Butterfield then develops the *ExtraValues/DefApp* strategy proposed by Everettians. Here one retains the Schrödinger evolution of a strictly isolated system and allows an indefinite macrorealm, but posits extra values to secure definite appearances (measurements) by appealing to decoherence. In its simplest version, this strategy ascribes a wavefunction, Ψ (specifically, a pure state), to the universe as follows: Ψ is a superposition corresponding to numerous different "macrorealms" (or "worlds" or "branches") that evolves according to the Schrödinger equation. According to Butterfield, the Everettians now propose a "breath-taking main idea": all these macrorealms are actual. He devotes much of the remainder of his essay to the meaning of macrorealms from the perspective of both "many-minds" and "many-worlds" interpretations, how macrorealms evolve over time, and what he sees as the attendant vagueness in these approaches.

His primary focus is the work of Saunders and Wallace, including their appeal to anthropocentrism, their definition of 'world' as relative state, their use of decoherence, and their arguments against a precise definition of 'branch'. A central concern is that of identity over time. Philosophers treat change of properties over time in two rival ways: either as properties of objects that persist self-identically over time (i.e., objects that "endure"), or as properties of objects having stages or temporal parts (i.e., objects that "perdure"). Similarly, for Everettians, we could say that the pointer has different properties (e.g., positions) in different worlds, or that there are multiple pointers similar to one another ("copies") but with different properties in different worlds. He discusses the pros and cons of the "copy" choice most Everettians make, and he endorses the analogy between "worlds" as conceived by Everettians and instants of time represented by a "block universe" conception of time.

In his final section, Butterfield discusses the dynamics of the universe as a whole and of its subsystems. Everettians find the deterministic evolution of the universe as a whole to be compatible with the indeterministic evolution of its open subsystems and with the "almost deterministic" evolution of isolated subsystems. He closes by suggesting that, although the relativistic invariance of the universe as a whole is clear, a lacuna remains concerning the relativistic invariance of its subsystems.

Michael Redhead's essay is based on the assumption that nonlocality as instantaneous causal action-at-a-distance is to be avoided since it violates "the spirit of special relativity." He therefore undertakes a meticulous examination of a variety of proofs of nonlocality in the quantum mechanical treatment of many-particle entangled states, seeking to detect and assess their assumptions.

Redhead starts with the assumption that relativity is more than a phenomenological invariance principle; instead it is grounded in the causal structure of spacetime. Specifically, Redhead claims that relativity entails the Philosophically Grounded Invariance Principle (PIP), which asserts that causal influences cannot operate outside the light-cone. If, alternatively, relativity entailed the First Signal Principle (FSP), it would disallow faster than light signals, where "signals" are controllable causal processes. David Bohm's interpretation of quantum mechanics is both deterministic and, to many scholars, consistent with relativity since it does not allow superluminal signaling—although it does allow superluminal causal processes. But since Redhead believes that relativity entails PIP, and since PIP → FSP, Redhead claims that Bohm's interpretation violates relativity. He also objects to it on theological grounds, since its determinism does not allow "room" for incompatibilist divine action. He therefore turns to indeterministic approaches involving nonlocality in both nonrelativistic and relativistic quantum mechanics.

He begins with John Bell's analysis of the (nonrelativistic) EPR argument, which delineated between two meanings of nonlocality: a) action-at-a-distance between individual particles, and b) nonseparability, in which at least some properties cannot be attached to individual particles. Bell's argument, in turn, rested on assumptions involving joint probability distributions and determinism, both of which Redhead explores in detail. He then discusses algebraic proofs of nonlocality that seek to demonstrate that local hidden-variable theories are self-contradictory.

Next, Redhead turns to the search for a relativistic EPR argument. First, he reviews the problems encountered in seeking to translate the nonrelativistic EPR argument into a relativistic context, paying particular attention the need to reformulate the "reality criterion" (e.g., every element of physical reality must have a counterpart in the theory). A relativistic EPR argument must employ a relativistic wavefunction and must not depend on the existence of absolute time-ordering for space-like events. Redhead describes in detail one proposal for a relativistic reality criterion by Ghirardi and Grassi and its reliance on the truth of certain classes of counterfactual statements. Ghirardi and Grassi's argument involves a distinction between what Redhead describes as OM-Loc, that the outcome of a measurement cannot be influenced by performing nonlocal measurements, and ER-Loc, that elements of reality cannot be created by performing nonlocal measurements. Ghirardi and Grassi claim to show that relativity and quantum mechanics are in "peaceful coexistence," but to do so they must also claim that violating ER-Loc is more serious than violating OM-Loc. Redhead disagrees, but offers a further assumption which he calls the "Principle of Local Counterfactual Definiteness (PLCD)." With this he shows that Ghirardi and Grassi's relativistic reformulation of the EPR argument is less general than they suggest; it is in fact limited to deterministic systems.

In his concluding section Redhead first argues that nonlocality seems unavoidable for any reconstruction of quantum mechanics which is both realist, i.e., in which all observables have sharp values at all times, and deterministic. We either turn to a stochastic hidden-variable framework or seek to understand correlations in terms of what Shimony describes as "passion-at-a-distance." In the anti-realist option pursued by Ghirardi and Grassi, Redhead challenges the claim that the existence of action-at-a-distance is not a valid deduction from the EPR argument, but he then rescues the claim by the additional assumption of determinism. He regards his results as closing further gaps in the peaceful co-existence argument, but the "mysterious harmony" of quantum correlations remains "spooky" even if it does not involve causal dependence. For the anti-realist, the role of measurement is to actualize potentialities. But when quantum mechanics is applied to cosmology, where there is nothing "outside" the universe to serve as a measuring device, the realist option may be preferred, and with it the notion of nonseparability.

Redhead's essay thus gives arguments for invoking either indeterminism or holistic nonseparability. The author sees these as having important theological implications: indeterminism is important for theories of divine action on particular occasions, while holism is an anti-reductionist thesis "which shows how every element of the universe has for its ground of being the totality of the whole, which pantheists would want to identify with God."

The overall aim of **Chris Clarke**'s essay is to show how a modification of the consistent-histories interpretation of quantum mechanics provides a natural setting for understanding human and divine action. For Clarke, religion is largely about finding the meaning of the "good life," and our aim is to help people live it. Hence we tell stories about the world we live in, some deriving from science, others from the great myths of religion. Weaving them together is important even though no part of the story represents a reality independent of ourselves and not all parts are equally supported by experiment. Clarke draws upon the phenomenology of Heidegger and Merleau-Ponty to argue that objective reality is located in the "second-person relationship" that lies between subject and object. Quantum theory plays an essential role in Clarke's understanding of "the interplay of self, society, and Other" through which the concreteness of the world emerges.

Clarke then turns to the central problem of quantum mechanics: if it is a generalization of, and not an alternative to, classical mechanics, should we not use it to describe all of physics in a unified way? Yet if we do so, the theory predicts that macroscopic phenomena will be superpositions of states "flagrantly at variance with our experience," as the Schrödinger's cat experiment vividly depicts. Bohr and Von Neumann avoided this problem by dividing the world into the quantum system and its classical environment, and they characterized these realms by two separate time developments. But Clarke's goal is an overall picture of the world that places the observer and the observed system on the same footing. To do so, Clarke focuses on the consistent-histories approach in which state reduction is unnecessary and only appearances are definite.

A "histories" approach links a sequence of preparation and measurement pairs such that each measurement becomes the preparation for the next. A "consistent" histories approach tries to rule out superpositions of macroscopically distinguishable states by considering only those histories whose probabilities obey classical logic. The approach was introduced by Robert Griffiths in 1984 and then extended to cosmology by Murray Gell-Mann and James Hartle, but problems were soon raised by Dowker and Kent. Clarke's hope is to reformulate the approach to avoid these problems and then relate it to human and divine agency. To do so, he focuses on how we might restrict the possibilities of future histories given a fixed and acceptable history up to the present, looking at unmodified dynamics within sets of histories. This leads Clarke to propose a specific definition of consistency in terms of logical exclusivity and the rules of quantum mechanics, the physical significance of which arises through decoherence. As it turns out, although all classically acceptable histories are consistent, not all consistent histories are acceptable; additional structure is needed to single out acceptable histories. According to Clarke, the past history from which the future is predicted can help provide such a structure. This approach does not divide the world into classical and quantum domains, nor does it involve the collapse of the wavefunction. Moreover, in this approach, the history of the world, as both contingent and governed by decoherence, accounts for why the universe will continue to be classical.

Clarke then turns to the issue of human agency and, by analogy, divine action. According to Mae-Wan Ho, living organisms exhibit coherence, maintaining phase relations between the quantum states of their constituents over considerable distances and times. If so, our experience might not obey classical logic, and the non-Boolean aspects of an organism's own history may be observationally detectable by an external observer. Quantum mechanics thus opens the possibility that we can share histories, at least momentarily. This provides Clarke with a way to engage the "other minds" problem. Drawing on Heidegger and Levinas he discusses the objective world in terms of the interrelation of beings-in-the-world. The "co-creation of the universe" then arises through the set of such intermeshing histories. The consistent-histories approach also allows us to move beyond the "determinism vs. random" debates about free will. Instead, decision-making involves a shift from one consistent Boolean logic to another. We experience this as creativity, though the shift appears random to others. Our free will is thus characterized by the simultaneous creation of volition and a framework of meaning which justifies this volition.

Clarke describes his experience of the divine as of a guidance that is immanent in the concrete flow of events and yet transcendent, not contained in any horizon. He uses the idea of "entrainment" in quantum theory, where previous events are realized in the present, to describe divine action as top-down entrainment, coordinating and informing the individual acts of will that it contains, rather than as a divine influence at the atomic level. This concept of divinity goes beyond pantheism, but coheres with panentheism and with a view of divinity embodied in the world as suggested by the second *mahavakya* of the Upanishads.

4.3 Part III: Theological Issues

Six essays now follow in Part III. These essays start with many of the philosophical and historical issues raised above and focus more specifically on the task of undertaking theological research in light of quantum mechanics.

Although it has been "spectacularly successful" in its predictive power, **John Polkinghorne** begins by stressing that we do not fully understand quantum theory. The central difficulty is the "measurement problem": how do determinate macroscopic states (i.e., particular results) come about when a measurement is made on apparently indeterministic quantum states? Viewing this as a "collapse of the wavepacket" only renames the problem, since such a collapse contradicts the dynamical (Schrödinger) equation under which the wavepacket evolves smoothly. Niels Bohr spoke dualistically about classical and quantum worlds that had to intermesh, but this does not really work in principle since there is only a single physical reality in which even the classical apparatus is made of quantum constituents. Polkinghorne also finds unsatisfactory a statistical interpretation of quantum mechanics which refrains from speaking about individual quantum processes, including consistent-histories approaches. He then outlines various groups of proposals for interpreting quantum theory which seem more promising.

The first group starts with quantum theory as it is and attempts to resolve the measurement problem by including "decoherence." According to the superposition principle, exclusive classical states (e.g., "here" or "there") are admitted together as a viable physical quantum state (e.g., "here" and "there"). Superposition gives rise to "interference" effects suggesting the wavelike aspect of quantum states. Why, then, don't we see superposition and interference in our everyday experience? Some have proposed that decoherence, which involves interactions between the quantum process and its radiative environment, helps solve the problem by rapidly minimizing all but one state and by canceling interference effects. The problem is that decoherence does not tell us why any particular outcome, and not one of the other possibilities, actually occurred.

The second group, "hoped-for physics," believes that the interaction with large systems brings about the collapse of the wavefunction. In Polkinghorne's opinion, the irreversibility of the behavior of macroscopic systems may provide a clue to how this happens, but it has not done so yet. A third group seeks what Polkinghorne calls "unknown new physics," where, for example, the amount of matter involved in the interaction determines whether the collapse occurs, or where quantum gravity plays a crucial role. David Bohm's work represents a fourth approach: "hidden new physics." For Bohm, there is no collapse of the wavepacket, but Bohm's approach offers no predictive advantages over conventional quantum physics. According to Polkinghorne, the choice between Bohm and Bohr has to be made on the basis of extra-scientific criteria, including metaphysical principles. Theologians who believe that divine action requires indeterministic physical processes have every right to prefer Bohr's conventional approach, as long as they do not claim that science alone supports their choice. The final group appeals to "unknown but 'nearby' metaphysics" in reaching out to additional factors in nature to solve the problem. These include the role of consciousness of the observer and the many-worlds and many-minds strategies, which accept the formalism of quantum mechanics but actualize all states in the quantum superposition.

All of these proposals seem to assign a special role (or "preferred basis") to spatial position in their formulation of and solution to the measurement problem. If so, this would imply a change in the way we think about physics, which, since its inception, treats all dynamical variables equally. They also limit their focus to laboratory measurements, and they may not be extendable to the wider context of natural physical processes. Polkinghorne then suggests that the standard classical account of macroscopic processes may need to be reconsidered. Complex classical systems point to the presence of a "pattern-forming causality of an holistic kind ('active information')." Perhaps the equations of classical physics are "downward emergent" approximations of a more complex account of macroscopic physics as well as "upward emergent" from quantum physics. Still, the unresolved complexities of quantum chaology pose a challenge to such an approach. Even the meaning of the term "quantum event" cannot be reduced either to occasional measurements or to the general unfolding of the wavefunction governed by the Schrödinger equation.

Finally, Polkinghorne turns to the theology of divine and human agency. Though autonomous in many ways, the metaphysical backing for such discourse should still take account of science, particularly where quantum mechanics and chaos theory suggest ontological openness. But many unresolved problems beset such attempts. Quantum theory may not be helpful because of the limited and episodic character of measurement events in which indeterminacy seems to hold. Moreover, although some quantum processes, such as gene mutation, may lead to macroscopic consequences, they do not seem to generate a basis for the "flexible actions of agents." Chaos theory provides a more "flowing character" for agency. Of course, chaos theory is normally framed within a deterministic, Newtonian, context, but it could be given a wider framework. The real problem is how to combine chaos theory with quantum mechanics, and in the process, solve issues in quantum chaology. Polkinghorne believes that the best hope for future progress will lie in an increased understanding of the nature and implications of quantum chaology.

According to **Michael Heller**, the evolution of concepts is a driving force in science. New concepts inherit much from their predecessors and yet are open to future generalizations. When they produce paradoxes and inconsistencies, a crisis arises which can be called a conceptual revolution or, more properly, an evolution within a conceptual framework. One example of a conceptual evolution is the origin of rational discourse about the world begun in sixth-century BCE Greece. Another is current research in quantum mechanics. A sign of such conceptual evolution is an increasing generalization in which the old concepts are restricted to a smaller domain of validity than they originally enjoyed. Both science and theology can be seen as attempts to catch reality in a net of concepts and theories, attempts which always fail. Still we should not remain silent, at least about God, since it is better to say something even if it is always tentative. All language demands interpretation, as quantum physics clearly shows. The breakdown of language in physics points to the need to generalize; perhaps theology could learn something by analogy from this fact. Thus the goal of this essay is to look at contemporary quantum theory and to derive from it a lesson for theology.

Heller's starting point is the fact that the main distinguishing feature of quantum mechanics is its noncommutativity; he seeks to show the degree of generalization already present in quantum theory by using the recently discovered noncommutative geometry. It not only clearly shows the generalizing mechanisms underlying the present theory, but it also points towards further possible generalizations. Heller explores the possibility that at its fundamental level, physics is modeled by noncommutative geometry. Quite independently of whether this hypothesis will prove true, he claims that we can learn a lesson from it. Heller analyzes a few concepts, such as causality, probability and chance, which are of great importance for philosophy and theology when they are transferred from their usual context to the environment of the "noncommutative world." The main characteristic of this world is its a-temporality and a-spatiality. It turns out, for instance, that in this a-temporal world authentic dynamics (albeit in a generalized sense) is possible.

Heller does not claim that concepts elaborated in noncommutative geometry can be used directly in theology. Instead he tries to draw consequences for theological discourse from the fact that even in physics some concepts undergo such drastic evolution that they distance themselves from our everyday linguistic intuition. He begins with an algebraic formulation of quantum mechanics based on general C^*-algebra; this formulation allows one to recover the more limited formulation in terms of Hilbert spaces. C^*-algebras that are relevant to quantum mechanics are noncommutative algebras, and it is this noncommutativity which is responsible, according to Heller, for all the peculiarities of quantum theory. Algebraic formulation also leads to the possibility of geometrizing quantum mechanics. The so-called "noncommutative spaces" are totally global in character; no local concept can be given any meaning. This in turn could lead to the unification of quantum mechanics and general relativity. The idea is that fundamental physics is based on a noncommutative geometry that is nonlocal; only at a higher level does the distinction between spatio-temporal geometry and physical dynamical processes arise. Even at this fundamental level, there can be an authentic, though generalized, dynamics. But here the distinction between singular and nonsingular is lost, undermining such ideas as the beginning of the universe and the concept of the individual. Instead, and unlike previous approaches in physics and philosophy, singularities are a part of our macroscopic perspective, but their distinctive character is meaningless at the fundamental level. Equally, nonlocal phenomena, such as those which the EPR experiment points to, are explained within the noncommutative approach.

In his closing sections, Heller shows how important theological concepts, such as causality, are reshaped by the noncommutative framework and its properties of timelessness and nonlocality. Causality becomes a "dynamical nexus" rather than a temporal ordering of cause and effect, a combination of a-temporal and nonlocal behavior that is fertile ground for thinking theologically about God as Creator.

Philip Clayton gives two reasons why constructive theology should engage in dialogue with quantum physics: it cannot afford the fideistic position that results from disengaging with science, and it should seek a more hypothetical, fallible, and revisionist method than traditionally allowed, thus opening itself up to the engagement without becoming fully relativistic. Clayton's method will be to consider an array of interpretive models in a specific scientific field. He will then look for areas of compatibility with theology, in the process revising both theology and science. Eventually, he will repeat this activity across a variety of disciplines. But why should physics provide constraints on how God might act? Clayton's response: it can do so if divine (or human) agency occurs in the physical world in conformity with physical law. This obviously holds for us, and it may indeed hold for the way God chooses to act. It thus becomes Clayton's "wager": the structure of the physical world sets parameters on, and tells us about the manner in which, God can act. He locates this position midway between those who give a purely subjective account of theology or who worry that quantum mechanics will not be helpful for divine action, and those who seek even stronger theological conclusions from science or possibly the convergence of science and religion.

Clayton then explores three quantum mechanical constraints on divine action. The first is the role of the observer. Minimalists focus on macroscopic measurements by an observer who is never within the quantum mechanical system being studied. Maximalists introduce subjectivity and consciousness in explaining a quantum experiment, despite the resistance of many physicists. Here Clayton finds another crucial issue at work: reductionism assumed by minimalists versus emergence and even dualism assumed by maximalists. The second issue is the "many-worlds" interpretation as represented by Hugh Everett, Bryce DeWitt, and others, compared with those who defend the irreducible role of subjectivity in nature, such as Eugene Wigner, John von Neumann, John Wheeler, Henry Stapp, and Roger Penrose. Both of these interpretations are deeply influenced by metaphysics: physicalists who accept a branching universe versus subjectivists who view quantum mechanics as evidence of mind as irreducible in nature. Clayton then turns to his third issue, indeterminism and free will. He reminds us how the early defenders of the Copenhagen view saw the free choice of an experimenter as playing an irreducible role in the outcome of the experiment. Despite counterarguments, Clayton claims that ontological indeterminism remains a significant factor in these debates: it seems a necessary condition for an incompatibilist view of free will, particularly if incompatibilist free choices are to be enacted in the world. In turn one can argue that God so created the world as to allow for human freedom.

Clayton then argues that questions like these three show that physics and philosophy lie on a continuum, particularly when the philosophical questions are closely connected with physics research. Even theology lies on this continuum, though it is further removed from physics research than from philosophy. Clayton now turns to three additional issues. First he considers Bernard d'Espagnat's ontology, terming it "Spinozistic Monism." Here the state vector expresses properties of a deeper, underlying reality which we can never describe in itself but which is manifested in what we observe and which can be understood as Being. Next Clayton engages critically those who interpret quantum physics in terms of Eastern mysticism, including Capra, Bohm, and Wilber. Though their stress on holism may be compelling, their metaphysical conclusions, like any others, are options not directly supported by physics. Finally Clayton turns to theistic metaphysics, considering both classical theism and panentheism. Theism asserts that the world as it appears to us is real and that it has its origin in an ultimate principle called spirit. The divine spirit is an active principle in this world and is in many ways personal. Classical theism has advantages over the preceding views, but it can become problematic if it places too great a distance between God and the world; the analogy with human agency breaks down for a fully disembodied view of God. To Clayton, panentheism avoids some of these difficulties, particularly as it understands the world to be within God even while, as with classical theism, God is more than and distinct from the world. Here each physical event can be an expression of divine agency in a "top-down" manner which does not violate physical law. It also provides a metaphysics that coheres nicely with some of the interpretations of quantum physics previously discussed, particularly those which stress holism, veiled reality, interconnection, and interdependence.

Uncertainty regarding the meaning of "the acts of God" pervades modern theology, according to **Thomas Tracy**. Critical historical and literary techniques have deepened the problem of interpreting biblical texts and the connection they make between story and history, while the natural sciences have changed the intellectual context of interpretation by offering an account of nature without appeal to transcendent causes. On the one hand, scientific methods do not rule out divine action, and scientific findings are not inconsistent with it. Ironically, theologians from deists to liberals such as Schleiermacher, Bultmann, and Kaufman, have worked with a closed causal picture of the world that they feel is authorized by science. They have taken this to be incompatible with divine action in the world, leaving either a God who only sets the world's initial conditions or whose actions violate the laws of nature. But contemporary natural science does not necessarily lead to a deterministic metaphysics. Tracy cites two possible responses. First, a theologically sufficient account of God's particular actions in history might actually be developed that still limits God to being the creator of history as a whole. Second, God can be said to act in particular cases without intervention in history if one can defend an indeterministic interpretation of natural causes. It is here that quantum physics might be relevant.

Though Tracy's focus is on the second response, he starts with an extended treatment of the first one since he does not want to underestimate its resources and since he explicitly assumes it as background for the second response. Here God's fundamental action is the free intentional act of creating the world, which continuously gives being to the created world in its entirety but which cannot be understood by analogy with human agency. Moreover, God gives to created things active and passive causal powers, so that God's action is direct in causing their existence, but indirect in acting through them and their powers to produce results in the world. Thus even though God acts *uniformly* in all events, we can affirm God's objectively special action in two ways: particular events may reveal God's overall purposes, and they may play a special causal role in shaping history. It is interesting to note that, in identifying this second way, Tracy is making an important addition to the typology developed in previous CTNS/VO publications and republished above, where only the first way, called "subjectively special action," was discussed.

If, however, the structures of nature are on some level(s) indeterministic, God can act to determine the outcomes of natural processes without disrupting their intrinsic causal properties. Here God could be thought of as acting in all such chance events or in just some of them, though the latter generates conceptual puzzles. Moreover, the extent of ontological chance in nature will influence the extent of God's action in nature. Indeterminism also plays a role in "incompatibilist" accounts of free human action. Here again God could be thought of as acting in all human acts, as John Calvin and Aquinas seemed to imply, or as empowering people to make their own choices. Both options raise further issues, including the problem of evil and the ultimate redemption of the world. Indeed, faith in God's redemptive action in history provides "a compelling theological reason" to argue for a noninterventionist account of divine action and thus an indeterministic interpretation of nature.

A number of challenges, however, face any attempt to use quantum physics for such an account. First, quantum physics can be interpreted in a variety of different ways including the Copenhagen interpretation, Bohmian nonlocal hidden-variable determinism, many-worlds determinism, and so on. While it is legitimate, even unavoidable, to prefer one of these on theological grounds, we should stress that others are available and their theological use in each case is tentative and provisional. A second challenge is the measurement problem found in some of these interpretations. Does this overly limit the occasions of divine action, or is "measurement" more universal in nature than some interpreters suggest? And how do the worlds of quantum processes and observable objects relate? A third challenge is to show that indeterministic transitions associated with measurement can produce a difference in the course of the everyday world. Laboratory equipment, of course, involves precisely this sort of "amplification," but so do natural processes, such as vision and genetic mutation. In conclusion, Tracy stresses the primary importance of God's creating and sustaining the world, and within this, God's indirect action through created causes and, possibly, God's direct noninterventionist action at points of underdetermination in natural processes.

George F.R. Ellis lays out a thoroughgoing critique of reductionism and a complex ontology for reality as a whole, drawing from his previous publications and offering new reflections in light of quantum theory. Against reductionism, Ellis claims that nature is hierarchically structured, with emergent levels of order and meaning, as well as bottom-up and top-down action, occurring throughout the hierarchy. He begins with classical physics, chemistry, and biology, where reductionism is framed in terms of micro-to-macro relations of bottom-up deterministic causality. But Ellis notes that quantum processes give rise to the regularities of the classical world, and that they can have macroscopic results in the classical world. Ellis's examples include amplifiers such as instruments (e.g., photomultipliers), biological organs (e.g., the eye) and processes (e.g. genetic mutations expressed in the organism); coherent implementation of micro-effects; "essentially quantum effects at the macrolevel" (e.g., superconductivity); and quantum entanglement (e.g. Bose-Einstein condensate). These examples undermine the reductionist claim that the properties of parts entirely determine the properties of wholes.

He then gives numerous examples of macro-to-micro relations indicative of top-down action in physics (e.g., nucleosynthesis), in biology (e.g adaptation, expression of genetic information), and in human volition (e.g., intentions that lead to actions). Quantum physics provides several ways to understand how these phenomena can occur at the level of physics: interaction potentials, experiments and the collapse of the wavefunction, state preparation, decoherence, and the arrow of time. These features, according to Ellis, discredit reductionism in several ways. In hierarchically structured systems there is top-down action as well as bottom-up action. The outcome, even when there is determinism and mechanism at the microlevel, is partially effected by the context of boundary conditions, macro-constraints, and macroinfluences. Here systems thinking, based on synthesis, is needed as well as reductionistic analysis. Quantum entanglement provides another crucial argument against reductionism. Not only do cooperative effects between constituents of entangled states modify their individual behavior, but entanglement makes it hard to speak in terms of independent properties of constituents parts. Quantum uncertainty further undermines reductionism not only in microsystems but also in macrosystems when micro-uncertainties are amplified. Thus simplistic ideas of reductionism, that "we are nothing but the sum of particles controlled by forces at the microlevel," do not hold. They must be replaced by more sophisticated views integrating both bottom-up ("microcausation"), bottom-bottom ("co-operative"), and top-down ("context-setting") interactions. Still, while Ellis sees reductionism as wrong in principle, reductionism in practice is legitimate. He offers criteria for when it is suitable (e.g., when bottom-up causation dominates and microcomponents maintain their individual properties) and when it is not suitable (e.g., when either top-down causation is involved or when cooperative phenomena change the behavior of component parts). He closes this section with comments on two issues related to quantum theory and ontology: the possibility of chaotic/fractal structures in quantum processes, and the status of the theoretical and mathematical terms in quantum theory (e.g., are they human invention or Platonic realities?).

Ellis then develops an elaborate ontology to describe the many different aspects of material, human, and ideational reality, building on the works of Popper, Eccles, Penrose and others. In his hierarchical structure, ontological status and phenomenological laws of behavior are assigned to higher, as well as lower, levels. His ontology includes: WORLDS 1 (the physical world), 2 (individual and communal consciousness), 3 (Aristotelian possibilities), 4 (Platonic abstract realities), and 5 (underlying purpose). He offers for their foundation and ultimate context, WORLD 0 (God), and he describes the complex ways these WORLDS relate to each other. He concludes with a discussion of God's action in the world, drawing from his previous publications with Nancey Murphy. Divine action is kenotic, revealing God's purpose and the ethical core of God's nature through personal religious experience. Divine action, in turn, requires an openness in physical processes such that God's action has real causal effects in the physical world. The ontological nature of quantum uncertainty provides such openness. "The outcome of quantum measurements are fully under God's control, while being apparently random to humans..." Such effects at the quantum level can, in turn, affect the macrolevel without calling on chaos theory or getting "entangled in the problem of quantum chaology." He defends his view of divine action in light of problems such as quantum randomness, the existence of micro-to-macro effects, the suggestion that such divine action would be episodic, a possible violation of the conservation of energy, and other challenges.

Robert John Russell develops and extends the thesis that, if one interprets quantum mechanics philosophically as pointing to ontological indeterminism, then one can construct a robust bottom-up, noninterventionist, approach to objective, mediated, direct divine action. In this approach, God's indirect acts at the macroscopic level, understood as both general and special providence, arise in part from God's direct action at the quantum level. God sustains the deterministic time-development of elementary processes governed by the Schrödinger equation, and God also brings about irreversible interactions that are not described by the Schrödinger equation (i.e., "measurements") and that can lead to specific macroscopic effects. Thus divine action ultimately results in the regularities of our everyday world, which we attribute to general providence and describe by the laws of physics, and in specific macroscopic events which we view as acts of special providence.

Russell begins with clarifications and comments on methodology. His thesis does not explain how God acts or constitute an argument that God acts, but merely shows theological coherence of his theory of divine action with natural science. It is neither an epistemic nor an ontological "God of the gaps" argument. It does not reduce God to a natural cause; instead, God's action is hidden from science. It does not propose that God alters the wavefunction between measurements or other such views. It opts for a bottom-up approach, since this seems the best way to discuss God's action during the billions of years from the early universe to the evolution of primitive organisms on Earth; when sentient life is considered, bottom-up and top-down approaches should be combined. Finally, he responds to two questions. First, why should we take quantum mechanics seriously if it will one day be replaced? His response is that our alternative, classical physics, is wrong as a fundamental theory, and its depiction of nature as a closed causal system has already been thoroughly explored theologically. Second, how can we use quantum mechanics theologically if it can be given multiple philosophical interpretations? His response is that every scientific theory is open to multiple interpretations and that this poses a problem for all theological engagements with science. The key is that constructive theology can take a "what if" strategy, exploring the implications of one particular interpretation to its fullest without incurring the foundationalist problems of natural or physico-theology. Here he explicitly works within the Copenhagen indeterminist approach.

Next Russell turns to the measurement problem from the perspective of the Copenhagen interpretation. He distinguishes between the time development of the wavefunction governed by the deterministic Schrödinger equation, and the irreversible interactions between a quantum system and other systems to which the Schrödinger equation does not apply. Such interactions are routinely called "measurements," but he claims their scope is much wider than usually acknowledged. It includes: micro-macro (e.g., the absorption of a photon by the retina), micro-meso (e.g., the capture of an electron by an interstellar dust particle), and irreversible micro-micro (e.g., proton-proton scattering in the presence of heavy nuclei) interactions, though it does not include reversible micro-micro interactions (e.g., proton-proton scattering in free space). The phrase "the collapse of the wavefunction" is used loosely to suggest "what happens" during a measurement, where the inapplicability of the Schrödinger equation and thus intrinsic unpredictability is taken as pointing to ontological indeterminism. The term "quantum event" can be defined as referring to this collection of ideas related to irreversible interactions.

Turning to theological issues, Russell first argues that quantum statistics (Bose-Einstein and Fermi-Dirac), as well as the Schrödinger time evolution and irreversible interactions, together lead to the classical world we interpret via general providence. He weighs arguments for viewing special divine action as either "ubiquitous" or "episodic" and concludes that "pervasive" is more helpful. He proposes that the spatial and temporal characteristics of the wavefunction and its collapse in an irreversible interaction point to divine action as both global and local. Finally he discusses scientific and theological challenges raised by special relativity, suggesting that we need a richer theological conceptuality of "time and eternity."

He then discusses four crucial theological issues. First, does God act providentially in all quantum events (Murphy's option), or only in some (Tracy's option)? Russell prefers Murphy's option, though there are advantages and disadvantages to both. Second, if divine action and mind/brain top-down causality are both operative in acts of human volition, how do we avoid what Russell calls "somatic overdetermination"? After examining the responses by Murphy, Tracy, and Ellis, Russell suggests that God acts in all quantum events until the rise of life and consciousness, after which God limits God's action, leaving room for top-down, mind/brain causality. Third, why doesn't God act to minimize suffering, disease, death, and extinction in nature? Again, after assessing responses by Murphy, Tracy, and Ellis, Russell proposes that we can give a more persuasive response to theodicy if we move from creation theology (and thus providence) to a trinitarian theology of redemption, particularly as developed by Wolfhart Pannenberg. This, in turn, leads to Russell's fourth issue, which he sees as the crucial challenge to the theology-and-science discussion today: the meaning and intelligibility of the resurrection and eschatology in light of physics and cosmology.

Russell's essay includes an appendix on philosophical problems in quantum mechanics, including a proposed "architecture of philosophical issues," a discussion of Bell's theorem, and a comparison of nonlocality and (in)determinism in Bohm and Bohr's interpretations of quantum mechanics.

I. SCIENTIFIC AND HISTORICAL CONTEXT

Scientific Background

Abner Shimony
Raymond Y. Chiao
Michael Berry

Historical Antecedents

Ernan McMullin

THE REALITY OF THE QUANTUM WORLD

Abner Shimony

1 Introduction

We live in a remarkable era in which experimental results are beginning to elucidate philosophical questions. In no domain have the results been more dramatic than in quantum mechanics. The theory has been confirmed magnificently since the 1920s as its predictions of atomic, molecular, nuclear, optical, solid-state, and elementary-particle phenomena were shown to be accurate. Yet in spite of these successes the bizarre and counterintuitive character of quantum mechanics has led some investigators, including Einstein, to believe quantum-mechanical descriptions of physical systems are incomplete and in need of supplementation. Recent experiments show that this opinion is very likely wrong. The experimental results reveal more clearly than ever that we live in a strange "quantum world" that defies comfortable, commonsense interpretation.

Here are a few of the new, strange findings we must begin to accept. First, two entities separated by many meters and possessing no mechanism for communicating with each other nonetheless can be "entangled": they can exhibit striking correlations in their behavior, so that a measurement done on one of the entities seems instantaneously to affect the result of a measurement on the other. The finding cannot be explained from a classical point of view, but it agrees completely with quantum mechanics. Second, a photon, the fundamental unit of light, can behave like either a particle or a wave, and it can exist in an ambiguous state until a measurement is made. If a particle-like property is measured, the photon behaves like a particle, and if a wavelike property is measured, the photon behaves like a wave. Whether the photon is wave- or particle-like is indefinite until the experimental arrangement is specified. Finally, the notion of indefiniteness is no longer confined to the atomic and subatomic domains. Investigators have found that a macroscopic system can under some circumstances exist in a state in which a macroscopic observable has an indefinite value. Each of these findings alters drastically the way we perceive the world.

2 Quantum Basics

An understanding of these experiments and their philosophical implications requires some familiarity with the basic ideas of quantum mechanics. Essential to any discussion of the theory is the concept of the quantum state, or wavefunction. The quantum state specifies all the quantities of a physical system to the extent that it is possible to do so. The caveat at the end of the preceding sentence is crucial, because according to quantum mechanics not all quantities of a system have simultaneously definite values. The familiar Heisenberg uncertainty principle, which asserts that the position and the momentum of a particle cannot be simultaneously definite, is perhaps the best-known instance of this proposition.

What the quantum state of a system does provide unequivocally is the probability of each possible outcome of every experiment that can be done on the system. If the probability is one, the outcome is certain to occur; if the probability is zero, the outcome is certain not to occur. If, however, the probability is a number between zero and one, then it cannot be said in any individual case what the outcome will be. All that can be said is what, on the average, the outcomes of a specified experiment carried out on a large number of replica systems will be.

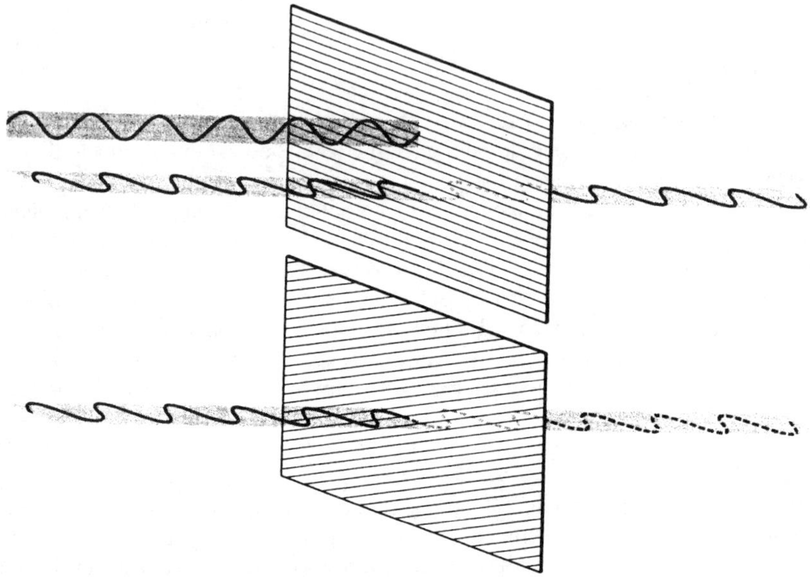

Figure 1. INDEFINITENESS of a quantum system is illustrated for a photon. A sheet of polarizing film transmits all light incident on it at a right angle if the light is linearly polarized along a certain direction in the film called the transmission axis (*hatching*). This polarization state of the photon is represented by the darker wavy line at the top. The film blocks all light incident on it at a right angle if the light is linearly polarized perpendicular to the transmission axis (*wavy gray line at top*). Now suppose a photon is linearly polarized at some angle to the transmission axis between zero and ninety degrees (*bottom*). Then whether or not the photon will be transmitted is indefinite; the probability of transmission is a number between zero and one (the square of the cosine of the angle).

Imagine, for instance, that measurements are made on a photon. The quantum state of the photon is fixed if three quantities are known: the photon's direction, its frequency and its linear polarization (the direction of the electric field associated with the photon). A suitable apparatus for measuring polarization is a sheet of polarizing film. The film is idealized so that it transmits all light incident on it at a right angle if the light is linearly polarized along a certain direction in the film called the transmission axis. The film blocks all light incident on it at a right angle if the light is linearly polarized perpendicular to the transmission axis.

Various experiments can be performed by rotating the polarizing film in different ways. If the photon is linearly polarized along the transmission axis, there is a probability of one that it will be transmitted. If the photon is linearly polarized perpendicular to the transmission axis, the probability that it will be transmitted is zero. A further implication of quantum mechanics, going beyond what has been said so far, is that if the photon is linearly polarized at some angle to the transmission axis between zero and ninety degrees, the probability of transmission is a number between zero and one (specifically, the square of the cosine of that particular angle).

If the probability is, say, one-half, then out of one hundred photons linearly polarized at the corresponding angle to the transmission axis fifty will be transmitted on the average.

Another basic idea of quantum mechanics is the superposition principle, which asserts that from any two quantum states of a system further states can be formed by superposing them. Physically the operation corresponds to forming a new state that "overlaps" each of the states from which it was formed. The concept can be illustrated by considering two quantum states of a photon in which the direction of the photon's polarization in the first state is perpendicular to the direction of the photon's polarization in the second. Then any number of states can be formed in which the photon's polarization points at some angle between the two perpendicular directions.

3 Implications of Quantum Mechanics

From these two basic ideas alone—indefiniteness and the superposition principle—it should be clear already that quantum mechanics conflicts sharply with common sense. If the quantum state of a system is a complete description of the system, then a quantity that has an indefinite value in that quantum state is objectively indefinite; its value is not merely unknown by the scientist who seeks to describe the system. Furthermore, since the outcome of a measurement of an objectively indefinite quantity is not determined by the quantum state, and yet the quantum state is the complete bearer of information about the system, the outcome is strictly a matter of objective chance—not just a matter of chance in the sense of unpredictability by the scientist. Finally, the probability of each possible outcome of the measurement is an objective probability. Classical physics did not conflict with common sense in these fundamental ways.

Even more startling implications flow from quantum mechanics if the system consists of two correlated parts. Suppose two photons fly apart in opposite directions. One possible quantum state of the pair of photons is the state in which both photons are linearly polarized along a vertical axis. Another possible state is the one in which they are both linearly polarized along a horizontal axis. There is nothing particularly bizarre or surprising about either of these two-photon quantum states, beyond the peculiarities of the single-photon states mentioned above. But if the superposition principle is brought into play, strange effects can occur.

In particular, by using the superposition principle one can form a quantum state that contains equal amounts of the vertically polarized state and the horizontally polarized state. This new state will figure prominently in what follows, and so it will be given a name, Ψ_0 (since the Greek letter psi is commonly used to represent a quantum state). The properties of Ψ_0 are most peculiar indeed. Imagine, for instance, inserting in the paths of the photons polarizing films with vertically oriented transmission axes. Because Ψ_0 contains equal amounts of the vertically and horizontally polarized states, there is a probability of one-half that both photons will be transmitted through their respective films and a probability of one-half that both will be blocked. What cannot happen is that one photon will be transmitted and the other will be blocked. In other words, the outcomes of the linear-polarization experiments on the two photons are strictly correlated.

The results will be the same if the polarizing films are oriented at an angle of forty-five degrees with respect to the horizontal: either both photons will be transmitted or both will be blocked. It simply cannot happen that one photon will be transmitted and the other will be blocked. In fact, it does not matter what the orientations of the films are as long as they match each other; the outcomes of the linear-polarization experiments are strictly correlated for an infinite family of possible experiments. (Of course, no more than one of the experiments can actually be carried out.) Somehow the second photon of the pair "knows" whether to pass through its polarizing film in order to agree with the passage or nonpassage of the first photon, even though the two photons are well separated and neither has a mechanism for informing the other of its behavior. In this kind of situation, then, quantum mechanics challenges the relativistic concept of locality, which holds that an event cannot have effects that propagate faster than light (and, in particular, instantaneous effects-at-a-distance).

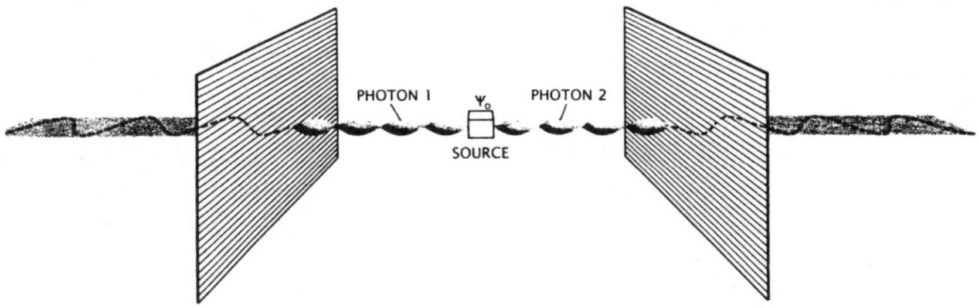

Figure 2. CORRELATIONS between the polarizations of two photons occur when the photons are in a special state called Ψ_0 (after the letter psi in the Greek alphabet). The state can be formed by superposing the state in which both photons are linearly polarized along a vertical axis with the state in which they are both linearly polarized along a horizontal axis. The state Ψ_0 contains equal amounts of the vertically polarized state and the horizontally polarized state. Now imagine that polarizing films with horizontally oriented transmission axes are inserted in the paths of the photons. Since Ψ_0 contains equal amounts of the two states, there is a fifty percent probability that both photons will be transmitted through their respective films and a fifty percent probability that both will be blocked. What cannot happen is that one photon will be transmitted and the other will be blocked: the outcomes of the linear-polarization experiments are strictly correlated. In fact, it does not matter what the orientations of the films are as long as they match each other: somehow the second photon of the pair "knows" whether to pass through its polarizing film in order to agree with the passage or nonpassage of the first photon, even though the photons are well separated.

4 Hidden Variables?

It must be emphasized that all the peculiar implications that have been drawn so far—objective indefiniteness, objective chance, objective probability, and nonlocality—depend crucially on the premise that a system's quantum state is a complete description of that system. A number of theorists have maintained, however, that the quantum state merely describes an ensemble of systems prepared in a uniform manner, and that this is why good predictions can be made about the statistical results of the same experiment performed on all members of the ensemble. At the same time, the argument goes, the individual members of the ensemble differ from one another in ways not mentioned by the quantum state, and this is the reason the outcomes of the individual experiments are different. The properties of individual systems that are not specified by the quantum state are known as hidden variables.

If hidden-variables theorists are correct, there is no objective indefiniteness. There is only ignorance on the part of the scientist about the values of the hidden variables that characterize an individual system of interest. Moreover, there is no objective chance and there are no objective probabilities. Most important, the quantum correlations of well-separated systems are no more surprising than the concordance of two newspapers printed by the same press and mailed to different cities.

In 1964 John S. Bell of CERN, the European laboratory for particle physics, showed that the predictions of local hidden-variables models are incompatible with the predictions of quantum mechanics. Reflection on some hidden-variables models of David Bohm of Birkbeck College London and Louis de Broglie led Bell to prove the important theorem that no model that is local (in a carefully specified sense) can agree with all the statistical predictions of quantum mechanics. In other words, there are physical situations in which the predictions of quantum mechanics disagree with those of every local hidden-variables model.[1]

The idea of Bell's theorem can be grasped, at least in part, by returning to consider the quantum state Ψ_0. As noted above, the results of linear-polarization experiments performed on a pair of photons in this state must be strictly correlated when the angle between the transmission axes of the two polarizing films is zero degrees (as it is when both axes are aligned vertically). It should not be surprising to learn, therefore, that for the state Ψ_0 there is always at least a partial correlation between the outcomes, no matter what the angle between the transmission axes is. (Specifically, if one of the photons is transmitted through its polarizing film, then the probability that the other photon will be transmitted through its film is the square of the cosine of the angle between the two transmission axes.)

Consequently a hidden-variables model that agrees with all the statistical predictions of quantum mechanics must assign quantities to each pair of photons in the ensemble in a delicate way in order to guarantee the strict or partial correlations at every angle between the axes. But the condition of locality requires that the quantities assigned to each photon in a pair must be independent of the orientation of the polarizing film on which the other photon impinges and independent of the other photon's passage or nonpassage. It is this locality condition that makes quite impossible the delicate adjustments that would be necessary for reproducing all the correlations, strict and partial, implied by Ψ_0.

5 Some Crucial Experiments

Bell's theorem shows that in principle one can determine experimentally which is correct: quantum mechanics or the local hidden-variables models. It was important to do such a test because, in spite of the immense body of confirming evidence for quantum mechanics at the time Bell proved his theorem, the very points where quantum mechanics is without equivocation irreconcilable with common sense had not yet been probed.

In 1969 John F. Clauser, then at Columbia University, Michael A. Horne of Boston University, Richard A. Holt, then at Harvard University, and I proposed a

---

[1] See Bernard d'Espagnat, "The Quantum Theory and Reality," *Scientific American* November, 1979: 128–40.

design for the requisite test. Pairs of photons with correlated linear polarizations were to be obtained by exciting atoms to an appropriate initial state; the atoms would subsequently return to the unexcited state by emitting two photons. Filters and lenses would ensure that when the photons flew off in opposite or virtually opposite directions, one photon would impinge on a polarization analyzer and the other would impinge on another analyzer. By switching between two orientations of each analyzer and recording the number of photon pairs transmitted in each of the four possible combinations of orientations of the two analyzers, measurements of correlations of transmissions of the photons of a pair could be made.

We suggested that either calcite crystals or piles of glass plates serve as the polarization analyzers, since each of them is much more efficient than an actual polarizing film in blocking photons polarized perpendicular to the transmission axis. Photodetectors placed behind the analyzers would detect a fixed fraction of the photons passing through the analyzers. If two photons, one at each detector, were registered within twenty nanoseconds (billionths of a second) of each other, the probability would be quite high that they were emitted by the same atom. Since the lenses would collect the two photons over a finite angle, the quantum state would not be exactly the Ψ_0 state discussed above but a modified state Ψ_1, which also leads to correlations that cannot be reproduced by any local hidden-variables model.

The experiment was done by Stuart J. Freedman and Clauser at the University of California at Berkeley in 1972, by Edward S. Fry and Randall C. Thompson at Texas A. & M. University in 1975, and by other groups after that. Most of the experimental results agree with the correlation predictions of quantum mechanics and disagree with the hidden-variables models. Moreover, the reliability of the dissenting experiments is suspect because of subtle weaknesses in their design.

Yet until very recently all the experiments had a loophole that allowed staunch defenders of hidden-variables models to maintain their hopes: the polarization analyzers were kept in their respective orientations for intervals of a minute or so, which is ample time for the exchange of information between the analyzers by some

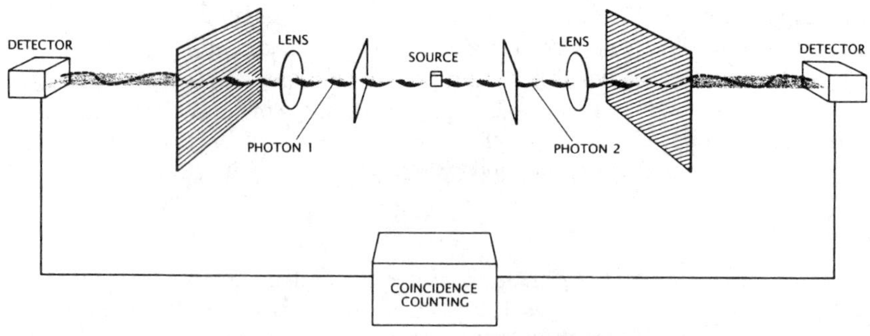

Figure 3. SEARCH FOR CORRELATIONS between members of pairs of photons was carried out in the 1970s by a number of investigators. The photon pairs were emitted in energy-state transitions of calcium and mercury atoms; each photon impinged on a polarization analyzer. Quantum mechanics predicts there must be delicate correlations in the passage or nonpassage of the photons through their analyzers, even though the photons have no apparent means of communicating with each other. The experiments mainly confirmed quantum mechanics, but they had a loophole: the orientations of the two analyzers were fixed before the photons were emitted. Consequently it was possible that information was somehow exchanged between the analyzers.

hypothetical mechanism. As a result the defenders could contend that the special theory of relativity does not imply the validity of Bell's locality condition in the physical situation of the experiments. But then these experiments would not serve as decisive tests between quantum mechanics and the local hidden-variables models.

To block this loophole, Alain Aspect, Jean Dalibard, and Gérard Roger of the Institute of Optics of the University of Paris did a spectacular experiment in which the choice between the orientations of the polarization analyzers is made by optical switches while the photons are in flight. In their experiment, which required eight years of work and was completed only in 1982, each switch is a small vial of water in which standing waves are periodically generated ultrasonically. The waves serve as diffraction gratings that can deflect an incident photon with high efficiency. If the standing waves are turned on, the photon will be deflected to an analyzer that is oriented one way; if the standing waves are turned off, the photon will travel straight to an analyzer that is oriented another way.

The switching between the orientations takes about ten nanoseconds. The generators that power the two switches operate independently, although (unfortunately for the complete definitiveness of the experiment) the operation is periodic rather than random. The distance between the analyzers is thirteen meters, so that a signal moving at the speed of light (the highest speed allowed by the special theory of relativity) takes forty nanoseconds to travel between them. Consequently the choice of orientation for the first polarization analyzer ought not to influence the transmission of the second photon through the second analyzer, and the choice of orientation for the second analyzer ought not to influence the transmission of the first photon through the first analyzer. The experimental arrangement is thus expected to satisfy Bell's locality condition. It follows that—according to Bell's theorem—there should be some violations of the quantum-mechanical predictions of correlations in the experimental outcome.

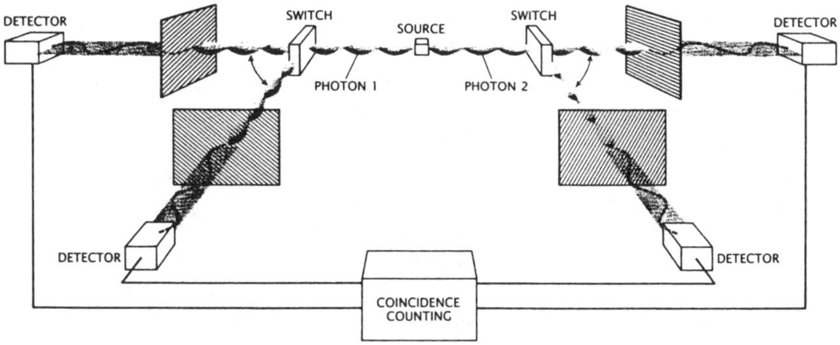

Figure 4. RAPID SWITCHING between orientations of polarization analyzers as photons flew was the hallmark of the experiment done by Aspect and his colleagues, which was completed in 1982. When a switch was "on," a photon was deflected to an analyzer that was oriented one way; when the switch was "off," the photon traveled straight to an analyzer that was oriented another way. The time required for light to travel between the analyzers was greater than the time required to switch between orientations, so that the choice of orientation for each analyzer could not influence the observation made at the other analyzer. (Unfortunately for complete definitiveness, however, the switching was periodic rather than random.) The experiment confirmed quantum mechanics; it would appear that the strange implications of the theory must be accepted.

In point of fact, however, the experiment yielded just the opposite result. The correlation data agree within experimental error with the quantum-mechanical predictions that are calculated on the basis of the quantum state Ψ_1. Moreover, the data disagree by more than five standard deviations with the extreme limits allowed, according to Bell's theorem, by any of the local hidden-variables models.

Even though the experiment of Aspect and his colleagues is not completely definitive, most people believe the prospects of overthrowing the results by future experiments are extremely small. It seems unlikely that the family of local hidden-variables models can be salvaged. The strange properties of the quantum world—objective indefiniteness, objective chance, objective probability and nonlocality—would appear to be permanently entrenched in physical theory. One of the strangest properties of the quantum world is nonlocality. Can the fact that under some circumstances a measurement on one photon apparently instantaneously affects the result of a measurement on another photon be capitalized on to send a message faster than the speed of light? Fortunately for the special theory of relativity the answer to the question is no. An underlying assumption of that theory—that no signal can travel faster than light—is preserved.

6 Nonlocality

Here is a brief argument that shows why. Suppose two people want to communicate by means of a device similar to the one for testing local hidden-variables models. Between the observers a source emits pairs of correlated photons. Each observer is provided with a polarization analyzer and a photodetector. The observers are free to orient the transmission axes of their analyzers any way they choose.

Suppose the observers agree to align the transmission axes vertically. Then every time a pair of photons is emitted there will be a strict correlation in the outcome: either both photons will pass through the analyzers or both will be blocked. But the strict correlation is of no value for each observer in isolation from the other. The first observer will note that half of the time photons pass through the first analyzer, on the average, and half of the time they are blocked. The second observer will note the same thing for the second analyzer. In other words, each observer in isolation sees only a random pattern of transmissions and blockages.

Now imagine that the first observer tries to encode some information and send it to the second observer by changing the orientation of the first polarization analyzer. Depending on the orientation of that analyzer, there will be either a strict or a partial correlation between the outcomes of the events at each detector. Once again, however, each observer will note that on the average half of the time photons pass through the analyzer and half of the time they are blocked. In general, no matter what the orientations of the analyzers are, each observer in isolation sees only a random (and statistically identical) pattern of transmissions and blockages. The quantum correlations between the photons can be checked only by comparing the data accumulated at the two detectors. Hence the attempt to exploit the quantum correlations to send messages faster than light cannot succeed.

In this sense there is a peaceful coexistence between quantum mechanics and relativity theory, in spite of quantum-mechanical nonlocality. For this reason it would be misleading (and wrong) to say that nonlocality in the quantum-mechanical sense is a reversion to action-at-a-distance, as in the pre-relativistic gravitational theory of Newton. It is tempting to characterize quantum-mechanical nonlocality as "passion-at-a-distance," not with any pretension to provide an explanation for the strange correlations, but only to emphasize that the correlations cannot be exploited to exert a controlled influence more rapidly than a light signal can be sent.

7 The Delayed-Choice Experiment

Another test, called the delayed-choice experiment, which was proposed in 1978 by John Archibald Wheeler, then at Princeton University, also reveals the strangeness of the quantum world. The basic apparatus of the experiment is an interferometer, in which a light beam can be split and recombined. A pulse of light from a laser is fired at the beam splitter, which is oriented in such a way that half of the light passes through the splitter and half is reflected at right angles to the direction of the incident pulse. If the light from the two paths is subsequently recombined, an interference pattern can be detected, which demonstrates the wave-like quality of light.

Now suppose the pulse of laser light is attenuated so severely that at any time there is only one photon in the interferometer. In this case two different questions can be asked about the photon. Does the photon take a definite route so that it is either transmitted or reflected by the beam splitter, thereby exhibiting a particle-like property? Or is the photon in some sense simultaneously transmitted and reflected so that it interferes with itself, thereby showing a wave-like property?

An answer was recently supplied by Carroll O. Alley, Oleg G. Jakubowicz, and William C. Wickes of the University of Maryland at College Park and independently by T. Hellmuth, H. Walther, and Arthur G. Zajonc of the University of Munich. Both groups found that a photon behaves like a particle when particle-like properties are measured and that it behaves like a wave when wave-like properties are measured. The remarkable novelty of the results is that the experiment was arranged so that the decision to measure particle-like or wave-like properties was made after each photon had interacted with the beam splitter. Consequently the photon could not have been "informed" at the crucial moment of interaction with the beam splitter whether to behave like a particle and take a definite route or to behave like a wave and propagate along two routes.

The length of both routes in the interferometer was about 4.3 meters, which a photon can traverse in roughly 14.5 nanoseconds. Obviously, this does not allow enough time for an ordinary mechanical device to switch between measuring particle-and wave-like properties. The feat was made possible with a switch called Pockels cell, which can be actuated in nine nanoseconds or less. A Pockels cell contains a crystal that becomes birefringent when a voltage is applied across it: light polarized along one axis of the crystal propagates at a velocity different from that of light polarized along the perpendicular direction. Given the proper choice of voltage and configurational geometry, light polarized in one direction when it enters the cell will emerge polarized in the perpendicular direction. The Pockels cell was inserted in one of the two routes the photon could take after interacting with the beam splitter (see figure 5 below).

A piece of polarizing film was the other essential element needed to switch between measurements of particle-like and wave-like properties. Light emerging from the Pockels cell impinged on the film. If the cell was "on," the polarization of the light was such that the polarizing film reflected the light into a photodetector, thereby answering the question of which route and confirming the photon's particle-like properties. If the cell was "off," the polarization of the light was such that the polarizing film transmitted the light, which then was combined with the contribution coming from the other route; interference effects confirmed the photon's wave-like aspect.

Both groups of investigators have reported results that are in excellent agreement with quantum mechanics. Their work shows that the choice between the two questions can be made after an individual photon has interacted with the beam splitter of an interferometer.

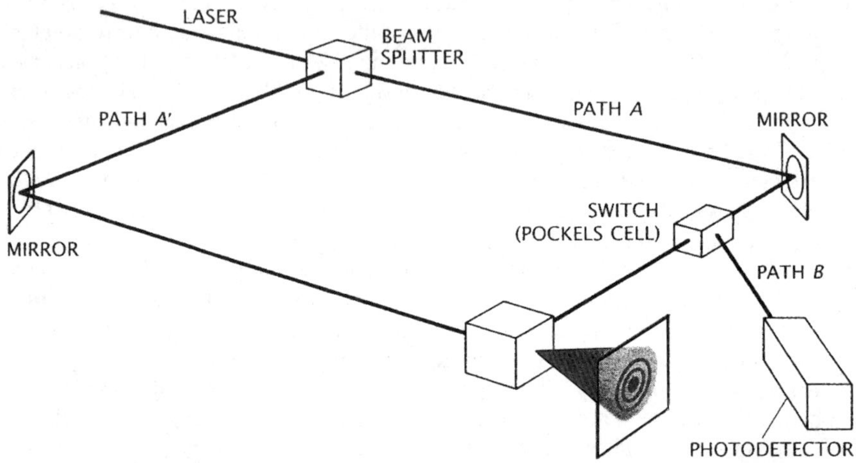

Figure 5. The Delayed-Choice Experiment is another test that reveals the strangeness of the quantum world. A photon impinges on a beam splitter. Two questions about the photon can be asked. Does the photon take a definite route so that it is either transmitted or reflected by the beam splitter, thereby exhibiting a particle-like property? Or is the photon in some sense both transmitted and reflected so that it interferes with itself, exhibiting a wave-like property? To find out, a switch is positioned in one of the two paths the photon can take after interacting with the beam splitter (*here, path* A). If the switch is on, the light is deflected into a photodetector (*path* B), thereby answering the question of which route and confirming the photon's particle-like properties. If the switch is off, the photon is free to interfere with itself (*paths A and A'*) and produce an interference pattern, demonstrating the photon's wave-like properties. Results from the experiment show that a photon behaves like a wave when wave-like properties are measured and behaves like a particle when particle-like properties are measured. Remarkably, the switch was triggered after the photon had interacted with the beam splitter, so that the photon could not have been "informed" whether to behave like a particle and take a definite route or to behave like a wave and propagate simultaneously along two routes.

How are the results of the delayed-choice experiment to be interpreted? It is worthwhile first to disclaim one extravagant interpretation that has sometimes been advanced: that quantum mechanics allows a kind of "reaching into the past." Quantum mechanics does not cause something to happen that had not happened previously. Specifically, in the delayed-choice experiment quantum mechanics does not cause the photon to take a definite route at time zero if twelve nanoseconds later the Pockels-cell switch is turned on, and it does not cause the photon to take both routes, in wave-like fashion, if the switch is off.

A more natural interpretation is that the objective state of the photon in the interferometer leaves many properties indefinite. If the quantum state gives a complete account of the photon, then that conclusion is not surprising, since in every quantum state there are properties that are indefinite. But the conclusion does raise a further question: How and when does an indefinite property become definite? Wheeler's answer is that "no elementary quantum phenomenon is a phenomenon until it is a registered phenomenon." In other words, the transition from indefiniteness to definiteness is not complete until an "irreversible act of amplification" occurs, such as the blackening of a grain of photographic emulsion. Students of the foundations of quantum mechanics disagree about the adequacy of Wheeler's answer, however. The next experiment shows why the question is still open.

8 Quantum Indefiniteness in the Macroscopic World

In 1935 Erwin Schrödinger proposed a famous thought experiment. A photon impinges on a half-silvered mirror. The photon has a probability of one-half of passing through the mirror and a probability of one-half of being reflected. If the photon passes through the mirror, it is detected, and the detection actuates a device that breaks a bottle of cyanide, which in turn kills a cat in a box. It cannot be determined whether the cat is dead or alive until the box is opened.

There would be nothing paradoxical in this state of affairs if the passage of the photon through the mirror were objectively definite but merely unknown prior to observation. The passage of the photon is, however, objectively indefinite. Hence the breaking of the bottle is objectively indefinite, and so is the aliveness of the cat. In other words, the cat is suspended between life and death until it is observed. The conclusion is paradoxical, but at least it concerns only the results of a thought experiment.

It is now more difficult to dismiss the paradoxical nature of the conclusion, because something similar to Schrödinger's thought experiment has recently been achieved by a number of groups of investigators including Richard F. Voss and Richard A. Webb of the IBM Thomas J. Watson Research Center in Yorktown Heights, Lawrence D. Jackel of the AT&T Bell Laboratories, Michael H. Devoret of Berkeley, and Daniel B. Schwartz of the State University of New York at Stony Brook. Their work has relied to a certain extent on calculations that were done by Anthony J. Leggett of the University of Illinois at Urbana-Champaign and Sudip Chakravarty at Stony Brook, among other investigators.

The experimental apparatus consists of an almost closed superconducting ring. A thin slice of insulating material (called a Josephson junction) interrupts the ring, but an electric current can circulate around the ring by "tunneling" through the insulator. The current generates a magnetic field.

The quantity that is of interest in the system is the magnetic flux through the ring, which (when the field is uniform) is equal to the area of the ring multiplied by the component of the magnetic field perpendicular to the plane of the ring. If the ring were uninterrupted, the flux would be trapped within the ring, but the insulator allows the flux to slip from one value to another. With modern magnetometers the flux can be measured with fantastic accuracy. The fact that the flux arises from the motion of enormous numbers of electrons (on the order of 10^{23}) in the superconducting ring justifies speaking of the flux as a macroscopic quantity. There is now good evidence that states of the superconducting ring can be prepared in which the flux does not have a definite value—a quantum-mechanical feature that had previously been established only for observables of microscopic systems.

Figure 6. A MACROSCOPIC SYSTEM can under some circumstances exist in a state in which a macroscopic variable has an indefinite value: indefiniteness is not limited to microscopic systems, such as the photon. The system shown here is a superconducting ring that does not quite bend back on itself. A thin slice of insulating material separates the two ends of the ring from each other, and an electric current is made to circulate around the ring by "tunneling" through the insulator. The current generates a magnetic field. If the ring were continuous, the magnetic flux through the ring (the area of the ring multiplied by the component of the magnetic field perpendicular to the plane of the ring) would be trapped at a fixed value, but the insulator allows the flux to slip from one value to another. Surprisingly, the flux does not have a definite value.

To understand how this indefiniteness is demonstrated experimentally, it is necessary to know that for each value of the flux the ring has a certain potential energy. Ordinarily one would not expect that the flux through the ring could change spontaneously from one value to another, because a potential-energy barrier separates adjacent values of the flux from each other. Classical physics forbids the transition between two such values of the flux unless some external source of energy, typically thermal, is supplied to traverse the barrier between them. In quantum mechanics, on the other hand, the barrier can be tunneled through without any external source of energy. The groups of investigators mentioned above have shown that the flux does change between two values, and that the change cannot be entirely accounted for thermally; the observed tunneling must be at least partially quantum mechanical, particularly at very low temperatures. But quantum-mechanical tunneling rests essentially on the indefiniteness of the flux, which thus cannot be localized definitely at or close to one value or another (see figure 7 below).

The experimental demonstration of quantum indefiniteness in a macroscopic variable does not ipso facto contradict the statement by Wheeler quoted above, but it does show that amplification from a microscopic to a macroscopic level does not in itself exorcise quantum-mechanical indefiniteness. The emphasis in Wheeler's statement about an "irreversible act of amplification" must be placed on the word "irreversible." The conditions for the occurrence of an irreversible process are far from settled in contemporary theoretical physics. Some students of the subject (including me) believe new physical principles must be discovered before we can understand the peculiar kind of irreversibility that occurs when an indefinite observable becomes definite in the course of a measurement.

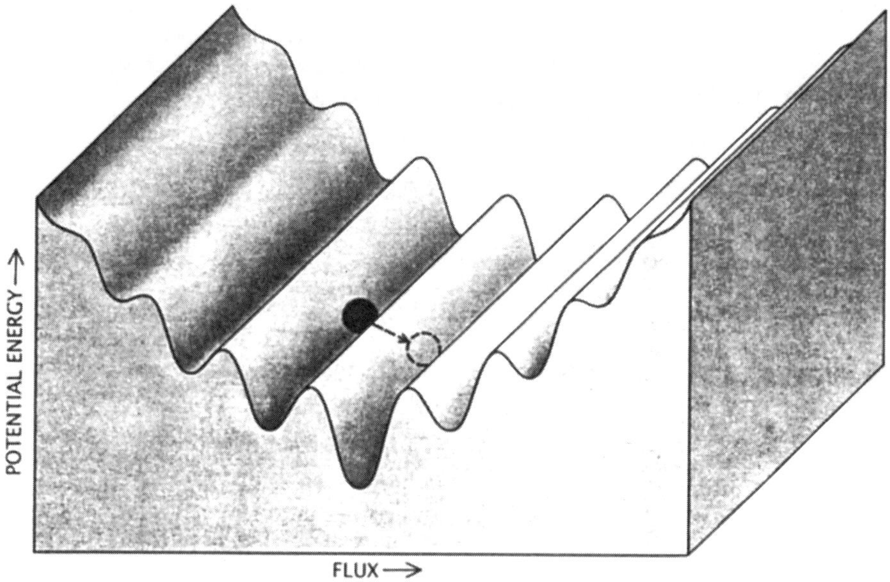

Figure 7. INDEFINITENESS in the system is depicted schematically. Each value of the flux through the superconducting ring has a certain potential energy associated with it. Ordinarily one would not expect that the flux through the ring could spontaneously change from one value to another, because a potential-energy barrier separates the adjacent values of the flux from each other. The barriers can be thought of as hills, and the state the system is in can be represented as a ball residing in a valley among the hills. According to classical physics, a transition between two values separated by a barrier requires outside energy (to push the ball over the hill). Quantum mechanically, however, the barrier can be tunneled through without any external source of energy. Tunneling is essentially a manifestation of the indefiniteness of the flux.

9 A Concluding Note: Neutron Interferometry and the Aharonov-Bohm Effect

The strangeness of the quantum world continues to be explored. Still other experiments have recently been performed or are now under way; two classes of these experiments should be mentioned here, even though there is no room to discuss them in detail. In the neutron-interferometer experiments of Helmut Rauch and Anton Zeilinger of the Atomic Institute of the Austrian Universities, Samuel A. Werner of the University of Missouri at Columbia, and Clifford G. Shull of the Massachusetts Institute of Technology and their associates, the wavefunction of a neutron is split by a sheet of crystal and recombined by one or two other sheets. The interference effects exhibited in the recombination demonstrate a number of remarkable properties, including the indefiniteness of the neutron's route through the interferometer.

Finally, R.G. Chambers of the University of Bristol, G. Möllenstedt of the University of Tübingen, and Akira Tonomura of Hitachi, Ltd., have confirmed by electron interferometry the remarkable Aharonov-Bohm effect, in which an electron "feels" the presence of a magnetic field that is in a region where there is zero probability of finding the electron. This is a striking demonstration of a kind of nonlocality different from, although remotely related to, the nonlocality exhibited by correlated photon pairs. A thorough understanding of the relation between the two kinds of nonlocality as well as the many other perplexing issues raised by experiments probing the nature of the quantum world awaits further work.[2]

[2] For further reading, see John Archibald Wheeler and Wojciech H. Zurek, eds., *Quantum Theory and Measurement* (Princeton, N.J.: Princeton University Press, 1983); Susumu Kamefuchi et al., eds., *Foundations of Quantum Mechanics in the Light of New Technology* (Tokyo: Physical Society of Japan, 1984); John C. Polkinghorne, *The Quantum World* (Princeton, N.J., Princeton University Press, 1985); Alastair I.M. Rae, *Quantum Physics: Illusion or Reality?* (Cambridge: Cambridge University Press, 1986).

QUANTUM NONLOCALITIES: EXPERIMENTAL EVIDENCE

Raymond Y. Chiao

τὰ πάντα ἐν αὐτῷ συνέστηκεν
("in him all things hold together," Colossians 1:17b)

1 Introduction

The purpose of this essay is to point out that certain experiments already show that the world as we know it possesses counterintuitive, nonlocal actions-at-a-distance of at least three kinds. Before beginning the discussion, let me say a word about methodology. It is important to emphasize that the scientific method is based primarily on an inductive, *a posteriori* reasoning whose starting point is experience, i.e., experiment, and is only secondarily based on a deductive, *a priori* reasoning whose starting point is a set of seemingly self-evident propositions about the world, i.e., axioms. Perhaps our theology should similarly also be based on experience.[1] Having had a conversion experience, in which I converted from atheism to the faith of Christ, I fully realize that the knowledge of God that gives life cannot be obtained by reason alone, but by a personal revelation of the Spirit. Nevertheless, I shall limit myself here to what we can learn, by scientific means alone, from laboratory experiments concerning the quantum nonlocalities of the universe in which we live.

By "quantum nonlocality," I shall mean quantum actions-at-a-distance, which have no classical explanation. By "actions-at-a-distance," I shall mean physical effects, events, or conditions, or more precisely, the *correlations* between such effects, events, or conditions that are separated by a space-like interval from each other, i.e., so that a light signal could not have passed between them. Such actions do not in fact violate special relativity, since, as we shall see, none of these actions-at-distance can be utilized to reverse the order of cause and effect, i.e., to send a signal that links a cause to its effect faster than light, and thus backwards in time. We shall see that it is the uncontrollable, probabilistic nature of quantum events, e.g., radioactive decays, whose fundamental uncontrollability originates from the uncertainty principle, which prevents any violation of relativistic causality. Nevertheless, some very counterintuitive quantum nonlocal effects have been observed to occur.

The three explicit kinds of quantum nonlocalities to be discussed here occur in:

1. the Aharonov-Bohm effect
2. the Einstein-Podolsky-Rosen effect
3. the tunnel effect

All three nonlocal effects stem from the superposition principle of quantum mechanics and quantum field theory, i.e., quantum interference. This fundamental principle states that an arbitrary superposition of two physically allowable states $|\psi_1\rangle$ and $|\psi_2\rangle$,

$$|\psi\rangle = c_1|\psi_1\rangle + c_2|\psi_2\rangle, \tag{1}$$

[1] Owen Thomas, "Theology and Experience," *Harvard Theological Review* 78 (1985): 179–201.

where c_1 and c_2 are two arbitrary complex numbers, is again a physically allowable state. The nonlocal concept of "state" or "wavefunction," which requires the initial specification of information about a physical system *everywhere at once*, is the source of nonlocalities in the quantum world. This kind of stipulation of the initial conditions of the system is necessitated by the first-order nature of the time-dependent Schrödinger equation. Note that the notion of "simultaneity" implicitly contained in the words "at once" in the stipulation of "everywhere at once" is in tension with the fact that simultaneity is frame-dependent in relativity. Also, note that it is difficult in practice to specify the necessary information about a system "everywhere," since for any finite-sized laboratory, the relativistic light-cone structure of spacetime restricts our knowledge of the world to within finite event horizons. (We are neither omnipresent nor omniscient!) These tensions between quantum theory and relativity still remain in relativistic quantum field theory (see Redhead's essay in this volume), but as far as we know, they do not result in any open contradictions.

The first and the third nonlocalities involve single-particle interference, but the second involves two-particle interference, i.e., a superposition state, or an "entangled state," of two particles. I shall not try to distinguish here between these three kinds of nonlocalities on theoretical grounds, but shall simply present some relevant experimental facts. At present, I do not have any helpful insights into their differences or their commonality, apart from the fact that they all stem from the superposition principle.

2 Nonlocality of the First Kind: The Aharonov-Bohm Effect

Consider the electron interference experiment around a torus performed by Tonomura and et al. (see fig. 1 below).[2] A torus has the important topological property of dividing space into an interior and an exterior that are physically separated from each other by the surface of the torus. An electron emerges from a source S and proceeds along two possible paths either through the middle or around the outside of the torus to a detector D, where an interference between the two possible paths occurs. The first path α passes from S through the hole at the center of the torus, to D, whereas the second path β passes from S outside of the outer diameter of a torus, to D. Neither path penetrates through the surface of the torus into its interior.

Now the interior of the torus is filled with a highly permeable ferromagnet, which traps almost all of the magnetic flux lines inside the torus. The surface of the torus is then coated with a superconductor, so that when cooled well below the superconducting transition temperature, the Meissner effect sets in, and all of the magnetic flux lines are bottled up inside the interior of the torus, i.e., inside the ferromagnet. Thus any stray magnetic field lines, which could have leaked to the outside, and could therefore have intersected with the possible paths of the electron, become exponentially small. Hence the electron experiences essentially no Lorentz force arising from the magnetic field of the ferromagnet. Furthermore, the surface of the superconductor is overcoated with a thick copper layer so that the wavefunction of the electron decays exponentially into the interior of the copper layer, and hence into the interior of the torus.

---

[2] Akira Tonomura et al., "Evidence for Aharonov-Bohm Effect with Magnetic Field Completely Shielded from Electron Wave," *Phys. Rev. Lett.* 56 (1986): 792–5.

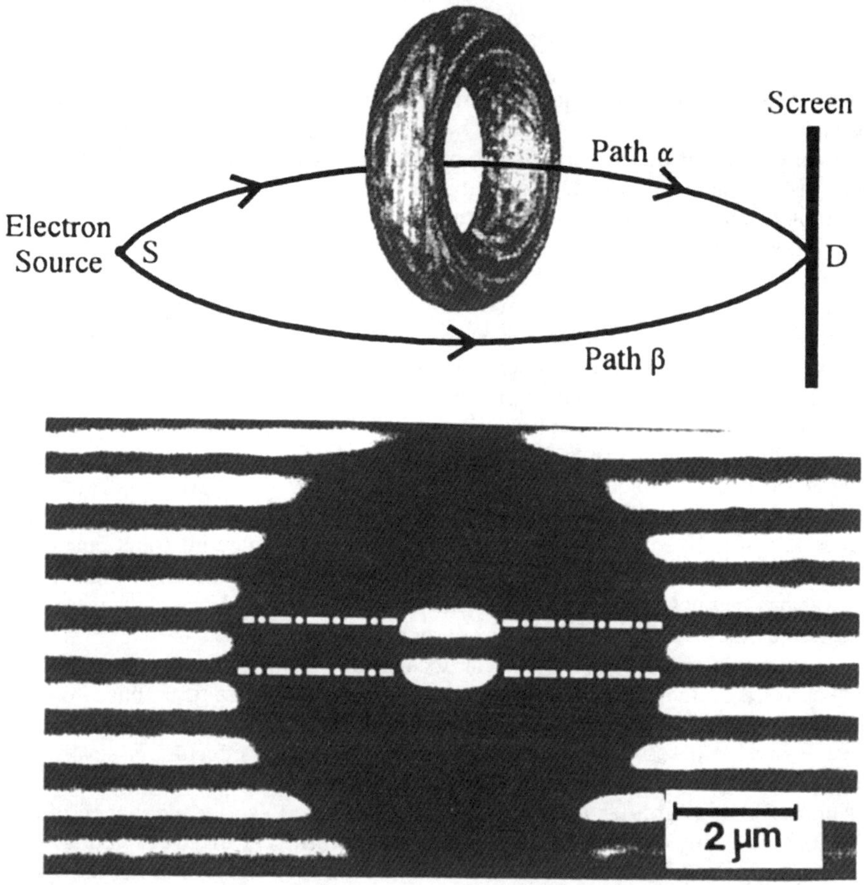

Figure 1. Schematic of an experiment to measure the Aharonov-Bohm phase shift in an electron interferometer. A source S emits an electron, which can travel either along path α through the hole of a ferromagnetic torus coated by a superconductor, or along path β around the outside of the torus, to a detector D. The observed interference fringes are shown below the schematic.[3]

In this way, the space of the electron interferometer is separated into two mutually exclusive regions. In the first region outside the surface of the torus, the electron wavefunction is allowed to propagate, but the magnetic field is forbidden from penetrating to the outside by the Meissner effect. In the second region inside the surface of the torus, the magnetic field is allowed to exist, but the electron wavefunction is forbidden to penetrate to the inside by the copper layer. Due to their mutual spatial separation, the electron never *locally* experiences, at any point along its possible paths, any nonvanishing magnetic field (and hence it never feels any Lorentz force), and the magnetic field never *locally* interacts with, at any point in its nonvanishing domain, the electron. Nevertheless, the phase shift of the electron

[3] Interferogram kindly furnished to the author by A. Tonomura.

wavefunction $\Delta\phi$, observable through the shift of the interference pattern detected by the detector D, is determined by the magnetic flux inside the torus as follows[4]:

$$\Delta\phi = \frac{e}{\hbar c}\oint_C \mathbf{A}\cdot d\mathrm{x} = \frac{e}{\hbar c}\Phi, \tag{2}$$

where e is the electron charge, $\hbar$ is Planck's constant divided by 2π, c is the speed of light, C is the closed circuit formed by paths α and β, $\mathbf{A}$ is the vector potential, $d\mathrm{x}$ is a line element of C, and Φ is the magnetic flux contained inside the torus, quantized in units of $2\pi\hbar c/2e$. Hence there is predicted to be a nonvanishing phase shift of π, which was observed by Tonomura et al. (see fig. 1 above).[5]

The Aharonov-Bohm effect demonstrates the existence of a quantum nonlocality of the first kind, a nonlocality that is topological in nature, as is apparent from the fact that the topology of the torus used in the above experiment plays a central role in it. The effect is nonlocal, in the sense that it is a global, topological feature of electromagnetic interactions, which has no classical, local explanation (remember that in this experiment, the electron never feels any local force, such as the Lorentz force). This nonlocality of the first kind is a result of a basic principle of quantum physics known as "the principle of local gauge invariance." The fact that the phase of the wavefunction in quantum mechanics can be locally altered without observable effects necessitates the existence of a nonintegrable phase factor (and not the existence of the vector potential itself, which is not a gauge-invariant quantity), which leads to the Aharonov-Bohm effect.[6] This fundamental invariance principle lies at the basis not only of the electromagnetic interactions, but also of the weak and the strong interactions of high-energy particle physics.

Although there exists no classical, local explanation of the Aharonov-Bohm effect based on the Lorentz force, we can explain the Lorentz force based on the Aharonov-Bohm effect (see fig. 2 below). Thus nonlocal quantum effects lie at a deeper level of description of the world than local classical effects. Classically, the trajectory of an electron that is injected into a region with a uniform magnetic field is curved into a circular arc. This can be understood as a consequence of the Aharonov-Bohm phase shift, which causes a tilting of the wavefront of the propagating electron's de Broglie wave. This wavefront is the surface normal to the local tangent of the electron's classical trajectory. Now apply Huygens's principle: two secondary wavelets that are emitted from points A and B on the original wavefront allow us to reconstruct the next wavefront of the propagating electron after an infinitesimal incremental step. Let us choose on the original wavefront the two secondary source points A and B that are located symmetrically on either side of the central point C, which represents the original central position of the electron on its trajectory, where the probability of finding it is a maximum.

The next wavefront of the propagating electron is determined by the following geometric construction: draw the common envelope, i.e., a surface that is the common tangent to all the secondary wavelets, including the points A' and B', which emanate from A and B. This envelope is the reconstructed wavefront, and the mid-

[4] Yakir Aharonov and D. Bohm, "Significance of Electromagnetic Potentials in the Quantum Theory," *Phys. Rev.* 115 (1959): 485–9.

[5] Tonomura et al., "Evidence for Aharonov-Bohm Effect."

[6] Tai Tsun Wu and C.N. Yang, "Concept of Nonintegrable Phase Factors and Global Formulation of Gauge Fields," *Phys. Rev. D* 12 (1975): 3845–57; Gordon Belot, "Understanding Electromagnetism," *Brit. J. Phil. Sci.* 49 (1998): 531–55.

point C' is the new central point through which the electron trajectory now passes. Note that in the presence of a magnetic field, the trajectory of the electron from C to C' must no longer be rectilinear: there now exists a nonvanishing Aharonov-Bohm phase shift between the secondary wavelets emanating from points A and B that is proportional to the enclosed area of the trapezoid $AA'B'B$ times the magnetic field, i.e., proportional to the magnetic flux through the shaded area of the figure. This phase shift can be calculated starting from equation 2. Therefore, as a consequence of the Aharonov-Bohm effect, the reconstructed electron wavefront is tilted, and its associated trajectory is no longer rectilinear but is now curved into a circular arc in the presence of the magnetic field, just as one would expect classically on the basis of the Lorentz force law. In fact, we can quantitatively derive this force law from the Aharonov-Bohm phase shift (see fig. 2). From this example, we can see underneath the local effects of classical physics to the deeper, nonlocal effects of quantum physics from which they spring.

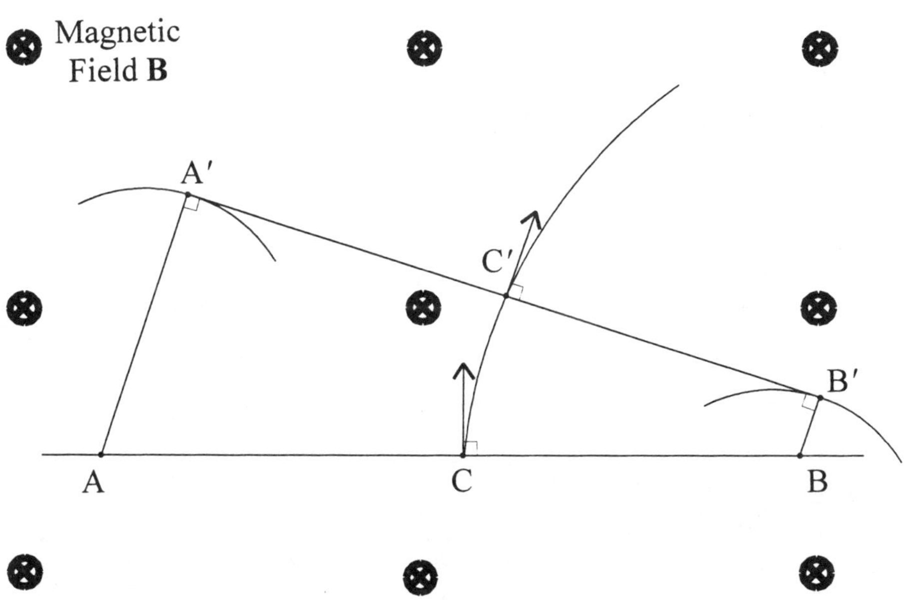

Figure 2. The Lorentz force law comes out from the Aharonov-Bohm effect. Consider an electron plane wave with a momentum **p** propagating in a uniform magnetic field **B**. Apply Huygens's principle. The secondary wavelets emitted at points A and B acquire a phase shift that is proportional to the shaded area of the trapezoid $AA'B'B$, after propagation for a short amount of time Δt. This implies that the reconstructed wavefront will be tilted with respect to the original wavefront by a small angle $\Delta\theta = \Delta p/p = \Delta\phi/kl$, where $\Delta\phi$ is the Aharonov-Bohm phase given by equation 2, $k = p/\hbar$ is the wave number of the plane wave, and l is the distance between A and B. In the short wavelength limit, a simple calculation shows that the trajectory CC' of the electron will now be curved as if by an effective force of magnitude $F = dp/dt = evB/c$, i.e., the Lorentz force.

3 Nonlocality of the Second Kind: The Einstein-Podolsky-Rosen Experiments

Einstein, Podolsky, and Rosen[7] pointed out that two particles emitted from a common source can exhibit nonlocal correlations upon measurement, even long after they have been separated by a large distance from each other, and therefore are no longer able to interact significantly with each other by means of any long-range force. For example, consider two spin-1/2 atoms emitted in a decay into opposite directions from a common spin-0 source prepared in the Bohm singlet state,[8] which is a superposition of two product states, i.e., an "entangled state,"

$$|S_{total} = 0\rangle = \frac{1}{\sqrt{2}}\left\{|S_{z1} = +1/2\rangle|S_{z2} = -1/2\rangle - |S_{z1} = -1/2\rangle|S_{z2} = +1/2\rangle\right\}, \quad (3)$$

such as occurs when a diatomic silver molecule photodissociates into two silver atoms (see fig. 3). Note that as a result of the *superposition* of the two product states, this state is intrinsically *nonfactorizable*. This means that the results of spin measurements on the two silver atoms, even when they are separated by a large distance from each other, are *not independent of each other*, but rather are *contingent* on the conditions of the measurement made on the other atom. For example, when a measurement of one atom by a detector located at the left end of an experimental apparatus after a Stern-Gerlach analyzer yields a definite result corresponding to a spin state $|S_{z1} = +1/2\rangle$, then conservation of the total spin angular momentum of the system necessarily requires that immediately upon this measurement, the other atom at the right end must definitely be in a spin state with $|S_{z2} = -1/2\rangle$. Einstein, Podolsky, and Rosen argued that a combination of the principles of realism and locality requires that this result must be *independent* of whether or not any measurement is actually made on the other atom at the right end of the apparatus. Note, however, that the choice of the quantization axis (i.e., the *z*-axis of the Stern-Gerlach magnet) is up to the experimenter, and that the action of turning the angle of the Stern-Gerlach analyzer by the experimenter at left end of the apparatus immediately forces a change in the *conditions* for the possible outcomes of measurements at the right end, which may be a long distance away.

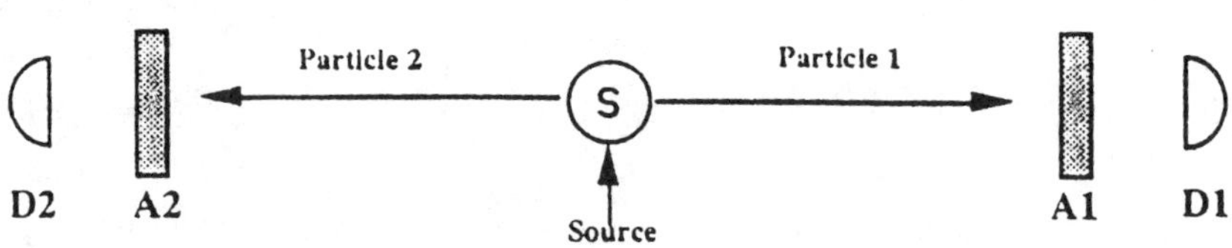

Figure 3. Schematic of a typical Einstein-Podolsky-Rosen experiment. The source S emits two particles in an entangled state, such as two silver atoms in the Bohm singlet state, equation 3. A1 and A2 represent two analyzers, such as two Stern-Gerlach apparatuses, and D1 and D2 represent two detectors.

Einstein, Podolsky, and Rosen argued starting from the principle of realism as expressed in their statement,

---

[7] Albert Einstein, B. Podolsky, and N. Rosen, "Can Quantum-Mechanical Description of Physical Reality Be Considered Complete?" *Phys. Rev.* 47 (1935): 777–80.

[8] David Bohm, *Quantum Theory* (Englewood Cliffs, N.J.: Prentice-Hall, 1951), 614.

> If, without in any way disturbing a system, we can predict with certainty (i.e., with probability equal to unity) the value of a physical quantity, then there exists an element of physical reality corresponding to this physical quantity,[9]

which, when combined with the principle of locality, requires that each particle in the pair possesses a local, definite value of spin *before* any measurement has been made. For example, according to their argument, the particle on the right side of the above experiment must have had a definite value of spin, namely $S_{z2} = -1/2$, even *before* we have made any measurements on it. Since standard quantum theory does not allow for any such "elements of physical reality," Einstein, Podolsky, and Rosen concluded that quantum theory must be incomplete.

Bell's theorem[10] provides a crucial insight into the meaning of the Einstein-Podolsky-Rosen-type experiments. *Locality* is again one of the main assumptions that goes into the proof of this theorem: the outcomes of measurements in a finite spacetime region A cannot be affected by any actions, specifically the choice by the experimenter of settings of measuring devices (e.g., the direction of the axis of the Stern-Gerlach magnet, in a well-separated spacetime region B).[11] Following the lead of Einstein, Podolsky, and Rosen, Bell combined this locality assumption with the assumption of *realism*, i.e., that particles possess definite physical properties, such as their spin, *independent* of any acts of observation, detection, or measurement. He thereby obtained heuristically a *strong separability* condition, i.e., the *factorizability* of joint probabilities of space-like separated coincidence detections. This factorizability is the most important step in the proof of Bell's theorem, but it is in direct contradiction with the *nonfactorizability* of the superposition states such as the entangled states given by the above Bohm singlet state (eq. 3), and hence in contradiction with the *nonseparability* of such states in quantum theory.

Many experiments, mainly using photons, including ours at Berkeley, such as in the Franson experiment[12] (see below), have shown that the inequalities that follow from Bell's theorem are violated (apart from technical caveats concerning detection loopholes, etc.), and that the predictions of quantum theory, which also violate these inequalities, are confirmed. An important implication of the observed violations of Bell's inequalities is that, in contradiction to Einstein, Podolsky, and Rosen's belief, all properties of particles such as position, momentum, energy, spin, etc., do not have any physical meaning before measurement, and, in this sense, all these properties *do not even exist until they are observed.* This reminds us of Bishop Berkeley's famous aphorism: *Esse est percipi* ("To be is to be perceived"). Berkeley believed that the universe exists because God is its Observer.

These observational results imply that some or all of the seemingly self-evident assumptions used in the proof of Bell's theorem are inconsistent with experiment. Setting aside the question of which assumptions should be abandoned in light of the

---

[9] Einstein, Podolsky, and Rosen, "Can Quantum-Mechanical Description of Physical Reality Be Considered Complete?", 777.

[10] John S. Bell, "On the Einstein Podolsky Rosen Paradox," *Physics* 1 (1964): 195–200.

[11] "Well-separated" means that all points of A have space-like separations from all points of B. Thus no light signals can propagate from any point in B to any point in A, and vice versa, and therefore they must be causally disconnected.

[12] Paul G. Kwiat, A.M. Steinberg, and R.Y. Chiao, "High-Visibility Interference in a Bell-Inequality Experiment for Energy and Time," *Phys. Rev. A* 47 (1993): R2472–5.

observed violations of Bell's inequalities,[13] the point here is simply that nonlocal correlations of space-like separated events (e.g., simultaneous "clicks" in distant Geiger counters) have been observed, and that these nearly perfect correlations have been smoothly changed to nearly perfect anti-correlations by the local action of the experimenter (e.g., by turning a knob that changes the orientation of an analyzer axis). There thus exists a type of *action-at-a-distance*, in which the experimenter's *local action* (freely choosing the experimental settings) changes Bohr's *global conditions* of the experiment,[14] changing the nature of the observed correlations-at-a-distance continuously from essentially perfect correlations to essentially perfect anti-correlations. (However, no signals can be sent faster than light, since only *after* correlating the otherwise completely random pattern of single-detector count rates does this action-at-a-distance by the experimenter become apparent.) This kind of action-at-a-distance stems from the *nonfactorizability* of entangled states, such as the Bohm singlet state, which implies a kind of *nonseparability* of the world.[15]

As an example of this second kind of nonlocality, let us review the Berkeley version of the Franson experiment.[16] The light source that we used produced two photons that are prepared in an entangled state of energy, by means of spontaneous parametric down-conversion in a nonlinear crystal (see fig. 4).[17] This was also the light source that was later used in our tunneling-time experiments.[18] A well-collimated, monochromatic beam of light from an ultraviolet laser is incident on a nonlinear crystal of potassium dihydrogen phosphate. During the process of parametric down-conversion inside the crystal, a rainbow of many colors is generated in conical emissions around the ultraviolet laser beam, in which one parent ultraviolet photon breaks up into two daughter photons, conserving energy and momentum in the process. This down-conversion process is very much like radioactive decay, in which a parent particle decays spontaneously into two daughter particles. Two pinholes are used to select out the two daughter photons on opposite sides of the laser beam, whose energies are approximately one-half of that of the parent photon.

Thus the two daughters share the original energy of the parent photon approximately equally. These two photons are "twins," in that they are observed to be born

---

[13] However, for a more expanded discussion on this point, see the articles in this volume by Cushing and Redhead, as well as Raymond Y. Chiao and J.C. Garrison, "Realism or Locality: Which Should We Abandon?", *Foundations of Physics* 29 (1999): 553–60.

[14] In his response to Einstein, Podolsky, and Rosen, Bohr proposed the idea of "an influence on the very conditions which define the possible types of predictions regarding the future behavior of the system"; Niels Bohr, "Can Quantum-Mechanical Description of Physical Reality be Considered Complete?", *Phys. Rev.* 48 (1935): 696.

[15] Paul C. Davies, *God and the New Physics* (New York: Simon & Schuster, 1983); Bernard d'Espagnat, *Veiled Reality* (Reading, Mass.: Addison-Wesley, 1995).

[16] Kwiat et al., "High-Visibility Interference."

[17] David C. Burnham and D.L. Weinberg, "Observation of Simultaneity in Parametric Production of Optical Photon Pairs," *Phys. Rev. Lett.* 25 (1970): 84–7.

[18] Aephraim M. Steinberg, P.G. Kwiat, and R.Y. Chiao, "Measurement of the Single-Photon Tunneling Time," *Phys. Rev. Lett.* 71 (1993): 708–11; Aephraim M. Steinberg and R.Y. Chiao, "Subfemtosecond Determination of Transmission Delay Times for a Dielectric Mirror (Photonic Band Gap) as a Function of the Angle of Incidence," *Phys. Rev. A* 51 (1995): 3525–8; Raymond Y. Chiao and A.M. Steinberg, "Quantum Optical Studies of Tunneling and Other Superluminal Phenomena," in *Proceedings of the Nobel Symposium No. 104* on "Modern Studies of Basic Quantum Concepts and Phenomena," E.B. Karlsson and E. Brändas, eds., *Physica Scripta* T76 (1998): 61–6.

simultaneously inside the crystal, and hence to cause two simultaneous "clicks," which are always registered *in coincidence* in the two equidistant Geiger counters used to detect them. However, due to the finite size of the two pinholes that subtend a significant fraction of the spectrum of the rainbow, these two photons possess a broad frequency spectrum, and hence a broad energy spectrum. From Heisenberg's energy-time uncertainty principle, $\Delta E \Delta t \geq \hbar/2$, it follows as a consequence of their broad spectrum that the two twins must both be produced in a wavepacket of very short time duration. Each of these two "twin" wavepackets contain one and only one photon. The short duration of these wavepackets makes it possible to make a precise timing measurement in the coincidence detection of one down-converted photon relative to the detection of its twin, a feature that will become important later in the discussion of our tunneling-time measurements.

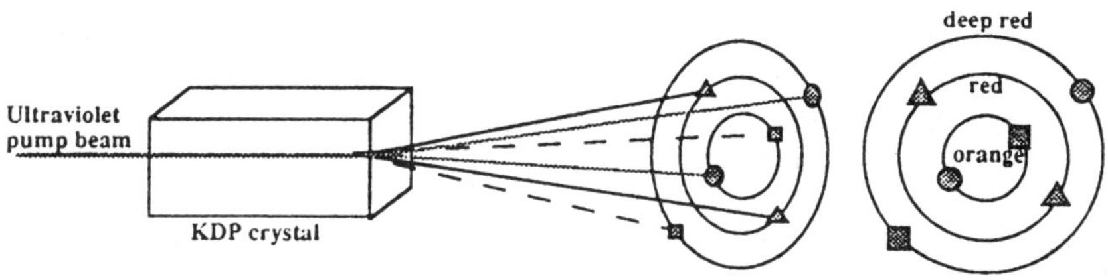

Figure 4. The parametric down-conversion light source produces correlated pairs of photons in a rainbow of conical emissions around an ultraviolet laser pump beam. Matching shapes (circles, triangles, and squares) indicate correlated photon pairs that would be detected in coincidence. Rings indicate different colors of the rainbow. A "parent" ultraviolet photon decays in this down-conversion process into a pair of "daughter" photons, symbolized by a matching pair of shapes. In the Berkeley experiments, two daughter photons of approximately equal energy (represented here by the pair of triangles) were selected out by means of two pinholes (not shown) for coincidence detection.

The parent ultraviolet photon has a very sharp spectrum in energy, since it was produced by a monochromatic laser, but for the above reasons the two twin down-converted daughter photons each possess a very broad spectrum, arising from the finite size of the pinholes. However, the *sum* of the two energies of the two individual photons must be very sharp in value, and must be exactly equal, upon each coincidence detection, to the very sharp energy of the parent photon, since energy must be conserved during the down-conversion process. Hence the twins exhibit a strict anti-correlation in the amount of energy they possess upon coincidence detection: whenever one twin possesses an energy greater than half the parent's energy, the other twin must necessarily possess less than half the parent's energy, such that the sum of their two energies exactly equals the sharp value of the energy of the parent. This is true no matter how far apart they are in space when they are detected. Therefore the outgoing quantum state of the two photons is the entangled state:

$$|2 \text{ photons}\rangle = \int_0^\infty \int_0^\infty dE_1 dE_2 \delta(E_0 - E_1 - E_2) A(E_1, E_2) |E_1\rangle |E_2\rangle, \tag{4}$$

where E_0 is the sharp energy of the parent ultraviolet laser photon, and E_1 and E_2 are the unsharp energies of the two down-converted daughter photons.

The wavelength of the ultraviolet laser that produced the parent photons in the Berkeley experiments was 351 nanometers. The crystal was cut with an optic axis oriented so that the two twin daughter photons at around the same red wavelength of 702 nanometers emerged at a small angle of a few degrees relative to each other. We used two millimeter-sized pinholes to select out these two twin photons. The size of these pinholes limits the bandwidth of the light that passes through them, and the resulting single-photon wavepackets have temporal widths around 20 femtoseconds or a bandwidth of around 6 nanometers in wavelength.

One striking consequence of the above entanglement of the energies of the two daughter photons that we observed in an earlier experiment is the following "collapse" phenomenon: a sharp measurement of the energy of one twin, *A*, by means of a very narrow-band filter (a Fabry-Perot interferometer) placed in front of the detector of twin *A*, forces a nonlocal "collapse of the wavepacket" of the other twin, *B*, (using the "collapse" language of Heisenberg) to a wavepacket of a very long duration that corresponds to the sharpness of the energy measurement of twin *A*. Thus the "collapsed" duration of the wavepacket of twin *B* matches the very narrow bandwidth of the filter through which twin *A* passed, but through which twin *B* has never passed! A subsequent measurement using coincidence detection of the duration of the "collapsed" wavepacket of twin *B* by means of a Michelson interferometer, whose arm length difference matches the much expanded length of the "collapsed-in-energy" wavepacket of twin *B*, now exhibits interference fringes, indicating that the wavepacket of twin *B* has indeed increased dramatically in its duration, so that it can overlap with itself at the central beam splitter of the Michelson interferometer, and thereby interfere with itself. This occurs even when twins *A* and *B* are distantly separated from each other.[19]

The Franson experiment is a modification of the above experiment, in which the Fabry-Perot and the Michelson interferometers used in the detection of the two twins are replaced by two identical unbalanced Mach-Zehnder interferometers, as originally proposed by Franson (see fig. 5).[20] For this modified experiment, one can formulate a Bell inequality. We have observed a violation of this inequality, which implies that the twin photons do not possess any definite energy (or color), nor do they possess any definite time of emission (or birthday), prior to their detection (i.e., prior to the registration of the "clicks" in the detectors). Thus, in a sense, these properties are created *ex nihilo* during the act of measurement.

Each twin photon enters its own, physically separated, Mach-Zehnder interferometer through an initial beam splitter, after which there is a choice for it either to take a long path or a short path to a final beam splitter, before detection. The long arm of each of the two Mach-Zehnder interferometers can be continuously scanned with respect to the short arm, so that the path length difference for each of the photon twins can be separately changed at will by the experimenter. When one photon is detected *independently* of its twin, we observed no interference fringes (i.e., maxima interleaved with minima in the "singles" detection count rate), when the path length difference of its own interferometer is being scanned. However, when this photon is detected *in coincidence* with its twin, interference fringes now appear

---

[19] Raymond Y. Chiao, P.G. Kwiat and A.M. Steinberg, "The Energy-time Uncertainty Principle and the EPR Paradox: Experiments Involving Correlated Two-photon Emission in Parametric Down-Conversion," in *NASA Conference Publication 3135*, D. Han et al., eds. (Washington, D.C.: NASA, 1991), 61–71; Paul G. Kwiat and R.Y. Chiao, "Observation of a Nonclassical Berry's Phase for the Photon," *Phys. Rev. Lett.* 66 (1991): 588–91.

[20] James D. Franson, "Bell Inequality for Position and Time," *Phys. Rev. Lett.* 62 (1989): 2205–8.

during this scanning process. More remarkably, these interference fringes still appear, with no change in intensity, when only the path length difference of its *twin's* interferometer, *through which it has never passed*, is being scanned.

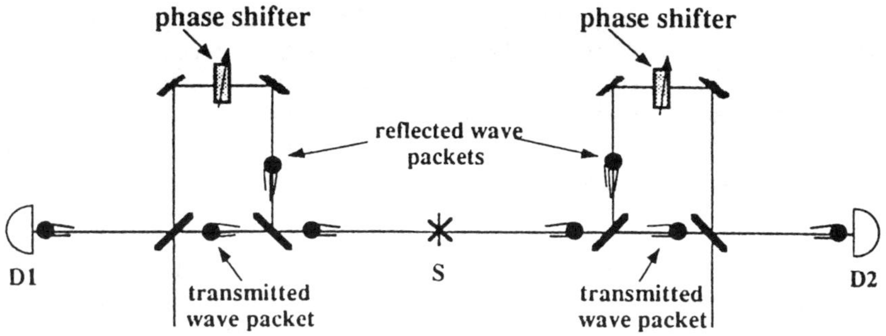

Figure 5. The Franson experiment. The interference of two spatially separated, but entangled, photons in two unbalanced Mach-Zehnder interferometers.

It is easy to understand why no interference fringes are observed when one photon is detected independently of its twin: the width of its wavepacket, which is typically on the order of 10 micrometers (corresponding to its typical 30 femto-second duration), is much smaller than the difference between the long and short paths of its interferometer, which is typically on the order of a meter. Hence the transmitted and reflected portions of the wavepacket miss each other by a wide margin at the final beam splitter, and no interference fringes in the single count rates, which would have arisen from the overlap of these portions of the wavepacket, are observed (see fig. 6).

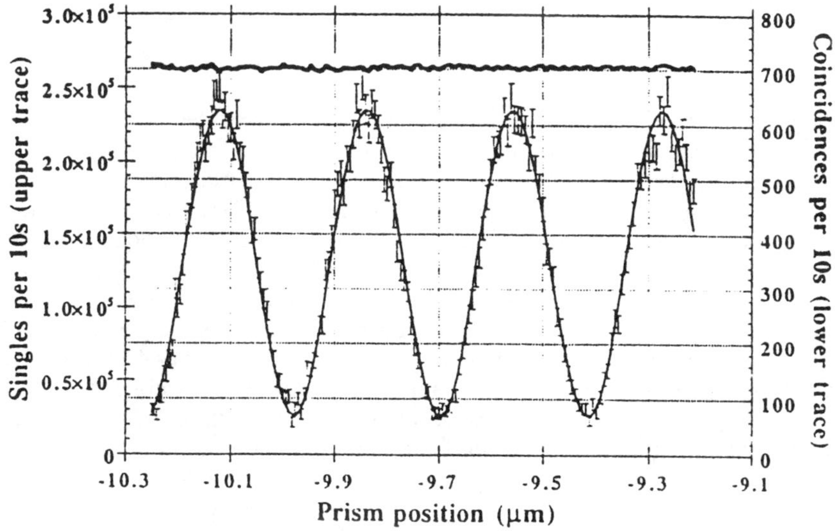

Figure 6. Interference fringes seen in the Berkeley Franson experiment in coincidence detection. Upper trace: the singles rate exhibit no interference fringes. Lower trace: The coincidence count rate between the two detectors for the two photon twins of figure 5, as a function of the path length difference between the long and short arms of one of the two unbalanced Mach-Zehnder interferometers, exhibit fringes with very high visibility (i.e., contrast).

However, there exists a second-order (i.e., two-photon) interference effect that we observed in coincidence detection: the coincidence count rate exhibits interference fringes, with a very high contrast observed between maxima and minima (see fig. 6 above). The easiest way to calculate this effect is to employ Feynman's rules for interference: if two or more *indistinguishable* quantum processes result in the same final outcome, add the complex probability amplitudes for these processes, and then take the absolute square of the result to obtain the probability of this outcome. Otherwise, if they are *distinguishable*, add their probabilities. Applying these rules to the two Mach-Zehnder interferometers, we note first that when a coincidence "click" occurs at the two detectors at opposite ends of figure 5, there is no way to tell whether the photon twins both have passed through their individual interferometers along the short paths, or both have passed through the long paths, since both possibilities end up with the same final outcome, namely, simultaneous "clicks" at the two Geiger detectors that register in a coincidence counter. Hence the indistinguishable processes here are (i) the "short-short" process, where both photons take the short paths, and (ii) the "long-long" process, where both take the long paths. The distinguishable processes are (iii) the "short-long" process, where one photon takes the short path and its twin takes the long path, and (iv) the "long-short" process, where the opposite occurs. In principle and also in practice, we are able to reject these distinguishable processes by using a sufficiently large path length difference between the long and the short paths, and also by using a sufficiently narrow coincidence timing window in our electronics, so that the time ordering of the two "clicks" in the two detectors for processes (iii) and (iv) become clearly distinguishable. We are thus left with the two indistinguishable processes (i) and (ii), for which we must first add the *amplitudes* before taking the absolute square. Hence the probability of coincidence detection is given by

$$|\text{short}\cdot\text{short}+\text{long}\cdot\text{long}|^2=|1\cdot 1+\exp(i\phi_2)\cdot\exp(i\phi_2)|^2=2(1+\cos(\phi_1+\phi_2)), \quad (5)$$

where ϕ_1 and ϕ_2 are the phase differences arising from the difference between the long and short arms of the left and right interferometers, respectively. This Feynman calculation predicts a 100% visibility, i.e., minima that consist of perfect nulls, of the interference fringes. However, a Bell inequality for this experiment (when certain reasonable auxiliary assumptions, such as the fair sampling assumption, are made) implies that sinusoidal fringes can have at most $1/\sqrt{2}$ or 70.7% visibility. The data in figure 6 above show a violation of the Bell inequality by sixteen standard deviations. This implies that *local realistic* theories, in which particles are assumed to be objects possessing definite properties, such as energy or time of emission, which exist before any measurements have been made, are inconsistent with experiment. In fact, these properties must have come into existence, that is, must have been created *ex nihilo* upon their observation, detection, or measurement. This again reminds us of Bishop Berkeley's aphorism: *Esse est percipi*.

To sum up, in the Franson experiment two twin photons that were so far apart when they were detected that they could not possibly have communicated with each other, still behave in correlated ways at the two final beam splitters, as if they knew about each other's behavior, and also about the experimenter's action, at a distance. We have demonstrated that the behavior of the two distantly separated photons at the two final beam splitters (whether they decide to be transmitted or reflected at these beam splitters) can be changed, controllably and smoothly, from a nearly perfect correlated behavior to a nearly perfect anti-correlated one, by a mere turning of a knob by the experimenter, who smoothly changes the path length of one and only one of the interferometer arms. Seemingly this should affect the path length of the photon that passes through it, but not the behavior of the twin that does not. However, this is *not* the case. The twin seems to know about the experience of its sibling.

4 Nonlocality of the Third Kind: The Photon Tunneling-time Experiments at Berkeley

The third manifestation of nonlocality occurs in the phenomenon of quantum tunneling,[21] in connection with the question: How long does it take for a particle to tunnel through a barrier? It is very important in posing this question to state precisely at the outset the *operational* definition of the quantity being measured. For the tunneling time for our experiments, this definition is based on the following *Gedanken* experiment (see fig. 7 below). Suppose, as discussed above, that a single parent particle (a photon) decays into two daughter particles (two photons). Suppose further that these two daughter particles have the same speed in the vacuum (i.e., *c* in our case), and that they were detected by means of two detectors placed at equal distances from the point of decay. There result simultaneous "clicks" at the two equidistant detectors, e.g., two Geiger counters, which could then be registered in a coincidence counter. We shall interpret the time that a "click" occurred at a Geiger counter, as registering the time of arrival of a particle at that detector.

Now suppose we place a tunnel barrier in the path of one of the daughter photons. (The other daughter photon continues to travel unimpeded through the vacuum.) Of course, this would greatly diminish the coincidence count rate. However, whenever a tunneling event does occur, a single photon tunnels through the barrier as an *indivisible* unit, and the *difference* in the time of arrival of the two daughters, as measured by the difference in the time of the "clicks" of the two detectors, constitutes a precise definition for the tunneling time.

The question concerning the superluminality of the tunneling process can now also be precisely stated. Does the "click" of the detector that registers the arrival of the photon that traversed the tunnel barrier go off *earlier* or *later* (on average) than the "click" of the detector that registers the arrival of the photon that traversed the vacuum? If the tunnel barrier had simply been a thin piece of transparent glass, then the answer would obviously be "later," since the group velocity for a photon inside the glass would be *less* than the speed of light, and the group delay for the photon traversing the glass relative to that of the vacuum would be *positive*. However, if, as some tunneling-time theories predict, the tunneling process is superluminal, then the counterintuitive answer would be "earlier," since the effective group velocity for a photon inside the tunnel barrier would be *greater* than the speed of light, and the group delay for the photon traversing the barrier relative to that of the vacuum would be *negative*. Hence it is the *sign* of the relative time between the "clicks" in the two detectors that determines whether tunneling is subluminal or superluminal.

The reader may ask why relativistic causality is not violated by the superluminality of the tunneling process, if it should indeed be superluminal. It has been

---

[21] Steinberg et al., "Measurement of the Single-Photon Tunneling Time"; Steinberg and Chiao, "Subfemtosecond Determination of Transmission Delay Times for a Dielectric Mirror"; Chiao and Steinberg, "Quantum Optical Studies of Tunneling and Other Superluminal Phenomena."

shown[22] that special relativity does not forbid the *group* velocity to be faster than c; only Sommerfeld's *front* velocity must not exceed c. For example, here in the context of the tunneling process, a superluminal group velocity arises because a Gaussian wavepacket can be *reshaped* due to a selective transmission of its earlier parts by the tunnel barrier. By this *causal* reshaping process, the peak of the transmitted wavepacket can be moved forwards earlier in time relative to the peak of the incident wavepacket, without any violation of special relativity. Hence by the Born interpretation of the wavefunction, the "click" of the tunneling particle at its Geiger counter will most probably occur *earlier* than that of its twin that traversed the vacuum (the peak of a wavepacket represents the point of maximum probability for a "click" to occur). Also remember that the moment of tunneling of the signal photon is not under the experimenter's control due to the uncertainty principle. However, any genuine signal must be a *controllable* signal.

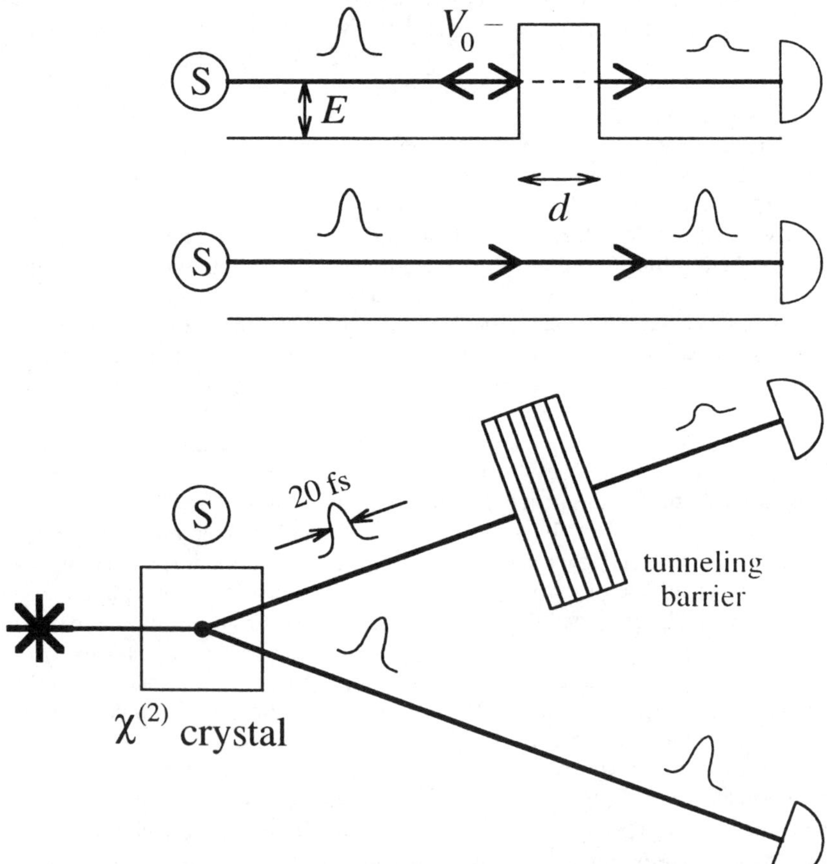

Figure 7. *Gedanken* experiment to measure the tunneling time by means of two particles simultaneously emitted from a source S and detected by two equidistant detectors.

[22] Raymond Y. Chiao and A.M. Steinberg, "Tunneling Times and Superluminality," in *Progress in Optics*, vol. XXXVII, E. Wolf, ed. (Amsterdam: Elsevier, 1997), 347–406.

Presently, the best detectors for photons (silicon avalanche photodiodes operated in the Geiger mode, i.e., as Geiger counters) have picosecond-scale response times, which are still not fast enough to detect the femtosecond-scale time differences expected in our tunneling-time experiment. It was therefore necessary to utilize a two-photon (Hong-Ou-Mandel) interferometer, which has a femtosecond-scale temporal resolution for measuring the time difference between the travel times of the two photons traversing the two arms of the interferometer. By placing the tunnel barrier in one of these arms, a precise measurement of the delay due to tunneling could then be performed. The tunnel barrier used in our experiments was a dielectric mirror in which periodic layers of alternately high and low index media produce a photonic band gap at the first Brillouin zone edge. The problem of photon propagation in this periodic structure is analogous to that of the Kronig-Penney model for electrons propagating inside a crystal. In particular, near the midgap point on the first Brillouin zone edge, there exists due to Bragg reflection inside the periodic structure an evanescent (i.e., exponential) decay of the transmitted wave amplitude, which is equivalent to tunneling. One important feature of this kind of tunnel barrier is the fact that it is nondispersive near midgap, resulting in little distortion of the tunneling wavepacket.

4.1 Tunneling Time Theories

A strong motivation for performing experiments to measure the tunneling time was the fact that there were many conflicting theories for this time.[23] It suffices here to list the three main contenders:

1. Wigner time (or "phase time" or "group delay")
2. Büttiker-Landauer time (or "semiclassical time")
3. Larmor time (with Büttiker's modification)

The Wigner time calculates how long it takes for the peak of a wavepacket to emerge from the exit face of the tunnel barrier relative to the time the peak of the incident wavepacket arrives at the entrance face. Since the peak of the wavepacket in the Born interpretation is the point of highest probability for a "click" to occur (see the above *Gedanken* experiment), it is natural to expect this to be the relevant time for our experiments. This calculation is based on an asymptotic treatment of tunneling as a one-dimensional scattering problem (the tunneling amplitude is viewed simply as the forward scattering amplitude from the potential barrier), and it utilizes the method of stationary phase to calculate the position of the peak of the tunneling wavepacket. The physical idea behind this calculation is simple: the peak of the wavepacket is where its strongest frequency component, and all of its nearby components, undergo constructive interference. Hence the phase of every one of these components must be approximately the same. The result is also simple: this tunneling time is proportional to the derivative of the phase ϕ of the tunneling amplitude with respect to the frequency, and hence the energy E, of the particle (see eq. 6 below).[24]

---

[23] See the following reviews: Eivind H. Hauge and J.A. Støvneng, "Tunneling Times: A Critical Review," *Rev. Mod. Phys.* 61 (1989): 917–36; Rolf Landauer and Th. Martin, "Barrier Interaction Time in Tunneling," *Rev. Mod. Phys.* 66 (1994): 217–28; Chiao and Steinberg, "Tunneling Times and Superluminality."

[24] Eugene P. Wigner, "Lower Limit for the Energy Derivative of the Scattering Phase Shift," *Phys. Rev.* 98 (1955): 145–7.

The Büttiker-Landauer time is based on a different *Gedanken* experiment. Suppose that the height of the tunnel barrier is perturbed sinusoidally in time. If the frequency of the perturbation is very low, the tunneling particle will see the instantaneous height of the barrier, and the transmission probability will adiabatically follow the perturbation. However, as one increases the frequency of the perturbation, at some characteristic frequency the tunneling probability will no longer be able to follow adiabatically the rapidly varying perturbation. It is natural to define the tunneling time as the inverse of this characteristic frequency. The result is again simple[25]: for opaque barriers, this tunneling time is the distance traversed by the particle (i.e., the barrier width d) divided by the absolute value of the velocity of the particle $|v|$. (In the classically forbidden region of the barrier, this velocity is imaginary, but its characteristic size is still given by its absolute value).

The Larmor time is based on yet another *Gedanken* experiment. Suppose that the tunneling particle had a spin magnetic moment (e.g., the electron). Suppose further that a magnetic field were applied to the region of the barrier, but only to that region. Then the angle of precession of the spin of the tunneling particle is a natural measure of the tunneling time. However, Büttiker noticed that in addition to this Larmor precession effect, there is a considerable tendency for the spin to align itself either along or against the direction of the magnetic field during tunneling, since the energy for these two spin orientations is different. The *total* angular change of the tunneling particle's spin divided by the Larmor precession frequency is Büttiker's Larmor time.[26]

One consequence of the Wigner time is the Hartman effect[27]: the Wigner tunneling time τ_W "saturates" for thick barriers, i.e, in the limit of large d, where the barrier transmission becomes exponentially small (in the "opaque" barrier limit), it approaches a limiting value consistent with the uncertainty principle,

$$\tau_w = \hbar \frac{\partial \phi}{\partial E} \rightarrow \frac{\hbar}{\left(E(V_0 - E)\right)^{1/2}} \geq \frac{\hbar}{2V_0}, \tag{6}$$

where E is the energy of the tunneling particle and V_0 is the height of the rectangular potential barrier. Thus the resulting time delay for tunneling becomes independent of d, for sufficiently large d (but not so large that significant distortions of the wavepacket occur). The apparent superluminality of tunneling is a consequence of this effect, since as d is increased, there is a point beyond which the saturated value of the tunneling time is exceeded by the vacuum traversal time d/c, and the particle appears to have tunneled faster than light.

By contrast, the Büttiker-Landauer theory predicts a tunneling time $d/|v|$, which increases linearly with d for opaque barriers, as one would expect classically. For a nonrelativistic electron incident on a rectangular barrier with a height $V_0 << mc^2$, the effective velocity $|v|$ is always less than c. Nevertheless, for the periodic structure that we used in our experiment, the effective velocity $|v|$ at midgap is infinite, which is a behavior even more superluminal than that predicted by the Wigner time. This fact

25 M. Büttiker and R. Landauer, "Traversal Time for Tunneling," *Phys. Rev. Lett.* 49 (1982): 1739–42.

26 M. Büttiker, "Larmor Precession and the Traversal Time for Tunneling," *Phys. Rev. B* 27 (1983): 6178–88.

27 Thomas E. Hartman, "Tunneling of a Wave Packet," *J. Appl. Phys.* 33 (1962): 3427–33.

makes it easy to distinguish experimentally between these two theories of the tunneling time. However, we hasten to add that the Büttiker-Landauer time may not apply to our experimental situation, since the *Gedanken* experiment on which it is based is quite different from the one relevant for our experiment.

Büttiker's Larmor time predicts a tunneling time that is independent of d for thin barriers, but that asymptotically approaches a linear dependence on d in the opaque barrier limit, where it coincides with the Büttiker-Landauer time. In our first experiment it was impossible to distinguish experimentally between this time and the Wigner time. Only in our second experiment could these two theories clearly be distinguished from one another.

Spontaneous parametric down-conversion was the light source used in our experiments, as described before. We used two pinholes to select out only twin photons with approximately the same energy. Their wavelengths were approximately 702 nanometers. The size of these pinholes determined the bandwidth of the light that passed through them, and the resulting single-photon wavepackets had temporal widths around 20 femtoseconds and a bandwidth of around 6 nanometers in wavelength.

The tunnel barrier consisted of a dielectric mirror with eleven quarter-wave-length layers of alternately high index material (titanium oxide with $n = 2.22$) and low index material (fused silica with $n = 1.45$). The total thickness of the eleven layers was 1.1 micrometers. This implied an *in vacuo* traversal time across the structure of 3.6 femtoseconds. Viewed as a photonic bandgap medium, this periodic structure had a lower band edge located at a wavelength of 800 nanometers and an upper band edge at 600 nanometers. The transmission coefficient of the tunneling photon tuned near the midgap (700 nanometers) was 1%. Since the transmission had a broad minimum at midgap much larger in width than the wavepacket bandwidth, there was little wavepacket distortion. The Wigner theory predicted at midgap a tunneling delay time of around 2 femtoseconds, or an effective tunneling velocity of $1.8c$. The Büttiker-Landauer theory predicted a zero tunneling time at midgap, which implies an infinite effective tunneling velocity.

To achieve the femtosecond-scale temporal resolutions necessary for measuring the tiny time delays associated with tunneling, we brought together these two photons by means of two mirrors, so that they impinged simultaneously at a beam splitter before they were detected in coincidence by two Geiger-mode silicon avalanche photodiodes. Since the two photons are bosons, they always pair off into the same directions. There resulted an extremely narrow null in the coincidence count rate as a function of the relative delay between the two photons, a destructive interference effect first observed by Hong, Ou, and Mandel.[28] The narrowness of this coincidence minimum, combined with a good signal-to-noise ratio, allowed a measurement of the relative delay between the two photons to a precision of ±0.2 femtoseconds. This allowed an accurate measurement of the delay for the photon to traverse through the tunneling barrier.

To check the sign of the tunneling delay relative to that for an equal distance in the vacuum, which is crucial for the interpretation of the superluminality of the tunneling process, we tilted the mirror towards Brewster's angle for the substrate ($\theta \approx 56°$), where there is a very broad minimum in the reflection coefficient as a function of angle. Near Brewster's angle this minimum is so broad that it is not very

---

[28] C.K. Hong, Z.Y. Ou and L. Mandel, "Measurement of Subpicosecond Time Intervals Between Two Photons by Interference," *Phys. Rev. Lett.* 59 (1987): 2044–6.

sensitive to the difference between the high and low indices of the successive layers of dielectrics. Thus to a good approximation, the reflections from all layers vanish simultaneously near this angle. Hence the Bragg reflection responsible for the band gap disappears, and the evanescent wave behavior and the tunneling behavior seen near normal incidence also disappears. The dielectric mirror should then behave like a thin piece of transparent glass with a *positive* delay time relative to that of the vacuum.

Another reason for tilting the mirror is that one can thereby distinguish between the Wigner time and Büttiker's Larmor time, as they differ considerably in the region near the band edge, which occurs near Brewster's angle. This can be seen in figure 8, where there is a considerable divergence as the band edge is approached between the solid line representing the theoretical prediction of the Wigner time and the long-dashed line representing that of Büttiker's Larmor time. The data points in figure 8 seem to rule out Büttiker's Larmor time (although again we hasten to add that this theory may not apply to our experiment). The agreement with Wigner's theory is better, but there are discrepancies that are not completely understood.

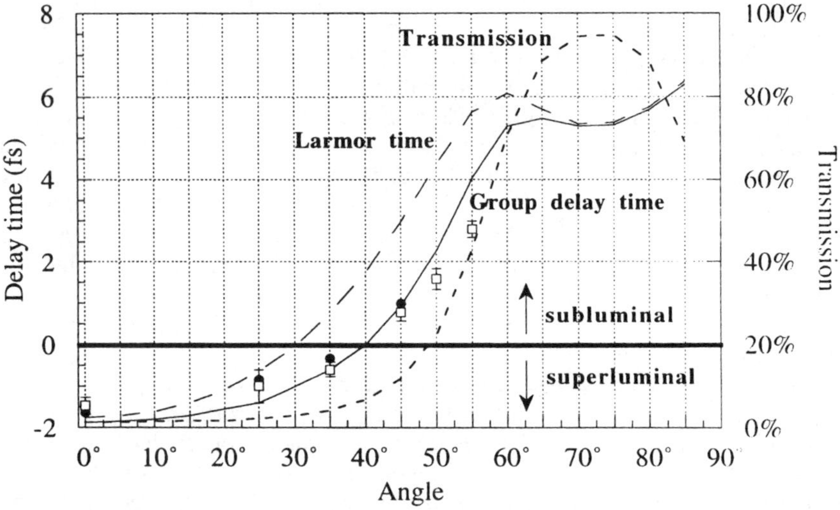

Figure 8. Temporal shifts of the minima seen in data plotted as a function of incidence angle, compared with the theoretical predictions of the Wigner or group delay time (solid curve) and of Büttiker's Larmor time (long-dashed curve). The transmission versus incidence angle is shown as the short-dashed line. All curves are for *p*-polarization. Solid circles and open squares denote data from two different tunnel barrier samples.

Other experiments confirming the superluminality of tunneling have been performed in Cologne, Florence, and Vienna.[29] The Cologne and Florence groups

[29] A. Enders and G. Nimtz, "Zero-Time Tunneling of Evanescent Mode Packets," *J. Phys. I* (France) 3 (1993): 1089–92; A. Ranfagni, P. Fabeni, G.P. Pazzi, and D. Mugnai, "Anomalous Pulse Delay in Microwave Propagation: A Plausible Connection to the Tunneling Time," *Phys. Rev. E* 48 (1993): 1453–60; Ch. Spielmann, R. Szipöcs, A. Stingl, and F. Krausz, "Tunneling of Optical Pulses Through Photonic Band Gaps," *Phys. Rev. Lett.* 73 (1994): 2308–11.

performed microwave experiments, and the Vienna group performed a femtosecond laser experiment. All these groups have confirmed the Hartman effect. One of these groups[30] has claimed to have sent Mozart's fortieth symphony at a superluminal speed of $4.7c$ through a microwave tunnel barrier 114 mm long consisting of a periodic dielectric structure similar to our dielectric mirror. However, the further interpretation that their experiment represents a violation of relativistic causality, is in our opinion unfounded.[31]

An experiment indicating the simultaneous existence of two different tunneling times was performed in Rennes.[32] In frustrated total reflection, the tunneling of photons through an air gap occurs between two glass prisms when a light beam is incident upon this gap beyond the critical angle. The Rennes group observed in frustrated total internal reflection both a lateral *displacement* of the tunneling beam of light and an angular *deflection* of this beam. These two effects could be interpreted as evidence for two different tunneling times that simultaneously occurred in the same tunneling barrier. The lateral displacement is related to the Wigner time, and the angular deflection is related to the Büttiker-Landauer time. As evidence for this, they cited the saturation of the beam displacement (the Hartman effect), and simultaneously a linear increase of the beam deflection, as the gap was increased. Recently, these results were confirmed by experiments in which superluminal tunneling times were directly measured in frustrated total internal reflection using femtosecond laser pulses.[33]

The experiments at Berkeley and elsewhere thus indicate that the tunneling process in quantum physics is superluminal. According to special relativity, this implies a nonlocality of a third kind, in which an observer moving at a sufficiently fast speed (close to the speed of light) past the tunnel barrier would infer that a single tunneling particle exists at two places at the same time, namely at both the entrance and exit faces of the tunnel barrier simultaneously. In our opinion, this does not imply that one can "communicate" faster than c, despite claims to the contrary by Heitmann and Nimtz.[34] The group velocity cannot be identified as the signal velocity of special relativity, by which a cause is connected to its effect. Rather, it is Sommerfeld's front velocity that exclusively plays this role, and no causal loop paradoxes can arise.[35]

There have also been recent discoveries in contexts other than in quantum tunneling, in which superluminal phenomena make their appearance, such as in atomic media with inverted populations, but we shall not discuss them here.[36]

---

[30] W. Heitmann and G. Nimtz, "On Causality Proofs of Superluminal Barrier Traversal of Frequency Band Limited Wave Packets," *Phys. Lett. A* 196 (1994): 154–8.

[31] Chiao and Steinberg, "Tunneling Times and Superluminality."

[32] Ph. Balcou and L. Dutriaux, "Dual Optical Tunneling Times in Frustrated Total Internal Reflection," *Phys. Rev. Lett.* 78 (1997): 851–4.

[33] J.J. Carey, J. Zawadska, D.A. Jaroszynski, and K. Wynne, "Noncausal Time Response in Frustrated Total Internal Reflection?", *Phys. Rev. Lett.* 84 (2000): 1431–4.

[34] Heitmann and Nimtz, "On Causality Proofs of Superluminal Barrier Traversal."

[35] John C. Garrison, M.W. Mitchell, R.Y. Chiao, and E.L. Bolda, "Superluminal Signals: Causal Loops Paradoxes Revisited," *Phys. Lett. A* 245 (1998): 19–25.

[36] Raymond Y. Chiao, "Faster-Than-Light Propagations and Their Applications," in *Proceedings of the 8th Rochester Conference on Coherence and Quantum Optics* (New York: Plenum, to be published).

5 Philosophical Implications

We have seen that there exists empirical evidence for three kinds of quantum nonlocalities in the world. What are the implications of these discoveries in quantum physics for philosophy and theology? In this section, I shall discuss some provisional philosophical implications, and in the next and final section, offer some tentative theological reflections.[37]

It may be objected that one should keep completely separate from each other physical truths, on the one hand, and nonphysical (or "metaphysical") speculations, on the other hand, including those of philosophy and theology. However, I believe that one cannot maintain such a dualistic worldview, in which truth is fundamentally separated into materialistic and nonmaterialistic compartments. Instead, intellectual integrity demands that we integrate our understanding into a single, undivided whole.

Here I shall suggest a neo-Berkeleyan viewpoint, in which quantum nonlocalities may shed new light on the question of the nature of divine action in the world. Such action may not only be local, but indeed may also be nonlocal. These nonlocalities can occur not only throughout space, but also throughout time. To illustrate the latter point, consider the following *Gedanken* experiment: suppose that an individual atom were placed in a Paul trap, so that it is initially in an excited state, and is about to emit a photon. Suppose further that a distant observer (say, a billion light years away from the atom) were to set up a powerful telescope to observe the light from this atom. Quantum mechanics tells us that the moment the observer sees a flash of light go off in his eye due to the emitted photon—or a "click" goes off in his Geiger counter, registering the arrival of this photon—the wavefunction of the entire system (atom plus photon) "collapses" into a state where the momentum vector of the photon is definitely directed along the line of sight that connects the atom to the detector in which the photon has been registered.

Let us apply the principle of the conservation of momentum to this atom-photon system. We conclude that the atom must have recoiled precisely in the *opposite* direction from the direction of the detected photon, i.e., against the line of sight. This atomic recoil occurred a billion years in the past, in the frame of the observer, and yet the observer's detection of the photon will have occurred a billion years in the future, in the frame of the atom. Hence the detection of the photon and of the direction of its momentum, as observed *in the future* of the atom, is precisely correlated with the emission event and the direction of the recoil of the emitting atom, as observed *in the past* of the observer. This is an example of an Einstein-Podolsky-Rosen effect involving the entanglement of past and future events in the universe. Most importantly, we should remind ourselves that as a result of the observed violations of Bell's inequalities, no properties of the particles composing the system, such as the recoil momentum of the atom, can actually exist until the moment an observation has actually been made. This forcefully reminds us of Bishop Berkeley's dictum: *Esse est percipi*.

---

[37] For a fuller discussion of the philosophical and theological implications of quantum physics, see the earlier articles by John Polkinghorne, "The Quantum World," 333–42, Robert J. Russell, "Quantum Physics in Philosophical and Theological Perspective," 343–74, Chris J. Isham, "Creation of the Universe as a Quantum Process," 375–405, all in *Physics, Philosophy, and Theology: A Common Quest for Understanding*, Robert J. Russell, William Stoeger, S.J., and George Coyne, S.J., eds. (Vatican City State: Vatican Observatory, 1988), and, of course, the other articles in this volume.

One should also mention here Wheeler's "delayed choice" *Gedanken* experiment.[38] Not only does the observer *create* by his observations the properties of the system being observed, but he is also able to *choose*, as the experimenter, the settings of the apparatus in such a way as to influence the behavior of a photon in its past (e.g., whether it behaves as a particle or as a wave). Not only can the observer, by his own *free will*, make such a choice of experimental settings, but he can *delay* his choice until the last moment before the "click" of a detector occurs, so that this change of settings in the present could influence the photon's behavior in the distant past. One version of this "delayed choice" experiment has been suggested and performed.[39] This experiment, and also the Einstein-Podolsky-Rosen experiments discussed above, show that the free will of the experimenter, at the highest levels of reality, can influence the quantum phenomena that occur at the deepest levels of reality.

6 Theological Reflections

To start off my theological reflections on the implications of these quantum phenomena, I shall assume as a basic principle that the universe we live in bears witness to the Creator who created it (Psalm 19:1ff; Romans 1:20). It is therefore not my objective here to intrude on the role of divine revelation in the Scriptures, which is of course essential for saving faith, but to explore what we can learn from what Galileo called the "book of nature," i.e., physics, and especially from the recent findings of quantum physics. In particular, the transcendence and immanence of the Creator with respect to creation may acquire new meaning in light of the findings of quantum physics. As suggested above, let us generalize Berkeley's philosophical principle to a "neo-Berkeleyan point of view," in which God is the Observer of universe, in the quantum sense of "observer."

This generalization starts from small systems, such as the atom-photon system considered in the previous section, in which an observed-created reality is seen to occur upon every elementary act of measurement, and ends up with large systems—in particular, with the entire universe. In this viewpoint, every elementary, individual quantum event, as in the above illustration, is a result of a creative act of the universal Observer, in which all properties of all particles come into existence upon their observation, in continual acts of *creatio ex nihilo*, which constitute a kind of *creatio continua* occurring everywhere at once. Thus the existence of the universe itself is *contingent* upon the continual observations of the Creator. The idea of *contingency of existence*, in the sense of the utter dependency of the universe for its properties and existence at each moment upon its Creator, is thereby introduced via quantum physics into philosophy and theology. This neo-Berkeleyan viewpoint does not imply a kind of solipsism, but rather suggests a kind of contingent realism. Furthermore, this viewpoint suggests a new meaning of the *immanence* of the Creator with respect to creation, since God is acting as the Observer *everywhere at once* in the universe. Thus God is omnipresent, omniscient, and omnipotent.

---

[38] John Archibald Wheeler and Wojciech H. Zurek, eds., *Quantum Theory and Measurement* (Princeton, N.J.: Princeton University Press, 1983), 182.

[39] Y.H. Shih and C.O. Alley, "New Type of Einstein-Podolsky-Rosen-Bohm Experiment Using Pairs of Light Quanta Produced by Optical Parametric Down Conversion," *Phys. Rev. Lett.* 61 (1988): 2921–4, and references therein.

Moreover, from this viewpoint, the universe as a whole, at its moment of creation in standard Big Bang cosmology, is Georges Lemaître's "primeval atom," i.e., the entire universe at its beginning is to be conceived as a single quantum object. Thus the beginning of the universe is itself an elementary, individual quantum event, out of which emerged all particles in an entangled state. Since there exists no *physical* "environment" outside the universe with which it could have interacted, the decoherence of its initial quantum state could not have occurred: the universe as a whole is a closed quantum system, which is in a pure state with zero entropy. All parts of the universe must therefore have been initially entangled with every other part. The neo-Berkeleyan viewpoint proposed above, when applied to the universe as a whole, suggests the following: the entire universe as an individual quantum system came into existence in a singular observational act of *creatio ex nihilo* by the Creator. This suggests a new meaning of the *transcendence* of the Creator with respect to the creation, in that God is the *transcendent* Observer of the universe, and therefore its Source, at the moment of its creation.

One possible difficulty with this neo-Berkeleyan viewpoint arises from the "quantum Zeno" paradox. In this paradox, if an observer decides to observe extremely frequently the evolution of a quantum system, each observation projects the system into a fixed eigenstate of the observable being measured (e.g., the position of a particle). In the standard von Neumann postulate for quantum measurement, the system after the measurement is projected into a given eigenstate of the observable being measured, and thus the state of the system is "prepared" in this eigenstate from a previous measurement for the next measurement. It therefore follows that if measurements or observations are performed at infinitesimal time intervals of the position of a particle, such observations would freeze the motion of the particle, and no evolution would be possible, i.e., the particle would not be able to move at all. This is similar to Zeno's paradox, in which the division of time into infinitesimal intervals would seem to preclude any motion at all. Since we empirically observe that the universe does indeed change with time, this would imply that if the neo-Berkeleyan viewpoint is correct, God would not be performing *continuous* observations at infinitesimal time intervals. Closely connected with this is a quantum *nondeterminism* at a fundamental level, which would allow room for free will and divine action.

The neo-Berkeleyan viewpoint suggests a "top-down" approach, in which large-scale, emergent phenomena, such as consciousness and free will, would be integrated with small-scale quantum phenomena through the process of quantum measurement. Similar ideas were expressed by Wigner,[40] who suggested that *human* consciousness, or the *human* mind, is directly involved in the final step of quantum measurement, i.e., in the "collapse of the wavefunction." Here, however, I suggest that ultimately it is God's consciousness or mind, instead of human consciousness, that is involved at the ultimate step in quantum measurement. This proposed *theocentric* viewpoint avoids the difficulty of Wigner's *anthropocentric* viewpoint, which arises from the fact that human observers are limited in space and time, whereas quantum events are occurring everywhere and at all times in the universe.

In another extension of the interpretation of quantum mechanics, Wheeler suggested a kind of self-consistent feedback-loop process, in which we as observers are emergent *co-participants* in the creation of the universe, by the mere fact that we

---

[40] Eugene P. Wigner, "Remarks on the Mind-Body Question," in *Quantum Theory and Measurement*, J.A. Wheeler and W.H. Zurek, eds., 168–81.

are able to look back and observe the beginning of the universe.[41] Again, I would replace human observation by divine observation. This is not to deny the possibility of a "bottom-up" viewpoint, in which the deepest-level quantum phenomena form the basis for the emergence of the highest-level phenomena, eventually culminating, for example, in the emergence of human consciousness and free will. However, for the reasons given above, I believe that in addition to these feedback-loop processes, which are *internal* to the universe, and which result in emergent observers within the material universe, there must exist a transcendent Observer who is *external* to the universe and who gives concrete reality to all properties, even to the very existence of the universe through divine observation.

In conclusion, I believe that the relationship between the observer and the observed in quantum physics, through analogous reasoning, may illuminate the relationship of the Creator to the creation. The neo-Berkeleyan viewpoint introduced here suggests not only a continual *creatio ex nihilo qua creatio continua* by an immanent Creator, but also a singular *creatio ex nihilo* by a transcendent Creator. Moreover, the above Einstein-Podolsky-Rosen effects imply a quantum *nonseparability* of the world, which ties together the universe nonlocally as a whole. This reminds us of the words of the Apostle John,

> All things came into being through him, and without him not one thing came into being that has come into being (John 1:3).

and of the words of the Apostle Paul,

> All things have been created through him and for him.... and in him all things hold together (Colossians 1:16,17).

From the context of these and other similar passages of the New Testament, we infer that "all things" refers to the universe. Not only are all distant parts of the universe woven together throughout space, but also its future and its past are entangled throughout time, as if the universe were one seamless garment.

Acknowledgment. I would like to thank Bob Russell and the Center for Theology and the Natural Sciences for the invitation to contribute to this volume, and George Coyne, S.J., for his hospitality at the Vatican Observatory. I also thank Paul Kwiat and Aephraim Steinberg for their work on the experiments described herein, David Burnham and John Frame for an early reading of the manuscript, John Garrison, Colin McCormick, Marlan Scully, and Anton Zeilinger for their insightful comments, and all the participants of the Wye and the Vatican Observatory meetings for many stimulating and helpful discussions. I would especially like to thank Akira Tonomura for supplying me with the electron interferogram shown in figure 1.

[41] John A. Wheeler, "Law Without Law," in *Quantum Theory and Measurement*, Wheeler and Zurek, eds., 182–213.

CHAOS AND THE SEMICLASSICAL LIMIT OF QUANTUM MECHANICS (IS THE MOON THERE WHEN SOMEBODY LOOKS?)

Michael Berry

1 Introduction

The question of how chaos in classical mechanics gets reflected in the quantum mechanics of the microworld is slippery and subtle.[1] On the one hand, it is known from observations, computations, and mathematical theorems that Newtonian mechanics often predicts unpredictability: some systems of forces are so unstable that nearby trajectories separate exponentially fast.[2] This is chaos—the celebrated "sensitivity to initial conditions." Chaos is, strictly speaking, a phenomenon defined in the limit of long times, for systems that are confined to a restricted region of space (for example, a particle bouncing in a billiard table that is the shape of a stadium). One of the hallmarks of chaos is the lack of periodicity or quasi-periodicity in the time-development of any dynamical variable (for example, the x component of position or the y component of velocity); the mathematical statement of this is that these variables have a continuous spectrum.

On the other hand, the corresponding quantum systems always have discrete (that is, quantized) energy levels, and since the energy spectrum governs the time-development of any dynamical quantity such evolution cannot be chaotic. This is the origin of the often-repeated, and strictly true, assertion that there is no chaos in quantum mechanics. Two awkward questions immediately arise. The first stems from the fact that classical mechanics is supposed to be the limit of quantum mechanics in situations where Planck's constant h can be neglected. This looks like a contradiction, and we ask: What kind of limit is it where a system is not chaotic for any finite value of h, yet is chaotic when $h = 0$? The second question is connected with the recognition that strictly speaking all systems—even our orbiting moon—obey the laws of quantum mechanics, and yet they are often observed to evolve chaotically (even the moon is slightly chaotic, and other astronomical bodies much more so). We ask: What is the origin of the observed chaos in our macroscopic but strictly quantum world?

My purpose here is to try to answer these questions in the simplest way, that is without formalism. First, though, I draw attention to the fact that the relation between chaos and quantum physics, important enough as it is in its own right, can also be regarded as a microcosm of much larger questions. Our scientific understanding of the world is a patchwork of vast scope; it covers the intricate chemistry of life, the sociology of animal communities, the gigantic wheeling galaxies, and the dances of elusive elementary particles. But it is a patchwork nevertheless, and the different areas do not fit well together. This is the notorious problem of reduction: we might

---

[1] Michael V. Berry, "Semiclassical Mechanics of Regular and Irregular Motion," in *Les Houches Lecture Series*, vol. 36, G. Iooss, R.H.G. Helleman, and R. Stora, eds. (Amsterdam: North-Holland, 1983), 171–271; idem, "Quantum Chaology (The Bakerian Lecture)," *Proc. Roy. Soc. Lond. A* 413 (1987): 183–98; Martin Gutzwiller, *Chaos in Classical and Quantum Mechanics* (Berlin: Springer, 1990).

[2] Heinz-Georg Schuster, *Deterministic Chaos: An Introduction* (Weinheim: VCH Verlagsgesellschaft, 1988).

believe that a living organism is described completely by the quantum mechanics of all its molecular (atomic, electronic, nucleonic, quarkish...) parts, but where in its byzantine wavefunction, governed by the Schrödinger equation, is the spark of life? Even if we knew, how could that quantum description extend beyond living cells and individual creatures, to the displays of peacocks and the flocking of birds?

Even within physical science, reduction between different levels of explanation is problematic—indeed, it is almost always so. Chemistry is supposed to have been reduced to quantum mechanics, yet people still argue over the basic question of how quantum mechanics can describe the shape of a molecule. The statistical mechanics of a fluid reduces to its thermodynamics in the limit of infinitely many particles, yet that limit breaks down near the critical point, where liquid and vapor merge, and where we never see a continuum no matter how distantly we observe the particles; the critical state is a fractal, and much theoretical physics of the last forty years has gone into getting a clear understanding of it. The geometrical (Newtonian) optics of rays should be the limit of wave optics as the wavelength becomes negligibly small, yet we shall see that the reduction (mathematically similar to that of classical to quantum mechanics) is obstructed by singularities, even when there is no chaos. (A singularity is a place where the smoothness of a mathematical quantity or geometrical pattern is disrupted, for example by the quantity becoming infinite; I will give examples later.)

What follows should not be misconstrued as anti-reductionist. On the contrary, I am firmly of the view, beautifully expressed recently,[3] that all the sciences are compatible and that detailed links can be, and are being, forged between them. But of course the links are subtle, and my emphasis will be on a mathematical aspect of theory reduction that I regard as central, but which cannot be captured by the purely verbal arguments commonly employed in philosophical discussions of reduction.

My contention here will be that many difficulties associated with reduction arise because they involve *singular limits*.[4] These singularities have both negative and positive aspects: they obstruct the smooth reduction of more general theories to less general ones, but they also point to a great richness of borderland physics between the theories. So, indeed, the problem to be considered here, of chaos in the quantum world, is an example of something grander. But examples are important. To paraphrase the numerical analyst Beresford Parlett: "Only wimps study only the general case. Real scientists study examples."

Before getting started, I must make four points. The first is that although we will be dealing with the limit $h \rightarrow 0$, at first encounter this seems meaningless. Planck's constant cannot be set equal to zero because it is a constant of nature with a fixed value. And yet, h has dimensions, so its "fixed" value can be anything, depending on the system of units used. In fact, by "the semiclassical limit $h \rightarrow 0$" I will mean "situations where the dimensionless variable obtained by dividing h by any classical quantity with the same dimensions (action) is negligibly small."

The second point is that discussions of chaos have been confused by conflating two different questions. The first is: How does classical behavior (e.g., chaos) emerge in the semiclassical limit? If the limit $h \rightarrow 0$ were not singular, this emergence would be trivial: simply solve the quantum problem, which involves h,

---

[3] Edward O. Wilson, *Consilience: The Unity of Knowledge* (New York: Knopf, 1998).

[4] Michael V. Berry, "Asymptotics, Singularities and the Reduction of Theories," in *Proc. 9th Int. Cong. Logic, Method., and Phil. of Sci.*, D. Prawitz, B. Skyrms, and D. Westerståhl, eds. (Amsterdam: Elsevier, 1994), 597–607.

then set $h = 0$. But we shall see that the true answer is much more subtle. This first question is important for the intellectual coherence of our description of the physical world, yet in a sense it is backward-looking, since we know the answer: the classical world has to emerge somehow. The second question, more productive of new physics, is: What nonclassical phenomena emerge as $h \to 0$? This sounds like nonsense, and indeed if the limit were not singular the answer would be: no such phenomena.

The third point is that in discussions of chaos and quantum physics there always lurks a second limit, in addition to $h \to 0$. This is the limit of long times, $t \to \infty$. As mentioned above, it is only in the long-time limit that classical chaos emerges: it takes infinitely long time to verify that a trajectory explores all the possibilities compatible with conservation of energy (and similar quantities), or to determine that two initially infinitesimally close trajectories separate exponentially fast. But the semiclassical and long-time limits do not commute, and therein lies much of the difficulty and subtlety associated with our subject.

The fourth point is that there are two questions that were the legitimate focus of attention by most of the other participants at the CTNS–Vatican Observatory Conference, but that I will not discuss. One is the relation, if any, that these considerations of quantum chaology bear to theological matters; I am unqualified (and also, as a nonbeliever, unmotivated) to do that. The other is the relation, if any, between quantum chaology and the interpretation of quantum mechanics (measurement, wavepacket reduction, etc.); I have some sympathy with the opinion mischievously expressed recently that such interpretation is unnecessary.[5]

2 Singular Limits: When One Plus One Does Not Make Two

To see that singularities in the semiclassical limit are unavoidable, consider the simplest possible situation[6] (where there is no chaos) of two beams of quantum particles, or waves of light, + and −, traveling in opposite directions, and coherent (e.g. created from a common source). The associated waves—complex functions of position and time—are

$$\psi_{\pm}(x,t) = \exp\left\{2\pi i\left(\pm\frac{x}{\lambda} - \nu t\right)\right\}, \tag{1}$$

where the wavelength λ and frequency ν are related to the momentum p and energy E of the quantum particles by the de Broglie and Planck laws:

$$\lambda = h/p, \quad \nu = E/h. \tag{2}$$

Each wave has intensity $|\psi|^2 = 1$.

By linearity (one of the principles of quantum mechanics), the resultant wave is the sum (superposition) of the two, so the (observable) intensity is

$$I(x) = \left|y_+ + y_-\right|^2 = 4\cos^2\left(\frac{2\pi x}{\lambda}\right). \tag{3}$$

[5] Christopher Fuchs and Asher Peres, "Quantum Theory Needs no 'Interpretation'," *Physics Today* (2000): 70-71.

[6] Michael V. Berry, "Some Quantum-to-Classical Asymptotics," in *Chaos and Quantum Physics,* vol. 52, *Les Houches Lecture Series*, M.-J. Giannoni, A. Voros, and J. Zinn-Justin, eds. (Amsterdam: North-Holland, 1991), 251–304.

This is the simplest representation of interferometry, where x would be distance measured on an observation screen. The oscillatory function (3) describes the interference fringes familiar since their explanation in terms of superposition by Young two centuries ago, and its mathematization by Fresnel (they were studying light, but nowadays the same interference is routinely observed for all quantum particles). Its values oscillate (figure 1) between zero and 4, with a spatial period of $\lambda/2$.

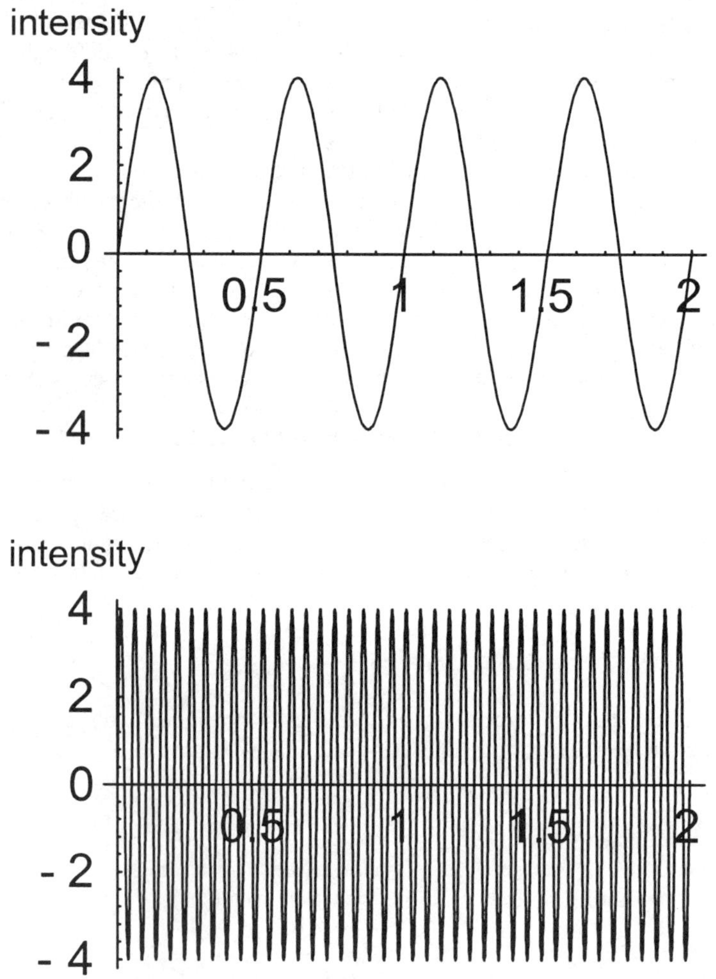

Figure 1. Graphs of the intensity interference function (3) for (top) $\lambda = 0.5$, (bottom) $\lambda = 0.05$, illustrating the approach to the singular limit $\lambda = 0$, where the intensity takes all values between 0 and 4.

Now we ask: what is the classical, or geometrical-optics, limit of this wave? By (2), $h \to 0$ corresponds to fixing the momentum and letting $\lambda \to 0$. Classically (or in geometrical optics) there is no interference, so the intensity resulting from adding two waves, each with unit intensity, should be $I = 2$. But it is not! As $h \to 0$, the intensity (3) oscillates faster and faster (figure 1, bottom), and in the limit it attains all values between zero and 4 in arbitrarily small ranges of x. In fact, (3) has a powerful singularity at $\lambda = 0$ (or $h = 0$), obstructing the smooth passage to the limit. We used to tell ourselves, and our students, that this singular function, and others like it, are mathematical curiosities, of no interest in physics. Now we see that the singularity arises unavoidably when we try to bridge the gap between two of our major theories, in the simplest possible case. For wave intensities, $1 + 1 = 2$ is false, all the way to the classical limit.

And yet we know that, in words attributed to Lord Rayleigh, two candles are twice as bright as one, that is, when waves are irrelevant, one plus one is indeed two. The only way to extract this result from (3) is by *averaging*. Any external influence that compromises the purity of (3) will have the effect of blurring (3), and in the classical limit this function is infinitely sensitive to blurring. Two possible influences are finite x-sensitivity (window size) of the detector, and nonmonochromaticity of the beams (finite spreads in wavelength or momentum). Now the average of the function $\cos^2$ is 1/2, so

$$\langle I(x)\rangle = 4\left\langle \cos^2\left(\frac{2\pi x}{\lambda}\right)\right\rangle = 4 \times \frac{1}{2} = 2, \tag{4}$$

which is of course the classical result.

We need a slightly expanded version of (4). The average is over the phase difference χ between the two waves. Imagine the external influences result in a gauss-distributed χ with spread Δ; then

$$\langle \cos^2\chi\rangle = \frac{1}{\Delta\sqrt{2\pi}}\int_{-\infty}^{\infty} \mathrm{d}\chi \cos^2\chi \exp\left\{\frac{\chi^2}{2\Delta^2}\right\} = \frac{1}{2}\left(1 + \exp\left\{-2\Delta^2\right\}\right). \tag{5}$$

Phase randomness is very efficient: for phase spread $\Delta = \pi/2$, $\langle\cos^2\chi\rangle = 0.5003$; for $\Delta = \pi$, $\langle\cos^2\chi\rangle = 0.500000001$.

So, the correspondence principle does hold, but there has been a price to pay: because of the singularity, classical mechanics emerges not directly, but only after averaging over phase-scrambling effects that can be ascribed to the environment—that is, influences external to the pure quantum wave initially calculated—in modern parlance, decoherence effects. All this to get 1 + 1 = 2![7]

3 Decoherence and the Emergence of Chaos

Now, what about chaos? To fix our ideas, consider Hyperion, the sixteenth satellite of the planet Saturn. This body is a potato-shaped rock about as big as New York City ($R \approx 142$km). To the best of our knowledge, Hyperion is unique in that its rotation is *chaotic*.[8] Under resonant forcing from the big moon Titan, in combination with Saturn's gravity and its own nonsphericity, Hyperion tumbles erratically. The

[7] I cannot resist recalling that Bertrand Russell and Alfred North Whitehead in their classic work, *Principia Mathematica*, which constructed mathematics from its basis in logic—before Gödel proved that enterprise is doomed—derived 1 + 1 = 2 as a theorem somewhere in the middle of volume 2.

[8] Jack Wisdom, S.J., and F. Mignard, "The Chaotic Rotation of Hyperion," *Icarus* 58 (1984): 137–52; G.J. Black, P.D. Nicholson, and P.C. Thomas, "Hyperion: Rotational Dynamics," *Icarus* 117 (1995): 149–61.

instability has a time-constant (for exponential forgetting of its initial state) of about 100 days, which is long compared with both its orbital period (21.3 days) and its instantaneous rotation period (about 5 days).

Now consider the rotation of Hyperion, regarded as a quantum object. Its quantized angular momentum J consists of $2\pi J/h \approx 2 \times 10^{58}$ Planck units (see Appendix A for the basis of the estimates given in this section). From this enormous number, together with the correspondence principle—and with the same modeling as we use for the classical dynamics, where the only forces come from Saturn and Titan—the quantum predictions for the motion ought to coincide with the classical to very high accuracy. Let us see.

It helps to think of the classical motion on the phase space represented by the angular momentum vector $\mathbf{J}$ (whose magnitude is $J = |\mathbf{J}|$); this is the surface of a sphere with polar coordinates θ, ϕ. The motion of the vector $\mathbf{J}$ can be regarded as that of a mechanical system with a single degree of freedom, whose "coordinate" is the azimuth angle ϕ and whose "momentum" is $p = J\cos\theta$; therefore phase-space area is area on the sphere, divided by J. For Hyperion, the total available phase-space area is $4\pi J^2/J \approx 10^{58}h$.

If we knew the motion precisely, Hyperion's rotation would be a single point moving on the sphere. But even classically our measurements are not perfectly precise, so the state of Hyperion is a little blurry patch, tiny on the sphere but with an area many times h; call its boundary $B(0)$, evolving into $B(t)$. Chaotic evolution means that the different points on B separate exponentially, so that $B(t)$ develops into a convoluted curve, with tendrils exploring the whole phase sphere. The area enclosed by $B(t)$ remains constant (this is Liouville's theorem); however B's perimeter length grows exponentially, as $L(t) \approx L(0)\exp(t/T_c)$, where for Hyperion the chaos time is $T_c \approx 100$ days.

Classically, the convolutions get ever more intricate. For a while, the quantum state representing Hyperion follows these convolutions; this is the spreading of the wavepacket, chaos-style. But quantum physics prevents the intricacy—that is, the chaos—from developing forever, because phase-space fine structure is limited by Planck's constant h. Any smaller structure must be blurred. After some time T_q, the areas of typical tendrils get finer than h; thereafter, classical and quantum evolutions are different. This is the quantum suppression of classical chaos.[9] At first thought, T_q for astronomical objects must be enormously long, probably irrelevant to any observations on time scales relevant to astronomers—remember that gigantic 10^{58}. Not so. A simple estimate (Appendix A) gives

$$T_q \approx T_c \log(J / h). \tag{6}$$

For Hyperion, this is

$$T_q \approx 100 \text{ days} \times \log(10^{58}) = 37 \text{ years} \tag{7}$$

[9] Giulio Casati, B.V. Chirikov, J. Ford, and F.M. Izraelev, "Stochaistic Behavior of a Quantum Pendulum Under a Periodic Perturbation," in *Stochastic Behavior in Classical and Quantum Hamiltonian Systems*, Lecture Notes in Physics, vol. 93, Giulio Casati & Joseph Ford, eds. (Berlin: Springer, 1979), 334–52; Michael V. Berry, N.L. Balazs, M. Tabor, and A. Voros, "Quantum Maps," *Ann. Phys. N.Y.* 122 (1979): 26–63; H. Jürgen Korsch and M.V. Berry, "Evolution of Wigner's Phase-Space Density Under a Non-integrable Quantum Map," *Physica D* 3 (1981): 627–36.

—an astonishingly short time.[10]

Now, the chaotic rotation of Hyperion was discovered less than 37 years ago, but nobody thinks that after the year 2020 astronomers will begin to see quantum effects—specifically the calming of Hyperion's instability and its replacement by some more moderate multiply periodic motion. And indeed, the predicted quantum suppression of chaos is itself suppressed, and classicality restored, by another quantum effect. This is *decoherence*. It arises from the fact that Hyperion is not isolated.

At the very least, seeing Hyperion involves photons from the sun, arriving unpredictably. We would never think to include these photons in the calculation of Hyperion's dynamics, because their energy is so minuscule: 4×10^{-19}J for a visible photon, as compared with $2 \times 10^{+19}$J for Hyperion's rotational energy. On the other hand, their energy is vast in comparison with the energy spacing 10^{-39}J of Hyperion's quantum rotational levels, so a quantum effect from these photons is conceivable.

The proper way to calculate this effect would be to enlarge the physical system to include not just Hyperion in the field of Saturn and Titan but also the sun's radiation field, and formally eliminate the radiation variables (by tracing over them), to get a modified quantum mechanics involving only Hyperion's dynamics. In recent years powerful formalisms have been developed for doing this,[11] and have been applied to the situation where there is classical chaos.[12] Rather than repeat these arguments here, I will first estimate the effects of decoherence in a bare-handed way, by treating the environment as a random time-dependent modification of Hyperion's Hamiltonian—a kick from each impact of a photon from the sun. In Appendix B I give a slightly more formal treatment, but still incorporating what John Polkinghorne calls "the patter of photons" as an external influence, rather than a constituent of a composite system.

We begin by noting that the quantum suppression of chaos involves the interference of waves and therefore their phases (in some situations this is analogous to Anderson localization in solids[13]). Therefore the suppression can be easily—and quickly—destroyed by decoherence from external influences. The change in Hyperion's angular momentum resulting from the random impact (impulsive torque) of a single visible photon (wavelength λ) is $\Delta J \approx (h/\lambda)R$. This impact will change the total phase of Hyperion's wavefunction. And, with even slight convolutions of B, so that there is more than one classical contribution to the wavefunction for some azimuth values ϕ, it will also (by about the same amount) change the phase difference between these contributions. The change (associated with the component of $\Delta\mathbf{J}$ parallel to $\mathbf{J}$, which spins up Hyperion and swells the sphere and therefore also the convolutions of B on it) is

$$\Delta J / h \approx (h/\lambda)R/h = R/\lambda \approx 10^{11} \qquad (8)$$

---

[10] This was first calculated in a similar way by Ronald Fox, private communication.

[11] Roland Omnès, "Consistent Interpretations of Quantum-Mechanics," *Rev. Mod. Phys.* 64 (1992): 339–82; idem, "General Theory of the Decoherence Effect in Quantum Mechanics," *Phys. Rev. A* 56 (1997): 3383–94.

[12] Wojciech H. Zurek and J.P. Paz, "Decoherence, Chaos and the 2nd Law," *Phys. Rev. Lett.* 72 (1994): 2508–11; idem, "Quantum Chaos—A Decoherent Definition," *Physica D* 83 (1995): 300–8; Wojciech H. Zurek, "Decoherence, Chaos, Quantum-Classical Correspondence, and the Algorithmic Arrow of Time," *Physica Scripta* 76 (1998): 186–98.

[13] Shmuel Fishman, D.R. Grempel, and R.E. Prange, "Chaos, Quantum Recurrences, and Anderson Localization," *Phys. Rev. Lett.* 49 (1982): 509–12.

This uncontrollable phase shift between the waves associated with any two classical contributions, acting according to (5), is overwhelmingly sufficient to induce decoherence, and so suppress the suppression of chaos—and just from a single photon. In the stream of sunlight (flux ≈ $10 W/m^2$), photons arrive at a rate of one every 10^{-30}s, so decoherence occurs extremely fast. Stretching this argument, we can estimate the decoherence time T_d for the stream of photons from the sun to disturb the phase by about 2π radians:

$$T_d \approx 10^{-53} \text{s} \tag{9}$$

Do not take the precise value of this number seriously. The foregoing is very far from a proper calculation of the decoherence effect for Hyperion. But the extreme smallness of T_d should be taken very seriously. It shows that an external disturbance so small that its effect on a system's dynamics is negligible (one photon, or a tiny fraction of a photon) can nevertheless dramatically and effectively instantaneously influence the kinematics of its quantum state—and more strongly as the classical limit is approached. Decoherence—the generalization of the 1 + 1 = 2 calculation of the previous section—emerges as a dominant physical effect, allowing classical chaos to persist.[14]

I chose Hyperion's chaos to dramatize the way in which uncontrolled environmental influences can induce quantal decoherence. In fact, when applied in such extreme situations decoherence has nothing to do with chaos *per se*, but to classicalization: since $T_d << T_c$, the destruction of quantum interference occurs long before chaos develops. So the claim sometimes made, that chaos amplifies quantum indeterminacy, is misleading. The situation is more subtle: chaos magnifies *any* uncertainty, but in the quantum case h has a smoothing effect, which would suppress chaos if this suppression were not itself suppressed by externally-induced decoherence, that restores classicality (including chaos if the classical orbits are unstable).

The calculation in Appendix B shows this decoherence-induced classicalization more clearly for the illustrative example of Hyperion; very recently, a more general and formal argument for the universality of the phenomenon has been given.[15] At the opposite extreme from Hyperion are the humble two-state systems, where decoherence, and the related inhibition of quantum transitions through the quantum Zeno effect, can be explored in detail.[16]

---

[14] Zurek and Paz, "Decoherence, Chaos and the 2nd Law"; idem, "Quantum Chaos—A Decoherent Definition"; Zurek, "Decoherence, Chaos, Quantum-classical Correspondence, and the Algorithmic Arrow of Time."

[15] D. Braun, F. Haake, and W. Strunz, "Universality of Decoherence," *Phys. Rev. Lett.* 86 (2001): 2913–7.

[16] Robert A. Harris and L. Stodolsky, "Quantum Beats in Optical Activity and Weak Interactions," *Phys. Lett. B* 78 (1978): 313–17; idem, "Two-State Systems in Media and 'Turing's Paradox'," *Phys. Lett. B* 116 (1982): 464–9; Robert A. Harris and R. Silbey, "On the Stabilization of Optical Isomers Through Tunneling Friction," *J. Chem. Phys.* 78 (1983): 7330–3; Peter Pfeifer, "Chiral Molecules: A Superselection Rule Induced by the Radiation Field," Thesis, Chemistry Dept., ETH Zürich, 1980; Saverio Pascazio, M. Namiki, G. Badurek, and H. Rauch, "Quantum Zeno Effect with Neutron Spin," *Phys. Lett. A* 179 (1993): 155–60; H. Nakazato, M. Namiki, S. Pascazio, and Y. Yamanaka, "Quantum Dephasing by Chaos," *Physics Letters A* 222 (1996): 130–6; Michael V. Berry, "Some Two-State Quantum Asymptotics," in *Fundamental Problems of Quantum Theory*, D.M. Greenberger and A. Zeilinger, eds., *Ann. N.Y. Acad. Sci.*, 255 (1995): 303–17; P. Facchi, S. Pascazio, and A. Scardicchio, "Measurement-Induced Quantum Diffusion," *Phys. Rev. Lett.*

4 Emergent Semiclassical Phenomena

The example of Hyperion addressed the first of the two commonly conflated questions mentioned near the end of the Introduction, namely, the emergence of classical effects—specifically, chaos—as $h \rightarrow 0$. Now, with two examples, we consider the more interesting question of what nonclassical effects emerge and persist in the limit $h \rightarrow 0$. The first does not involve chaos; the second does. Both of these emergent phenomena are delicate, in the sense that their detection requires magnification that increases as h gets smaller; this contrasts with more familiar examples of macroscopic quantum phenomena (superfluidity, superconductivity, the impenetrability of matter in bulk…).

Corresponding to a quantum scattering wavefunction (for example of a beam of atoms striking a target of other atoms, or electrons in an electron microscope) or to a field of diffracted light waves (for example sunlight encountering a raindrop), is a *family* of trajectories. These are the orbits of massive particles, or light rays. (Even a localized wavepacket corresponds to a family, rather than a single trajectory, because its localization cannot be perfect.)

Now, a family of trajectories possesses a holistic property not inherent in any individual trajectory, namely *focusing*. In three dimensions, focusing usually occurs on surfaces, and in two dimensions, on lines. These are the envelopes of the family of trajectories. They are called *caustics*. Caustics are the singularities of ray (geometrical, classical) theory.[17] On them, the ray intensity rises to infinity. They are the brightest places in the field. Examples are rainbows (angular focusing of sunlight refracted and reflected by raindrops, and their analogues in atomic and molecular scattering) and the patterns of light dancing on the bottom of a swimming-pool. In our eyes, microscopes and cameras, the images of each point are also formed by focusing, but of a very special kind, namely the nongeneric points fashioned by evolution or the lensmaker's art, rather than the surfaces and lines occurring when there is no symmetry. Caustics are important.

But they are nonexistent! Close examination, on the scale of the wavelength, dissolves the singularities, and reveals that they are decorated by intricate and beautiful interference patterns. These can be understood by an application of the same mathematics, developed in the 1960s and 1970s, that enables the different caustic singularities to be classified. This is the celebrated and (in other contexts) notorious *catastrophe theory*.[18] The *diffraction catastrophes*[19] are emergent semiclassical (more generally, short-wave) phenomena, with many applications throughout mathematics as well as physics (they describe the supernumerary arcs

83 (1999): 61–4.

[17] Michael V. Berry, "Singularities in Waves and Rays," in *Les Houches Lecture Series*, vol. 35, R. Balian, M. Kléman, and J.-P. Poirier, eds. (Amsterdam: North-Holland, 1981), 453–543; John F. Nye, *Natural Focusing and Fine Structure of Light: Caustics and Wave Dislocations* (Bristol: Institute of Physics Publishing, 1999).

[18] Vladimir I. Arnold, *Catastrophe Theory* (Berlin: Springer, 1986); Tim Poston and Ian Stewart, *Catastrophe Theory and Its Applications* (London: Pitman, reprinted by Dover, 1978).

[19] Nye, *Natural Focusing and Fine Structure of Light*; Michael V. Berry, "Waves and Thom's Theorem," *Advances in Physics* 25 (1976): 1–26; Michael V. Berry and C. Upstill, "Catastrophe Optics: Morphologies of Caustics and Their Diffraction Patterns," *Progress in Optics* 18 (1980): 257–346.

sometimes seen just inside the main rainbow, that were impossible to explain using ray theory). Although they are contained in wave equations (Schrödinger's, Maxwell's), they were not derived by the reductionist approach of starting with solutions of wave equations, but by top-down methods, starting with the caustic singularities of ray physics; only later were the full connections with wave equations elucidated and consilience established.

Before getting to the second example, I digress to point out that the story of caustics extends deep into wave physics, which in a way nicely illustrates the role of singularities. Wave physics dissolves the caustic singularities of geometrical optics. But wave physics has singularities of its own, complementary to caustics; these are the singularities of *phase*—the new quantity introduced in the generalization from rays to waves. The singularities are the *dislocation lines*[20] (so called because of their close analogy with dislocations in crystals), where the phase is singular and the intensity is zero. Dislocations are fine-scale features of waves; they are now beginning to be investigated in detail in optical experiments with interfering laser beams.[21] They are optical singularities in the approximation where light is regarded as a scalar wave. But light is not a scalar wave; its vector nature allows for *polarization*. And in the passage from scalar to vector, the dislocations are dissolved, and replaced by yet other singularities, of the pattern of field vectors; these are lines again, but where the polarization is purely circular or purely linear.[22] At each stage, generalization to a deeper theory dissolves the singularities of the old theory, and replaces them by new ones.

Now we return to chaos, with our second example, and consider the conventional quantum mechanics of isolated systems, mindful of the fact that as the classical limit is approached the criteria for effective isolation become more stringent. An isolated bounded system possesses discrete energy levels. For fixed energy, the semiclassical limit corresponds to highly excited states, so we expect these to reflect the nature of the corresponding classical trajectories, in particular their chaology. And so it turns out, in a remarkable example of an asymptotically emergent phenomenon associated with a singular limit.

First, there is an obvious way in which the classical limit is singular. Between the classical energy continuum and the quantum discontinuum is a logically unbridgeable gap. In philosophers' parlance, the two theories are incommensurable. In practice, the incommensurability is inconsequential, because any spectroscope has finite resolution, and under sufficiently semiclassical conditions the discrete levels (spacing h^D for systems with D degrees of freedom) cannot be discriminated. Beyond

---

[20] Nye, *Natural Focusing and Fine Structure of Light*; John F. Nye and M.V. Berry, "Dislocations in Wavetrains," *Proc. Roy. Soc. Lond. A* 336 (1974): 165–90; John F. Nye, "The Motion and Structure of Dislocations in Wavefronts," *Proc. Roy. Soc. Lond. A* 378 (1981): 219–39; Michael V. Berry and M.R. Dennis, "Phase Singularities in Isotropic Random Waves," *Proc. Roy. Soc. Lond. A* 456 (2000): 2059–79.

[21] Marat S. Soskin, *Singular Optics*, SPIE Series, vol. 3487 (Bellingham, Washington: SPIE, 1998); Mikhail Vasnetsov and K. Staliunas, *Optical Vortices* (Commack, New York: Nova Science Publications, 1999).

[22] John F. Nye, "Polarization Effects in the Diffraction of Electromagnetic Waves: The Role of Disclinations," *Proc. Roy. Soc. Lond. A* 387 (1983): 105–32; idem, "Lines of Circular Polarization in Electromagnetic Wave Fields," *Proc. Roy. Soc. Lond. A* 389 (1983): 279–90; John F. Nye and J.V. Hajnal, "The Wave Structure of Monochromatic Electromagnetic Radiation," *Proc. Roy. Soc. Lond. A* 409 (1987): 21–36.

this, we can imagine studying the details of the semiclassical spectrum by performing a mental magnification to keep the average level spacing unity.

Now it is possible to ask whether the thus-magnified spectrum is qualitatively different for a classically integrable (regular, nonchaotic) system and for a chaotic one. For example, the much-investigated quantum dots can be caricatured as quantum billiards, i.e., electrons moving freely in planar regions with specular reflection at the boundary walls; square or circular boundaries generate classically regular motion; a stadium-shaped wall generates chaos. (Other chaotic systems that have been studied experimentally include hydrogen atoms in very strong magnetic fields, and the vibrations of molecules whose atoms are anharmonically coupled.)

These spectra display *universality*. There is a sense in which the spectra of all quantum systems whose classical mechanics is chaotic are the same, and similarly for integrable systems.[23] The universality does not extend to the positions of the individual levels: of course these are different for different systems in each class. It is to the *statistics* of the levels that universality applies. Statistics are natural in the semiclassical limit, since there are many levels in a classically small energy interval. Spectral statistics has become a richly elaborated subject,[24] of which the merest sketch must suffice here.

Consider one of the simplest statistics of the magnified spectrum (where the mean spacing is unity), namely the probability $P(S)$ that neighboring levels have spacing S. For almost all integrable systems with more than one degree of freedom,

$$P(S) = \exp(-S) \text{ (integrable);} \tag{10}$$

this is the negative exponential characteristic of Poisson-distributed "events" (levels); in particular, the most probable spacing is $S = 0$, indicating clustering of the levels.[25] It is a little strange at first encounter that regular classical motion corresponds to a random quantum spectrum, but it is so.

For chaotic classical motion, the level spacings distribution is very different, and closely approximated by the Wigner distribution[26]

$$P(S) = \tfrac{1}{2}\pi S \exp(-\tfrac{1}{4}\pi S^2) \text{ (chaotic).} \tag{11}$$

This is the statistic characteristic of the eigenvalues of random matrices.[27] Note that $P(0) = 0$, indicating repulsion of levels, in contrast to the clustering of (10).

---

[23] Berry, "Quantum Chaology (The Bakerian Lecture)"; Michael V. Berry and M. Tabor, "Level Clustering in the Regular Spectrum," *Proc. Roy. Soc. Lond. A* 356 (1977): 375–94; Oriol Bohigas and M.-J. Giannoni, "Chaotic Motion and Random Matrix Theories," in *Mathematical and Computational Methods in Nuclear Physics*, Lecture Notes in Physics Series, vol. 209, J.S. Dehesa, J.M.G. Gomez, and A. Polls, eds. (Berlin: Springer, 1984), 1–99; Oriol Bohigas, M.-J. Giannoni, and C. Schmit, "Characterization of Chaotic Quantum Spectra and Universality of Level Fluctuation Laws," *Phys. Rev. Lett.* 52 (1984): 1–4.

[24] Fritz Haake, *Quantum Signatures of Chaos* (Berlin: Springer, 1991).

[25] Berry and Tabor, "Level Clustering in the Regular Spectrum."

[26] Berry, "Semiclassical Mechanics of Regular and Irregular Motion"; idem, "Quantizing a Classically Ergodic System: Sinai's Billiard and the KKR Method," *Ann. Phys.* 131 (1981): 163–216.

[27] Charles E. Porter, *Statistical Theories of Spectra Fluctuations* (New York: Academic Press, 1965); Madan L. Mehta, *Random Matrices and the Statistical Theory of Energy Levels* (New York: Academic Press, 1967).

I have simplified matters. There are actually three universality classes for classically chaotic quantum spectra. Equation (10) applies when there is time-reversal symmetry, when the particles have integer spin (bosons), or where spin plays no part (as in electrons in quantum billiards). The other universalities apply when there is time-reversal symmetry for fermions,[28] and when there is no time-reversal symmetry.[29]

It is one thing to identify universality classes, by computer experiments inspired by intuition, or (now) by laboratory experiment, but quite another to provide a theoretical explanation of the universality. However, this has now been achieved, both in broad outline and in many details.[30] It turns out that the quantum spectral universality is begotten by a universality of the classical mechanics, connected with the distribution of very long periodic orbits.[31] And in one of those unexpected connections that make theoretical physics so delightful, the quantum chaology of spectra turns out to be deeply connected to the arithmetic of prime numbers, through the celebrated zeros of the Riemann zeta function: the zeros mimic quantum energy levels of a classically chaotic system. The connection is not only deep but also tantalizing, since its basis is still obscure—though it has been fruitful both for mathematics and physics.[32]

It is important to emphasize why spectral universality is an emergent nonclassical semiclassical phenomenon. It is nonclassical because it is a property of discrete energy levels, which have no classical counterpart. It is semiclassically emergent because only as $h \rightarrow 0$ are there many levels in a classically small interval—it is impossible to calculate statistics using only the ground state. And only the singular

---

[28] Rainer Scharf, B. Dietz, M. Kus, F. Haake, and M.V. Berry, "Kramers' Degeneracy and Quartic Level Repulsion," *Europhys. Lett.* 5 (1988): 383–89.

[29] Thomas H. Seligman, J.J.M. Verbaarschot, and M.R. Zirnbauer, "Spectral Fluctuation Properties of Hamiltonian Systems: The Transition Region Between Order and Chaos," *J. Phys. A* 18 (1985): 2751–70; Michael V. Berry and M. Robnik, "Statistics of Energy Levels Without Time-Reversal Symmetry: Aharonov–Bohm Chaotic Billiards," *J. Phys. A* 19 (1986): 649–68; Michael V. Berry and R.J. Mondragon, "Neutrino Billiards: Time-Reversal Symmetry-Breaking Without Magnetic Fields," *Proc. Roy. Soc. Lond. A* 412 (1987): 53–74.

[30] Michael V. Berry, "Semiclassical Theory of Spectral Rigidity," *Proc. Roy. Soc. Lond. A* 400 (1985): 229–51; Jonathan P. Keating, "Periodic-Orbit Resummation and the Quantization of Chaos," *Proc. Roy. Soc. Lond A* 436 (1992): 99–108; Eugene B. Bogomolny and J.P. Keating, "Gutzwiller's Trace Formula and Spectral Statistics: Beyond the Diagonal Approximation," *Phys. Rev. Lett.* 77 (1996): 1472–75.

[31] Nathan Argaman, F.M. Dittes, E. Doron, J.P. Keating, A. Kitaev, M. Sieber, and U. Smilansky, "Correlations in the Actions of Periodic Orbits Derived from Quantum Chaos," *Phys. Rev. Lett.* 71 (1992): 4326–9; Jonathan P. Keating, "Long-Time Divergence of Semiclassical Form-Factors," *J. Phys. A.* 27 (1994): 6605–15.

[32] Michael V. Berry and J.P. Keating, "The Riemann Zeros and Eigenvalue Asymptotics," *SIAM Review* 41 (1999): 236–66; idem, "A New Approximation for Zeta(1/2 +it) and Quantum Spectral Determinants," *Proc. Roy. Soc. Lond. A* 437 (1992): 151–73; Michael V. Berry, "Semiclassical Formula for the Number Variance of the Riemann Zeros," *Nonlinearity* 1 (1988): 399–407; Jonathan P. Keating, "The Riemann Zeta-Function and Quantum Chaology," in *Quantum Chaos*, G. Casati, I. Guarneri, and U. Smilansky, eds. (Amsterdam: North-Holland, 1993), 145–85; Jonathan P. Keating and M. Sieber, "Calculation of Spectral Determinants," *Proc. Roy. Soc. Lond. A* 447 (1994): 413–37; Eugene B. Bogomolny and J.P. Keating, "Random Matrix Theory and the Riemann Zeros I: Three- and Four-point Correlations," *Nonlinearity* 8 (1995): 1115–31; idem, "Random-Matrix Theory and the Riemann Zeros II: *n*-point Correlations," *Nonlinearity* 9 (1996): 911–35.

nature of the semiclassical limit allows such nonclassical phenomena to emerge without paradox.

Acknowledgment. I am grateful for the hospitality of the A.D. White Professors-at-Large Program at Cornell University, where the first draft of this paper was written.

Appendix A

Caricaturing Hyperion as a homogeneous sphere with radius R (= 142km) and mass M (= 1.77×10^{19}kg), its moment of inertia is $I = 2MR^2/5$, and from its rotational angular speed $\Omega_H \approx 2\pi/(5 \text{ days})$ follows the angular momentum

$$J = I\Omega_H = \frac{2}{5}MR^2\Omega_H, \tag{A1}$$

and thence the number $N = 2\pi J/h = 2 \times 10^{58}$ given in the text, of Planck units in Hyperion's angular momentum.

For the estimate of T_q (equations 6 and 7), we imagine the boundary of the patch representing Hyperion's classical state evolving into tendrils each of area h. Since the total area of the angular-momentum sphere (phase space) is $4\pi J$, the number of such tendrils is about N, and their total perimeter is $L(T_q) \approx N \times$ (radius of J sphere) $\approx N\surd J = N^{3/2}\surd h$; the initial perimeter is $L(0) \approx \surd J = \surd(Nh)$. Solving the chaos equation

$$L(T_q) = L(0)\exp(T_q / T_c) \tag{A2}$$

for T_q now gives (6) and (7).

To estimate the spacing of Hyperion's rotational eigenenergies E_N, we use

$$E_N = \frac{J^2}{2I} = \frac{N^2h^2}{8\pi^2 I}, \tag{A3}$$

whence

$$\Delta E_N \equiv E_{N+1} - E_N = \frac{Nh^2}{4\pi^2 I} = \frac{Jh}{2\pi I} = \frac{h\Omega_H}{2\pi}. \tag{A4}$$

This gives $\Delta E_N \approx 10^{-39}$J as stated in the text.

Finally, to estimate the decoherence time T_d we start with the energy of sunlight reaching unit area of Hyperion each second. This is about $P = 10\text{W/m}^2$, since 1000W/m^2 reaches the earth and Saturn is about ten times further from the Sun. The total power reaching Hyperion is thus $\pi R^2 P$; thence the number of photons (frequency ν) reaching hyperion in T_d is $N_p \approx \pi R^2 P T_d / h\nu$. These give random impacts, so the total phase change is $\surd N_p$ times the value in (8), and this must be of order unity, leading to

$$T_d \approx \frac{hc\lambda}{\pi R^2 P} \tag{A8}$$

and thence to (9).

Appendix B

My aim here is to use a simple model to get an expression for the "decoherence factor" that describes explicitly how impacts from photons "classicalize" Hyperion's rotation, by causing the density matrix, in the representation corresponding to Hyperion's angular position, to become diagonal very rapidly.

On the very small time scales associated with the onset of decoherence, Hyperion's chaotic tumbling is irrelevant, so it suffices to treat the rotation by a single angle θ, with associated angular momentum component J. Let a photon with momentum p strike Hyperion with impact parameter b ($-R \le b \le R$), and let the reflectivity be r. This will change J by

$$\Delta J = pb(1 - r) \tag{B1}$$

A classical Hamiltonian incorporating these impacts (denoted by subscripts i) is

$$H = \frac{J^2}{2I} + \theta \sum_i \Delta J_i \delta(t - t_i) \tag{B2}$$

(H is not periodic in θ, but in the present context θ can be considered as living on the covering space consisting of the whole real line.)

Quantally, the variables θ and J become operators $\hat{\theta}$ and $\hat{J}$, and the evolution of a state $|\psi(t)\rangle$ over the interval $\tau_i = t_{i+1} - t_i$ ($\approx 10^{-30}$s) between the successive impacts is determined (from the Schrödinger equation corresponding to (B2)) by the unitary evolution operator

$$U_i = \exp\{-\mathrm{i}\hat{\theta}\Delta J_i / \hbar\} \exp\{-\mathrm{i}\hat{J}^2 \tau_i / (2I\hbar)\}. \tag{B3}$$

We apply this in the position representation, for wavepackets that are classically narrow (corresponding to an accurate specification of Hyperion's rotation) and correspond to rotation with average angular momentum J, yet wide enough to neglect the spreading of the wavepacket during the tiny interval τ_i between photon impacts. These conditions are very easy to justify for a planetary wavepacket, as can be confirmed by exact calculations on a Gaussian model. Then the evolution is, approximately,

$$\begin{aligned} \psi(\theta, t_{i+1}) &= \exp\{-\mathrm{i}\theta\Delta J_i / \hbar\} \langle\theta| \exp\{-\mathrm{i}\hat{J}^2 \tau_i / (2I\hbar)\} |\psi(t_i)\rangle \\ &\exp\{-\mathrm{i}\theta\Delta J_i / \hbar\} \exp\{-\mathrm{i}\hat{J}^2 \tau_i / (2I\hbar)\} \psi(\theta, t_i) \end{aligned} \tag{B4}$$

Expressing this in terms of the density matrix $\hat{\rho}_i = |\psi(t_i)\rangle\langle\psi(t_i)|$ gives, for the evolution after N impacts,

$$\langle\theta|\hat{\rho}_{i+N}|\theta'\rangle \approx \exp\left\{-\mathrm{i}(\theta - \theta') \sum_{m=i}^{i+N} \Delta J_m / \hbar\right\} \langle\theta|\hat{\rho}_i|\theta'\rangle. \tag{B5}$$

Now comes the central step: averaging the exponential factor over the random impacts. In a time T during which there are many impacts (which can still be macroscopically extremely short), the exponent is a Gaussian random variable. With this observation, and introducing the average photon frequency ω, the *decoherence factor* can be written

$$\begin{aligned} F(\theta, \theta', T) &= \left[\exp\left\{-\mathrm{i}(\theta - \theta') \sum_{i=1}^{N} p_i b_i (1 - r_i) / \hbar\right\}\right]_{\mathrm{av}} \\ &= \exp\{-A\omega T(\theta - \theta')^2\}, \end{aligned} \tag{B6}$$

where the dimensionless *decoherence amplification constant* A, expressed in terms of the power P striking unit area of Hyperion, is

$$A \approx \frac{PR^4}{hc^2} \approx 10^{38}. \tag{B7}$$

The enormous value of A, together with (B6), ensures that in even the shortest macroscopically significant time T the off-diagonal elements of the density matrix are negligible for any difference $\theta - \theta'$ in Hyperion's angular position that is large enough to conceivably be measured.[33]

---

[33] For a more general version of (B6), see equation (3) of Braun, Haake, and Strunz, "Universality of Decoherence."

FORMALISM AND ONTOLOGY IN EARLY ASTRONOMY

Ernan McMullin

1 Introduction

A point often overlooked in the proliferating discussions of the relevance of recent advances in physical theory to theological concerns is that such discussion almost necessarily takes for granted that these advances bear on *ontology*, on the underlying structure of the world. They presume, then, a *realist* interpretation of the theories they discuss. But such an interpretation cannot by any means be assumed; anti-realist criticisms of the realist thesis have multiplied in the last few decades. Were one to adopt an anti-realist reading of physical theory generally, construing theory, in effect, as no more than a convenient way of saving the phenomena for essentially practical ends, the dramatic advances in physical theory of the past century could have little or no relevance for the theological elaboration of religious belief. To serve theological ends, theory has to declare (or at least strongly suggest) how the world is.

The conference from which the essays gathered in this volume derive dealt with developments in quantum theory, much the most enigmatic domain of recent science. But can these developments carry with them the consequences for traditional theological concerns that many have claimed they do? Some of the most energetic critics of scientific realism are drawn from the ranks of philosophers of quantum mechanics. Were one to take the side in this debate of such critics as Bas van Fraassen and Arthur Fine, one would scarcely anticipate that there could be any potential implications of quantum theory for theology in the first place.

But the challenge is even more pointed than this. Even if one *were* to adopt a broadly realist position in regard to the implications of theory, a further problem arises, one that is more or less peculiar to mechanics. How is one to *interpret* the quantum formalism in ontological terms? Interpretation poses little difficulty in most other scientific fields. But, somehow, interpretation has always been a problem in the domain of mechanics. And that problem has become much more acute with the advent of quantum mechanics. Though the quantum formalism, with its innumerable varieties of application, can claim to be one of the most successful in empirical terms in the entire history of the natural sciences, there is a quite striking level of disagreement as to how it ought be interpreted, disagreement, that is, as to what it tells us about the microstructure of the familiar world of our experience. This inevitably must affect any further effort to relate the findings of quantum theory to issues of theological concern.

Two related sorts of difficulty face those who claim quantum theory as warrant for a particular ontological claim. First, the quantum formalism itself seems to yield two different ontological interpretations, two interpretations that could hardly be more different. One, the so-called "Copenhagen" interpretation, claims the parameters of the microworld to be in important respects indeterminate and the basic form of action at that level to be indeterminist. The other, associated in particular with David Bohm, would, on the contrary, take the microstructure to be as determinate and the action as determinist as both were in classical mechanics. How is it possible that basically the *same* formalism can sustain two translations into ontology that are so radically at odds?[1] And how, in the circumstances, is one to decide between them, since by definition they predict equally well, being interpretations, effectively, of two formalisms mathematically reducible to one another?

Yet from the perspective of the theologian, the implications of the rival interpretations are significantly different. Throughout the Newtonian era, the determinist thrust of the mechanics of the day was believed to raise special difficulties for the notions of free will and moral responsibility that are altogether central to Western theology. Though the indeterminism proclaimed so confidently as a major consequence of the new quantum mechanics was welcomed by some who argued that it opened the way to reconcile physics and human freedom once more, it was soon realized that indeterminism in neural agency may pose as great a challenge for freedom of action as does determinism, though for quite different reasons. So what *are* the consequences of quantum theory for this key component of standard theological doctrine?

Further complicating efforts to assess the relevance (or lack of it) of quantum theory for theology generally is the troubling strangeness of the ontologies associated with it, whether for example that of Bohr or that of Bohm. From the realist perspective, the Bohr interpretation abandons the objective measurement-independent ontology of particles with well-defined velocities and positions of classical physics; it also requires, in effect, a mysterious sort of "entanglement" between pieces of apparatus at unlimited distances from one another. Bohm, on the other hand, retains the objective status and the determinate/determinist features of the earlier mechanics but at the cost of introducing, among other things, an additional highly unintuitive form of action. If the aim is to find analogues in the quantum domain that would illuminate the ways in which the Creator acts within the natural order, it is not at all clear what sort of illumination one can expect from a quarter itself so ontologically problematic.

I am not going to address these difficult issues directly, leaving this task to other essays in this collection. My aim is a more modest one. The problem of relating formalism to ontology, which weighs so heavily upon quantum theory, and hence also on all efforts to develop plausible links between that theory and contemporary theological issues, is in no sense new. Indeed, it is almost as old as the kind of inquiry that could in a very broad sense be characterized as "scientific," meaning that it contains significant elements of the practices that we group together today under the label "natural science." My plan is to investigate the early history of astronomy in order to show that similar difficulties in making the step from formalism to ontology arose right from the beginning. It was in astronomy that mathematical formalism first came to be successfully employed to describe observed physical regularity. But reasoning from this to a plausible causal substructure that would explain why the regularity holds (the sort of inference that would today be called retroduction or abduction[2]) encountered obstacles that modern-day interpreters of the quantum formalisms would surely find familiar.

---

[1] The complicated story of how these interpretations came to be, of how the first of them, associated in particular with Niels Bohr, came to be regarded as the canonical, indeed the only, interpretation, is well told in James T. Cushing, *Quantum Mechanics: Historical Contingency and Quantum Hegemony* (Chicago: University of Chicago Press, 1994).

[2] See Ernan McMullin, *The Inference That Makes Science* (Milwaukee: Marquette University Press, 1992).

Tracing these obstacles in some detail through several contexts in pre-Copernican and Copernican astronomy should prove of some interest in its own right. But it should also serve to bring out how problematic the inference from successful formalism to appropriate ontology has always been in the explanation of motion. The moral is, of course, that one must be exceedingly cautious in advancing claims for a particular interpretation of a successful predictive/descriptive formalism in mechanics. It is no accident, then, that the critics of scientific realism in recent philosophy of science draw their inspiration so often from mechanics. Though a realist perspective is more difficult to articulate in that domain than in any other part of natural science, the prospects, as we shall see, are not entirely bleak.

2 The Matter of Interpretation

Before embarking on the historical studies that constitute the body of the essay, it will be helpful to clarify first the notions of *formalism* and *interpretation* employed. A mathematically expressed formalism, $s = 1/2at^2$, for example, tells us nothing whatever about the world, as it stands. It can be manipulated by means of the grammar, the syntax, of mathematics. Theorems like $ds/dt = at$, or $d^2s/dt^2 = a$, can be derived from it. But to read it as the law of falling bodies, for example, some of its terms have to be *interpreted* as measurable physical quantities. Mathematics of itself cannot provide the needed semantics. Linking terms to world is a complex process involving specific measurement techniques as well as theoretical specification of how the outcome of these techniques is to be understood. Terms like 'data' and 'observation' mask the complexity of this link. Observation, in its original meaning of sense-experience, plays only a minor (if indispensable) role in securing this primary link between pure mathematical formalism and the physical world it is meant to portray.

There are, therefore, two quite different processes of interpretation involved when one speaks in a general way of the problem of linking formalism to ontology in natural science. The first is the association of mathematical terms with specific physical meanings so that the formalism can convey empirical data. This is already a form of interpretation and at times by no means an unproblematic one. A glance at the internal debates among logical positivists during their brief period of philosophical dominance will recall how convoluted and how crucial this first level of interpretation can be, especially for those empiricists for whom it is the only relevant level.

The product of this interpretation is a formalism that is no longer, strictly speaking, a *pure* formalism. Nor is it merely a mathematical formalism, though it is often so described. Kepler's laws, Boyle-Charles's law, and the like, employ mathematical formalism but they are themselves something much more. They make truth claims about the world, specifically about the world as "observed." They might best, perhaps, be described as *physical* formalisms, conveying in a mathematical syntax the results of systematic observation.

It is with this type of formalism, already the product of the first level of interpretation, that we are concerned here. It is the relation of *physical* formalism to ontology that poses the much more controversial issues that divide scientific realists from nonrealists and divide realists themselves into different camps. The Copenhagen interpretation of quantum mechanics is already a *second* level of interpretation. Quantum theory itself is a *physical* formalism in the sense defined above. In order to relate it to the data of measurement, it must already be at least partially interpreted in physical terms. But as it stands, it leaves the second level of interpretation open; it does not directly specify underlying ontological or causal structure. That is where Bohr, Bohm, and the rest come in.

A second ambiguity in the usage of the term 'interpretation' in the quantum context arises from the claim that there are two "interpretations" of the quantum formalism. Why not two different *theories*, as James Cushing recommends?

> The (physical) interpretation refers to what the theory tells us about the underlying structure of these phenomena (i.e., the corresponding story about the furniture of the world—an ontology). Hence, *one* formalism with *two* different interpretations counts as *two* different theories.[3]

Customary usage, however, would suggest that there is only *one* theory here: physicists speak of "quantum theory" and make a point, it would seem, of referring to the Bohr/Bohm alternatives as "interpretations," not as two different quantum theories. Yet Cushing would appear to have a point: if we look to theory to tell us something about the underlying structure of the world, the Bohr and Bohm accounts could hardly be further apart and in that sense should count as very different theories.

The original usage of the term 'interpretation' in regard to the task of finding an acceptable ontology to go with the new quantum formalism of the 1920s was intended, presumably, to underline the difficulty of that task in general and the particular difficulty in characterizing an ontology of such a totally unfamiliar sort. The interpretation here was emphatically *after* the fact. It was highly tentative. And it seemed, in a sense, optional: those engaged in specific applications of quantum theory did not need to concern themselves, apparently, with the further issue of interpretation. The subsequent efforts of the leading Copenhagen figures to present Bohr's as the only *possible* interpretation[4] ought, it might seem, to have led them to characterize their understanding of the quantum formalisms as *the* quantum theory. But there was still a measure of disagreement among them about just *how* the ontology should be characterized. And there was the ghost of Einstein hovering over their efforts to legitimate the "Copenhagen" interpretation. In Einstein's eyes, at least, this latter did not deserve the further term of approbation "theory." Then came Bohm and Bell and a direct challenge to the claim that the Copenhagen interpretation is the only viable one, that it, so to speak, *is* quantum theory.

In natural science generally, furthermore, the specification of an ontology eventually tends to drive further development of the formalism (think of plate tectonics or molecular biology). This has not, so far as I can tell, been the case in quantum mechanics to any notable extent (though that may change, and in changing, provide a reason, finally, to favor one of the ontologies over the other). Whether these be the reasons or not, where customary usage is concerned, "interpretation" the Bohr and Bohm ontologies remain. In this essay, I will be using the term in the inclusive sense of proceeding from formalism to causal substructure, with no suggestion, necessarily, of the proposed ontology's being particularly problematic.

The language of "interpretation," "substructure," "ontology," ought to serve as a reminder once again that this whole discussion makes sense only in a realist perspective. Cushing's remark above that an "interpretation" of quantum theory refers to "underlying structure" brings out the fact that the Bohr/Bohm debate counts

---

[3] James T. Cushing, in this volume, 100.

[4] Chronicled in a lively way in Mara Beller, *Quantum Dialogue: The Making of a Revolution* (Chicago: University of Chicago Press, 1999); see also Cushing, *Quantum Mechanics*, passim.

for something only if one assumes the legitimacy of the move to ontology that each side implicitly relies on.[5] It would not matter either way which of the two "interpretations" were to be preferred, if the ambition of theory must be strictly limited to saving the phenomena.

3 Greek Astronomy

Our review may begin from the very earliest uses of mathematical formalism to order the regularities found in our experience. In practice, this was to order celestial phenomena, the only domain where regularity and recurrence were, at first sight, at all evident. Notable among these efforts was the complex practice that built up around the omen-list, *Enuma Anu Enlil,* in Mesopotamia beginning from the second millennium BCE. This list matched appearances in the sky with major happenings on Earth, reflecting the religious beliefs of the region. It evidently motivated the growth of systematic observation of celestial phenomena, the use of mathematics (first developed independently, perhaps, for commercial purposes) to schematize the observations, and the creation of long-term records from which, for example, the planetary periods could eventually be calculated. The extant Mesopotamian records from the seventh century BCE onwards are quite extensive—baked clay tablets fortunately, are exceptionally durable!—and testify to a sophisticated practice enabling the prediction of eclipses, first and last appearances of planets and constellations, and the like.[6] Two methods of approximation were developed that eventually enabled the construction of elaborate ephemerides. System A and System B (as they have come to be called) employed the equivalent of arithmetical step-functions and zigzag functions, respectively.

The goal of these physical formalisms was entirely practical. There was no attempt to follow the orbits of the planets in the sky. Orbits were without significance as omens, and in any case the Babylonian omen-guardians lacked the geometrical methods that would have been required to trace them. Nor was any question raised, so far as we can tell, about the physical nature of the heavenly bodies or about the causes of their motions. The physical formalisms, so successful from the predictive standpoint, remained uninterpreted in causal terms. Since the skies seem to have served as a sort of message-board of the gods, questions of a more physicalist kind about the specifics of the planetary motions would, presumably, have been unlikely to arise.

But the same was by no means true in the Greek-speaking world next door, where notions like *nature* and *cause* were in the process of development and where observed regularities invited further explanation. The phenomena of the heavens, a mixture of the regular and the apparently irregular, posed an especially intriguing

---

[5] There was a leaning for a time towards instrumentalism, notably on the part of Heisenberg, in the early stages of the development of quantum mechanics. And some would still characterize Bohr's position as at bottom anti-realist. But on balance, it seems clear that Bohr was making a wide variety of quite specific realist claims (e.g., about the effect of measurement on previously undetermined properties). However, this is an issue I leave for others to debate.

[6] See, for example, Otto Neugebauer, *The Exact Sciences in Antiquity* (New York: Dover, 1969); Frank R. Hodson, ed., *The Place of Astronomy in the Ancient World* (Oxford: Oxford University Press, 1974). For further references, see Ernan McMullin, "The Goals of Natural Science," *Proceedings and Addresses of the American Philosophical Association* 58 (1984): 37–63.

challenge. Prompted, it was said, by a hint from Plato that the seeming irregularities of the planetary motions ought to be reducible to a set of circular motions, the only kind of motion that gave promise of the eternal continuance proper to the motions of the heavenly bodies, Eudoxus of Cnidos (c. 408 BCE – c. 355 BCE) proposed to view each planet's observed motion as the resultant of a set of four independent circular motions (for the Sun and Moon, three only) around the Earth as center. The most striking feature of the observed planetary paths was the occasional occurrence of a looping or "retrograde" motion. Eudoxus' formalism could, very broadly at least, account for this, as well as for the planets' motions in latitude. His achievement would lay a permanent impress on Greek astronomy.

There is some disagreement as to whether he attempted to interpret his formalism in terms of an underlying physical structure. Specifically, did he interpret the circular motions (all that the formalism of itself required) as being embodied in rotating concentric spheres? G.E.R. Lloyd remarks:

> The theories of Eudoxus and Callippus were purely mathematical constructions. Neither astronomer said anything about the mechanics of the heavenly movements, about the nature of the concentric spheres, or about how movement was transmitted from one to the others.[7]

J.L.E. Dreyer agrees:

> As Eudoxus made no attempt to connect the movements of the various groups of spheres with each other, it seems probable that he only regarded them as geometrical constructions suitable for computing the apparent paths of the planets.[8]

D.R. Dicks supposes Eudoxus to have been talking about concentric spheres, but regards these as no more than "mathematical abstractions," on the grounds that he did not speculate about the material of which they were composed or the kinds of connections that existed between them or the powers responsible for their motion.[9] This, then, would appear to have been a physical formalism without any suggestion of ontology.

These arguments from silence seem far from convincing, most especially because none of Eudoxus' own works have survived and we are entirely dependent (as is all too usual with Greek astronomers) on excerpts and summaries in later Greek writers. Furthermore, his system was clearly of ontological inspiration, beginning with the separate circular motions themselves. It would have been difficult to visualize the combining of circular motions around a common center, each on a different plane, without interpreting them in physical terms as embodied in rotating concentric spheres each with a different axis but all axes intersecting at the center of the Earth.

Whether or not Eudoxus *did* propose an underlying physical structure of this sort, his contemporary, Aristotle, did not hesitate to do so, drawing on the proposed formalism. By then, Callippus had added seven more circles (or spheres) he thought necessary to save the phenomena. Aristotle took the whole discussion in a different direction by regarding the formalism as a stepping stone to a causal explanation of *why* the planets move as they do, to a cosmic ontology, taking the form of a single interlocking system of spheres. (Eudoxus had treated each planet's set of motions as

---

[7] G.E.R. Lloyd, *Early Greek Science: Thales to Aristotle* (London: Chatto and Windus, 1970), 92.

[8] J.L.E. Dreyer, *A History of Astronomy* (New York: Dover, 1953 [1906]), 91.

[9] D.R. Dicks, *Early Greek Astronomy to Aristotle* (London: Thames and Hudson, 1970), 258.

autonomous.) Only such a causally unified system could, Aristotle said, provide an explanation proper. In order to achieve this unification, however, he had to add twenty-two extra "counteracting" spheres in order to prevent the motions distinctive of each planet from being communicated downwards to the planets below it. Because the "axles" of each sphere are carried on the next outer sphere, the motion of each sphere is the resultant of the motions of *all* the spheres outside it, with the exception that the movement of rotation of each sphere around its own "axle" (which cannot be explained in this way) has to depend on an active mover-principle associated with that sphere. Aristotle's system requires fifty-five spheres in all (fifty-six if the sphere of the "fixed" stars be included).[10] And it employs two different modes of causal explanation, both familiar from our ordinary experience, one mechanical and the other teleological.

This was an immensely complicated interpretation of the Eudoxan formalism. What gave it its apparently overwhelming appeal? After all, it survived for two thousand years; it still had its defenders in Galileo's day. Its long popularity was certainly not due to its merits in saving the appearances. Though the original Eudoxan model had some modest success in that regard, pointing in a fruitful direction, at least, it was beset with difficulties from the beginning. Historians point out that though it could account, very roughly at least, for the retrograde motions of Jupiter and Saturn, it could not be made to yield the corresponding motions for Mars and Venus. Nor could it, in general, get the shapes and widths of these motions right.[11] More evidently, it could not account for the well-known periodic variations in planetary brightness as well as the variations in the apparent size of the Moon, both understood even then to signify variations in distance from the Earth. A concentric geocentric model could not, in principle, admit such variations. Further, it could not, in practice, predict the planetary positions within the limits of observational accuracy of the time. All in all, its "saving of the phenomena" seems at best to have been of a promissory sort.[12]

Where, then, did its appeal lie? Clearly, in its claim to *explain* the planetary motions in plausible causal terms. This overrode, and long continued to override, its shortcomings in actually saving the phenomena. Aristotle and his successors could even suggest some of the properties the carrying spheres would have to possess: they would have to be solid in order to carry and be carried; they would have to be transparent ("crystalline") in order that the light of stars and planets would be visible on Earth; they would have to be incorruptible in order that their motions might be unchanging in their regularity... The ontology of the spheres could, then, take on a life of its own, almost independent of the mathematical complexities of the formalism—yet not really independent. The warrant for belief in the existence of

[10] Later followers of Aristotle, particularly the Aristotelians of the Christian Middle Ages, attempted to reduced the awkward multiplicity of fifty-six independent First Movers by supposing that the movers of the spheres were themselves animated by the desire to imitate as far as possible the state of rest of the single Unmoved Mover. Though the general analysis of motion in Book VIII of the *Physics* seems to lean towards a single First Mover (in a tone not entirely assured, however; see 259a 7–14), once Aristotle turns to the specifics of the Eudoxan astronomy in the *Metaphysics*, he insists that there must be as many additional Movers, themselves unmoved by another, as there are spheres (1073a 30–40).

[11] See Lloyd, *Early Greek Science*, 82–92; Dreyer, *A History of Astronomy*, 87–107.

[12] It was also, in the absence of an adequate spherical geometry, extremely difficult to manipulate. For an overall assessment, see Larry Wright, "The Astronomy of Eudoxus: Geometry or Physics?", *Studies in the History and Philosophy of Science* 4 (1973): 165–172.

these spheres, quite unfamiliar in kind as they were, still lay in large part in the Eudoxan formalism and its admittedly limited success in saving the phenomena of planetary motion. Aristotle could, however, be relatively cavalier about the exact number of the spheres; this could be left, he says, to the "mathematicians." And he could, effectively, hand over to these latter the messy and complex task, at once observational and mathematical, of wrestling with the phenomena directly.

The mathematicians eventually did just that, but with a reversal of the Aristotelian priorities. Putting the criteria of saving the phenomena first, some of them sought a mathematical formalism that would better accord with the totality of the known phenomena, even at the sacrifice of the ontological plausibility that Aristotle had prized. Apollonius and Hipparchus, mathematicians of genius, introduced two innovations that greatly increased the flexibility of the formalisms available to the astronomer. First was the eccentric, which in its simplest form meant shifting the center of the planet's rotation a short distance from the Earth. It meant, of course, abandoning the fitted spheres of the Aristotelian cosmos, but with it one could save the known inequalities in length of the four seasons.[13] More radical and more flexible was the epicyclic model, where the planet would revolve on a circle (the epicycle) whose own center would be carried on a second circle (the deferent) around the Earth as center. And the eccentric model itself could be given added flexibility by allowing its center of rotation to revolve in a small circle around the fixed Earth. The ontological preference for uniform circular motion was thus retained. But any further tie between formalism and ontology was now more difficult to divine. It was not easy to find a plausible ontological interpretation for the epicycle and the movable eccentric, especially one that would retain the sort of explanatory force in which Aristotle's concentric model had excelled.

But there was another puzzle of a new kind, one that presaged, faintly at least, the sort of problem about interpretation that plagues current quantum mechanics. It was indeed, as Hipparchus is reported to have said, something to marvel at.[14] With a proper setting of the initial parameters, the epicyclic and the movable eccentric formalisms could be shown to be mathematically equivalent, so that they yielded *exactly* the same predictions regarding the planetary motions. But how could this be possible? How could two radically different ontologies—recall that for the astronomers of the day, the component circular motions were real—yield, in principle, exactly the same phenomena?[15] The tie between formalism and ontology seemed to be broken. The criterion of saving the appearances was not going to be enough in this case to reveal the true nature of the planetary motions.

Reflecting on this dismaying situation, Adrastus would later (c. 100 CE) comment:

---

[13] See Olaf Pedersen and Mogens Pihl, *Early Physics and Astronomy* (London: Macdonald, 1974), 84–6.

[14] Theon of Smyrna in his *Astronomia*, quoting Adrastus (whose own work is lost) quoting Hipparchus (whose own work is likewise lost). See Pierre Duhem, *To Save the Phenomena*, E. Doland and C. Maschler, trans. (Chicago: University of Chicago Press, 1969), 8.

[15] Two reservations need, however, to be noted. First, the equivalence does not hold for the original simpler fixed eccentric. And second, the eccentric formalism is properly applicable only to the movements of the outer planets, which is probably why astronomers came to prefer the epicyclic alternative for most purposes; see Dreyer, *A History of Astronomy*, 156.

> No matter which hypothesis is settled on, the appearances will be saved. For this reason, we may dismiss as idle the discussions of the mathematicians, some of whom say that the planets are carried along eccentric circles only, while other claim they are carried by epicycles, and still others that they move around the same center as the sphere of the fixed stars.[16]

Mathematical astronomers whose task it was to fit geometrical figures to the observed celestial phenomena were, in Adrastus' view, wasting their time debating the relative merits of the two formalisms if they thought they could in this way establish the *true* circular motions involved in the observed movements of the planets.[17] Instead, they should turn to properly *physical* considerations. They could, for example, argue that the epicyclic formalism should be preferred because it could be modeled physically by means of a small sphere rolling between two concentric spheres. According to Adrastus, Hipparchus had much earlier come to the same conclusion, on the grounds that the epicyclic formalism "conforms better to nature," whereas the eccentric formalism, lacking a plausible ontology, saves the phenomena "only accidentally."[18] Adrastus' moral was thus clear: when two physical formalisms associated with two different and incompatible ontologies save the phenomena equally well, a decision between them is to be reached by determining which of the ontologies can be made to yield the better causal explanation.

I will follow this story no further, though it would be instructive to see what Ptolemy made of the situation and even more instructive to trace the uneasy division that prevailed throughout the medieval period between Ptolemaic mathematical astronomy (whose complex formalism saved the phenomena admirably) and a physical astronomy that stubbornly relied on the merits of the superior explanatory force and overall conformity to physical principles of the Aristotelian alternative. The tendency among natural philosophers was to place their ontological bets on the latter and to attribute the success of the former in saving the phenomena rather to the ingenuity of the mathematicians than to its merits as a clue to ontology. But it is time to pass on to a new historical context where the relation between formalism and ontology was once again in question and where a solution of the sort suggested earlier by Adrastus and others was once more offered, this time in much more confident detail.

4 The Reality of the Sun's Motion: Copernicus

At one level, the Copernican formalism could be described as a mathematical transform of the Ptolemaic one, involving nothing more than a shift of reference-frame from Earth to Sun. It would follow that there would in that case be no way to

---

[16] Theon, *Astronomia*, chap. 26, quoted in Duhem, *To Save the Phenomena*, 9–10. The third alternative mentioned here was, of course, the Aristotelian one. It was not, however, by any means mathematically equivalent to the other two.

[17] When Duhem, referring to this passage, takes the speaker to imply the impossibility of "astronomy's" ever discovering "the true hypothesis," it was clearly only mathematical astronomy he had in mind. In the pages that followed, he was at some pains himself to point to the fact that Theon and Adrastus clearly held that astronomy employing arguments belonging to physics could, in principle, reach the truth about the real motions. In an extended critique of Duhem's tendency to discover his own special brand of instrumentalism in the works of Greek astronomical commentators, Lloyd appears to misinterpret Duhem in this instance; "Saving the Appearances," *Classical Quarterly* 28 (1978): 202–22; 218.

[18] Theon, *Astronomia*, chap. 34; Duhem, *To Save the Phenomena*, 9.

distinguish between the two on the grounds of saving the phenomena, the traditional criterion of the mathematical astronomer. Where observed planetary motions were concerned, at least, the predictions afforded by the two formalisms would be the same. However, matters were not quite as simple as this; for one thing, Copernicus made use of new observations to improve the accuracy of his computations, enabling him to construct new and improved tables.[19]

More important in his own eyes, he found a way to eliminate the hated "equant-point" that Ptolemy had, reluctantly, introduced in order to achieve the match between formalism and observed planetary positions that the eccentric-and-epicycle techniques alone, flexible though they were, seemed unable to bring about.[20] Because the adoption of the equant meant abandoning the principle of uniformity of circular motion—which, as we have seen, was inviolable on ontological grounds for earlier Greek astronomers—it had been a source of uneasiness for mathematical astronomers in the Ptolemaic tradition long before the time of Copernicus. But Copernicus was the first to work out an effective alternative that would do away with the equant, while still retaining the techniques of epicycle and eccentric that marked him as an inheritor of the older mathematical tradition. His formalism would have to save the phenomena, of course, but if he were to claim the reality of the terrestrial motions it postulated, it would have to do significantly more. And it is that "more" that concerns us here.

The heliocentric system he was proposing was one that could, as he reminded the readers of his *De Revolutionibus*, claim Greek ancestry.[21] But it was probably not from this source that his inspiration to shift the center of motion from Earth to Sun had originally come. The puzzling tie between the motions of Venus and Mercury and that of the Sun had been noted by astronomers from earliest times. Venus never strays further in angular distance than about 45° from the Sun; Mercury stays even closer, never more than about 25° away. But the Ptolemaic epicyclic formalism had brought to light an even more striking link, this time between the (mean) Sun's motion and the motions of *all* the planets. The line joining each of the superior planets (Mars, Jupiter, Saturn) to the center of its epicycle rotates so as to remain at all times parallel with the distant line joining Earth to the (mean) Sun. It follows from this that the epicyclic period of each of these planets is the same and is thus

[19] Anton Pannekoek, *A History of Astronomy* (New York: Interscience, 1961), 193.

[20] The equant was defined by Ptolemy as the center relative to which the epicycle center moves uniformly, i.e., sweeps out equal angles in equal times as it moves around the deferent circle. The equant is placed in such a way that the center of the deferent circle is half-way between it and the Earth, which is, like the equant, eccentrically located. This construction entails that the motion of the epicycle-center is nonuniform relative to its own orbit on the deferent circle. See Dreyer, *A History of Astronomy*, 197.

[21] He attributes elements of it to Greek sources, notably to the "Pythagoreans." Indeed, Copernicus's work was routinely described as "Pythagorean" in the century that followed its publication; the decree of the Congregation of the Index in 1616 suspending the *De Revolutionibus* "until it be corrected" referred to "the false Pythagorean doctrine." But the "motion around the fire" mentioned by Philolaus, the Pythagorean to whom Copernicus alludes, was almost certainly *not* a motion around the Sun (Dreyer, *A History of Astronomy*, 40–52). Heracleides of Pontus, not a Pythagorean, to whom Copernicus also refers, did propose a daily rotation of the Earth on its axis. But it was Aristarchus of Samos (a reference to whom Copernicus deleted for some reason in the final version of his manuscript; see A. Rupert Hall, *The Revolution in Science 1500–1750* (London: Longman, 1983), 63) who suggested (c. 280 BCE) a yearly revolution of the Earth around the Sun.

roughly equal to the period of the Sun around the Earth, i.e., one year. Something similar is true of the inferior planets (Venus and Mercury): the center of the planet's epicycle always lies on the line joining Earth and Sun, once again ensuring a one-year period, this time for the center of the planet's epicycle as it orbits the Earth.[22] Ptolemy himself noted these strange coincidences. In his *Planetary Hypotheses*, he remarks that each of the planets seemed to have one of its two epicycle-related periods determined in advance.[23]

It would not have required a very large leap of mathematical intuition on Copernicus's part—one wonders if it would not also have occurred to Ptolemy and been set aside—to realize that this epicyclic tie between Sun and planets could be very simply explained by drawing on the long-familiar equivalence of epicycle and movable eccentric. All that was needed was to place the Sun on the movable eccentric corresponding to the planetary epicycle, with the Earth as its center of motion.[24] The same eccentric would suffice for all five of the planets. Then the motion could be mathematically transferred from the Sun's frame of reference to that of the Earth, making the Sun the center of rotation for all of the planets, now including Earth. This would, at one stroke, eliminate the one-year coincidences, the puzzling links between the motions of planetary bodies that seemed on the face of it to be independent of one another. It would, in short, *explain* why it was that each planet's motion, when analyzed in epicyclic terms, had a component with the same period as that of the Sun in its apparent orbit around the ecliptic.

Furthermore, it would also explain why Venus and Mercury, viewed from the next outer planet, Earth, could never appear on the other side of the sky from the Sun; their motions would *necessarily* be linked to the apparent motion of the Sun. The shift of perspective here was not just a matter of simplifying a mathematical formalism, of reducing the number of circles needed. It provided an *explanation* for objective features of the planetary motions that Ptolemy had to leave as simply given, as puzzling coincidence.

In the Preface to the *De Revolutionibus*, Copernicus explicitly assumes the status of "philosopher," someone who, on the basis of causal reasoning, can claim truth for his constructions. He knew very well that this went beyond what his Ptolemaic forebears had been permitted to claim. In chapter 10 of Book I, he laid out his case. Clearly, he could not appeal simply to the traditional criterion of saving the phenomena, partly because his advantage in that regard was not great,[25] but much more because it would not carry conviction with those who sought to know what the *real* motions were. Nor could he rely on a claim to have simplified the formalism in certain respects; this was a pragmatic, but surely not an epistemic, consideration. Indeed, it was not entirely clear that he *had* simplified it, overall.[26]

---

[22] Arthur Berry, *A Short History of Astronomy* (New York: Dover, 1961 [1898]), 116–7; Alexandre Koyré, *The Astronomical Revolution*, R.E.W. Maddison, trans. (London: Methuen, 1973), 103–4.

[23] James Evans, "Ptolemaic Planetary Theory," in *Encyclopedia of Cosmology*, Norriss S. Hetherington, ed. (New York: Garland, 1993), 513–26; 525.

[24] This preliminary would be the Tychonic solution. For details, see Hall, *The Revolution in Science 1500–1750*, 61–5.

[25] Though he could claim some advantages even there. The elimination of the lunar epicycle, for example, removed a serious anomaly in the Ptolemaic system that entailed a much larger periodic variation in apparent lunar size than was observed to be the case.

[26] Historians of science seem quite divided on whether or not the Copernican formalism

Instead, he could offer the sort of argument appropriate to the *philosophical* (physical) astronomer, an argument that called on cause rather than on merely saving the phenomena. Unlike Ptolemy, he could provide a *cause* for the numerous features of the earlier system that appeared there as odd coincidence. He could explain the ties, already described, between the solar motions and those of the planets, so prominent a feature of the Ptolemaic formalism and yet entirely unexplained there. He could, in addition, explain why it was that the outer planets are invariably at their brightest (and thus, nearest) "at the time of their evening rising" (i.e., when in opposition to the position of the Sun), and dimmest (farthest) "at the time of their evening setting" (i.e., when in conjunction). Ptolemy could, in principle, impose constraints of this sort on the parameters of his system, but he could give no reason of any sort for why they should be needed.

But what impressed Copernicus most—"the chief point of all," he calls it—was an argument of a somewhat different kind. It was still causal, but it was directed to establishing order and harmony where none was perceived before. This evidently counted for more, in his eyes, even than explaining what had previously seemed odd coincidence. He could now, for example, specify the exact order of the planets from Mercury outward to Saturn. Prior to that, the ordering of Mercury and Venus relative to the Sun had been debated; the Ptolemaic formalism of itself lacked the resources to settle the issue. Furthermore, the distances of the planetary orbits from Earth were entirely indeterminate in the Ptolemaic system unless several ontological postulates were added, namely, embedding the epicycles in spheres that rolled within two containing spheres, with each set of containing spheres contiguous to the sets on either side of it. In the Copernican dispensation, on the other hand, the average distances of the planets from the Sun could easily be calculated, with the distance of Earth to Sun as unit. Finally, the planetary periods would now lengthen steadily with the distance of the planet from the Sun, with the sphere of the fixed stars, now understood to be immensely distant from the outermost planets, at rest. Far more appropriate that it should be at rest than rotating in almost incomprehensibly rapid motion, as Ptolemy had to suppose it did.

Another kind of ordering was that afforded by the retrograde motions, which had led the Greeks to invent elaborate schemes of sphere or epicycle. There was now "a wonderful commensurability," "a sure bond of harmony." If the Earth is the third planet outwards from the Sun, it will "lap," or be "lapped" by planets moving at a different angular velocity relative to the Sun. This will *necessarily* produce retrograde motions in the orbits of the planets as observed from Earth. Assigning this

---

was simpler than the Ptolemaic one, notably in regard to the number of circles required. Otto Neugebauer declares that the common view that the Copernican formalism was simpler "is obviously wrong. The choice of the reference-frame has no effect whatever on the model, and the Copernican models require about twice as many circles as the Ptolemaic models and are far less elegant and acceptable"; *The Exact Sciences in Antiquity*, 2nd ed. (Providence, R.I.: Brown University Press, 1957), 204. Arthur Berry, at the other extreme, states that Copernicus needed only thirty-four circles. (He is evidently drawing on Copernicus's own estimate at the end of his early work, the *Commentariolus*; in *Three Copernican Treatises*, E. Rosen, trans. (New York: Dover, 1959), 90). And Berry contrasts this with Fracastoro's count of seventy-nine for the Ptolemaic system in 1538 (*A Short History of Astronomy*, 121). But Copernicus was forced to add many minor circles in the *De Revolutionibus* (see M.P. Lerner and J.P. Verdet, "Copernican Cosmology," in *Encyclopedia of Cosmology*, 71–92). And there were by the 1500s many different versions of the Ptolemaic formalism. The accepted view among recent historians seems to be that there was not a marked advantage in this respect for either formalism.

as cause also allows the magnitudes and frequencies of the retrograde motions to be estimated. Copernicus notes, with great satisfaction, that he can show why the retrograde motion is "greater in Jupiter than in Saturn and smaller than in Mars" and that it is more frequent "in Saturn than in Jupiter, less often in Mars and Venus than in Mercury." "All these things," he concludes, "proceed from the same cause, which resides in the movement of the Earth."

What gave this discovery of a hitherto hidden harmony such weight for Copernicus was undoubtedly a theological premiss: one would *expect* such harmony in the work of "the most orderly Workman of all" (Preface), "the best and greatest Artist" (I, 10). The heliocentric formalism could surely not (one can imagine him saying) be a mere matter of calculational convenience, as philosophers had so often in the past taken the constructions of the mathematical astronomers to be. Harmony of so striking a kind could only be a testimony of reality. For it to *be* harmony, the Earth's motions would have to be real.

This emphasis on harmony has led many historians of science to characterize the argument of this crucial chapter of the *De Revolutionibus* as "aesthetic." In his influential work, *The Copernican Revolution*, Thomas Kuhn remarks:

> As Copernicus himself recognized, the real appeal of sun-centered astronomy was aesthetic rather than pragmatic. To astronomers the initial choice between Copernicus's system and Ptolemy's could only be a matter of taste, and matters of taste are the most difficult of all to define or debate.[27]

I do not for a moment agree that Copernicus would have recognized that the preference he asserts for the heliocentric option was only a matter of taste. He saw the harmony he had discovered by setting the Earth in motion to be a positive sign that he had hit on the way Creation *really* is. There was, on the other hand, nothing harmonious about its only rival, the Ptolemaic system. He was not arguing that the more beautifully ordered system ought to be accepted as true just *because* it was beautiful. It was, in his eyes, because the Creator would ensure such order, such beauty, in his handiwork. A causal argument of this kind would have carried *some* weight, at least, with the astronomers of the generations between Copernicus and Newton, even with those who would not have been sympathetic to Copernicus's ardent neo-Platonism.

But, as we have seen, Copernicus's causal arguments for the reality of the Earth's motions are of two kinds. The second is concerned with harmony and its likely Cause. But the other needs no theological supplementation. It offers a choice between explaining something puzzling and leaving it as brute fact. As argument, it is not coercive; some have more tolerance for brute fact than others! But it *does* carry legitimate epistemic weight as one form of retroduction, arguing from effect back to cause. The Copernican retroduction from puzzling apparent coincidence to a "natural" explanation (as he calls it) is, however, of a special kind that is not as straightforward as the usual type of retroduction in natural science. Much depends in the Copernican case on just *how* puzzling the coincidence appears, how loudly it cries out for explanation, how unsatisfactory it would seem to leave it unexplained when a reasonable explanation is at hand. Kuhn does allow that the Copernican system is "far more natural" than its alternative. But he allows no epistemic weight to this. It does have an "aesthetic advantage," and such advantages, he allows, are not negligible: they do influence scientists, as they surely did Copernicus. But in the

---

[27] Thomas S. Kuhn, *The Copernican Revolution* (New York: Random House, 1957), 172.

end, he asserts, Copernicus's claims to have explained what Ptolemy left unexplained "do not 'prove' a thing. The Ptolemaic system explains these phenomena as completely as the Copernican."[28]

The Ptolemaic system did *not*, however, explain the coincidental features described earlier. It did, of course, *incorporate* these features, the puzzling ties between the Sun's motions and the planetary motions, for example. But it did so only because it was programed to do so. It could not assign any *reason* why the system should exhibit these features in the first place, features that seemed to point to some further cause. Kuhn equates prediction (or saving the phenomena, in the older phrase) and explanation, not an uncommon identification at the time he was writing, the heyday of the so-called "deductive-nomological" model of explanation associated with logical positivism. But not every prediction is explanatory and not all explanations yield predictions, so the critics of this reductive account of explanation finally prevailed.

Copernicus can be counted an opponent of the positivist tenet: he is saying, equivalently, that to "save" phenomena is not necessarily to explain them causally. The phenomena that indicate a tie between the nearest approach of the outer planets to Earth and the simultaneous position of the Sun can be "saved" by simply including this tie (if it can be consistently done) in the constraints set on the parameters specifying the formalism. But this, Copernicus is insisting, does not explain why the planetary system should exhibit this rather striking constraint in the first place. Successfully assigning a *cause* for this latter is clearly a step towards determining the truth of the matter.

Kuhn is, admittedly, not convinced. As far as he is concerned, "Copernicus's system is neither simpler nor more accurate than Ptolemy's."[29] And these criteria, empirical accuracy and simplicity, are the ones, he implies, that carry conviction, then as now, with working scientists, the former for epistemic reasons, the latter for pragmatic ones.[30] He concludes that, apart from a "limited and perhaps irrational" brand of astronomer with neo-Platonic sympathies, the Copernican argument would have appealed to the average astronomer of that time, if indeed it would have appealed at all, on aesthetic grounds and on those alone.[31]

Kuhn could have called in his support here the historical fact that, apart from his young colleague, Rheticus, no other astronomer was persuaded by Copernicus's arguments for the reality of the Earth's motion until Kepler came along, seventy years later. Does this mean that the arguments were, as Kuhn and many others have painted them, appeals to the aesthetic sense only? Why did they not convince, if they were indeed of epistemic value, as I am arguing here? There seems to me to be a simple answer to this. Copernicus could not come up with a physics that would counter the powerful Aristotelian arguments for the immobility of the Earth at the center of the universe, as well as the equally powerful Aristotelian objections

---

[28] Kuhn, *The Copernican Revolution*, 178. For a fuller discussion, see Ernan McMullin, "Rationality and Paradigm-Change in Science," in *World Changes: Kuhn and the Nature of Science*, P. Horwich, ed. (Cambridge: MIT Press, 1993), 70–6.

[29] Kuhn, *The Copernican Revolution*, 171.

[30] In fairness, it must be added that in his later work, Kuhn abandoned this quasi-positivist account of theory-assessment, advancing instead a far more nuanced account of the values that normally guide theory-choice in science. See "Objectivity, Value Judgement, and Theory Choice," in his *The Essential Tension* (Chicago: University of Chicago Press, 1977), 320–39.

[31] Kuhn, *The Copernican Revolution*, 181.

pointing out the absence of the effects one should see were the Earth really to be in motion. Galileo's *Dialogue on Two Chief World Systems* was still a century away. Furthermore, there were the biblical references to the Sun's motion and the Earth's immobility, just then taking on an increasingly literalist cast thanks to the Reformation debates regarding the interpretation of Scripture. The professional astronomers of the day would have been wary, and not just on pragmatic grounds, of an interpretation of the new astronomy that would put it at odds with the Scriptures, literally taken. And there was the declaration set at the beginning of the *De Revolutionibus*, certain to reinforce that reaction by its denial of any realist intent to the heliocentric formalism contained in the work to which it was an apparently (though not really) authorized Foreword.

In those circumstances how likely was it that Copernicus's arguments in favor of a realistic construal of his innovative formalism would get a hearing? Not very likely. My contention here is not that they were of sufficient weight that they ought to have been accepted, despite the range of arguments arrayed against them. Rather, I am urging the more modest thesis that Copernicus's arguments for his heliocentric ontology had more epistemic merit than is commonly allowed them, and that, were it not for the weighty objections arrayed against the heliocentric thesis, these arguments might well have prevailed.

5 The Reality of the Earth's Motion: Kepler

Fifty years later, these arguments found a receptive listener in a mathematical astronomer of genius. In his *Mysterium Cosmographicum* (1596), Johannes Kepler set out to explain the Copernican ordering and relative spacing of the planets by means of neo-Platonic speculation involving the five perfect geometrical solids. In those intervening years, no astronomer of any consequence had come to the defense of the Copernican ontology, though the Copernican formalism had rapidly gained ground as a practical aid among professionals in the field. Kepler, however, expressed unqualified support for Copernicus's realist assertions about the Earth's motions and the Sun's immobility. What inspired his confidence? First, of course, was his own theologically inspired belief in the harmony to be expected in the work of an all-powerful and all-wise Creator, and his consequent sympathy for the harmony arguments of the *De Revolutionibus*. Second was his conviction that the biblically based objection to the Copernican theses was misguided. This conviction would later find eloquent expression in the Introduction to his *Astronomia Nova*, where he argued persuasively against taking the biblical text as an authority in matters astronomical.[32]

But as he tells us in the Preface to the *Mysterium Cosmographicum*, he had already in his student days at Tübingen, under the inspiration of his teacher, Michael Maestlin, "collected together, little by little... the advantages which Copernicus has mathematically over Ptolemy."[33] And he lays these out in the opening chapter of the work. Not only does Copernicus save the phenomena far better (he alleges) than

[32] For a comparison of Kepler's arguments with those advanced some years later by Galileo in his Letter to the Grand Duchess Christina, see Ernan McMullin, "Galileo on Science and Scripture," in *The Cambridge Companion to Galileo*, Peter Machamer, ed. (Cambridge: Cambridge University Press, 1998), 271–347.

[33] Johannes Kepler, *Mysterium Cosmographicum: The Secret of the Universe*, A.M. Duncan, trans. (New York: Abaris, 1981), 63.

Ptolemy did,[34] but "what is far more important is that... only Copernicus magnificently gives the explanation, and removes the cause of wonder, which is not knowing causes."[35] Ptolemy simply "could not account at all for a number of outstanding features" of the planetary phenomena, even though he could incorporate the phenomena themselves in his formalism. For instance, he could not give "the *reasons* for the number, extent, and durations of the retrogressions, and why they agree precisely, as they do, with the positions and mean motions of the Sun."[36] Copernicus can explain these so well that their cause must necessarily lie in the physical order that he postulates.[37] A "very strong argument for the Copernican arrangement of the spheres" can also be found in the orbit followed by the comet of 1577.

More striking again is the fact that in Ptolemy's system "the three eccentrics of the Sun, Venus, and Mercury have equal times of revolution." The answer to this puzzle is that these bodies "do not really revolve [in these eccentrics] but instead of them, the Earth does so on its own."[38] And Kepler recalls the planetary brightness patterns that Copernicus had already singled out: "The ancients rightly wondered why the three superior planets are always in opposition to the Sun when they are at the bottom of their epicycles, but in conjunction when they are at the top." For Copernicus, this is no longer a source of wonder. One final argument, this time for the diurnal rotation of the Earth, is the elimination of the incredibly distant and even more incredibly swift-moving outer sphere. Kepler asks "whether it is easier for it to happen and to be believed" that the relatively tiny Earth is rotating once a day or to suppose that "the complete universe goes with ten distinct motions (as there are ten mutually independent circles) with inconceivable rapidity, and is subject to nothing but that small point [the Earth] which alone is motionless."[39]

A few years later, when Kepler was trying to enter the service of the astronomer, Tycho Brahe, his prospective employer required him to write a response to a polemic by a certain Nicolaus Reimarus Ursus (Baer) accusing Brahe of plagiarism and dismissing the hypotheses of mathematical astronomers as no more than "fictitious suppositions" that reveal nothing of the form of the world. In late 1600, Kepler composed his *Apologia pro Tychone contra Ursum*. But due to the deaths in quick succession of both protagonists, Tycho and Ursus, the *Apologia* unfortunately remained unpublished, "unfortunately" because its analysis of hypothetical reasoning was more sustained than almost any other for a long time after.[40]

---

[34] Kepler later acknowledged, in the introduction to his *Rudolphine Tables*, that this was not the case. See Eric J. Aiton's commentary in *Mysterium Cosmographicum*, 235.

[35] *Mysterium Cosmographicum*, 75.

[36] Ibid., 75–7; my emphasis.

[37] Aiton remarks in this connection that for Kepler, "the quality of simplicity provides the ground for the physical truth of the Copernican hypothesis" (ibid., 235). But though Kepler does, time and again, extol the virtue of simplicity, his principal point here is that Copernicus, unlike Ptolemy, can provide causal arguments for phenomena that Ptolemy has to leave unexplained. This criterion is inadequately described as "simplicity."

[38] Ibid., 81.

[39] Ibid., 85.

[40] The *Apologia* was not published until 1858. An English translation and commentary by Nicholas Jardine, titled *The Birth of History and Philosophy of Science* (Cambridge: Cambridge University Press) appeared in 1984. Jardine argues that this little work marked the beginnings of the now familiar joint discipline named in its title.

What Kepler was concerned most to challenge was Ursus' claim that because different hypotheses in astronomy often can save the phenomena equally well, there can be no grounds for attributing *truth* to such hypotheses. Kepler's first response is that a false hypothesis may save the phenomena for a time "by chance" but it will ultimately "betray itself." Two hypotheses whose ontologies *really* differ may seem to yield the same results, but they will ultimately separate in that regard.[41] A second way to decide on the merits of rival and apparently equivalent hypotheses is to call on "physical" considerations, that is, considerations other than the "geometrical" phenomena, i.e., the observed positions and motions of the heavenly bodies. Thus, for example, when Copernicus is able to show the *cause* of the outer planets being nearest to Earth when in opposition to the Sun, this immediately favors his system over that of a rival who cannot provide such a cause. Or again, Copernicus can eliminate the equant on the grounds that it is at odds with the nature of things. Even where two ontologically different hypotheses coincide in the "geometrical" realm (i.e., they share, effectively, the same mathematical formalism), each "will have its own peculiar corollary in the physical realm."[42] But because astronomers are, in their everyday work, used to dealing with "geometrical" phenomena only, they are liable to claim equivalence between hypotheses when a wider acquaintance with "related sciences" would right away show that the supposed equivalence does not hold.

Kepler's painstaking analysis of Brahe's protracted and unprecedently accurate observations of Mars would, over the next few years, afford a type of "physical" consideration that would be more effective in discriminating between the rival world-systems than any available to Copernicus. Kepler was able to show that Mars's orbit, assuming the Copernican perspective, could not be rendered accurately by the combinations of circles to which his predecessors had been restricted, and that it was best understood as being elliptical in shape. The problem of finding an ontology to go with the formalism describing planetary motion was at one stroke completely transformed. Brahe's analysis of the orbit of the comet of 1577 had already undermined belief in the carrier spheres through whose spaces the comet had passed, apparently unimpeded. Now the need for such spheres had passed, for a second reason. But what should take their place? What sort of ontology did the new formalism point to?

Though most of Kepler's epochal work, the *Astronomia Nova* of 1609, is devoted to showing in numbing detail how he had arrived at his conclusion about Mars's orbit, the broader aim of the work is made clear by its full title: *New Astronomy Based upon Causes; or Celestial Physics treated by means of Commentaries on the Motions of the Star Mars*. This was not to be a work in mathematical astronomy of the traditional sort, one that would set out to save the phenomena of the planetary motions, as we perceive them, but would lay no further claim to revealing the real planetary motions and their causes. It would, on the contrary, proceed in two opposite directions at once. It would draw on physical (causal) considerations in finally establishing the parameters of the Martian orbit. And it would, on the other hand, make use of these orbital parameters to infer to the physics that could make such an orbit possible. (There is an obvious similarity of procedure here to that followed by Aristotle in his own appropriation and enlargement of the Eudoxan formalism.) Kepler admits that his argument will "mingle the

[41] Ibid., 140.

[42] Ibid., 141–2.

probable with the necessary" but that, he says, should not be surprising, since physics (unlike mathematics) of its nature involves an element of conjecture.[43]

One thing was already clear. The planets sailed through their orbits unsupported. Some other form of causal explanation was needed, one of an obviously less conventional sort. Kepler had two initial suggestions to offer, both with a history behind them. One was a soul of some sort moving the planet; Aristotle had already postulated such an *anima motrix* as the cause of each sphere's rotation on its axis. Might not such a soul, especially if it were to be endowed with intelligence, undertake the admittedly more complex task of guiding each planet in its orbit? On the other hand, there was a precedent for supposing that there might be an attraction of some sort between distant bodies; the correlation between the lunar position and the terrestrial tides had for long led to the supposition of some sort of influence of the Moon on earthly waters. Even more immediate was the example of magnetic phenomena. How the attraction was to be exerted was not clear, since there was a general disinclination to allow the possibility of direct action-at-a-distance.[44] The ontology that would account for planetary motions for which the new planetary formalism would serve as warrant had to be in part, at least, invented.

A second notable feature of Kepler's resolution in the *Astronomia Nova* to grapple directly with the problem of linking a descriptively successful formalism to the more straightforward ontological issue, in this case of establishing the reality of the Earth's motion, was that he faced not a choice between two clear-cut alternatives, as Copernicus had, but a more difficult choice between three, the new candidate being Tycho Brahe's geoheliocentric compromise between Ptolemy and Copernicus. The Tychonic system retained the Earth at rest, but set the planets other than the Earth in motion around the Sun, and the Sun in its turn in motion around the Earth, carrying the planets with it. From the perspective of today, Brahe's system posed a much more testing challenge to the Copernican system because it could be construed (as the Ptolemaic system could not) as a simple choice in kinematic terms, at least, of an alternative reference-frame, holding all else in the Copernican system of relations the same. To complicate matters further for Kepler in his role as defender of the Copernican alternative, Tycho Brahe had been his patron and the source of the data on which his great work on Mars had depended. Kepler confides that Tycho on his deathbed, knowing that his protégée was "of the Copernican persuasion," had asked him to include Tycho's own alternative in his planned work, together with those of Copernicus and Ptolemy.[45]

One feature that leapt out of Kepler's elegant new rendering of Mars's orbit was the fact that the planet moved fastest when nearest the Sun, slowest when farthest from it. The same could be shown to be true for the other planets, and indeed, as Kepler notes, was already mathematically implicit in Ptolemy's equant device.[46] The implication was evident: the Sun was responsible for, or at least somehow related causally to, the planets' motions. And since the same speeding and slowing was true of the Earth, according to the Copernican construal, it suggested that the Earth too

---

[43] Johannes Kepler, *New Astronomy*, William H. Donahue, trans. (Cambridge: Cambridge University Press, 1992), 47.

[44] Ernan McMullin, "The Explanation of Distant Action: Historical Notes," in *Philosophical Consequences of Quantum Theory*, J. Cushing and E. McMullin, eds. (Notre Dame: University of Notre Dame Press, 1989), 272–302.

[45] Kepler, *New Astronomy*, 157.

[46] Ibid., 53.

somehow derived its motion from the Sun. The geocentric alternative would be to suppose that, because of this slowing and speeding, the source of motion for Sun and planets somehow lay in the Earth. Kepler considered this alternative to be "absurd," given the relative dimensions of Earth and Sun.

Taking the Earth to be in motion meant a rejection of Aristotle's physics of natural motion. Kepler proposed a new "theory of gravity" (his name for it) in the light of the Copernican reordering. Gravity, he suggests, is a disposition on the part of all bodies to draw together with other bodies of the same kind. Thus the Earth can be said to attract a stone, no matter where the Earth itself may be situated or how it may be moving. In the same way, the Moon's attraction of the waters of Earth is responsible for the tides. The strength of the attraction is proportional to the bulk (*moles*) of the attracting body.[47] This "grasp" of the Earth on the bodies close to its surface, akin to a "magnetic force," explains why, he says, despite the Earth's rapid rotation, these bodies (clouds, projectiles, etc.) show no effects of the Earth's rapid rotation. Thus the objections to the Earth's motion brought with such force by Aristotelian critics could be undercut by this new physics of attraction. There is no need to postulate a soul or "animate faculty," as Copernicus did, to explain this lack of effect of the Earth's motion on nearby bodies. Such a faculty is superfluous; "a corporeal faculty, which we call gravity, or the magnetic faculty" is all that is needed.[48]

What should be emphasized here is that this new physics, entirely different from the traditional Aristotelian one, was prompted by the attempt to specify an ontology that would match the new formalism describing the shape of the planetary orbits, as well as to satisfy Kepler's intuitions about the causal primacy of the Sun. Despite its unintuitive character—the notion of attraction clashed with long-held intuitions about the necessity of contact action to explain motion—it seemed to Kepler the best, if not the only, way to understand the key features of the elliptical orbits.

Lacking the later notion of inertial motion, he still had to find a cause for the onward motions of the planets in their orbits, and this cause should somehow be related to the Sun. His solution was to suppose the Sun to be rotating, swinging the force-carrying "species" it emitted into a sort of "whirlpool" that would urge the planets onward in their paths. Though this "species," like that of light, was immaterial, it could act on material bodies; its intensity would depend on the emitting body and on the distance between Sun and planet. The fact that the periods of the planets lengthened with distance from the Sun served as corroboration.

There is no need to go further into this elaboration of the new "celestial physics," into the ingenious use, for example, of Gilbert's hypothesis that the Earth is a giant magnet, to explain why the distances of the planets from the Sun vary as they do, as their (supposed) magnetic poles alter position relative to the Sun's (supposed) poles, or the later extension directly to the Sun-planet relationship in his *Epitome Astronomiae Copernicanae* (1618–21) of the magnetic analogies of attraction and repulsion and his repudiation of the alternative he had earlier considered that an intelligence might be guiding the planet in its mathematically intricate path. The motion of the planets, he finally decided, was to be attributed solely to the Sun at the center of the system.[49] But Kepler could never devise an entirely coherent causal

[47] Ibid., 55–6.

[48] Ibid., 58–9.

[49] For a summary account, see "Kepler's Dynamics of Planetary Motion," 280–5 in McMullin, "The Explanation of Distant Action." For a fuller treatment, see Bruce Stephenson, *Kepler's Physical Cosmology* (Princeton: N.J.: Princeton University Press, 1987); Gerald

account of planetary motion, one that would lend itself to quantification and test. What counted, however, in spite of this was his unwavering conviction that the future of astronomy must lie in fashioning such an account, in discovering an ontology that would match the mathematical-physical formalism that had finally reduced the planetary orbits to simple ellipses.

As it was, however, the *Astronomia Nova* had already made impressive progress in that direction. Kepler skillfully organized the work as a choice between three different ontologies, Ptolemaic, Copernican, and Tychonic, represented by approximately equivalent mathematical-physical formalisms that initially, at least, could be seen to save the phenomena reasonably well. Then came the assault on the Ptolemaic system, on the *ad hoc* features that Copernicus had already drawn to attention, on its fatal dependence on the principle of uniform circular motion, on its inability to make causal sense once the orbs are taken away and one has to resort to imaginary points as centers of attraction.

The Tychonic system, as Tycho had originally formulated it, relied on the epicycle/eccentric formalism that the elliptical orbits had definitively superseded. It required carrier spheres that intersected with one another in ways that made no sense in mechanical terms. But Kepler knew that the system could, in principle, be reformulated mathematically so as to be equivalent to the heliocentric system as far as saving the phenomena of planetary position was concerned. His arguments therefore had to prescind from this latter criterion entirely and focus on the implausibility in dynamic/causal terms of the vast system of planets and stars swinging around the relatively tiny Earth. The prospects of finding plausible "physical considerations" that would support such an ontology seemed to Kepler sufficiently remote to rule out the Tychonic option definitively.

But the Copernican system, as it had originally been devised, also had to be (gently) dismantled. Kepler still counted himself a "Copernican," but this was in recognition both of Copernicus's basic heliocentric claim and of the formalism that had provided him with the indispensable framework from which his own analysis could begin. Of that formalism of epicycles, eccentrics, and Tusi couples very little remained, once Kepler's analysis had reached its conclusion. In retrospect, it seems fair to say that all was now ready for Newton, once Galileo's two laws of motion and Descartes's principle of rectilinear inertia had been put in place.

Galileo's *Dialogue on Two Chief World Systems* (1632) is given pride of place in most accounts of how the claim that the Earth moves came to be, if not finally accepted, at least regarded as plausible. And one can see why. Galileo wrote a simple and accessible vernacular prose, avoided technicalities as far as possible, and laid out a convergence of arguments in favor of the Copernican system in skillful rhetorical form. Kepler did none of these. Galileo could call on the spectacular new telescopic discoveries for which he could claim the entire credit, discoveries that had all of Europe buzzing. In particular, he could show conclusively with their aid that the Sun, not the Earth, had to be Venus's center of rotation, and if so, then plausibly the center for the other planets also. And he could effectively challenge the Aristotelian objections to the Earth's motion by breaking down the traditional distinction between the physics of Earth and that of the skies, and by arguing persuasively for the imperceptibility of shared motion.

Holton, "Johannes Kepler's Universe: Its Physics and Metaphysics," *American Journal of Physics* 24 (1956): 340–51.

Where his argument fell short was in its attempt to show that the Earth rotates on its axis as well as sharing in the motion around the Sun of the five planets. Demolishing objections was one thing; providing positive arguments in favor was something else. The title of his work spoke of *two* world-systems, and one of these he had fairly well managed to undermine. But as Kepler had earlier emphasized in the *Astronomia Nova*, there are three, not two, world-systems to choose from. Though Galileo was fully aware of the Tychonic alternative, he passed it over in silence in the *Dialogue*, referring to Tycho as though he were just another defender, with Aristotle and Ptolemy, of the geocentric thesis.

Galileo, however, realized that in order to achieve the demonstration of the Earth's motions that his critics were calling for, he would have to produce a *causal* argument of some sort. This he believed he had in the phenomenon of the tides; he proposed that the tidal motions could only be explained by postulating a double motion for the Earth. But there were at least two immediate objections to this proposal. One was that the correlation between the tides and the position of the Moon had been known since antiquity, Kepler having already formulated a possible explanation of the tidal phenomena in terms of an attraction exerted by the Moon on the waters of Earth. But much more embarrassing, Galileo had already relied, in his response to the Aristotelian objections to the Earth's motion early in the *Dialogue*, on the claim that the rotation of the Earth would not have the disturbing effects on objects on the Earth's surface that the critics claimed, because these would share in the Earth's motion and thus the effect of the rotation would be imperceptible. (He was, in effect, relying on a dubious principle of circular inertia, as Galileo allowed his critic Simplicio to point out in the *Dialogue*.) But if this argument were correct, then there could not be tidal effects of the sort he was now proposing, in what was being presented as the clinching argument at the end of the *Dialogue* for the reality of the Earth's motions.

Galileo, however, is not really part of our story. Our concern in this essay has been with the problem of relating formalism to ontology. And Galileo, unlike Kepler, did not have a formal model of the planetary motions to offer. Though he was made aware of Kepler's claims regarding the planetary orbits as early as 1612,[50] he evidently never took them seriously, arguing strongly for the traditional principle of uniform circular motion and indeed going so far as to remark in the *Dialogue* that "the law of revolution and the structure of the orbit of each planet" and specifically that of Mars, have not yet been determined.[51] Galileo was not a mathematical astronomer in the tradition of Copernicus and Kepler. He had not struggled with the age-old astronomical problem of fitting geometrical figures to the change in observed planetary positions. His concern was that of the natural philosopher, eager to give a causal account of how the cosmic system *really* is. Without a formalism

[50] His friend, Federico Cesi, in a letter to Galileo on July 21, 1612, remarked that Kepler's ellipses were by then a matter of common knowledge; *Opere* (Florence: Barbèra, 1901–11), 365–7.

[51] Galileo Galilei, *Dialogue Concerning the Two Chief World Systems*, Stillman Drake, trans. (Berkeley, Calif.: University of California Press, 1953), 455. How could Galileo have set aside such a crucial and strongly supported claim, one that could have immensely strengthened his own case? An intriguing question, but it would lead us too far afield to pursue it. See the note by the editor, Owen Gingerich, in the new edition of Max Caspar, *Kepler*, C.D. Hellman, trans. (New York: Dover, 1993), 137. Among the more interesting contributions to the debate is Alexandre Koyré, "Attitude Esthétique et Pensée Scientifique," *Critique* 11 (1955): 835–47.

describing the planetary motions to serve as a clue to ontology, he was forced to look for physical effects on Earth of the reputed motions. Still implicitly adhering to the epicycles and eccentrics of the Copernican formalism, deprived of the solid orbs that might translate these in causal terms, deriding Kepler's notion of attraction as a vain fancy, and setting aside the explanation of gravity as a problem for later thinkers, Galileo left himself without the means to formulate the physics of planetary motion that he needed in order to make the kind of case for the Earth's motions that he so passionately sought. That would have to be left to Newton.

6 The Elusiveness of Mechanical Agency

Did Newton finally succeed, then, in linking an empirically successful formalism of planetary motion to the ontology that underlay it? This would require another essay.[52] Had he established the reality of the Earth's motions? Though some then (and more surprisingly, even some now) would insist that a more direct experimental proof was still needed (stellar aberration, James Bradley, 1728; the Foucault pendulum, 1851), most would agree that the cumulative case for the heliocentric system built up in the *Principia* (1687) was sufficient to warrant reasonable belief. But the associated ontology was another matter. Had Newton, in fact, produced an ontology, an account of the physical causes responsible for the planetary motions? He used terms like 'force' and 'attraction' that certainly seemed to suggest such an ontology. But he also pointedly refused to go beyond the "mathematical notion of these forces, without considering their physical causes and seats."[53] And it was not clear how he *could* go any further, given that he rejected the possibility of action-at-a-distance and that he had himself showed that the presence of an ether with mechanical properties would undermine the stability of the planetary system. But could there be any other alternative explanation for how gravitational attraction could work, one that would make the term 'attraction' something more than the mere name for a disposition on the part of a body to move in a certain way? He toyed for a time with the idea of "active principles" of a nonmaterial kind operating in the space between bodies, influenced perhaps by alchemical analogies. But he could not link these effectively with the gravitational formalism.

Here, then, was a highly successful physical formalism that appeared, on the face of it, to lack any possible causal ontology, or at the very least, to require the admission of an ontology of an altogether new and counterintuitive sort. The affinity between this situation and the current predicament in quantum mechanics is obvious. In both cases, the formalism outpaced the resources of interpretation in terms of previously familiar ontological categories, forcing unwelcome choices.

The most obvious moral of this long history, stretching from the very beginnings of astronomy down to the time of its blossoming in the seventeenth century, is therefore that the project of *interpretation* in mechanics is of its nature a highly problematic one. Finding a formalism that will successfully describe the regularities of a particular system's motions ("saving the phenomena") is one thing, but translating that formalism into ontological terms that will convert that formalism into a theory, an explanatory account in the fullest sense, has almost always been a

---

[52] See chap. 4, "How Is Matter Moved?", in Ernan McMullin, *Newton on Matter and Activity* (Notre Dame, Ind.: University of Notre Dame Press, 1978), 75–109.

[53] "Definition VIII," Isaac Newton, *Principia*, A. Motte and F. Cajori, trans. (Berkeley, Calif.: University of California Press, 1934), 5.

contentious matter. Mechanical agency has all along proved to be an uncommonly elusive quarry. Seeking to explain the motions of the planets through contact action, as common sense seemed to demand, led Aristotle to postulate spheres composed of a mysterious quintessence unlike anything encountered in our terrestrial experience. When the weight of evidence forced the abandonment of the spheres, Kepler had to invoke the analogy of magnetism to postulate an attraction between planets and Sun that sounded suspiciously like action-at-a-distance. Commenting critically on Newton's use of explanatory terms like 'attract' and 'draw' to describe the gravitational relations between Sun and planets, Berkeley argued that the most that Newton was entitled to claim was a *description* of planetary motions in terms of law-like regularities. He had not discovered the nature of the agency responsible for the motions, a topic that, as far as Berkeley was concerned, lay outside the scope of physics in any case.

Puzzling though these accounts of the agency responsible for planetary motion were, each did eventually gain acceptance for an appreciable time. And the effect of that was to press back, step by step, the boundary of acceptability in ontology where the explanation of motion was concerned. No longer would such explanation be tightly bound by the requirements of intuition or the analogies of familiar experience. Yet these constraints did not entirely lose their force either, and so the struggle to find acceptable interpretations of descriptively successful mechanical formalisms has never quite been resolved. There is something about the nature of mechanical agency (in part, perhaps, its character as ultimate in the explanatory order) that renders the interpretation in ontological terms of successful formalisms in mechanics problematic.

What warranted physical formalisms like those of Copernicus, Kepler, and Newton in the first place was, of course their success in saving the phenomena. But as Kepler saw, that was not enough: it would not distinguish a Ptolemy from a Copernicus. Which, he asked, gave a better causal explanation? Equally important, which proved itself over the course of time? Which demonstrated the resources to meet anomaly, to direct inquiry, to furnish novel predictions (I am filling in the open spaces in Kepler's analysis at this point). What must be overcome is the suspicion that successful saving of the phenomena may merely be a testimony to the ingenuity of the original innovator in fitting formalism to the phenomena to be saved. A postulated ontology cannot forever lie idle, the product of an interpretation whose original warrant lies only in the successes of the formalism, not in that of the ontology itself. An interpretation of a mechanical formalism, whether that of Kepler, of Newton, or of Bohr, must in the long run make itself at home as, perhaps in some expanded sense, explanatory. And in addition, it must over the course of time show its worth as imaginative resource.

7 Coda

What does quantum mechanics, and even more particularly quantum field theory, have to offer to those who are seeking clues to a deeper understanding of the Creator's agency in bringing to be and sustaining the created universe? This essay did not tackle that controverted issue directly. Its goal, in the context of this collection of essays, was less direct. There was a time when mechanical agency was thought to be of a simple push-pull sort; in that perspective, the action of the Creator, particularly in the context of human free choice, seemed deeply puzzling. But, gradually, progress in astronomy and later in mechanics more generally has forced the realization that mechanical agency is far more complex than that, indeed that it stretches almost unbearably the very boundaries of the human imagination to grapple with the ontologies that now suggest themselves.

Our foray into early astronomy has shown that inference from a successful formalism to an ontology that could plausibly support it has always forced an expansion of intuitive notions of mechanical agency, an expansion that has still quite evidently some way, perhaps a very long way, to go. The temptation of course might be to abandon the realist perspective in mechanics entirely at this point.[54] But the extraordinary coherence of the advances in quantum theory over the past few decades would argue against such surrender. The link with ontology is still there, but the ontology itself becomes ever more elusive and more difficult to grasp in terms of the agencies familiar from ordinary experience.

Are there theological morals to be drawn from the specifics of current quantum ontology, such as they are? For reasons that ought to be evident by now, I would hedge in the attempt to discover such morals with reservations. But having said that I would want to add that a more general theological moral *can*, indeed, be drawn from the long efforts to fashion an ontology that would be adequate to the increasingly sophisticated formalisms of astronomy and mechanics. What constitutes agency at the most basic level in the explanation of motion is immensely more complicated and more mysterious than earlier generations could ever have imagined. If this is the case, how much more complex, how much more mysterious, ought one expect the agency to be that brings into being a galactic universe and within it, beings who can pose questions such as these? And how correspondingly hesitant ought one to be in proposing analogies linking one domain to the other? Hesitant but not entirely without hope, however, as I trust my colleagues in this volume may be able to show.

[54] Heisenberg, for one, advocated such a move in the 1970s, an abandonment of the quest for underlying causal structure in terms of "particles" (the "Democritean" motif that has guided modern physics to this point) in favor of a "Platonic" acceptance of the formalism of mathematical symmetries as itself the ontology; in "Tradition in Science," in *The Nature of Scientific Discovery*, Owen Gingerich, ed. (Washington: Smithsonian Press, 1975), 219–36, 235.

II. PHILOSOPHICAL INTERPRETATIONS OF QUANTUM MECHANICS

William R. Stoeger, S.J.
James T. Cushing
Jeremy Butterfield
Michael Redhead
Chris Clarke

EPISTEMOLOGICAL AND ONTOLOGICAL ISSUES ARISING FROM QUANTUM THEORY

William R. Stoeger, S.J.

1 Introduction

In attempting to speak of divine action in light of quantum mechanics, it is important first to give a careful account of the epistemological and ontological implications of quantum theory. Without these we cannot usefully investigate the channels of influence and action available or unavailable to God. It is only after arriving at a reasonably secure purchase on the essential features of reality as revealed by quantum mechanics and by other disciplines, as well as a critical account of the access or lack of access we have to it—including a description of the relationships between the knowledge we have from quantum theory and physical reality itself—that we can begin to speculate in a worthwhile way on how God's action in nature and in experience can be modeled. If, for instance, we discover from our philosophical analysis of quantum physics that independently existing reality is in fact significantly "veiled" from our probing eyes, as some astute investigators maintain,[1] then our capability for discriminating or modeling divine action with any precision or in any falsifiable manner will be extremely limited. We cannot probe the modes of divine influence in physical reality as it is in itself unless we have some at least indirect access to it, and unless our access is not irretrievably convolved with the representational systems we construct to describe our interactions with it. We can only investigate divine action through those channels for which we have some confidence that our descriptions faithfully represent processes and features in reality, both as we know it and as it functions independently of our knowing it. If such channels are available, the partial knowledge we acquire through them may be enough to constrain our descriptions of divine action in interesting and insightful ways.

Determining the epistemological and ontological implications of quantum theory involves first sifting out the various interpretations according to their adequacy and fruitfulness and then studying what the survivors imply concerning our knowledge of physical reality and its characteristics insofar as we can know them. But what are the canons of adequacy according to which we can winnow out the most acceptable interpretations?

This is an enormous task. It has been the focus of many books and research projects. I cannot mount such a attempt here. What I shall try to do is to present what I consider the principal criteria for interpretation, summarize the central features of quantum physics, and apply the interpretational criteria to them to see what epistemological and ontological conclusions emerge. In doing this I shall presume a basic grasp of quantum mechanics and of the principal points at issue in its interpretation (these are adequately presented and discussed elsewhere in this volume). I shall also avoid indulging in critical reflections on either hidden-variable or many-worlds interpretations of quantum theory. There is neither time nor space

[1] Most notably Bernard d'Espagnat. See, for example, his masterful and detailed analysis, *Veiled Reality: An Analysis of Present-Day Quantum Mechanical Concepts* (Reading, Mass.: Addison-Wesley, 1995).

to include discussion of them here. And, though we cannot rule out these approaches, I consider them less adequate than others, either because of the difficulties in meshing them with special relativity, their lack of fruitfulness (e.g., in quantum field theory), or the somewhat extreme and counterintuitive philosophical positions to which some seem to lead. Thus, as I begin, some of the preliminary sifting has already been accomplished. I shall focus on those interpretations that are directly related to the Copenhagen position, to consistent histories or logics, and to those that incorporate decoherence as a partial answer to the measurement problem.

In this essay I shall first discuss the criteria I judge crucial in selecting adequate interpretations of quantum theory. Then I shall go on to give a brief description of the principal distinguishing features of quantum theory that need interpretation: nonseparability, quantization, objective uncertainty, complementarity, objective chance, correspondence, entanglement, measurement, and decoherence. Following this, I shall summarize the arguments in favor of Copenhagen-like interpretations (including those of Bernard d'Espagnat and Roland Omnès). On this basis, I shall next indicate the key epistemological and ontological consequences of this family of interpretations. I shall then briefly discuss the reciprocal influence of these consequences on interpretations of quantum mechanics themselves, and the issues that arise in attempting to understand the partially "veiled" character of reality and the "weak objectivity" of quantum physics. Finally, within this context I shall revisit the concept of "laws of nature," which I have discussed before, and draw some preliminary conclusions regarding what we can or cannot say about indirect divine action at the quantum level and about the openness of reality at the quantum level to direct divine influences.

2 Criteria of Adequacy for Interpretations of Quantum Theory

If we are going to prefer some interpretations of quantum theory to others, on what justifiable bases can we do so? In many ways this is the key question in this discussion. Strictly speaking, it cannot be divorced from the epistemological implications of quantum theory, which I discuss more thoroughly only later. There is a dialectic between an interpretation of quantum theory and its epistemological implications—we can complete the formulation of an interpretation only when we have distilled the epistemological implications, and we can express the epistemological implications only when we have a definite interpretation. They go hand in hand, and therefore must be mutually adjusted.

In fact, with closer scrutiny it becomes clear that there are several levels of interpretation, each important in its own way. First, there is the basic interpretation at the level of the physical theory itself (e.g., the wavefunction is interpreted so that its square indicates the probability of "finding"—measuring—the particle it represents with certain properties, such as position, momentum, and spin). This interpretation is based (justified) on the experiments that substantiate the fundamental aspects of the theory. Given this level of interpretation, there is still significant interpretative freedom left. For instance, how are entangled states to be understood in terms of the physics of the situation and in terms of measurements? How is wave-particle duality (complementarity) to be understood? How does the reduction of the wavefunction in measurement—which strictly speaking is outside basic quantum theory, but now is partially understandable in light of decoherence—take place? How does quantum physics mesh with classical physics (correspondence)? What are complementarity and quantum nonseparability (the Einstein-Podolsky-Rosen experiment and the Bell inequalities) really telling us about "reality" and the way we interact with it? What does complementarity tell us about our knowledge of reality? And what does the wavefunction really represent: all aspects of the "reality, or just those that can be systematically measured (completeness)?

These questions actually represent three further levels of interpretation beyond that of the most basic (first level). The second category has to do with the consistency and coherence of interpretations within the confines of the theory itself, considering its full application to various complex situations and the experiments that validate them. The third category pushes interpretation to the level where the theory, strictly speaking, does not seem to apply fully and shows signs of needing a complementary theory (e.g., a theory of measurement) or demands a coherent connection with other physical theories (e.g., with classical mechanics, or with special relativity)—these relationships with other physical theories must and do significantly constrain interpretations of quantum theory. The fourth category brings us to the level of epistemological and ontological interpretations proper. To what extent does quantum theory give us knowledge of the underlying independently existing reality, and to what extent doesn't it? And what is this independently existing reality really like?

The "upper"-level interpretations significantly constrain these deeper philosophical ones, but I think it is reasonably apparent that they do not completely determine them. Considerations outside quantum theory—general philosophical principles such as the principle of sufficient reason—may be appealed to in order to complement the constraints coming from quantum theory and quantum experiments. Whether or not these philosophical principles themselves are adequately justified or are subject to modification in light of quantum physics, is another crucial question and requires further discussion. My main point in this regard is simply that more is needed to fix this level of interpretation than is available from what is given by quantum physics and its attendant theories.

The much larger question at this stage is, of course: How do we determine whether the characteristics we discern through quantum physics and the experiments that substantiate it in fact pertain to the underlying reality of nature and the world? Do they instead only pertain to our limited and episodic interaction with it? And how can we determine which of the features we have perceived represent reality as it is—and to what extent?

Before reflecting more deeply on these philosophical questions and criteria, a few further comments concerning upper-level (scientifically oriented) criteria are in order. Fundamental are both the internal consistency of interpretations of quantum theory in more complex situations (e.g., mixtures, not just pure states, or in its elaboration into quantum field theory) and its external consistency and correspondence with other crucially important physical theories (e.g., classical mechanics and special relativity). It goes without saying, of course, that these levels of consistency and correspondence need to be strongly supported by experiment and observation. Finally, there is the criterion of fruitfulness and clarity: to what extent has the given theory, together with its interpretation, generated further theoretical development and interpretational insight? It is precisely on the basis of these criteria that many researchers generally exclude hidden-variable and other strongly deterministic interpretations of quantum theory.[2] A great deal of further discussion on these points

---

[2] Some authorities on hidden-variable theories (e.g., James Cushing, *Quantum Mechanics: Historical Contingency and the Copenhagen Hegemony* (Chicago: University

would be valuable. But now I want to suggest other deeper, more philosophical criteria for interpretation that are needed to derive epistemological and ontological conclusions from quantum theory.

There are two important meta-philosophical criteria for the epistemological and ontological interpretation of quantum theory. First, it is imperative to assume the minimum concerning external reality and our knowledge of it. Though we have to go beyond what "the phenomena" give us in order to construct adequate theories (e.g., the wavefunction!), we need to respect "the phenomena" when we describe and explain them rather than add entities not needed to explain what we observe. Why? In order to avoid endowing external reality with features it may not have at the quantum level, at levels where classical physical and philosophical features and assumptions are unjustified. We want to allow the quantum "facts" to speak for themselves, even if that leads to counterintuitive and puzzling conclusions. In particular, we want to avoid presuming that all reality, or basic quantum reality, is like the macroscopic reality with which we are familiar. We also want to avoid imposing the divisions we make in thinking about and modeling reality on reality itself, or at least be critically aware that we often do this out of convenience and without adequate justification.

It goes without saying that we also should respect this principle of parsimony at the level of physics itself. We should avoid, as much as possible, introducing theoretical entities or fields or processes that are not demanded or warranted by experiment or observation, particularly those that express unjustified philosophical presuppositions that go beyond the phenomena, or those that inhibit us from coordinating quantum theory with other fundamental components of physics, such as special relativity.

Second, it is important to insist that quantum theory is telling us *something* about reality as it is, independent of our knowledge or measurement of it. This fundamental preliminary conclusion is reasonably well justified, and thus is not an assumption or a postulate. It is rather a belief or conviction that constantly receives challenge and support as we interact with reality on various levels and orient ourselves with respect to the great variety of contexts it presents. We bring this belief and conviction to our experience of the quantum world, and then find that it is gradually nuanced and reinforced as we develop an understanding and an intuition for the surprising phenomena it yields. Obviously, quantum theory is not telling us everything about the underlying reality it attempts to describe and explain, and it may be difficult or impossible for us completely to de-convolve what external reality is contributing from what our measurements and observations determine. But, though this is the case, we are justified in maintaining that quantum theory reflects some of the principal features of external reality in the microscopic regime to at least a limited degree. The primary justification for holding this is simply that when we interact with the world at the quantum level, it is just as recalcitrant and resistant—it manifests patterns of behavior that severely constrain how we can conceive it and manipulate it—as the more familiar classical levels of reality. We simply cannot

---

of Chicago Press, 1994), 188–91) point out that, while the overall formalism of Bohm's theory is not Lorentz-invariant and therefore grates with special relativity, the *observable* quantities are Lorentz-invariant. Therefore, according to them, the theory *is* consistent with special relativity. It is important to determine whether or not this type of consistency is enough to render it theoretically acceptable as an alternative, particularly in light of the critique d'Espagnat gives in *Veiled Reality*, 279–83, 352.

impose what we want on reality—there are certain definite ways the systems and entities that constitute them behave as they interact with one another and with macroscopic systems.

If we do not maintain that the principal characteristics of quantum phenomena provide at least limited access to features of quantum reality as it really is, then we deny that the phenomena we detect, measure, and observe have any relation to anything underlying them. With that refusal we would deny ourselves any epistemological or ontological insights from quantum physics. This would be a serious misstep, particularly in light of the compelling evidence that reality is strongly relational and interactive at every level, and that the very entities that inhabit it are themselves definitively constituted through the interrelationships and interactions of entities at more fundamental levels. This evidence comes to us from all branches of the natural sciences, including quantum physics. In a sense there is nothing to be known except what interacts with us, or interacts with those entities that interact directly with us (e.g., measuring apparatus).[3] It is true that in the case of quantum phenomena the knowledge we derive from them does not put us in direct contact with the underlying objects or states that interact to produce these phenomena—they are in some sense veiled. But our experimental and theoretical investigations and modeling certainly place significant constraints on the properties those underlying entities can have. And this is precisely the point! We find that, proceeding on this supposition, the models we have constructed actually work and successfully predict and explain other phenomena.

Some may protest that we cannot know anything concerning the validity of our knowledge of reality, since we do not have any purchase independent of our limited perspective and point of view on reality as it is in itself. We only know what is manifested to us by our interaction with it, which may so distort or cloud what reality is in itself as to generate manifestations completely unrepresentative of it. However, we know this position cannot be correct, if only from our engagement with different levels of reality. From our growing knowledge of the different realms of microscopic and sub-microscopic reality, as well as the more usual macroscopic realms—limited though it is—we are able to orient ourselves successfully, making worthwhile predictions and applications and cumulatively building an experimentally well-supported body of models and explanations concerning a growing network of phenomena. Even though our overall "take" on reality may be limited and partial (there is also good evidence for this—we do not understand everything and we only gradually grow in knowledge and understanding), there seems to be every reason to believe that there is a strong connection between how reality manifests itself to us and how it really is, the "noumena": we can fit our many perceptions and "takes" on reality together into an integrated and consistent whole that "works." There are many other ways we could try to understand quantum phenomena (e.g., classically descriptive ways) that do *not* work.

This viewpoint is strongly supported by Ernan McMullin's emphasis on the powerful role retroduction plays in the natural sciences.[4] Retroduction is the process

---

[3] The one category of known entities that *may* not directly interact with us is that of abstract entities or Platonic forms. However, I would maintain that such abstract entities can be known only by some direct or indirect interaction with us unless they are innate, and I would argue against the innate knowledge of such objects. I am grateful to Jeremy Butterfield for pointing out this possible exception.

[4] Ernan McMullin, "Structural Explanation," *American Philosophical Quarterly* 15

by which the scientist uses his or her well-educated imagination to construct hypotheses concerning hidden realities, causes, or relationships in order to explain observable phenomena, such as regularities, processes, structures, correlations, and relationships. These hypotheses then undergo rigorous testing and modification over a period of time, until the modified hypotheses reach a point at which they are confirmed well enough to provide the basis for further scientific investigation. The justification of retroduction is its success and fertility. The actual entities hypothesized may continue to be hidden in many ways, or only indirectly manifested. But, if their employment in a scientific theory successfully explains what we observe and provides a key to further knowledge and understanding, then there is a strong—but always provisional—warrant for affirming that these entities exist and have the characteristics the theory says they have. McMullin supports his emphasis on retroduction by demonstrating its importance in many key scientific discoveries of the distant and recent past.

Obviously, this does not rule out the fact that there are many significant features of quantum reality—just as there may be of any other level of reality—that completely escape our detection. However, other such features that may eventually come to light, either through future advances in physics or through advances in other disciplines, would have to be integrated in some way with the features of reality of that we already have some knowledge, just as the quantum phenomena and relativistic phenomena have had to be integrated with our understanding of classical physics (correspondence!). In fact, there may be phenomena of which we as yet have no inkling, which will be discovered by a future generation of physicists and which will be extremely important in further comprehending physical reality. There may also be aspects of reality that are not even in principle knowable or discernable by us, simply because of our epistemological limitations. Finally, "reality for us" may have features that are not functions of the fundamental underlying entities as they exist in themselves—independently of us—but only as they are manifested in interaction with one another at the lowest levels, such as those that effect decoherence, and at the higher levels, such as the highly organized observations by conscious organisms that lead to knowledge. But such features still at least indirectly manifest something about the way the underlying entities are in themselves—about the way reality at that level is in itself.

All these limitations and uncertainties, however, do not undermine the essential point that the patterns, regularities and processes—the laws of nature—discernable in quantum physics reflect the existence and some of the important characteristics of physical reality at the sub-microscopic level itself. Yes, some features of our perception and understanding of quantum reality may be artifacts of our models or our limited and imperfect understanding, and may not be attributable to external quantum reality—but some may be reflective of it. In this regard it is important to make provisional judgments and maintain a critical stance, searching for and remaining open to further evidence that will help distinguish artifact from genuine feature. And yes, there may be aspects of quantum reality—and of the full range of reality—we cannot describe because of our experimental and theoretical limitations. But again, this is not an argument against our possession of a limited "take" on reality.

(1978): 139–47; idem, "The Shaping of Scientific Rationality," in *Construction and Constraint: The Shaping of Scientific Rationality*, Ernan McMullin, ed. (Notre Dame, Ind.: University of Notre Dame Press, 1988), 1–47; idem, *The Inference That Makes Science* (Milwaukee: Marquette University Press, 1992).

Any knowledge we have of the world around us, at any level, will be limited and channeled through our interaction with it—our experience of it. As I have already mentioned, we cannot know anything that we do not interact with, except perhaps abstract entities or Platonic forms whose representations may be innate. (There may be such realities, but they are not the ones in which we are immersed and by which we are constituted.) Reality is not divorced from us—it is not radically non-relational, and therefore unknowable. Certainly, the least we must say is that the reality with which we are associated, with which we interact, must have characteristics consistent with its manifestations in the phenomena we measure and observe at the various levels of structure and organization, including those of the quantum realm. To deny this—that there is any essential relationship between the phenomena and the underlying realities that support them—is tantamount to the Cartesian fallacy of separating perception and knowledge from their objects.[5] At other levels and for other considerations, we may emphasize different characteristics of the reality we experience, realizing full well that they in fact do at least reflect, however imperfectly, properties of the underlying reality (e.g., that it is contingent, evolving, relational, ordered in certain ways, autonomous). We may have to rely on some of these characteristics as fully or partially justified presuppositions—justified on grounds other than those of quantum physics—when we interpret quantum phenomena at the epistemological and ontological levels. It is difficult to conceive that the reality we interact with should have these characteristics without the underlying reality possessing something corresponding to them.

At the basis of these assumptions lie two essential metaphysical presuppositions we cannot avoid: the principle of sufficient reason, or sufficient cause, and what we might call the principle of relationality. The principle of sufficient reason rests on the intuition that whatever we experience, detect, or observe demands some underlying and proportionate reason or cause. Though we must be careful with this at the level of quantum physics (there may be some aspects or specific properties of phenomena that may not have a sufficient cause, e.g., that a given atom of ^{235}U decays at this instant rather than two weeks from now), there is strong justification for maintaining that the phenomena themselves have underlying proportionate causal explanations. They exhibit regularities that strongly indicate the existence of more fundamental ordered causes.[6]

The point of this brief discussion is that, just because a particular macroscopically measured property of an event or observed entity does not have a "sufficient reason" in the underlying quantum world, does not mean that the event or episode as a whole does not demand one. There *is* a sufficient reason why the decay takes place, even though there is *no* sufficient reason why it takes places at the precise time and place it does—at least none that goes beyond what the statistics give. This failure is not cause for abandoning the principle of sufficient reason in quantum theory, but only its applicability to certain properties of events that have to do with the unique macroscopic realization ("objectification") of one possibility out of those given by the observed quantum system itself. At the quantum level there appears to be sufficient reason for all possibilities to be realized. At the macroscopic level only

[5] Wolfgang Smith, *The Quantum Enigma: Finding the Hidden Key* (Peru, Ill.: Sherwood Sugden & Co., 1995).

[6] This does *not* imply the existence of nonlocal hidden variables, but only that the phenomenon—the decay and the statistics it exhibits—demands an underlying cause, which quantum mechanics provides, even if the precise time of a given decay may not be predictable.

one possibility is realized—quantum theory and its extensions so far do not explain why that particular one is picked out, only the statistics of their occurrence. This is closely related to the measurement problem itself.[7]

It important to note that it would be impossible to engage in science or in any other inquiry without relying on the principle of sufficient reason in some form. The strongest justification for it is that it is extremely fruitful—it works. The principle of relationality essentially says that the reality of which we are a part and with which we interact is constituted by networks of relationships on many different levels among the real or putative objects we discern. Reality is not constituted by a collection of isolated entities. Entities, whatever they are, are always in relationship and interacting with other entities—and these relationships and interactions are essential to what the world is.[8]

3 Key Features of Quantum Physics

Among the distinguishing features of quantum reality are nonseparability, quantized properties, objective uncertainty, complementarity, objective chance, correspondence, entanglement, measurement, and decoherence. In listing these terms I have in some way—at least for several of them—already implied how they should be interpreted. In a brief essay such as this, there is no easy way of avoiding that.

Nonseparability, which in many instances implies a certain nonlocality, means that two or more apparently distinct entities (e.g., particles) may be space-like separated and yet inextricably linked to one another by sharing the same wavefunction. What this indicates is that the fundamental "wholes" or entities are not the localized particles or objects that we first discerned but rather the connected reality given by the wavefunction itself, which is not localizable. As Abner Shimony puts it, "we observe correlated actualization of potentialities over space-like intervals."[9]

Quantized properties refers to the fact that the values of many properties (e.g., energy, angular momentum, spin, and charge) of quantum systems as they manifest themselves in their interactions with one another and with experimental instrumentation are discrete and indivisible. Energy, angular momentum, and charge, for example, come in "packets" of a certain size, not in a continuous spectrum of sizes.

Objective uncertainty or indeterminacy describes the fundamental limit on our ability to simultaneously determine—measure—two noncommuting operators specifying a quantum entity (e.g., its position and its momentum, or its energy at a certain time). By using the adjective "objective" I have already presumed the standard interpretation that this uncertainty or indeterminacy is not due to our lack of knowledge or the imprecision of our measurements, but rather is an underlying indeterminacy in physical reality itself. Shimony refers to this characteristic of reality as "objective indefiniteness"[10]: independent of the observer, some eventualities do not have a definite truth or falsity.

---

[7] Future advances in quantum theory may explain this "objectification"—how just one of the classically allowed possibilities is selected—through interaction with environment, or through nonlinear corrections to the dynamics.

[8] Cf. George Ellis in this volume, and references therein.

[9] Abner Shimony, "Conceptual Foundations of Quantum Mechanics," in *The New Physics*, Paul Davies, ed. (Cambridge: Cambridge University Press, 1989), 373–95.

[10] Shimony, "Conceptual Foundations of Quantum Mechanics," 374–5; see also his discussion of this characteristic in sec. 3 of his essay in this volume.

Complementarity refers to the strange and counterintuitive property of a quantum system by which it manifests one realization in one type of measurement or observation and a different realization in a different type of experiment incompatible with the first. The two different manifestations are not simultaneously real,[11] according to the standard interpretation, and are not properties that the underlying quantum reality possesses independently of the measurements. The primary example here, of course, is that a given quantum system may sometimes manifest itself as a wave and at other times as a particle. Each can be part of a consistent history of the quantum system. But the two consistent histories are, as I have just mentioned, incompatible with one another.[12]

Objective chance signifies that the occurrence of many types of events, as well as their specific outcomes, in the quantum regime are possible but not certain. This characteristic is a property of the given physical situation itself, not of the experimenter's lack of knowledge. It is intimately connected with objective uncertainty and thus presupposes objective indefiniteness.[13] This objective uncertainty and indefiniteness resides at the level of macroscopic measurement of events of the quantum world—empirical reality[14]—not at the level of the underlying quantum world itself, where determinism—definite futures specified by every past configuration—reigns. In the projection of this quantum world into the macroscopic world of our experience, determinism fails. Only certain determined states are realized macroscopically, not all. Furthermore, quantum theory assigns a probability of measuring a certain result, and another probability of measuring each of the other alternatives. Microscopically—quantum mechanically—all the possibilities are given; but macroscopically, in the actual measurement, only one particular outcome is realized (see the brief discussion of the problem of measurement below). Thus, this probability, too, is rooted in the physical reality itself, not in the imperfect knowledge of the observer. It is sometimes referred to as "objective probability."[15]

At this point, a number of authors, including Werner Heisenberg and Shimony, invoke the concept of potentiality. They suggest that there are potentialities underlying all indefinite but probable events and properties of entities, with specific measurements or interactions serving to actualize those potentialities. However, others would consider this language too realistic.[16]

Correspondence indicates that the quantum description of reality, despite being so different from the classical description, must smoothly mesh with the classical description in the proper limits—when Planck's constant is taken to zero, or for very large quantum numbers. In other words, the classical and the quantum descriptions of reality cannot be contradictory. The latter is appropriate for the regime of the very small, and can also be used, in principle, in all macroscopic situations. However, for the latter the quantum description reduces to the classical, as long as decoherence and "objectification" (collapse of the wavefunction to a unique state) are admitted.

---

[11] Ibid., 393–4.

[12] Roland Omnès, *Understanding Quantum Mechanics* (Princeton, N.J.: Princeton University Press, 1999), 69ff., 169ff.

[13] Shimony, "Conceptual Foundations of Quantum Mechanics," 374.

[14] Cf. d'Espagnat, *Veiled Reality*, 183–4, 196.

[15] Shimony, "Conceptual Foundations of Quantum Mechanics," 374–5.

[16] D'Espagnat, *Veiled Reality*, 382–3, 421–3.

Entanglement, briefly, signifies the common situation in which the sum of two or more states is actualized, even though none of the component states is.

The issue of *measurement*—or as it is sometimes called, the problem of measurement—is central to quantum theory. When a measurement is performed, only one of the possible outcomes is realized, even though other outcomes were possible. The wavefunction involved typically represents a superposition of possible outcomes, each outcome having a probability of occurrence given by the square of the respective component of the wavefunction (probabilities can be verified statistically). However, only one of them is macroscopically realized. Strictly speaking, from the point of view of basic quantum theory, the superposition or combination of possible outcomes, and the interference among them, should be transferred to the measurement. We should measure the full superposition of outcomes, but we only actually detect one. Somehow, the wavefunction "collapses" to yield just one of the possible outcomes during the measurement procedure, i.e., during its interaction with a macroscopic system. Until relatively recently this phenomenon seemed to require either a process completely outside the framework of quantum theory, or a radical modification of quantum theory itself—particularly because of its linear character. To solve this problem completely may still require this.

However, now, with our understanding of *decoherence*, it seems that the measurement problem may have found part of its solution. Decoherence is the interactive process of a quantum system with the many degrees of freedom of the macroscopic environment whereby the interference terms among the different possible outcomes of a quantum measurement are very rapidly dissipated, thus yielding a set (technically, an improper mixture) of classical outcomes, each with its particular probability of being realized. There still is, however, considerable controversy on this point.[17] A key question in this regard pertains to the epistemological status of the extremely small probabilities of classically forbidden events that still remain as residues of quantum interference between possible outcomes. Do they represent actual, though rare, possibilities, or are they mere artifacts of the statistical model we are using and therefore not representative of the possibility of such classically forbidden events?

Moreover, decoherence by itself does not select among the several or many separate outcomes represented by the diagonal elements of the density matrix. This is what Omnès refers to as the problem of objectification—or of accounting for a unique outcome.[18] Omnès attempts to analyze the problem away, maintaining that the physics should not be expected to solve this problem as long as it explains probabilistically any outcome that occurs. Quantum theory, he argues, is basically probabilistic to begin with, from the perspective of macroscopic measurements of quantum systems. Further, the uniqueness of the datum is only known through empirical input—that is, the uniqueness of reality is not, according to Omnès, a matter of the relationships between theory and reality. Mensky, however, suggests

[17] Cf. for instance the differing views of Omnès, *Understanding Quantum Mechanics*, 242–56, and d'Espagnat, *Veiled Reality*, 177–89. Cf. also Michael B. Mensky, *Quantum Measurements and Decoherence: Models and Phenomenology* (Dordrecht: Kluwer, 2000), especially 189–97 (chap. 10).

[18] Omnès, *Understanding Quantum Mechanics*, 82–5, 242–8. See also Jeremy Butterfield, sec. 4, and John Polkinghorne, both in this volume.

that the full measurement problem, including objectification, can be resolved.[19] His version of the theory always treats finite open systems, which includes the environment. As a result, the output of any measurement is an element of the theory. From this point of view the problem arises from over-idealizing open systems by treating them as closed systems. Mensky speculates that this leads directly to a crucial role for consciousness. Other researchers, such as GianCarlo Ghirardi et al. (the Ghirardi-Rimini-Weber process and extensions and variations thereof[20]) attempt to account for objectification by appealing to nonlinear modifications of the dynamics, which in some cases involve interactions with the environment reminiscent of decoherence. However, no definitive conclusions have yet resulted from such investigations.

The primary insight with respect to decoherence that is crucial for our appreciation of the relationship between quantum reality and classical macroscopic reality is that, as systems at the quantum level interact with macroscopic systems, there is indeed what can be called objective chance, objective probability and objective indefiniteness (we should avoid ontologizing these characteristics at this point, but we also need to indicate that they are not merely subjective—the objective character of quantum entities is only realized when they interact with or are measured or detected by macroscopic systems, and not before that).[21] Reality at that quantum level is shadowy, uncertain, but endowed with potentiality for many different possibilities. It is in interaction with other systems—particularly with macroscopic systems—that some of these potentialities are actualized—but not all of them. I pointed out in section 2 that reality is relational. Here we see this in a striking way: actual events and entities only fully surface in the interaction and interrelationship of these shadowy potentialities with one another and especially with already actualized entities.

4 Acceptable Interpretations of Quantum Physics

In some ways I have already implied a standard Copenhagen interpretation of quantum physics by the way in which I have described the distinctive characteristics of the quantum world, except that I have, with many others, rejected privileging macroscopic systems. Here I would like to specify briefly, on the basis of the interpretational criteria I have described in section 2, the reasons for preferring Copenhagen-like interpretations of quantum mechanics, which include powerful extensions of it in the consistent-histories approach (though there are fine points of contention among exponents of these interpretative schemes). Before doing this, however, we should take note that, in some definite sense, *at the level of physics itself*, quantum theory does not need any further interpretation than is already present in the theory. It works very well in predicting and accounting for what we observe

---

[19] Mensky, *Quantum Measurements and Decoherence*, 194–7, and references therein.

[20] D'Espagnat, *Veiled Reality*, 292–7; Omnès, *Understanding Quantum Mechanics*, 248–56.

[21] Butterfield (private communication) denies that our experience of reality is tied to measurement or detection by macroscopic systems. However, it seems very much that it is. We do not experience the world of quantum reality as it is described by quantum theory. Otherwise, we would have no difficulty with measurement. Our everyday experience is very much mediated by the macroscopic world, although the hidden structures that underlie it are quantum.

and measure, whether or not we endow the pre-measured, pre-observed independent wavefunctions or quantum states themselves with any ontological status.

The Copenhagen, or standard, interpretation of quantum theory consists in taking seriously the objective character of chance, indefiniteness, and probability, as well as the notion of complementarity. A brief characterization of the essence of this interpretation would include the following two points[22]: (i) the indivisibility and completeness of the quantum of action implies that the interaction between the object measured and the detector is never negligible, that properties of a quantum system measured by incompatible experimental arrangements are not simultaneously real (complementarity), that the phenomena issuing from an interaction involving a given quantum system cannot be separated or divided into completely individuated objects or systems, and that the wavefunction contains all that there is to the quantum system (completeness); and (ii) measurement or interaction with a macroscopic object is the only process that endows the quantum systems with unambiguous meaning and properties.

Since the advent of this basic interpretation there has been a great deal of investigation and discussion regarding what "completeness" means, as well as regarding the privileged status of macroscopic objects. With our increased appreciation of the role decoherence plays in interaction and the measurement process, we now realize that macroscopic objects can and must also be considered as quantum systems, but as quantum systems whose collective interacting degrees of freedom render their behavior classical (correspondence). The epistemological and ontological weight we attribute to "completeness" is still a contentious issue, about which I shall have more to say at the end of this essay. It is perhaps the concept upon which our eventual epistemological and ontological conclusions will have principal impact.

If we attempt to take seriously the criteria of interpretation I have presented, and refuse to postulate either physical or metaphysical entities that are neither observed (directly or indirectly) nor needed, and if, at the same time we carefully examine how the theory meshes with other crucial physical theories, such as special relativity, and if, further, we appreciate its long-term fruitfulness, then the family of Copenhagen-like interpretations is by far the most satisfactory, as long as we include the radically quantum character of macroscopic systems. As I indicated above, the consistent-histories approach also falls into this category—it agrees with the essential features of the Copenhagen stance and in a sense is a marked improvement of that basic interpretation.[23] It can also be carried smoothly into relativistic quantum field theory

[22] Cf. for instance Shimony, "Conceptual Foundations of Quantum Mechanics," 393–5.

[23] A *history* is a sequence of time-ordered events expressed as projection operators into the relevant Hilbert space, thus referring to a definite physical system (Omnès, *Understanding Quantum Mechanics*, 137). A *family of histories* is a set of such histories, each of which expresses a possible sequence of events or operations for a given system in a particular context. It is necessary to construct a *complete* family of mutually exclusive histories (ibid., 138ff) and then define a probability (of each history) that satisfies the usual requirements of positivity, normalizability, and additivity. This can be done, as long as decoherence is taken into consideration. The consistent-histories approach can be interpreted in various ways, depending upon how one understands the mutual exclusivity of the histories in each complete family (many worlds, or one world) and how one understands the properties realized by the operators defining the events making up the histories (hidden variables or not). Understanding these in a way consistent with the Copenhagen interpretation is natural, as long as one includes decoherence and de-privileges macroscopic objects, which one should do in any case.

by the use of Wightman distributions and naturally defined time-ordering operators.[24] Hidden-variable theories and interpretations, as well as many-worlds interpretations, seem to be much less acceptable from the point of view of our basic criteria.

5 Principal Epistemological and Ontological Implications

Relying on the conclusions thus far reached, I would hold that (i) quantum phenomena do give us some limited, indirect purchase on external or independent quantum reality through the outcomes of their observable interactions with macroscopic systems, and that (ii) it therefore exists in some way as it is manifested to us independently of our knowledge of it, and is radically relational (i.e., it is characterized throughout by relations that do not supervene on the intrinsic properties of the relata[25]). Let me offer two conclusions at this point.

1. There is a reality that exists independently of our knowing it, and this reality is at least partially responsible for what we observe at the quantum and the classical levels of description.

2. Though we cannot know the objects or entities of this underlying quantum reality as they exist in themselves—they are "veiled" or partially hidden—we obtain some limited knowledge of them through their manifestations in interaction with our instruments of observation.[26] Furthermore, we can provisionally model what is knowable about them by using the theoretical apparatus of quantum physics—with the wavefunction, and the operators on Hilbert space, etc. However, we cannot say that the basic properties that characterize quantum reality as we observe it are actually possessed by the underlying quantum entities themselves.[27] We can only say that there are properties of the underlying quantum entities—entities we represent by wavefunctions—that yield the properties we observe in our measurements when we interact with them using a macroscopic apparatus. Those observed properties are, in a sense, projections of the underlying quantum properties into the world of

Chris Clarke, in this volume, maintains that the consistent-histories approach, though attractive, is inadequate, because it may allow histories that are consistent but unacceptably nonclassical. This *seems* to be at odds with Omnès's assessment of what a consistent history is, and with the effects of decoherence in enabling the additivity of history probabilities. Thus, more investigation, or clarification, of this facet of the consistent-histories interpretation is in order.

[24] Chris J. Isham, "Quantum Logic and the Histories Approach to Quantum Theory," *J. Math. Phys.* 35 (1994): 2157–85; Chris J. Isham, N. Linden, K. Savvidou, and S. Schreckenberg, "Continuous Time and Consistent Histories," *J. Math Phys.* 39 (1998): 1818–34; Kurt Just and W.R. Stoeger, "Combinability and Quantum Theory," in preparation.

[25] I am grateful to Jeremy Butterfield for formulating this in such a clear, felicitous way.

[26] This conclusion, which I share with d'Espagnat, does not, however, necessarily imply a fundamental monism (d'Espagnat's move). This would follow only if the underlying veiled reality were both radically undifferentiated and inclusive of all there is. We have some indications that it is not radically undifferentiated (reality is only partially veiled, and we see some structure in it), and I maintain that it is certainly not inclusive of all there is, but only of the fundamental physical systems that underlie the macroscopic world we experience. On this point, see also Philip Clayton's contribution to this volume, particularly sec. 4.

[27] Jeremy Butterfield (private communication) asks if this is true in any stronger sense than in the classical case (Kant's position!). I would argue against Kant that it is not as strong in the classical case as he maintains, and that it is even weaker in the quantum case—I maintain that the veil is only partial, not complete.

macroscopic experience through the interaction of quantum entities with macroscopic entities.

Thus, there must be at the quantum level, independent of our interaction with it, properties that underlie and generate the basic quantum properties we observe—the nonseparability, the objective uncertainty or indefiniteness, the objective chance, and the complementarity at the level of observation. We can then—and actually do, according to my epistemological interpretation—tentatively model (in the sense of designating something in our theory to stand for, but not necessarily describe or characterize) this underlying reality, which must have certain, however incompletely and indirectly known, characteristics and properties—using wavefunctions and quantum states in such a way that our model for the interactions with macroscopic measuring devices yields the properties we observe. We are blocked, however, from asserting that such underlying properties actually describe the properties of the underlying quantum entities as they are. The properties we observe are, nevertheless, consistent with the properties that they do have and with what we know of them through the quantum phenomena. As I have already stressed, there may well be other properties to which we do not yet have access or may never detect.

These are the principal epistemological and ontological conclusions we are justified in drawing from quantum physics on the basis of the criteria established. A cautionary nuance is appropriate at this point. There are experts in quantum theory and in the philosophy of quantum theory who would strongly insist that we cannot consider the wavefunction or the abstract quantum state as a "surrogate reality."[28] They would maintain that no wavefunction exists either before or after we conduct an experiment, because, according to them, the wavefunction is not an objective entity, and its evolution does not represent the evolution of a physical system. "It only gives the evolution of our probabilities for the outcomes of potential experiments on that system."[29] At the level of quantum theory itself this minimalist position makes a great deal of sense. Of course it is vital to stress that the wavefunction is not an objective reality. But, from a philosophical point of view, it certainly "stands for" or represents some hidden underlying "objective" reality. Since it gives the probabilities for outcomes of potential experiments on some system, it must in some way represent the system, however partially, indirectly, or inadequately. Again, this is strongly supported by the key role retroduction plays in scientific inference and understanding (see section 2).

Thus, one may say our knowledge of reality is "weakly objective."[30] By this I mean that we know that some independent reality exists and that it manifests itself to us through our interactions with it. However, we have no privileged purchase on that reality "as it is in itself"—so we cannot adequately or independently critique the correspondence of knowledge to it using observations and measurements. In particular, there may be—and probably are—many aspects of external reality we have no knowledge of, or have detected no manifestations of so far. In other cases, some of our conclusions about reality may be flawed and inaccurate, or simply the result of the projection of that reality into our limited representations, models, and

---

[28] Christopher A. Fuchs and A. Peres, "Quantum Theory Needs No 'Interpretation'," *Physics Today* 53.3 (March 2000): 70–1.

[29] Ibid., 71.

[30] The sense I give to this term is somewhat stronger than the sense d'Espagnat gives to it; he does not attempt to describe how empirical reality may give us some knowledge of the underlying independent reality, as I do here.

understandings. Nevertheless, from the patterns and regularities we have observed, we have strong reasons to believe that reality has told and is telling us definite things about itself that we can provisionally describe—a conclusion that McMullin would strongly support in light of his emphasis on retroduction in scientific knowledge (section 2). This is particularly true if we recognize that reality is profoundly relational and interactive, that the systems and entities that make it up exist only in interaction with one another, and that the basic potentialities are only actualized in relationship and interaction. Reality then is indeed "veiled," but not completely. There is some relationship between the empirical reality we know and the underlying physical reality of the naked quantum world that hovers beneath our observations.

Finally, we can already see that our epistemological and ontological interpretations of quantum physics depend upon the interpretations we give the components of quantum theory at the level of physics. This particular point is exemplified in Fuchs and Peres's denial of the objective reality of the wavefunction discussed above. We can also see this in the interpretation that now must be given to "completeness." Because of the veiled character of the underlying quantum reality, and the fact that we only know this reality through the mediation of its observable interactions with macroscopic systems, it becomes clear that the completeness attributed to the quantum of action, and by extension to the wavefunction, is epistemological completeness, not ontological completeness. It represents all the properties we can observe and therefore know from all the experiments we can potentially make on the quantum system. It does not represent everything about the underlying quantum reality itself—much less our unwarranted projections of classical categories upon it.[31]

6 The Laws of Nature and Divine Action

These epistemological and ontological conclusions concerning quantum physics strongly reinforce the point of view I have taken elsewhere on the ontological status of the laws of nature.[32] Previously I have stressed that (i) the laws of nature as we actually formulate them in the sciences are but imperfect and incomplete descriptions of the regularities, structures, relationships and processes that obtain in reality itself (the laws of nature as they actually function); and that (ii) the laws of nature as we describe them are descriptive, not prescriptive. Our conclusions concerning what we know about reality at the quantum level mesh perfectly with this point of view. The quantum laws of nature we have formulated to describe and explain quantum phenomena are indeed only imperfect and incomplete descriptions of what we observe and know. Obviously, they do not directly describe what goes on "behind the veil," at the level of the underlying quantum entities themselves. And finally, they only pertain to the observations, measurements, and considerations that are possible

---

[31] Ibid.

[32] William R. Stoeger, "Contemporary Physics and the Ontological Status of the Laws of Nature," in *Quantum Cosmology and the Laws of Nature: Scientific Perspectives on Divine Action*, Robert J. Russell, Nancey Murphy and Chris J. Isham, eds. (Vatican City State/Berkeley, Calif.: Vatican Observatory/Center for Theology and the Natural Sciences, 1993), 209–34; idem, "The Mind-Brain Problem, the Laws of Nature, and Constitutive Relationships," in *Neuroscience and the Person: Scientific Perspectives on Divine Action*, Robert J. Russell, Nancey Murphy, Theo C. Meyering, and Michael A. Arbib, eds. (Vatican City State/Berkeley, Calif.: Vatican Observatory/Center for Theology and the Natural Sciences, 1999), 129–46.

from the point of view of physics. In particular, they leave out properties that are philosophical or those that may be indicated by physics but are in principle not testable.

What relevance does this have for divine action? There can be various implications, depending on the theological or philosophical presuppositions we make with regard to divine action itself. Discussion of these is beyond the scope of this essay. Here I shall just point in a few possibly important directions.

First, God's universal creative action at the fundamental levels of reality must be partially realized "behind the veil," where there are certain regularities, processes, and relationships—certain laws of nature—to which we do not have significant access. Nevertheless, because of the interactive and relational character of nature at all levels, including the quantum level—and because the veil is only partial—God's universal creative and sustaining action, however it is effected "behind the veil," is constrained *to appear*, in terms of its *regular operation*, according to the characteristics we observe on *this side* of the veil. Because the veil *is partial*, some aspects of the appearance of God's universal action must reflect something of its basic functioning, even though there are obviously other aspects that will remain hidden. Thus, we can strongly constrain the regular *appearance* of the operation of divine creative action, but only in partial ways its *actual* regular operation.

In this regard I should note that I am speaking here only of those aspects of God's creative action that are manifest at the quantum level of physical reality. I am not saying anything about how it is manifested through other categories of the laws of nature, nor implying that, because these are the most fundamental in the *physical* or *scientific* sense, they are the most fundamental with regard to God's creative action considered philosophically or theologically. Further, the adjective "regular" contains an important restriction. Since the natural sciences, strictly speaking, deal only with recognizable patterns of phenomena—which can be confirmed by repeatable observations and experiments in controlled situations—they do not rule out isolated cases that appear to violate the laws of nature they embody, as long as such exceptions do not become regular. What is regularly true does not eliminate the possibility of infrequent deviation from the ordinary pattern, particularly if the context is different—and especially if a "higher-order" regularity (top-down causality) is acting to coordinate lower-level causes. Then, of course, the question is: By what means is such a higher-order causality being impressed? Such coordination by higher-level laws of nature may seem to violate lower-level laws, when we do not understand the higher-level causes and the way in which lower-level regularities can be influenced and directed by them.

Second, there are similar "veiled realities" at other levels of the natural world besides the quantum one—where we do not know the laws of nature as they actually function, not even approximately or provisionally. Consciousness (the hard problem within the complex of mind-body issues) and personal relationships are among these. Perhaps more to the point for the present discussion are the various aspects of the relationship between the transcendent ground of being and intelligent, conscious individuals and societies of individuals through experiences and manifestations of genuine transcendence. It is here that God's special action seems to be focused. In these cases, again, because the veil is partial, the appearance of God's activity must reflect something of its underlying operation. However it is concretely effected "behind the veil," both God's universal creative action and God's special action are constrained to appear according to the characteristics of inanimate, animate, personal, and social reality on *this* side of the veil. Furthermore, as Thomas Tracy suggests, it may be that God's special action in history and in personal relationships is a modality of God's relationship to the world as Creator,[33] so that special divine action should not be seen as completely separate from divine creative action.

Third, since special divine action is almost always effected in virtue of God's personal relationship with creatures capable of such interaction, it seems to involve top-down influences on the material causal structures of reality and thus to transcend what is or can be given simply by causes that we perceive and can adequately model by the sciences on this side of the various veils. Higher-level considerations and regularities having to do with persons and their relationships with one another frequently seem to render possible and even likely what we consider impossible from a strict physical or biological point of view, even in our ordinary human experience. So, from one point of view, it is not particularly surprising when the transcendent is involved. One may look at this either as higher-level relationships and regularities overriding or marshaling in novel ways the regularities of lower levels, or as an intervention or violation of the laws of nature at lower levels, *if* they are considered absolute ***under all circumstances***. But either way, such "violations" or "interventions" are not random, arbitrary, or gratuitous—they are done in virtue of God's love and care for those with whom God has special relationships, or those who are open to such invitations of the transcendent.

As a matter of fact, this happens in our relationships with one another, except that in this case we express our action as material beings on other material beings. The key problem in God's case is that we have no model for understanding how what is immaterial (God is spirit and therefore immaterial in some definite sense) can act on what is material. We do not even have an idea or concept of what an immaterial entity might be like that could mesh with our scientifically describable physical world. At any rate, if an immaterial entity—God—acts on a material system, then clearly one of the laws of nature, as we know them, will be contravened. The immaterial influence simply cannot be scientifically accounted for.

Human action is different—and acceptable—precisely because, though amounting to intervention in some sense, its influences can be accounted for physically and biologically. However, at a deep level human action is also mysterious. How does thought affect a material system? Or does thought—and decision—always involve a physical or neurological correlate?

With divine action, of course, there are many intermediate levels (e.g., prophets). However, at some remove—at some level—there eventually arises the principal question: How does God affect the material world in special cases in response to personal relationships with God? It seems that sensitivity to the transcendent on the part of the person or the community is always involved—a surrendering to the divine influence perceived outside oneself or welling up from within. But it is also true that a certain address by the transcendent is involved as well, in response to the pattern of openness and sensitivity on the part of the creatures.

If we are looking for the "causal joint" in special divine action, or for that matter in divine creative action,

> the direct causal nexus *is* the active, richly differentiated, profoundly immanent [because it is transcendent] presence of God in created beings and in their interrelationships—*is* at the same time their limited and specific participation—inclusion—in God's own existence and interrelationships as Trinity, which is utterly transcendent and immaterial but also radically open to and available for the realization of finite possibilities. The

[33] Thomas Tracy, in this volume.

presence of God in each entity constitutes the direct, the immediate, relationship of that entity with God, and therefore is the channel of divine influence in secondary causes.[34]

These are a few of my initial reflections on how all this relates to divine action. What I have primarily attempted to do in this essay is to provide some organized reflections on quantum physics and to describe its epistemological and ontological implications based on the preferred physical interpretations that have emerged. Personally, on these foundations, I conclude that God does act through the laws of nature as they function in reality—those upon which we have some limited purchase, as well as those that are unknown to us. Some of the latter lurk behind the veil that shrouds the quantum world from our macroscopically conditioned gaze, and others lurk behind similar veils through which the human sciences, philosophy, and theology can only catch fleeting glimpses of meaning and glory, "as through a glass darkly."

Acknowledgment. I am grateful for all the suggestions, questions, corrections and encouragement I have received from everyone at the conference. It has been a fruitful and stimulating exchange!

---

[34] William R. Stoeger, "Describing God's Action in the World in Light of Scientific Knowledge of Reality," in *Chaos and Complexity: Scientific Perspectives on Divine Action*, Robert J. Russell, Nancey Murphy and Arthur Peacocke, eds. (Vatican City State/Berkeley, Calif.: Vatican Observatory/Center for Theology and the Natural Sciences, 1995), 253–4.

DETERMINISM VERSUS INDETERMINISM IN QUANTUM MECHANICS: A "FREE" CHOICE

James T. Cushing

1 Introduction

The question of an "openness" (indeterminism, chance) versus a "closedness" (determinism) in the world can arguably be seen as *the* fundamental issue in relation to possibilities for a particular way of God's action in what is taken to be her creation.[1] Some suggest that quantum mechanics provides evidence for an "openness" or inherent indeterminism in the world. Hence, I take the issue of determinism versus indeterminism in quantum mechanics to be of particular importance for the theological discussion. Therefore, it is perhaps worth emphasizing that considerations of empirical adequacy and logical consistency alone do not force one to accept an indeterministic view of fundamental physical phenomena.[2] I appreciate that this present "contribution" of mine is largely "negative," or at best orthogonal to the general theological issues. I lay out a position that can quite reasonably be rejected because it does not fit in with certain metaphysical commitments one may have or with a theological project that one may wish to pursue.

To facilitate presenting an alternative version of quantum mechanics,[3] I suggest that, for purposes of this discussion, we consider the theory of quantum mechanics as having two conceptually distinct components: a formalism and an interpretation.[4] One could be a bit more refined and speak of a partially interpreted formalism (my "formalism") versus an *ontological* interpretation (my "interpretation").[5] It is in the

---

[1] For one prominent view of God's action in the world and its consonance with modern science, see John C. Polkinghorne, *Belief in God in an Age of Science* (New Haven, Conn.: Yale University Press, 1998). Furthermore, Owen Thomas informs me that a tension between free will and determinism exists only if one assumes an "incompatibilism" between the two. Since I am not competent to discuss this objection, I simply note it here.

[2] This is developed at length in James T. Cushing, *Quantum Mechanics: Historical Contingency and the Copenhagen Hegemony* (Chicago, Ill.: University of Chicago Press, 1994).

[3] There are, in fact, more versions, or interpretations, of quantum mechanics than the two I discuss in this paper. For an accurate and clear presentation of the main contenders, see David Z. Albert, *Quantum Mechanics and Experience* (Cambridge: Harvard University Press, 1992). However, the two I do consider are quite sufficient to make the points I wish.

[4] This is a distinction to which philosophers would rightly object as one that cannot be sharply maintained. Nevertheless, I believe that it will do as a "zeroth" approximation in the present context. Michael Heller has made a similar formalism-interpretation distinction in his "On Theological Interpretations of Physical Creation Theories" in *Quantum Cosmology and the Laws of Nature: Scientific Perspectives on Divine Action*, Robert J. Russell, Nancy Murphy and Chris J. Isham, eds. (Vatican City State: Vatican Observatory Publications; Berkeley: Center for Theology and the Natural Sciences, 1993), 91–102.

[5] Some authors take quantum theory to be a partially interpreted formalism, separate from its ontological interpretations. However, since we typically demand more of scientific theories than empirical adequacy, I include both the formalism and the interpretation as components of a theory. One can, of course, use either terminology—so long as we all understand what the terms mean. Nothing essential here turns on one choice or the other.

latter that the ontological determinism-vs.-indeterminism issue resides. After outlining each of these two components, I sketch the fundamentals of Bohmian mechanics and mention in passing how it happened historically that the standard, indeterministic (Copenhagen) version was accepted over a deterministic (de Broglie–Bohm) one. There follows a discussion of some apparent tensions between quantum mechanics and the special theory of relativity. This leads to consideration of Bell's theorem and its implications, the possible existence of a preferred frame, the distinction between "separability" and "locality," and the block universe question.

A superdeterministic picture of the physical universe does provide a straightforward and consistent picture of a god and its action in the world—albeit one not particularly compatible with a posited freedom of choice among individuals. In fact, given that some argue that evil is the price one must pay for freedom of action—and given all of the incredulity that can result from the evident magnitude of evil in our world—one can wonder whether or not the "data" of free choices are themselves unproblematical. In other words, if one is not struck by an evident existence of free will, then one might feel less than compelled to opt for an indeterministic version of quantum mechanics. In the end, of course, each may choose according to his or her predilections in this matter.

2 Quantum Mechanics: Formalism Versus Interpretations[6]

Let me make a basic distinction between two components of a scientific theory: its formalism and its interpretation. These are conceptually separable, even if they are often entangled in practice. It is sufficient for now to restrict my comments to a particular theory: quantum mechanics. By formalism, I mean a set of equations and a set of calculational rules for making predictions that can be compared with experiment (i.e., "getting the numbers right"). It is important to stress that I need only the fact that both standard quantum mechanics and Bohm's version use a common set of rules for predicting the values of quantities that can actually be *observed*. There may be additional quantities that have in-principle well defined values for Bohm (even though not for Copenhagen), but these are not accessible to observation (certainly not now and, arguably, never). The (physical) interpretation refers to what the theory tells us about the underlying structure of these phenomena (i.e., the corresponding story about the furniture of the world—an ontology). Hence, *one* formalism with *two* different interpretations counts as *two* different theories.[7]

I should at least outline what I have in mind by the Copenhagen interpretation, since I am here concerned mainly with the Bohmian alternative. The following is a bare-bones sketch, totally divorced from any history of the development and acceptance of the theory. Entire books have been written on the formalism of (nonrelativistic) quantum mechanics. Here I only outline in the briefest (if somewhat vague) form, the five postulates that are usually employed in making quantum-mechanical calculations:

---

[6] Some of this section and the next follow rather closely corresponding parts of Cushing, *Quantum Mechanics*, and of James T. Cushing, "It Is the Theory Which Decides What We Can Observe" in *Experimental Metaphysics*, Robert S. Cohen, Michael Horne and John Stachel, eds. (Dordrecht: Kluwer Academic Publishers, 1997), 13–23.

[7] The implications of this underdetermination of theory choice for critical realism is discussed in Cushing, *Quantum Mechanics*, chap. 11.

i. a state vector (e.g., ψ)—a vector (in a Hilbert space $\mathcal{H}$) representing the state of the physical system
ii. a dynamical equation (e.g., the Schrödinger equation)

$$H\psi = i\hbar\frac{\partial\psi}{\partial t} \tag{1}$$

giving the time evolution of the state vector ψ under the influence of the Hamiltonian H for the physical system ($\hbar$ being Planck's constant)
iii. a correspondence between (Hermitian) operators A on $\mathcal{H}$ and physical observables a. These physical observables a can take on only the eigenvalues a_j where

$$A\psi_j = a_j\psi_j \tag{2}$$

iv. average values for a series of observations of a given as $\langle\psi|A|\psi\rangle$
v. a projection postulate (either explicitly or effectively assumed) upon measurement

$$\psi = \sum_k c_k\psi_k \rightarrow \psi_j \tag{3}$$

where (for ψ normalized to unity) $|c_j|^2$ is the probability of obtaining the result a_j (that is, the chance of ending up in the eigenstate ψ_j)

Even if we accept these as a sufficiently adequate representation of the formalism of quantum mechanics, we still cannot give a simple and concise statement of *the* Copenhagen version of quantum mechanics because the term 'Copenhagen' has been used to refer to so many variations, all allegedly based on Niels Bohr's writing. While there has been much agonizing in the literature over exactly what constitutes the Copenhagen interpretation, I just mention a commitment to the completeness of the state-vector description, to a principle of complementarity, to probability that enters into quantum theory in an ineliminable and essentially nonclassical way (i.e., probabilities are not merely epistemic), and to a prohibition against any possible alternative causal description in a spacetime background. On the Copenhagen interpretation of quantum mechanics, physical processes are arguably, at the most fundamental level, both inherently indeterministic and nonlocal.[8]

3 A Summary of Bohmian Mechanics[9]

Although I have written more extensively elsewhere about John Bell's theorem and about Bohmian mechanics,[10] in this and the two following sections I shall give only

---

[8] See, however, my comments on nonseparability and nonlocality in section 5 below. Furthermore, I personally have no idea what the ontology of the Copenhagen interpretation might be. This goes far beyond, and is not really related to, the question of determinism versus indeterminism, nor even essentially to nonlocality. The real difficulty has to do with the (non-) existence of a definite physical reality at all times.

[9] The central ideas of Bohmian mechanics go back to the work of Louis de Broglie in the 1920s on his pilot-wave theory; see his *Non-Linear Wave Mechanics: A Causal Interpretation* (Amsterdam: Elsevier Publishing Co., 1960) for a retrospective summary of those developments. However, for simplicity I often refer to Bohm's theory or to Bohmian mechanics, since I shall be concerned with certain outgrowths of Bohm's ideas.

[10] See, for example, James T. Cushing and Ernan McMullin, eds., *Philosophical Consequences of Quantum Theory: Reflections on Bell's Theorem* (Notre Dame, Ind.:

a brief overview of some of the main issues on these topics as they relate to the concerns of this volume.[11]

As a preliminary, qualitative summary of Bohmian mechanics, the program can be characterized as an objective ("realistic"),[12] deterministic version of quantum mechanics—a "hidden-variables" theory, in which the additional variables (over and above the usual quantum-mechanical wavefunction $\psi(q)$, where $q = (q_j)$ are the "free" variables that are the arguments of ψ) are the positions ($Q = (Q_j)$) of the particles of the system.[13] Of central importance here is the question of *completeness* that separated Einstein and Bohr. That is, on the Copenhagen interpretation, a complete state description is given by $\psi(q)$, whereas for Bohmian mechanics a complete state description is given by $(\psi(q), Q)$. In Bohmian mechanics, the Schrödinger equation [eq. (1) above] governs the time evolution of the wavefunction $\psi(q)$, while the guidance condition [eq. (5) below] determines the time evolution of the positions Q.[14] As long as the *quantum-equilibrium* statistical distribution $P(x,t) = |\psi(x,t)|^2$ is introduced into Bohm's completely deterministic dynamics

University of Notre Dame Press, 1989), and Cushing, *Quantum Mechanics.*

[11] For complete, modern technical accounts of Bohm's theory, see Peter R. Holland, *The Quantum Theory of Motion* (Cambridge: Cambridge University Press, 1993); Detlef Dürr, S. Goldstein, and N. Zanghì, "Quantum Equilibrium and the Origin of Absolute Uncertainty," *Journal of Statistical Physics* 67 (1992): 843–907; idem, "Quantum Mechanics, Randomness, and Deterministic Reality," *Physics Letters A* 172 (1992): 6–12; idem, "Quantum Chaos, Classical Randomness, and Bohmian Mechanics," *Journal of Statistical Physics* 68 (1992): 259–70; idem, "Bohmian Mechanics as the Foundation of Quantum Mechanics" in *Bohmian Mechanics and Quantum Theory: An Appraisal*, James T. Cushing, Arthur Fine and Sheldon Goldstein, eds. (Dordrecht: Kluwer Academic Publishers, 1996), 21–44; Karin Berndl, M. Daumer, D. Dürr, S. Goldstein and N. Zanghì, "A Survey of Bohmian Mechanics," *Il Nuovo Cimento* 110B (1995): 737–50; Antony Valentini, "Pilot-Wave Theory of Fields, Gravitation and Cosmology" in *Bohmian Mechanics*, Cushing, Fine and Goldstein, eds., 45–66. W. Michael Dickson, in his *Quantum Chance and Non-Locality* (Cambridge: Cambridge University Press, 1998), gives a careful discussion of what nonlocality means in a completely deterministic setting. In Sheldon Goldstein, "Quantum Theory Without Observers—Parts One and Two," *Physics Today* 51.3 (1998): 42–6, and 51.4 (1998): 38–42, one finds a concise overview of the Bohmian project.

[12] By "realistic" here I mean an objectively existing (i.e., independent of observation) value for the positions of all particles at all times. (And, for Bohm—at least in the version I discuss here—particle position is the only observable.)

[13] The space of Bohmian mechanics (as for quantum mechanics generally) for the variable q is configuration space (typically of dimension 3N, where N is the number of particles in the system)—although all of the actual particle positions Q_j can be taken to label particle positions in physical three-dimensional space. Hence, one could see Bohmian mechanics as underwriting an ontology consisting of particles and forces only, all existing in physical three-space.

[14] In terms of the notations introduced in this paragraph, the Schrödinger equation and guidance condition would be written, respectively, as

$$i\hbar\frac{\partial}{\partial t}\psi(q,t) = H\psi(q,t), \quad \psi(q,t) = R(q,t)e^{iS(q,t)/\hbar}, \quad H = -\frac{\hbar^2}{2m}\nabla^2 + V(q)$$

$$\frac{d}{dt}Q_j \equiv v_j(Q,t) = \frac{1}{m}\nabla_j S(Q,t)$$

Here $V(q)$ is the usual classical potential and $\psi(q,t)$ has been written in polar form (which can always be done) in terms of its modulus $R(q,t)$ and its phase $S(q,t)$.

[eq. (5)], Bohmian mechanics is in principle empirically identical with standard quantum mechanics in all of its predictions for the outcome of any actual experiment or observation.

We can characterize the conceptual matrix of Bohm's theory as follows. As far as an ontology is concerned, it provides a picture story in a spacetime background (i.e., there are at all times actually existing particles that follow definite trajectories). This theory is completely deterministic and underpins an observer-independent, objective reality. Of course, this theory (and its ontology) is nonlocal: there *are* instantaneous, long-range influences between spatially separated stations. However, this ("benign") nonlocality cannot be used to send any message (let alone one faster than light). This is a result of the no-signaling theorem of quantum mechanics.[15] The measurement process, which is inherently many-body in nature, is basically an act of discovery—there is *no* quantum-mechanical measurement problem. There is a well defined criterion for a classical limit,[16] so that there is no *conceptual* mismatch between the classical and quantum domains. Finally, insofar as all measurements are ultimately *position* measurements and the usual quantum probability distribution holds (i.e., quantum equilibrium: $P = |\psi|^2$), Bohm's theory gives *complete* empirical equivalence with standard quantum mechanics.

In a bit more detail now, let me contrast the indeterministic, Copenhagen view of reality with that provided by Bohm's theory. His basic idea is the following. Beginning with the (nonrelativistic) Schrödinger equation [eq. (1) above] (which is *accepted, not derived*, here), he was able, *by a mathematical transformation alone*, to rewrite this dynamical equation as[17]

$$\frac{d\boldsymbol{p}}{dt} = -\nabla(V + U) \tag{4}$$

where $\boldsymbol{p}$ is given by the *guidance condition*

$$\boldsymbol{p} = mv = \nabla S \tag{5}$$

and U is the *quantum potential*

$$U \equiv -\frac{\hbar^2}{2m}\frac{\nabla^2 R}{R}, \qquad R = |\psi| \tag{6}$$

There is still the usual probability density $P = |\psi|^2$ governed by the continuity equation.[18]

---

[15] See, for example, Cushing, *Quantum Mechanics*, 195–6.

[16] There is a good deal more that remains to be said about this, in addition to the mere statement that $U = 0$ in eq. (4). On this point, see James T. Cushing and G.E. Bowman, "Bohmian Mechanics and Chaos" in *From Physics to Philosophy*, Jeremy Butterfield and Constantine Pagonis, eds. (Cambridge: Cambridge University Press, 1999), 90–107.

[17] For details, see, for example, Cushing, *Quantum Mechanics*, 60–2. For explicit definitions of S and of V, see note 14 above.

[18] Here I have glossed over the different emphases various Bohmians place on eq. (4) versus eq. (5) as being the fundamental equation of Bohmian mechanics. For a careful discussion of a tension that can be claimed to exist between these two schools, see Millard Baublitz and A. Shimony, "Tension in Bohm's Interpretation of Quantum Mechanics" in *Bohmian Mechanics*, Cushing, Fine and Goldstein, eds., 251–64. This difference in approach is also reflected in how "spin" is handled in Bohmian mechanics: simply as another degree of

One, of course, need not be particularly concerned about or interested in the mathematical manipulations leading to eqs. (4) through (6). What is relevant, though, for my purposes here is that the dynamics of quantum mechanics can be put into the "Newtonian" form $\boldsymbol{F} = m\boldsymbol{a}$ [eq. (4)] and given a causal interpretation in which microscopic particles, such as electrons, follow well-defined trajectories in spacetime. However, because of the influence of the quantum potential, these trajectories are extremely sensitive to the initial conditions ($\boldsymbol{x}_o$, $\mathbf{v}_o$) of the particles.[19]

We can consider P as the probability density of our ensemble of particles moving in the velocity field defined by ∇S. It is, in fact, just the quantum potential U (which is determined by the wavefunction ψ) that produces (or "transmits") the influence of the environment (and of the "measuring" apparatus) to the particle. Different observations or measurements correspond to different environments (e.g., a particular slit being opened or closed in a double-slit arrangement)—and hence to different wavefunctions ψ—so that different outcomes (or "trajectories") result. This truly reflects Bohr's concept of the holism of quantum phenomena and the spirit of his principle of complementarity. How a microsystem behaves depends upon its environment—an observed value is *contextual.*

All of the usual statistical interpretations of the standard Copenhagen theory are recovered from Bohm's theory provided *all* three of the following mutually consistent assumptions are made:

i. that the ψ-field satisfies Schrödinger's equation.
ii. that the particle momentum is restricted to $\boldsymbol{p} = \nabla S(\boldsymbol{x})$.
iii. that we do not predict or control the precise location of the particle, but have, in practice, a statistical ensemble with probability density $P(\boldsymbol{x}) = |\psi(\boldsymbol{x})|^2$. The use of statistics is, however, not inherent in the conceptual structure, but merely a consequence of our ignorance of the precise initial conditions of the particle.[20]

As stated previously, this last property ($P = |\psi|^2$) is often referred to as the quantum equilibrium condition. Notice that ψ plays two conceptually very different roles here: (a) it determines the influence of the environment on the microsystem ((i) and (ii)—the "dynamics") and (b) it determines the probability density P (iii). It is not a *logical* or an *a priori* necessity that the same function need play both of these roles. (Of course, complete predictive equivalence with standard quantum mechanics follows *only* if this is the case.) In fact, the *primary* conceptual role for ψ in Bohm's theory is (a). Bohm later gave a heuristic argument to show that, even if initially $P \neq |\psi|^2$, P would still be driven, through random interactions, to the equilibrium distribution $P = |\psi|^2$ (where it would remain by virtue of the continuity equation).[21]

---

freedom in the wavefunction (Dürr, Goldstein, and Zanghì, "Bohmian Mechanics as the Foundation of Quantum Mechanics"; Valentini, "Pilot-Wave Theory of Fields, Gravitation and Cosmology"; both in *Bohmian Mechanics*, Cushing, Fine and Goldstein, eds., 21–44, 45–66) or as the actual, mechanical, spin of a particle (David Bohm and Basil J. Hiley, *The Undivided Universe: An Ontological Interpretation of Quantum Theory* (London: Routledge, 1993), chap. 10).

[19] Strictly speaking, of course, it is the initial position alone that fixes the trajectory, since the guidance condition [eq. (4)] is a first-order equation.

[20] David Bohm, "A Suggested Interpretation of the Quantum Theory in Terms of 'Hidden' Variables. I & II," *Physical Review* 85 (1952): 166–93, 171.

[21] David Bohm, "Proof That Probability Density Approaches $|\psi|^2$ in Causal Interpretation of the Quantum Theory," *Physical Review* 89 (1953): 458–66. More recently, Antony

It is also worth pointing out here that, for Bohm, all observations ultimately reduce to position measurements. This is not unreasonable. After all, there *is* something special about *coordinate* space—we exist there (not, say, in momentum space) and all instrument outputs are ultimately readings in position space.

Finally, in Bohm's scheme, the classical limit corresponds to the quantum potential U (actually, the quantum force, which equals $-\nabla U$) being negligible compared to the classical potential. This is a precise and well-defined condition, not some heuristic limit such as $h \to 0$ (which is *not* under our control), which is often given a vague nod in standard presentations of quantum mechanics. On the other hand, we expect that, when ∇U is large (as happens, for example, in regions where $|\psi| = R$ is small and rapidly changing), then the motion can be irregular or chaotic, a type of "Brownian" motion.[22] Hence, small variations in the initial "hidden" parameters $\{\boldsymbol{x}_o\}$ can produce large variations in, say, the outcome of a scattering experiment. That is, averaging over these hidden variables leads to the statistical spread in a series of repetitions. One of the most beautiful aspects of Bohm's paper is his treatment of the measurement problem (which becomes a nonproblem).[23]

In summary, then, both the Copenhagen and Bohm theories of quantum mechanics operate with a common set of equations for making calculations and statistical predictions.[24] So, they must *always* yield identical results when applied to a common, well-formulated problem. This empirical equivalence is guaranteed only *modulo* no change in the formalism, of course. That is, one can argue that both Copenhagen and Bohm must make the same predictions for the numerical outputs of counter readings.

There is a story to be told about how it came to pass that the physics community around 1927 and thereafter accepted what is today the standard interpretation of quantum mechanics.[25] Since there is not space to do that here, let me simply quote

---

Valentini, in "Signal-Locality, Uncertainty, and the Subquantum H-Theorem. I," *Physics Letters A* 156 (1991): 5–11, has reconsidered the argument for an approach of a system to quantum equilibrium, along the lines of Boltzmann's *H*-theorem in classical statistical mechanics.

[22] See Cushing and Bowman, "Bohmian Mechanics," for a Bohmian perspective on quantum chaos.

[23] Bohm, "A Suggested Interpretation."

[24] Perhaps some may question whether or not the guidance condition [eq. (5) in the text] is part of standard quantum mechanics. Realize that $\boldsymbol{j} = P\boldsymbol{v}$ (where $P = |\psi|^2$ and $\boldsymbol{j}$ is the usual probability current density) and that $\boldsymbol{v} = \boldsymbol{j}/P$ is the guidance condition. This equation is there for both Copenhagen and for "Bohm," but only the latter gives it a literal meaning.

[25] See James T. Cushing, "Bohm's Theory: Common Sense Dismissed," *Studies in History and Philosophy of Modern Physics* 24 (1993): 815–42, and idem, *Quantum Mechanics*. For superb insights into the early history of quantum mechanics, see the extensive work of Mara Beller: "Matrix Theory Before Schrödinger: Philosophy, Problems, Consequences," *Isis* 74 (1983): 469–91; idem, "Born's Probabilistic Interpretation: A Case Study of 'Concepts in Flux'," *Studies in History and Philosophy of Science* 21 (1990): 563–88; idem, "The Birth of Bohr's Complementarity: The Context and the Dialogues," *Studies in History and Philosophy of Science* 23 (1992): 147–80; idem, "Einstein and Bohr's Rhetoric of Complementarity," *Science in Context* 6 (1993): 241–55; idem, "Rhetoric of Antirealism and the Copenhagen Spirit," *Philosophy of Science*, 63 (1995): 183–202; idem, "Criticism and Revolutions," *Science in Context* 10 (1997): 13–37; idem, *Quantum Dialogue: The Making of a Revolution* (Chicago, Ill.: University of Chicago Press, 1999). Jan Faye, in his *Niels Bohr: His Heritage and Legacy* (Dordrecht: Kluwer Academic Publishers, 1991), also explores the influence of Bohr's education and philosophical commitments on his

Bell on the choice by the scientific community of Copenhagen over de Broglie–Bohm.

> But in 1952 I saw the impossible done. It was in papers by David Bohm. Bohm showed explicitly how parameters could indeed be introduced, into nonrelativistic wave mechanics, with the help of which the indeterministic description could be transformed into a deterministic one. More importantly, in my opinion, the subjectivity of the orthodox version, the necessary reference to the "observer," could be eliminated.
>
> Moreover, the essential idea was one that had been advanced already by de Broglie in 1927, in his "pilot-wave" picture.
>
> But why then had Born not told me of this "pilot wave"? If only to point out what was wrong with it? Why did von Neumann not consider it? More extraordinarily, why did people go on producing "impossibility" proofs, after 1952, and as recently as 1978? When even Pauli, Rosenfeld, and Heisenberg could produce no more devastating criticism of Bohm's version than to brand it as "metaphysical" and "ideological"? Why is the pilot-wave picture ignored in the text books? Should it not be taught, not as the only way, but as an antidote to the prevailing complacency? To show that vagueness, subjectivity, and indeterminism are not forced on us by experimental facts, but by deliberate theoretical choice?[26]

That is, one can see the rejection of a deterministic pilot-wave representation of quantum phenomena, and the acceptance of an inherently indeterministic one, as something not forced on the scientific community by logic and empirical adequacy (although certainly *consistent* with these criteria), but as a choice made, finally, on other grounds. Similarly, I suggest, we should appreciate that one can choose to accept an indeterministic view of quantum mechanics for, say, theological reasons if one so wishes, but that is a different situation from claiming that quantum mechanics *necessarily* provides independent grounds for belief in the openness of the physical world at a fundamental level.

4 Objections to Bohm's Program and Responses

There is no shortage of objections to Bohm's theory. These commonly include charges that it is *merely* as empirically adequate as the standard version and, hence, makes no *new* predictions so that it is not scientifically interesting. Such a position certainly has a Whiggish air about it.

An old objection going back to Wolfgang Pauli asks *why* quantum equilibrium should obtain, simply as a happy coincidence. However, there are arguments that remove some of the mystery from this.[27]

It is often claimed that Bohm's theory is inherently nonrelativistic and cannot be extended to the relativistic domain—and hence, is a nonstarter, from a modern perspective. Well, first of all, there *are* Bohmian versions of quantum field theory that are completely equivalent predictively to standard relativistic quantum field theory.[28] As for the noncovariance of the basic dynamical equations of these Bohmian *predictively equivalent* theories, one can turn the question around (see the next section) and ask what the warrant is for demanding manifest Lorentz covariance of all of the equations of fundamental physics.

---

eventual formulation of the Copenhagen interpretation.

[26] John S. Bell, *Speakable and Unspeakable in Quantum Mechanics* (Cambridge: Cambridge University Press, 1987), 160.

[27] Dürr, Goldstein and Zanghì, "Quantum Equilibrium"; Valentini, "Signal-Locality I."

[28] Bohm, "A Suggested Interpretation"; Valentini, "Pilot-Wave Theory."

Recently in the physics literature, there have been claims and rejoinders about the "surreality" of Bohmian trajectories.[29] The upshot of all of this is that, while there are some unexpected surprises for the Bohmian perspective, there are no inconsistencies or conflicts with observation.

5 Tension Between Quantum Mechanics and Relativity

There is a vast literature on the subject of Bell's theorem.[30] Here I restrict my comments to fairly nontechnical matters and aim them at those for whom the subject is a relatively new one. It is important to appreciate that this famous theorem (i) has *nothing* to do (conceptually) with quantum mechanics and (ii) is a statement about the *world* in which we live (i.e., a statement about the way the world cannot *possibly* be). The basic question is: Could our world be both *objectively real* (i.e., observer independent; or, a definite truth of the matter at all times) and *local* (i.e., no *instantaneous* influences possible between spatially separated locations)? (Notice that *determinism* would provide *a*, but not *the* only, possible model for an objective reality.) The result, or upshot, of the theorem can be summarized symbolically as:

objective reality + locality + experiment → 1 > 2 (i.e., a logical contradiction).

So, a natural question is, What do we give up?[31] The three possibilities are:

i. an objective truth of the matter
ii. locality
iii. experimental results (loopholes)

Although the theorem itself is agnostic about this *choice*, my own opinion is that locality seems to be the problem, since both deterministic and stochastic models of reality can lead to Bell-type contradictions—once one assumes that observations made at one station cannot alter dispositions for (spacelike-separated) events (e.g., measurements) at another station.[32] (But, there is *much* more to be said here; for

---

[29] The original criticism along these lines was made by Berthold-Georg Englert, M.O. Scully, G. Süssmann, and H. Walther, "Surrealistic Bohm Trajectories," *Zeitschrift für Naturforschung* 47a (1992): 1175–86. Responses on behalf of Bohmian mechanics were then made by Detlef Dürr, W. Fusseder, S. Goldstein, and N. Zanghì, "Comment on 'Surrealistic Bohm Trajectories'," *Zeitschrift für Naturforschung* 48a (1993): 1261–2, and by Chris Dewdney, L. Hardy, and E.J. Squires, "How Late Measurements of Quantum Trajectories Can Fool a Detector," *Physics Letters A* 184 (1993): 6–11, with a reply again by Berthold-Georg Englert, M.O. Scully, G. Süssmann, and H. Walther, "Reply to Comment on 'Surrealistic Bohm Trajectories'," *Zeitschrift für Naturforschung* 48a (1993): 1263–4. The most recent salvo is by Yakir Aharonov, B.-G. Englert and M.O. Scully, "Protective Measurements and Bohm Trajectories," *Physics Letters A* 263 (1999): 137–46.

[30] To get some sense of this, see Cushing and McMullin, eds., *Philosophical Consequences*.

[31] Actually, Raymond Y. Chiao and J.C. Garrison, in "Realism and Locality: Which Should We Abandon?", *Foundations of Physics* 29 (1999): 553–60, point out that the concept of locality is typically formulated in a flat, four-dimensional space (e.g., one without, say, "wormholes" or "backdoor" connections between what appear to us to be far-separated stations). However, this possibility of resolving the conflict between locality and (objective) reality waits upon a specific model of a space that would still prove empirically adequate.

[32] A proof of a Bell inequality for an inherently probabilistic theory has been given, for example, by John F. Clauser and M.A. Horne, "Experimental Consequences of Objective Local Theories," *Physical Review D* 10 (1974): 526–35.

example, Michael Redhead quite correctly claims that attempts of general proofs[33] of nonlocality can break down in a general probabilistic context.[34]) The surprising fact is that *no* signaling is possible (with these instantaneous, long-range correlations). This avoids any conflict with the so-called first-signal principle of relativity. My interest here is *not* to review the most general discussions about the relation between Bell's theorem and various possible ontologies, but, rather, to examine that relation in the context of Bohm's theory.

When the results of Bell's theorem are brought to bear on quantum mechanics, we see immediately that no *local* hidden-variables theory is possible (i.e., one that can reproduce *all* of the predictions of quantum mechanics). However, as we have already seen in section 3, nonlocal, deterministic hidden-variables theories are "fine."

In an attempt to resolve any possible conflict, or even tension, with the special theory of relativity, there have been suggestions of distinguishing between "separability" and "locality"[35] and, in this same spirit, of taking seriously "relational holism."[36] Let me somewhat fancifully illustrate these differences with an example. Suppose there were a universe with just two dogs in it, say *A* and *B*, at spatially separated locations. Then, locality would forbid that anything dog *A* could do could *instantaneously* influence dog *B*. However, if one were to pinch dog *A*'s tail and his nose *immediately* itched, that would simply indicate the nonseparability (or relational holism) of that system—*not* of nonlocality.[37] However, it is not clear that such a distinction is useful physically.[38] Instantaneous influences-at-a-distance are arguably

---

[33] See, for example, Henry P. Stapp, "Locality and Reality," *Foundations of Physics* 10 (1980): 767–95; idem, "Bell's Theorem and the Foundations of Physics," *American Journal of Physics* 53 (1985): 306–17; idem, "Quantum Nonlocality," *Foundations of Physics* 18 (1988): 427–48; idem, "Quantum Nonlocality and the Description of Nature," in *Philosophical Consequences*, Cushing and McMullin, eds., 154–74.

[34] For counters to Stapp's arguments, see Rob Clifton, J. Butterfield and M.L.G. Redhead, "Nonlocal Influences and Possible Worlds—a Stapp in the Wrong Direction," *British Journal for the Philosophy of Science* 41 (1990): 5–58; W. Michael Dickson, "Stapp's Theorem Without Counterfactual Commitments: Why It Fails Nonetheless," *Studies in History and Philosophy of Science* 24 (1993): 791–14; Michael L.G. Redhead and P. La Rivière, "The Relativistic EPR Argument" in *Potentiality, Entanglement and Passion-at-a-Distance*, Robert S. Cohen, Michael Horne and John Stachel, eds. (Dordrecht: Kluwer Academic Publishers, 1997), 207–15. See also Arthur Fine, "Locality and the Hardy Theorem," in *From Physics to Philosophy*, Butterfield and Pagonis, eds., 1–11.

[35] Don Howard, "Einstein on Locality and Separability," *Studies in History and Philosophy of Science* 16 (1985): 171–201; idem, "Holism, Separability, and the Metaphysical Implications of the Bell Experiments" in *Philosophical Consequences*, Cushing and McMullin, eds., 224–53.

[36] Paul Teller, "Relational Holism and Quantum Mechanics," *British Journal for the Philosophy of Science* 37 (1986): 71–81; idem, "Relativity, Relational Holism, and the Bell Inequalities" in *Philosophical Consequences*, Cushing and McMullin, eds., 208–23.

[37] Some would refer to this use of the term "nonlocality" as signifying a vicious nonlocality, while nonseparability would be termed benign nonlocality. The first form of nonlocality would conflict with the first-signal principle of special relativity, while the second would not. For a discussion of these distinctions, see Dickson, *Quantum Chance*; Richard Healey, "Mining for Metaphysics," *Studies in History and Philosophy of Modern Physics* 30 (1999): 443–52.

[38] James T. Cushing, "Locality/Separability: Is This Necessarily a Useful Distinction?", in *Proceedings of the 1994 Biennial Meeting of the Philosophy of Science Association*, vol.

physical nonlocality, whatever one may wish to term them. In other words, it would seem reasonable to put the burden of proof or of argument on those who claim that these distinctions provide a fruitful and useful means of resolving the quantum-mechanics/relativity tension in some understandable way, as opposed simply to generating new terminology to paper over our ignorance.

Bohm's theory also provides an interesting perspective on the so-called "block-universe" question of whether or not special relativity is consistent with the possibility of becoming (the latter of which would seem necessary for a probabilistic view of the world).[39] A theory (such as Bohm's) that has a preferred frame for instantaneous actions surely allows for becoming. And, as we have seen, there is no observational conflict with relativity. However, of course, this theory is also completely deterministic. So, there is temporal becoming, but everything is predetermined. I mention this only as an aside.

In fact, rather than attempting to arrange some peaceful coexistence between quantum mechanics and relativity with terminological sleight of hand (as in the nonlocality/nonseparability gambit), it is really a conceptually fascinating possibility to countenance questioning the status of relativity (actually, Lorentz covariance) as a fundamental symmetry of nature.[40]

6 A Deterministic View of God

A completely deterministic universe certainly provides *a* representation of the way the world *might* be with a corresponding picture of a God completely in control. (Perhaps one can even see the preferred reference frame with its instantaneous connections as being reminiscent of Newton's idea of space as "God's *sensorium*."[41]) Just as with other scientific theories, this one offers a consistent story about a possibility for reality—but *truth* is another matter. As the quote from Bell in section 3 reminds us, in selecting an interpretation of quantum mechanics—and, here, I would add, in choosing a view of the physical world that supports our desired picture of God—we make a "deliberate theoretical choice."

One of the motivations for choosing an indeterministic version of quantum mechanics is the putative existence of free will, for which evil is the inescapable price.[42] A justification for the large amount of evil in the world becomes a genuine

---

I, David Hull, Micky Forbes, and Richard M. Burian, eds. (East Lansing, Mich.: Philosophy of Science Association, 1994), 107–16.

[39] For differing views on this, see Nicholas Maxwell, "Are Probabilism and Special Relativity Incompatible?", *Philosophy of Science* 52 (1985): 23-43; Howard Stein, "On Relativity Theory and Openness of the Future," *Philosophy of Science* 58 (1991): 147–67. See also the discussion in Chris J. Isham and John Polkinghorne, "The Debate over the Block Universe," in *Quantum Cosmology and the Laws of Nature*, Russell, Murphy, and Isham, eds., 135–44.

[40] Antony Valentini, "Signal-Locality, Uncertainty and the Subquantum H-Theorem. II," *Physics Letters A* 158 (1991): 1–8; James T. Cushing, "What Measurement Problem?", in *Perspectives on Quantum Reality*, Rob Clifton, ed. (Dordrecht: Kluwer Academic Publishers, 1996): 167–81.

[41] Isaac Newton, *Optics* (New York: Dover Publications, 1952), Book III, Pt. 1, Query 31, p. 403.

[42] For a quite different perspective on the problem of evil in the world, see Steven Weinberg, "A Designer Universe?", *The New York Review of Books* XXLVI (Oct. 21, 1999): 46–8.

difficulty. However, the concept of "free" choice is perhaps problematical, since it is not obvious to all that we do have such choices. Furthermore, one can marvel at the savage and inhospitable nature of the universe (for the long-term existence of humankind), as seen in modern cosmology, when one is attempting to understand the universe as the creation of a merciful God.

7 Conclusions

My only purpose in writing this brief *informational* essay has been to present in outline a position that allows one (if he or she so wishes) to put a completely deterministic gloss on physical phenomena, even at the quantum level. It has not been my intention to attempt to argue for the correctness, truth or desirability of such a position. It is not clear to me that the determinism-versus-indeterminism question is empirically decidable.[43] Of course, choices among theories in science are *not* made solely on the basis of logic and of empirical evidence ("objective" criteria) alone, but also involve consideration of other, "external" factors as well. Hence, to resolve the question of which ontological interpretation of quantum mechanics to choose, one might fruitfully employ other criteria (e.g., commitment to a certain view of relativity, theological *desiderata*, and the like).

Acknowledgment. I want to thank several people for making helpful comments on earlier drafts of this paper, but especially Jeremy Butterfield and John Polkinghorne.

[43] I appreciate that it is strictly speaking the theories, or equations, themselves that may or may not be said to have a mathematical property that can rightly be termed "deterministic," and that it is a long story to get from there to the determinism of the world itself. For some discussion of this point, see James T. Cushing, *Philosophical Concepts in Physics* (Cambridge: Cambridge University Press, 1998), 170–2.

SOME WORLDS OF QUANTUM THEORY

Jeremy Butterfield

1 Introduction

In this essay, I shall sketch some of the issues that arise in the interpretation of quantum theory from the perspective of a philosopher of physics. I will emphasize the measurement problem, which is widely agreed to be the theory's main interpretative problem. My over-arching aim will be to emphasize the difficulty of giving a satisfactory interpretation of quantum theory and to highlight the strange ontologies that emerge from various interpretations.

After some introductory remarks (sections 1 and 2), I present the measurement problem (3) and emphasize the role of decoherence (4). I then (5) distinguish four strategies for solving the measurement problem, briefly describing the first three. Finally (6–8), I present the strategy of Hugh Everett,[1] focusing particularly on the attractive versions of this strategy found in the work of Simon Saunders and David Wallace.[2] My rationale for concentrating on the Everettian strategy is that it aptly illustrates how radical our interpretations of quantum theory may have to be.

For the most part, I will cast my discussion in terms of elementary quantum theory, neglecting relativity (whether special or general) and quantum field theory. Indeed, my discussion will be philosophical, with almost no use of the technicalities of even elementary quantum theory. This restriction will not, I think, be misleading because most elements of the discussion—about the measurement problem and the strategies for solving it—still apply when relativity and quantum field theory are taken into account. The interpretive difficulties, in fact, are in general aggravated once we consider relativity and quantum field theory. The measurement problem, for example, is liable to be *more* severe once we accept, with relativity, that there is no absolute simultaneity, since one has to ask relative to which notion of simultaneity the wavepacket collapses. If the collapse is a physical process, one also has to define an appropriately relativistic dynamics. Furthermore, the Everettian strategy, unlike other strategies, does not propose to solve the measurement problem by changing the formalism of quantum theory—changes that, however satisfactory they might seem for elementary quantum theory, are difficult to adapt to relativity and quantum field theory. The Everettian strategy proposes instead a distinctive philosophical solution that one can reasonably expect to carry over smoothly to relativity and quantum field theory—if one accepts this solution for the nonrelativistic case.

---

[1] Hugh Everett, " 'Relative State' Formulation of Quantum Mechanics," *Reviews of Modern Physics* 29 (1957): 454–62.

[2] Simon Saunders, "Time, Quantum Mechanics, and Decoherence," *Synthese* 102 (1995): 235–66; idem, "Time, Quantum Mechanics, and Tense," *Synthese* 107 (1996): 19–53; idem, "Relativism," in *Perspectives on Quantum Reality*, R. Clifton, ed. (Dordrecht: Kluwer, 1996), 125–42; idem, "Time, Quantum Mechanics, and Probability," *Synthese* 114 (1998): 373–404; David Wallace, "Worlds in the Everett Interpretation," forthcoming in *Studies in History and Philosophy of Modern Physics*; idem, "Thoughts on Quantum Mechanics and Vice Versa: Fitting the Observer into the Interpretation of Quantum Theory," in preparation. The interpretations of David Deutsch, Murray Gell-Mann and James Hartle, Lev Vaidman, and Wojciech Zurek are similar, but I will not enter into details of their views.

The essay will have three general limitations. First, I am not qualified to say anything about possible connections between the interpretation of quantum theory and theology. Second, I will say nothing about the curious nonlocal phenomena that appear in quantum theory (noting only that Michael Redhead discusses these issues in his contribution to this volume). Finally, I will say next to nothing about three of my four broad strategies for solving the measurement problem because they involve less strange ontologies. In this regard they are more credible, but they face other difficulties. Jim Cushing's contribution to this volume discusses what is perhaps the best-developed example of a credible strategy, the pilot-wave interpretation.[3]

According to most accounts, especially in popular literature, quantum theory is indeterministic (the "collapse of the wavepacket"). To the theologian who naturally asks whether this might afford some scope for divine action in the world I say, "Beware," for this much-touted indeterminism is highly questionable. Only some of the four strategies accept it. In particular, the pilot-wave interpretation is a perfectly tenable and utterly deterministic interpretation. Similarly, the Everettian strategy postulates a fundamental determinism overlaid by an apparent indeterminism. Besides, some indeterministic interpretations give such detailed and thoroughly physical models of the collapse of the wavepacket that a theologian might well think twice about finding much scope for divine action in them.

I should also warn theologians against the opposite error of assuming that quantum theory is *merely* indeterministic. The assumption that it implies indeterminism and nothing more is false. Setting aside the pilot-wave interpretation, the indeterministic interpretations of quantum theory all involve a stranger ontology than the notion of indeterminism alone conveys. To put the point in terms of reality, quantum theory intimates a reality stranger than just indeterminism suggests.

I admit that it is frustratingly hard to say what this reality might be like, let alone for everyone to agree on such a description. Witness the effort that goes into articulating the various interpretations and the continuing controversy about which one is right. Nonetheless, I hope this essay will give some indication of the strange metaphysical possibilities that are "on the cards."

2 The Successes of Quantum Physics—And Its Open Problems

Before embarking on the measurement problem, I propose to set the stage in this section. First, I will emphasize how extraordinarily successful quantum theory has been in domains of application far beyond its original scope. The moral of this success is that we must take this theory extremely seriously as a description of the world. I will then offer some cautionary remarks about quantum theory's remaining problems, and about scientific realism and reductionism.

Examples of quantum theory's success are legion. To take just one, quantum theory describes the laser in a CD player. Although it was devised for systems of

---

[3] In this essay, I address only some aspects of the Everettian strategy, partly because I have discussed versions other than those of Saunders and Wallace elsewhere ("Worlds, Minds and Quanta," *Aristotelian Society Supplementary Volume* 69 (1995): 113–58; "Whither the Minds?" *British Journal for the Philosophy of Science* 47 (1996): 200–21), and partly because I agree for the most part with what Saunders and Wallace say about identity through time (see sec. 8 below) and with Saunders' interpretation of probability. Saunders casts his own views in terms of the consistent-histories approach, but his and Wallace's main ideas can be conveyed without relying on the technicalities of this approach. For a discussion of consistent histories, see Chris Clarke in this volume.

atomic dimensions (10^{-8} cm), the theory has turned out to work for scales much smaller (e.g., the nuclear radius of ca. 10^{-12} cm) and those vastly larger (e.g., superconductivity and superfluidity, involving scales up to 10^{-1}cm). Indeed, much of the history of twentieth-century physics is the story of quantum theory's basic postulates being applied ever more widely.

The point applies equally well when we look beyond present terrestrial physics. Quantum theory has been applied with extraordinary success to astronomy through the use of nuclear physics to develop an accurate and detailed theory of stellar structure and evolution, including the creation of nuclei in stars. On the cosmological scale, we now know a great deal about primordial nucleosynthesis.

And yet complacency, let alone triumphalism, is not in order! There is, of course, no valid inference from the empirical success of a theory—even one so broadly successful as quantum theory—to its literal truth. The most we might infer is that it is rational to believe the theory is true, or approximately so. But even this inference requires a strong form of that contentious doctrine, scientific realism. Moreover, physics itself contains hints that quantum theory is not true. Certain "clouds on the horizon" may prove as great a threat to the continued success of quantum theory as were the anomalies confronting classical physics at the end of the nineteenth century. Of course, people differ in what problems they find worrisome. As a philosopher, I worry about the various mysteries of interpreting quantum theory, but the wider physics community also recognizes at least two other clouds: the problem of reconciling quantum theory and general relativity, and the problems associated with the relation of quantum theory to special relativity, such as renormalizability and the rigorous existence of quantum field theories.[4]

Rather than ignore the clouds, we ultimately need to take account of them if we can. Otherwise, the debate over interpretation is likely to become isolated from the data of physics (to its own detriment) and unable to offer heuristic ideas to physics (to the detriment of physics). Here, however—for the reasons mentioned above—I will concentrate primarily on elementary quantum theory.

Finally, I need to make a few cautionary remarks about reductionism. Of course there is a strong sense in which particle physics as described by quantum field theory underlies atomic physics. In a broadly similar way, atomic physics underlies chemistry, which in a broadly similar way underlies biochemistry, and so on. But these inter-theoretic relations vary from one example to another and are in any single example often subtle and complicated (e.g., because they involve various kinds of limit[5]). Such examples do not provide neat cases of the traditional philosophical conception of theory-reduction as logical definitional extension (i.e., the deduction of one theory from the other, supplemented with judiciously chosen definitions of terms). Furthermore, to understand any single relation we may well have to take a stand on relevant interpretive issues. The relation of atomic physics to chemistry, for example, involves the transition from quantum to classical physics, and so is bedeviled by the measurement problem! In view of these complications, we must

[4] I discuss these clouds from a philosophical perspective in "The State of Physics: Halfway through the Woods," *Journal of Soft Computing* 5 (2001): 129–30.

[5] For further discussion of limits, especially the importance of singular limits, see Michael Berry in this volume; idem, "Asymptotics, Singularities and the Reduction of Theories," *Proceedings of 9th International Congress of Logic, Methodology, and Philosophy of Science*, D. Prawitz, B. Skyrms, and D. Westerstähl, eds. (Dordrecht: Kluwer, 1994), 597–607.

expect the "higher" levels to contain much (in terms of structures, concepts, or explanations) that is distinctive, giving them a high degree of "autonomy" from the "lower" levels.[6] "Outsiders to physics" should bear in mind that the greater the autonomy higher levels are taken to possess, the less pressure there will be to take lessons in ontology from physical theories of the lower levels.

In general, the idea that a "higher" level or theory might only approximate an underlying "lower" level is just one way among others that inter-theoretic relations can be subtle and complicated. Such approximations will nonetheless play a significant role in what follows, where all one can reasonably demand is that the classical should approximate the quantum. The details of the notion of approximation will vary. In some cases, such as decoherence (section 4), numerical predictions by classical theories approximate those of quantum theory; in other cases, such as Everettian interpretations (sections 6–8), classical concepts are used as approximations for grasping novel quantum concepts.

3 The Measurement Problem

I turn now to the measurement problem of quantum theory, which arises because quantum theory denies that objects such as electrons and atoms always have definite values for quantities like position, momentum, and energy. Essentially, the problem is that quantum theory apparently implies that such indefiniteness should also be endemic in the familiar macrorealm of tables and chairs. It is called the 'measurement problem' mainly because the argument that quantum theory implies an indefinite macrorealm is clearest for a measurement situation. Thus, if you use quantum theory to analyze a measurement of, say, the momentum of an electron, which quantum theory says has no definite momentum, you find that according to quantum theory the indefiniteness of the electron's momentum is transmitted to the apparatus's pointer so that it has no definite position!

First, let me present a brief summary of the formalism. Like other physical theories, quantum theory assigns states to systems. The state fully specifies the properties of the system. But the orthodox interpretation says that a quantum state is a catalogue of probabilistic dispositions (*Born-rule probabilities*). For each quantity (position, energy, momentum, etc.), the state defines a probability distribution on all possible values of the quantity. These states fall into two kinds.

1. *Pure States*. These are represented by vectors and written (in Dirac's notation) with angle-brackets, e.g. $|\psi\rangle$. Because they are vectors with length and direction, it makes sense to say that $|\psi\rangle$ is of length 1 or that two vectors are orthogonal (at right angles). More generally, the usual notions about vector spaces (e.g., basis and orthonormal basis) apply. The set of vectors representing the possible pure states of the system is called the *Hilbert space* of the system. For each pure state, there are physical quantities and values for each quantity such that the state ascribes probability 1 to some value for some quantity. But for each state, the great majority of quantities are ascribed a nontrivial probability distribution. To explain this, we first need some mathematical notions and jargon.

---

[6] Although I am happy to describe higher levels as 'emergent' from, rather than 'reduced' to lower levels, 'emergence' merely connotes some relation of dependence weaker than reduction. I am also suspicious of making emergence precise in terms of supervenience; see sec. 2 of Jeremy Butterfield and C. Isham, "The Emergence of Time in Quantum Gravity," in *The Arguments of Time*, J. Butterfield, ed. (London: Oxford University Press, 1999).

Each quantity is represented as a certain kind of function (*operator*) on the system's Hilbert space (namely, a self-adjoint operator). If the operator for a quantity sends a vector to a real-number multiple of itself, then we say the vector (or state) is an *eigenvector* (*eigenstate*) of the quantity, and the real number is the corresponding *eigenvalue*. Given a quantity, most vectors are not eigenvectors of it, but rather sums of eigenvectors (with different eigenvalues from each other); such sums are called *superpositions*. The set of eigenstates of a given quantity, for a given eigenvalue, is called an *eigenspace*, and it forms a subspace of the space of all vectors, just as a plane forms a subspace of ordinary three-dimensional space. If the subspace is one-dimensional (like a ray), the eigenvalue is called *nondegenerate*. If it is many-dimensional, the eigenvalue is *degenerate*.

Every quantity is "built" in a simple way from a special kind of quantity, called a *projector*. For any subspace, the function sending all vectors orthogonally into the subspace (so that each vector in the subspace is sent to itself) is the projector onto that subspace. The subspace is the projector's range, and is also its eigenspace for eigenvalue 1; the subspace of all vectors orthogonal to the given subspace is its eigenspace for eigenvalue 0. So projectors have just two eigenvalues, 1 and 0.

It turns out that every quantity is a linear combination (with real-number coefficients) of projectors onto mutually orthogonal subspaces, so that these subspaces are the quantity's eigenspaces and the real-number coefficients are its eigenvalues. The projectors onto these subspaces are called the quantity's *eigenprojectors* or *spectral projectors*. This implies that given a quantity we can find a set of its eigenvectors that are of length 1, are mutually orthogonal, and together form a basis (implying that any vector is a linear combination of them). This set is an orthonormal basis. Since it consists of eigenvectors of the quantity concerned, it is also called an *eigenbasis* of the quantity, but I shall usually just say "basis." Furthermore, if every eigenprojector of a quantity is one-dimensional (i.e. its range is a one-dimensional subspace)—or equivalently, if every eigenvalue of the quantity is nondegenerate—then the quantity's eigenbasis is essentially unique, and the quantity is called *maximal* (or *nondegenerate*). If an eigenvalue is degenerate, there is freedom in specifying an eigenbasis, for one can rotate a set of orthogonal eigenvectors spanning the eigenvalue's eigenspace.

Turning now to the physical interpretation of vectors as states that prescribe probabilities for quantities, an eigenstate of a quantity ascribes probability 1 to the corresponding eigenvalue. This is a special case of the general rule whereby the state's probability distribution for a quantity is derived from the geometry of the vector space. In the general case where the state is a superposition of the quantity's eigenstates, the probability prescribed by $|\psi\rangle$ that quantity Q takes value q is calculated as the square of the length of the vector obtained by projecting $|\psi\rangle$ down into Q's eigenspace for eigenvalue q. The convention that the original $|\psi\rangle$ is of length 1 secures (by a simple generalization of Pythagoras's theorem) that the probabilities for the various possible values add up to 1.

2. *Mixed States*. The second kind of state is more general and is represented by a weighted sum of pure states, with the weights adding up to 1. The probability distribution of any quantity thus becomes the weighted sum of the corresponding distributions of the pure states. Accordingly, mixtures are useful for calculating statistics for a heterogeneous population (*ensemble*) of systems in which different pure states occur in different proportions. A mixture is the linear combination of the projectors onto the rays (1-dimensional subspaces) that contain the mixture's component pure states, the coefficient of each projector being its corresponding

weight. As a linear function on the vectors, a mixture can be expressed as a matrix relative to any basis of the Hilbert space. This matrix is called a *density matrix*.

Mixtures, however, cannot always be interpreted straightforwardly in terms of a heterogeneous ensemble of systems (the *ignorance interpretation*). Surprisingly, for any composite system the vast majority of its pure states (the *entangled* or *non-factorizable* pure states) determine the state of each component system to be a mixture. In other words, for any entangled composite state, there is a set of pure states and associated weights for the component systems such that the correct probability distribution for any quantity on the component systems is given by the weighted sum of the distributions ascribed by the pure states. But beware! These mixtures cannot be given the ignorance interpretation without forcing the composite system to be a mixture. Thus, if we have an ensemble of composite systems in an entangled state $|\psi\rangle$, we cannot think of each single specimen of the component system as in one of the pure states in the mixture determined by $|\psi\rangle$. If it were, then the composite system would also be in a mixture (because a population of composite systems must be heterogeneous if a factor-population of component systems is). These curious mixtures are usually called *improper mixtures*. Ignorance-interpretable mixtures represent a heterogeneous population and are called *proper mixtures*.

For a given quantity, a (pure) superposition of its eigenstates and a mixture might ascribe the same nontrivial distribution. But pure states and mixtures are always distinct: any superposition and any mixture will ascribe different probability distributions from one another for some quantity—albeit perhaps a very arcane one. The numerical differences in the distributions, value for value, are *interference terms*.

Matters now become more contentious when we ask, What exactly are these probabilities of? The orthodox answer is that they are probabilities of measurement results. It is orthodox to say that the system has a value for a given quantity only when its state ascribes probability 1 to that value (the *eigenvalue-eigenstate link*). To a philosopher of probability, this claim will seem a howler—and it may well be wrong. But traditionally, it has been supported by two considerations. First, supposing there are values apart from those ascribed probability 1 (i.e., hidden variables), it has seemed impossible to design measurements that yield them reliably as results. Second, some proposals for such values, though simple and natural from a formal point of view, lead to outright contradiction (as in the theorems of John von Neumann, Andrew Gleason, Simon Kochen, and Ernst Specker). If we evade such contradictions by making the hidden variables "contextual," they will still need to be nonlocal in some sense in order to recover quantum theory's mysterious correlations between entangled but spatially separated systems (as in Bell's theorem).

More generally, orthodoxy says that the probability distribution ascribed by the state is not for values possessed at some time or other, but for values yielded as results (pointer-readings) on a measurement apparatus. It is curious for the interpretation of the quantum state to invoke something so extrinsic (and so vast in comparison) to the system, so variable from occasion to occasion, and so vague. Indeed, it is surely more than curious. It is unsatisfactory—unless you have an operationalist or instrumentalist view of theories. This predicament prompts us to analyze measurement from the point of view of quantum theory in hopes of vindicating this interpretation of state. The central concern here is to reconcile quantum theory's scarcity of values with the success of classical physics' opposite assertion that every quantity has a value for every system.

Many quantum measurements are, to a good approximation, repeatable. Whatever the initial state of the measured system, an immediate repetition of such a measurement always gives whatever result was first obtained. On the orthodox view, this motivates the claim that, in a measurement, the measured system goes into the eigenstate corresponding to the result obtained (the *projection postulate* or *collapse of the wavepacket*). The same line of thought also motivates a similar "collapse" of the apparatus to an eigenstate of pointer position. It is important to note that only in the collapse of the wavepacket does the alleged indeterminism of quantum theory occur. The state of an isolated quantum system evolves deterministically and continuously according to the Schrödinger equation (*unitary evolution*).

Thus we have arrived at the measurement problem. Quantum theory seems to veto such a reconciliation, and in a way that commits it to a manifestly false prediction. The quantum theoretic analysis of the interaction of, say, an electron that is indefinite for momentum with an apparatus for measuring momentum suggests that the electron's indefiniteness will be transmitted to the apparatus—leaving its pointer without any definite position! This has been proven for a wide range of quantum theoretic models of measurement, but we can confine ourselves to a simple model.

Consider a momentum measurement on an electron in a superposition of two momentum eigenstates: one for 1 unit of momentum, and the other for 2 units of momentum. Suppose we have a measurement apparatus or pointer, with ready-state $|r\rangle$, which reliably reads these eigenstates. The composite system will behave as follows:

$$|1\rangle|r\rangle \rightarrow |1\rangle|\text{reads "1"}\rangle \quad \text{and} \quad |2\rangle|r\rangle \rightarrow |2\rangle|\text{reads "2"}\rangle.$$

If the composite "electron + pointer" begins in the state on the left of each displayed evolution, it evolves by the Schrödinger equation in some fixed finite time to the state on the right. From this it follows (by the linearity of the Schrödinger equation) that measuring an electron initially prepared in a superposition yields (ignoring irrelevant normalizations):

$$\{|1\rangle + |2\rangle\}|r\rangle \rightarrow |1\rangle|\text{reads "1"}\rangle + |2\rangle|\text{reads "2"}\rangle.$$

But the final state on the right is not an eigenstate of pointer position. In fact, it determines the pointer state to be an improper mixture. So orthodoxy (more precisely, the eigenvalue-eigenstate link) declares that the pointer has no definite position!

There are two main ways to solve this problem. Either (i) we justify the collapse of the wavepacket for the pointer and replace the above final state by an eigen-state of pointer position, or (ii) we supplement the eigenvalue-eigenstate link's meager ascription of values by keeping the above final state but ascribing a definite position to the pointer without violating the no-hidden-variables theorems mentioned above.

In section 5, I will formulate these alternatives more sharply. For, in line with my remarks at the end of the previous section, a solution to the measurement problem need only recover classical physics' assertion of definite values as an approximation. For example, the first alternative (i) might only secure that the final state is approximately an eigenstate of pointer position, whereas the second alternative (ii) might ascribe a definite value not to pointer position, but to another quantity that is in some suitable sense "very close" to pointer position.

One might reply on behalf of orthodoxy that since quantum theory is meant to be indeterministic, and since the initial electron state was a superposition—thus ascribing probability 1 to no result—the final state on the right should *not* ascribe a definite position to the pointer. Rather, it should merely ascribe to the pointer probabilities for the two possible positions given by the electron's initial superposition—and the final state on the right surely does just that.

This reply is unobjectionable until the last step. There it slides from what it has, to what it needs in order to achieve a definite result. Specifically, it slides from orthodoxy's official interpretation that a *subsequent measurement* on the pointer would yield the results "reads '1'" and "reads '2'," with probabilities given by the electron's initial superposition, to what orthodoxy needs so that the pointer *now has* one of the two positions, i.e, a proper (i.e., ignorance-interpretable) mixture of two position eigenstates. But orthodoxy cannot get such a mixture—the final state is pure.

Thus we can summarize the measurement problem as the problem of justifying the replacement of a pure state by a proper mixture (or as the problem of eliminating the interference terms that distinguish such states). In order to formulate the alternative solutions more sharply, I first need to discuss decoherence.

4 Decoherence

There is a fallacious solution to the measurement problem (often repeated in the textbooks!) that bears on much significant recent work on decoherence. According to the solution, if the pure state of a composite system is entangled, then the state of each component system is a mixture. This seems to solve the measurement problem, for the final pure state above determines the state of the pointer to be just the right mixture of two position eigenstates!

This solution fails for two reasons. First, the mixture for the pointer is *not* ignorance-interpretable. It is an improper mixture, not a proper one. Although it gives the right statistics for subsequent measurements on the pointer, it does not secure a definite result (pointer position) at the end of each measurement. Again, we need to go beyond orthodoxy to get a definite result, either by justifying the collapse of the wavepacket or by ascribing extra values.

Second, the apparent success of this solution is an artifact of our simple model of measurement. Specifically, our measurement did not disturb the electron's momentum eigenstates. If it did disturb them, the final pure state of the composite system would *not* determine the pointer to be in the mathematically right mixture of two position eigenstates. (I say "mathematically" because the mixture is not interpretatively right—it is improper.) In fact, the final pure state would determine a mixture with the right weights, but of two eigenstates of some quantity different from position—the specific quantity depending on how exactly the electron was disturbed.

But unlike the first, this second reason is surmountable, and recent work on decoherence shows how.[7] Essentially, one needs to account for the fact that the pointer is not isolated but is continually interacting with its environment (e.g., air molecules or photons). Because the pointer's position is crucial to the physics of this interaction, the final state of the total system (electron + pointer + environment) determines the pointer to be in a mixture very close to the mathematically right mixture of position eigenstates no matter how much the measurement disturbs the

[7] This work has been carried out by many physicists over the last twenty-five years. See, for example, Domenico Giulini, et al., *Decoherence and the Emergence of a Classical World from Quantum Theory* (Berlin: Springer, 1996); cf. Berry in this volume.

electron. The process of obtaining this mixture is called *decoherence.*[8] Here, "very close to" means that the eigenstate components of the mixture are close to position eigenstates (i.e., close as vectors in Hilbert space, or—in physical terms—close if they ascribe very similar probability distributions to any quantity). To put it more precisely, the density matrix that expresses the final state in terms of position is nearly diagonalized, its off-diagonal terms—often called *tails*—being nearly zero.

In effect, pointer position "gets back into the act," not because of its initially intended role (registering the electron's momentum), but because of its importance in the physics of the pointer-environment interaction. The argument applies not only to apparatuses' pointers, but much more widely to all kinds of macrosystems that have a small number of macroscopic or slow degrees of freedom (like the position of the center of mass) that interact with an environment with very many microscopic or fast degrees of freedom. The macrosystem can even be tiny, such as a dust particle. Plausible models of many such cases lead to the macrosystem's state being very close to a mixture of position eigenstates.

Furthermore, the decoherence process is so efficient that in many models the macrosystem's state becomes amazingly close to the right mixture amazingly fast. Thus Erich Joos and Heinz Dieter Zeh[9] consider a dust particle (radius10^{-3} cm) in air and examine the interference terms distinguishing a superposition of two positions for the center of mass of the dust particle, which are just 10^{-4} cm apart, from the corresponding mixture. They show that these terms converge to zero like $\exp\{t/10^{-36}\text{ sec}\}$ and remain small for very long time-scales (10^{10} years)!

What do these results imply about how to solve the measurement problem? First, a warning: the above description of decoherence is simplified. A more precise description would acknowledge that there is in general no natural unique definition of 'macrosystem', since there will be rapid and efficient decoherence for a wide range of definitions. Similarly, there is in general no natural unique definition of the relevant 'final state', since decoherence does not "turn off" but continues to keep the interference terms (the tails) small.[10] Decoherence does not provide a *basis* of states of the macrosystem, for specifying a nonmaximal quantity—say the position of the center-of-mass of a pointer—leaves open a great deal about the pointer's other degrees of freedom (e.g., the positions of its individual atoms). Nor does decoherence even precisely specify a nonmaximal quantity. The requirement that interference terms be small (or soon be made small, and stay small, etc.) is of course vague, and so singles out only a vague range of quantities on the macrosystem.

Second, for all decoherence's efficiency, the mixture it provides is still not "interpretatively right"—it is improper. This point is not at all affected by the complications mentioned above. Nor is it contentious: some authors suggest that we

---

[8] The name arises from the use of 'coherence' to describe phenomena that manifest interference terms, such as when a state is a superposition rather than a mixture.

[9] Erich Joos and H.D. Zeh, "Emergence of Classical Properties through Interaction with the Environment," *Zeitschrift für Physik B* 59 (1985): 223–35.

[10] Because position values are a continuous rather than discrete set of real numbers, only approximate eigenstates of position are possible. One can argue that to recover a definite macrorealm one needs not just any such approximate eigenstates of position, but well-behaved states—with respect to other quantities such as momentum—that evolve according to approximately classical laws of motion. Indeed, there are such states (*coherent states*) that can be obtained by decoherence. In some models, at least, the components of the macrosystem's mixture are very close to coherent states.

should speak not of "decoherence," but of "diffusion (into the environment) of coherence." To solve the measurement problem, we must still choose between our initial alternatives: either justify the collapse of the wavepacket (applied to macrosystems) or ascribe extra values.

Third, however, decoherence is clearly relevant to the solution. Exactly how it is relevant will vary from one solution to another, but there are two general points important for my discussion of the Everettian strategy. As I remarked earlier, a solution to the measurement problem need only recover classical physics' assertion of definite values as an approximation. The astonishing numbers involved in decoherence liberate us from the traditional aim of obtaining macrosystems having definite values for *exactly* the familiar quantities like position. Once we see how rapid and ubiquitous decoherence is, we should be willing to accept that, rather than position itself, some other extremely close quantity is definite in value. Surely nobody can be so certain that *exactly* position, as against a "close cousin," is definite. Perhaps we should even go further and accept a vague range of quantities, the definiteness of any one of which would give a definite macrorealm. If we propose to ascribe extra values to position for macrosystems, then orthodox quantum theory on its own brings us close to our goal. Roughly speaking, we have only to go a bit further, from an improper mixture to a proper one. This will be what the Everettian strategy proposes.

5 Two Choices Yield Four Strategies

I turn now to a survey of strategies for solving the measurement problem. We have seen that one has to choose between justifying the collapse of the wavepacket and ascribing extra values to quantities. Let us make this a little more precise by introducing labels. We have to choose between:

Dynamics: secure a definite macrorealm by denying the Schrödinger evolution of a strictly isolated system, proposing instead new dynamical laws that can evolve a pure state (superposition) into a (proper, i.e., ignorance-interpretable) mixture.

ExtraValues: keep the Schrödinger evolution of a strictly isolated system and solve the measurement problem by ascribing to certain quantities some values beyond those prescribed by the eigenvalue-eigenstate link (while avoiding the various no-hidden-variable proofs).

But we must also make another choice independent of this one, so there will be four broad strategies to choose from. In my description of the measurement problem so far, I have assumed that the macrorealm was definite: macroscopic objects have definite values, if not for position then at least for very "close" quantities. But on reflection, this assumption is debatable. One could take the view that the evidence in its favor (the empirical success of classical physics with its assumption of definite values for all quantities) ultimately boils down to our *experience* of the macrorealm being definite.[11] Perhaps we should retrench and allow the solution to the measurement problem to become disjunctive: we need either to secure a definite macrorealm or to explain why it seems definite to us.

It is worth putting this point in terms of the general scientific enterprise of accounting for our experience of the world. The discussion thus far has envisaged

[11] Of course, one might retort that this view depends on a discredited foundationalist epistemology—but let us follow the line of argument.

that once a definite macrorealm is secured, an account will be given of why we experience it as definite. It is no problem to admit that such an account will be highly complicated (as shown by the complexity of theories in perceptual psychology) because nothing about it need be quantum-theoretic. Once a definite macrorealm is secured, some sort of nonquantum account (no doubt rooted in classical physics and physiology) can "take over the job" of accounting for experience. On the other hand, once we entertain the idea that what we experience as a definite macrorealm is actually indefinite, we will be forced to include some kind of quantum-theoretic weirdness in our account of experience.

Again, it is worth introducing labels for these two alternatives:

DefMac: solve the measurement problem by securing a definite macrorealm and expect some sort of classical psychophysics to account for why the macrorealm also *appears* definite to us.

DefApp: allow an indefinite macrorealm and solve the measurement problem by securing only that it *appears* definite—and so expect some sort of quantum psychophysics.

This choice is independent of the choice between *Dynamics* and *ExtraValues*. All four combinations are logically coherent strategies for solving the measurement problem; indeed, all are advocated in the literature. Of course, since the two choices and ensuing four strategies are cast in broad terms, there will be considerable variety within each strategy; and since the strategies are cast in rather vague terms, some views might perhaps be classifiable under more than one strategy.

The rest of this essay will concentrate on the Everettian version of the fourth strategy, which combines *ExtraValues* and *DefApp*. In particular, I will focus on an example of this strategy articulated by Saunders and Wallace. (Some Everettians, Saunders and Wallace included, could perhaps also be classified under *DefMac*.) In the remainder of this section I shall briefly review the other three strategies in order to help locate the Everettian strategy in the landscape of possible ontologies for quantum theory.

Dynamics & DefMac. Obvious examples of this strategy are the proposals of Nicolas Gisin, GianCarlo Ghirardi, Philip Pearle, and Ian Percival, all of whom propose precise equations for the "collapse of the wavepacket" of an isolated system in order to secure for the outcome of a measurement an ignorance-interpretable mixture with the correct (Born-rule) probabilities for its components.[12]

An assessment of these models might begin by asking about the prospects for relativistic versions. So far, the best-established models are nonrelativistic. These models also determine that the (pure) quantum state of an individual system at the end of a measurement is not an exact eigenstate of the preferred quantity that according to the model "controls the collapse." The state also has tails, small non-zero components for other values of the quantity. Is this a defect? Or is it enough to say that we simply learn that the quantity we originally preferred in setting up our model is not exactly the one that has definite values? (I myself answer "No" and "Yes," respectively.)

Dynamics & DefApp. The obvious examples of this strategy are the proposals of Wigner and more recently Stapp that mind (or consciousness) produces the

[12] For an exposition emphasizing interpretative issues, see GianCarlo Ghirardi et al., "Describing the Macroscopic World: Closing the Circle in the Dynamical Reduction Program," *Foundations of Physics* 25 (1995): 5–40.

collapse of the wavepacket.[13] Although they do not have well-developed models of the collapse, they suggest that the usual Schrödinger evolution, which holds throughout the physical realm, is broken only at the interface of brain and mind. Once the mind sees a measurement result, the measured system's (and pointer's) superposition is replaced by an eigenstate corresponding to the result seen. I admit to giving this sort of proposal very little credence, though it does have the merit of instantiating my general theme that we should be willing to be radical when interpreting quantum theory![14]

I should explain why in my taxonomy these proposals exemplify *DefApp* rather than *DefMac*, thus clarifying a contrast with the Everettian strategy. Admittedly, once consciousness has "done its stuff" and collapsed the wavepacket, the macrorealm on this account really is definite (for the quantities of which the collapsed state is an eigenstate). The components of the initial superposition corresponding to the values that are not seen by consciousness *really* disappear. Thus, it is not just a matter of the utter impracticality, thanks to decoherence, of confirming their existence by experiments whose statistics manifest the interference terms. I nonetheless classify these proposals as examples of *DefApp* because their version of definiteness is in general limited and temporary. After consciousness "stops looking," the usual Schrödinger evolution takes over again. This leaves no guarantee that the macrorealm will stay definite (in the favored quantities) as time goes on. This temporary definiteness of the macrorealm marks a major difference from the Everettian strategy, which as we shall see takes the definiteness of the macrorealm to be *always* an illusion.

So much for strategies that adopt *Dynamics*. I now move to those that adopt *ExtraValues*. Any such strategy must address two related issues about its postulated extra values: how they solve the measurement problem (either by *DefMac* or by *DefApp*), and how they evolve in time. Though I will emphasize the first question, the second is also crucial. The values must not evolve in such a way as to undermine the solution to the measurement problem they initially provided!

ExtraValues & DefMac. The obvious examples of this strategy are the pilot-wave interpretations initiated by de Broglie and Bohm between the 1920s and 1950s and developed by others since then.[15] According to these interpretations, a quantity is preferred *a priori* and thus always has a definite value (the preferred quantity is position in elementary quantum theory and field configuration in quantum field theory). The Schrödinger evolution of the quantum state prescribes a deterministic evolution of this quantity's definite value, solving the measurement problem (at all times!). Since Cushing provides a detailed account of these interpretations in this

---

[13] Eugene Wigner, "Remarks on the Mind-Body Problem," in *The Scientist Speculates*, I.J. Good, ed. (London: Heinemann, 1962); Henry Stapp, *Mind, Matter and Quantum Mechanics* (New York: Springer-Verlag, 1993).

[14] Butterfield, "Worlds, Minds and Quanta"; idem, "Quantum Curiosities of Psychophysics," in *Human Consciousness and Identity*, J. Cornwell, ed. (Oxford: Oxford University Press, 1996), 123–59, esp. 151–6.

[15] See, for example, David Bohm and Basil Hiley, *The Undivided Universe: An Ontological Interpretation of Quantum Theory* (London: Routledge, 1993); Peter Holland, *The Quantum Theory of Motion* (Cambridge: Cambridge University Press, 1993); and, especially for historical and philosophical issues, James Cushing, *Quantum Mechanics: Historical Contingency and the Copenhagen Hegemony* (Chicago: University of Chicago Press, 1994).

volume, I will only expand briefly on my earlier remark that they are perfectly tenable. Again, doing so will clarify a contrast with the Everettian strategy.

First, there are pilot-wave interpretations that can plausibly claim to be empirically adequate. They recover the well-confirmed empirical predictions of quantum theory and even relativistic quantum theories, including quantum field theories. However, they lack a feature that physicists tend to value highly: fundamental (as against phenomenological, or average) Lorentz covariance.

The Everettian strategy promises to retain fundamental covariance, but this is not to claim an indisputable advantage. The pilot-wave theorist can reply that quite apart from empirical adequacy, their interpretations also have some valuable *theoretical* features beyond being deterministic (a traditional desideratum for fundamental physics). For example, they account for quantum theory's curious nonlocal phenomena by appealing to a type of action-at-a-distance that is no more incomprehensible than Newton's gravity (though of course with the key difference that the strength of the quantum force does not decrease with distance). Rather than attempt to decide between the Everettian and pilot-wave theorist, it is enough here to note with Cushing that while one can legitimately reject pilot-wave interpretations—say, because one thinks the lack of fundamental Lorentz covariance is a disadvantage weightier than all the advantages—one cannot do so on empirical grounds.[16]

6 The Everettian Strategy

ExtraValues & DefApp. The Everettian strategy keeps the Schrödinger evolution of a strictly isolated system and allows an indefinite macrorealm, but it posits extra values to secure definite appearances. As we shall see, one can think of these values as in some sense perspectival (or even "mental"). Pioneered by Everett, this strategy encompasses several different interpretations.[17] Here I will focus on an interpretation elaborated by Saunders, which he calls "relativism" (or "relational quantum theory"), and a related interpretation of Wallace.[18] As I mentioned at the outset, Deutsch, Gell-Mann and Hartle, Vaidman, and Zurek have advocated similar interpretations.[19]

---

[16] Other examples of the *ExtraValues & DefMac* strategy include the modal interpretations proposed in the 1980s by such authors as Jeffrey Bub, Dennis Dieks, Richard Healey, and Simon Kochen. In fact, there are connections between these and pilot-wave interpretations; cf. Jeffrey Bub, *Interpreting the Quantum World* (Cambridge: Cambridge University Press, 1997). Whether these interpretations can succeed on issues such as empirical adequacy and Lorentz covariance is still an open question. The "state of the art" is represented by recent books such as Dennis Dieks and Pieter Vermaas, eds., *The Modal Interpretation of Quantum Mechanics* (Dordrecht: Kluwer, 1998) and Pieter Vermaas, *A Philosopher's Understanding of Quantum Mechanics: Possibilities and Impossibilities of a Modal Interpretation* (Cambridge: Cambridge University Press, 2000).

[17] For a fine survey, see Jeff Barrett, *The Quantum Mechanics of Minds and Worlds* (Oxford: Oxford University Press, 1999). I have discussed some of these interpretations under the banners "many worlds" and "many minds" in my "Worlds, Minds and Quanta," and "Whither the Minds?".

[18] Saunders, "Time, Quantum Mechanics, and Decoherence"; idem, "Time, Quantum Mechanics, and Tense"; idem, "Relativism," in *Perspectives on Quantum Reality*, Clifton, ed.; idem, "Time, Quantum Mechanics, and Probability"; Wallace, "Worlds in the Everett Interpretation"; idem, "Thoughts on Quantum Mechanics and vice versa."

[19] David Deutsch, *The Fabric of Reality* (London: Penguin, 1996); Murray Gell-Mann and James Hartle, "Quantum Mechanics in the light of Quantum Cosmology," in *Complexity,*

Although I do not have the space to discuss these various authors, I should mention that all seven share the following two main ideas (about which some other Everettians would disagree): (i) they secure the appearance of a definite macrorealm by appealing to decoherence to specify appropriate states of macrosystems (including tiny dust particles); (ii) they accept with equanimity that vagueness about decoherence—the smallness of an interference term (How small? How soon?)—makes for vagueness about what is definite. In order to secure definite appearances, they contend that it is perfectly reasonable to resort to approximate, indeed anthropocentric or perspectival, notions. In particular, it is enough to specify a vague range of quantities, the definiteness of any one of which would give definite appearances.

For the sake of clarity, in this section I will explain several basic ideas shared by all the Everettian interpretations. In the next two sections, I will distinguish between such interpretations as "many worlds" and "many minds."

At its simplest, the Everettian strategy proceeds as follows. One assumes that the entire universe has a presumably pure quantum state Ψ (the assumption could in fact be generalized to mixtures). Ψ always evolves according to the Schrödinger equation—after all, the universe is the only truly isolated system! Although Ψ can be expressed as a superposition of states in myriadly many ways, the measurement problem suggests it will be a superposition corresponding to many different definite macrorealms. As stated, this suggestion is vague, since *definite macrorealm* is vague. But the measurement problem suggests that however one makes it precise, or at least more precise (e.g., by appealing to decoherence), Ψ will have a component (i.e., a summand with nonzero coefficient) corresponding to many different definite macrorealms. At this point, the Everettians propose their breath-taking main idea: all these definite macrorealms are actual!

This idea needs much more clarification. Indeed, Everettian interpretations differ so greatly in how they clarify it that the common slogan "the various definite macrorealms are actual" is ambiguous. In this section I will give four clarifications that apply to all versions of the Everettian strategy. In sections 7 and 8 I will distinguish different Everettian interpretations. Section 7 concerns the ways in which Everettians use Everett's original idea of relative states to define their 'extra values' (the 'definite macrorealm'). Section 8 concerns the treatment of time, including the evolution of macrorealms over time. Though I think this order is optimal, it is not perfect because the various aspects of the discussion are of course related. In particular, postponing the treatment of time to section 8 means that until then we risk losing sight of the fact that: (i) 'macrorealm' should be defined so that macrosystems remain suitably stable over time; (ii) the Everettian strategy, unlike pilot-wave interpretations, allows a fundamentally relativistic dynamics.

Actuality. First, I should relate these different but actual macrorealms to the standard philosophical framework of possible worlds in Leibniz and modern modal metaphysics. Recall that while each such world is a logically possible total course of history, one such world is actual. Propositions including laws of nature (if such

---

Entropy and the Physics of Information, W. Zurek, ed. (Reading: Addison-Wesley, 1990), 425–59; Lev Vaidman, "On Schizophrenic Experiences of the Neutron, Or Why we Should Believe in the Many-Worlds Interpretation of Quantum Theory," *International Studies in Philosophy of Science* 12 (1998): 245–61; idem, "The Many-Worlds Interpretation of Quantum Mechanics," *The Stanford Encyclopedia of Philosophy* (2000), only on the Internet at http://plato.stanford.edu/.

there be) are made true at worlds. Let us assume from now on (with touching faith!) that the actual world obeys the laws of quantum theory, including the usual Schrödinger equation. More precisely, the actual world throughout all time is described by the sequence of states $\Psi(t)$ for all t. This assumption simply sets aside quantum theory-violating worlds, which are irrelevant for investigating the "world-picture" of quantum theory. On the Everettian account, the actual world contains very many macrorealms that are all definite (e.g., for the position of some pointer) but different from one another (e.g., regarding which position the pointer is in). So according to philosophers' standard usage of 'world', the name 'many worlds' is a misnomer.

However, some Everettians including Saunders prefer to say that the different macrorealms are all 'real', reserving 'actual' for the macrorealm 'we are in' (in a sense yet to be clarified!). I agree that this conforms at least equally well with everyday usage, but I have preferred to align my use of 'actual' with philosophers' current use of 'actual world'. (Some Everettians also use 'multiverse' for the one actual quantum cosmos, containing many macrorealms or 'worlds'.)

Branching. One might object that it is impossible for a pointer to have different values of its position or, more generally, for a system to have two or more values for a quantity. Such a constellation of different macrorealms cannot be in a single possible world. But the Everettian replies that she is postulating a hitherto-unnoticed parameter, relative to which objects have properties. Contradiction is avoided, since the object has the contrary properties relative to two different settings of the parameter. This reply is similar to the familiar way in which philosophers say that a change of properties over time involves no contradiction, since the object has the contrary properties relative to two different times. (The analogy with time will be developed at the end of section 7 and in section 8.)

The literature uses various words, such as 'perspective', 'branch', and 'world', for this new parameter—or rather, for its values—but none is dominant. In fact, none is ideal. 'Perspective' can sound too subjective and 'branch' suggests the indeterministic and thus possibly misleading metaphor of a tree-trunk splitting into branches over time (with the branches representing alternative possible histories). In this section and the next one, I shall for the most part use '(definite) macrorealm', world', and 'branch' synonymously, each being defined only at an instant. Section 8 will attempt to clarify these terms further.

I should, however, note a helpful terminological proposal by Wallace, who suggests using 'branch' for what I have so far labeled a '(definite) macrorealm'. Together with Saunders, he takes this notion to be approximate, to be defined by decoherence, but to be in part anthropocentric, to be temporally extended (not defined only at an instant), and to evolve stably. Because he also advocates that an arbitrary resolution of the universal quantum state Ψ is perfectly legitimate (though usually useless), he proposes to use 'world' for the corresponding aspects of the multiverse (i.e., of the philosophers' 'actual world'). On this view, worlds are precise and defined arbitrarily, but only for an instant. But beware! Saunders' and Wallace's views are very close, but Saunders tends to resist talk of 'branches' and 'worlds'. (There is even more variety of usage in the literature. For example, Vaidman uses 'world' for what is roughly Wallace's 'branch'.)

Links. Note that *relative to a branch*, the Everettian strategy retains the orthodox eigenvalue-eigenstate link. Precisely because a summand of Ψ is an eigenstate of a quantity (e.g., the position of some pointer), Everettians take the branch described by that summand to be definite for that quantity.

Indefiniteness. I have said previously that the Everettian strategy adopts *DefApp* rather than *DefMac*, and so has an indefinite macrorealm. This may now seem odd, since I have just said it involves a plethora of definite macrorealms. But this is a spurious conflict, a mere matter of words. We can either say there is one indefinite macrorealm that has perspectives or branches, relative to which there is definiteness, or we can say that there are many definite macrorealms. The original formulation of the measurement problem as a threat of indefiniteness prompts the first usage, while the Everettian approach prompts the second. In short, having many different definites (relative to different settings of a parameter) is one way of being indefinite.

Among the various Everettian proposals for how to make the term 'definite macrorealm' (or 'world' or 'branch') precise, some are mentalistic in a more radical sense than Saunders and Wallace (who do, remember, define branches in an anthropocentric way) would allow. They endorse *DefApp* more radically by explicitly invoking a notion of mind (or some very similar notion, e.g., sentience) and hence are called many-minds interpretations. For them, 'definite macrorealm' (or 'world' or 'branch') requires definiteness only of what appears to minds (or sentient beings). On these accounts, a macrosystem such as a pointer can be indefinite in position, even *within* a branch. All that is required is that *if* it appears to minds, it appears to them to be definite in position. This could be secured by ascribing definite values—relative to a branch—to appropriate quantities defined on brains (such as quantities that underpin states of perceiving a pointer to have a definite position).[20]

7 Defining the Branches

I now address what is perhaps the central question of this essay: How should the Everettian strategy define 'definite macrorealm' (or 'world' or 'branch')? This question is of course the analogue of the questions about which quantity is preferred that confronted the other views in section 5. Accordingly, it might be called the "preferred quantity problem." Unfortunately, it is more often called the "preferred basis problem," suggesting (wrongly) that the Everettian needs to specify not just a quantity, but a basis (or equivalently, a maximal quantity, i.e. a quantity all of whose eigenvalues are nondegenerate). Indeed, 'preferred quantity' also suggests that the eigenvalues, not just the orthogonal family of eigenspaces, matter to the Everettian strategy—which is presumably false. A phrase like 'preferred decomposition' might be even better, but I shall stick with the more familiar 'preferred quantity'.

This is a multi-faceted question, to which Everettians have given very different answers. Any full answer must of course include a discussion of how macrorealms (branches) evolve over time. But this topic can be postponed until section 8 because the common formal notion of *relative state*, introduced by Everett himself, can be explained independently of the treatment of time. (Indeed, the notion is itself needed for the discussion of time.) Before explaining the notion of relative state, and how different Everettians propose to fill it out, I need to draw attention to a general philosophical issue about vagueness.

7.1 Is Vagueness Acceptable?

One naturally requires a precise definition of 'macrorealm', 'world', or branch', which entails specifying a preferred quantity (as the pilot-wave and modal

[20] The conception of the macrorealm as indefinite within a branch is analogous to the Wigner-Stapp view discussed in sec. 5.

interpretations do). Since the Everettian proposes that each world is actual, what actuality itself consists of surely cannot be a vague matter. Yet it turns out that Everettian definitions of 'macrorealm', world', or 'branch' are usually vague.

Is a vague definition somehow acceptable? Saunders and Wallace answer "Yes." Saunders' reluctance to talk of worlds or branches, for example, reflects his belief that they are only vaguely definable (in part because they are anthropocentric notions). Wallace's "Yes" pertains to branches, which he defines vaguely because anthropocentrically, but not to worlds, which are precisely defined. As the more anthropocentric of the two, Saunders holds views similar to those of Gell-Mann and Hartle, both of whom relate 'branch' to the existence of stable records and thus to organisms or other complex macrosystems that exploit such records to survive—topics I will touch on again later.

To clarify this issue, I need to formulate two ways in which vagueness is acceptable. First, vagueness about precise matters is of course acceptable if it is simply a matter of having a working definition that one plans to make precise.[21] But vagueness can also be acceptable in a different way. If an explanandum is vague, the explanans can be vague, provided its vagueness does not undermine its explanatory role. Indeed, in this case vagueness would be an advantage if it corresponded to the vagueness of the explanandum, such that precise versions of the explanandum were satisfactorily explained by corresponding precise versions of the explanans.[22]

These points apply to the Everettian project of explaining the appearance of the familiar definite macrorealm (*DefApp*). First, the explanandum at issue is indeed vague—there is no precise agreement about exactly which systems in a "macrorealm" have definite values for exactly which quantities or about what an appearance of such requires. Rather, there is a cluster of approximate notions and doctrines relating them, including the example I have concentrated on: macrosystems having approximately definite positions. But one can instead emphasize the *appearance* of definiteness, and therefore consider a more anthropocentric explanandum (as with many minds). Nor do we *need* to have precise agreement about this explanandum. Recall (from section 2) that the empirical predictions of a superseded theory are often only approximately recovered by its successor.

Second, Everettians' explanans, i.e., their definitions of 'macrorealm', are often vague. As we shall see in more detail below, there have been two main Everettian approaches: (i) only what appears to minds is definite (many minds) and (ii) all macrosystems are definite (many worlds). Within the many-worlds approach, there have been two ways of understanding macrosystems. One way (ii-a) takes *apparatus* and *pointer-position* as primitive macroscopic notions, while the other (ii-b) defines macrosystems in terms of decoherence. Below I shall concentrate on (ii-b), an approach to which Saunders and Wallace belong. Suffice it to say here that (i) and (ii-a) clearly use vague notions, such as mind and apparatus, respectively.

---

[21] Of course one might complain that the various Everettian attempts over the past forty years to develop precise definitions have all been found wanting. A recent salient example, concerning the consistent-histories approach advocated by Saunders, can be found in Fay Dowker and A. Kent "On the Consistent-Histories Approach to Quantum Mechanics," *Journal of Statistical Physics* 82 (1996): 1575–646.

[22] "Corresponding vaguenesses" is familiar in the philosophy of language. Thus David Lewis, *Counterfactuals* (Oxford: Blackwell, 1973), chap. 4.2, defends his use of the vague notion of similarity in his proposed truth-conditions for counterfactual conditionals by maintaining that the kinds of vagueness correspond: resolving the vagueness of these conditionals' truth-conditions amounts to resolving the vagueness of similarity.

We saw at the end of section 4 that *macrosystem* and *decoherence* are equally vague notions.[23] In particular, how close to diagonal must the density matrix come (the size of the tails), and in which basis must the density matrix be nearly diagonal (the quantity selected by decoherence)? Broadly speaking, these kinds of vagueness *do* correspond to different ways of making precise the explanandum. Moreover, the ubiquity and astonishing efficiency of decoherence means that for all macrosystems the tails will very quickly become very tiny, and the selected quantity will be very nearly unique. Accordingly, this kind of vagueness will be unnoticeable by the standards of precision usual for macroscopic physics. So for (ii-b), Everettian vagueness is acceptable (at least while the physics of decoherence is still being explored) because it corresponds to vagueness in the explanandum, and here we seem to have met the proviso that the explanans' vagueness not undermine its explanatory role.

7.2 Relative States and Preferred Quantities

For any pure state of a composite system consisting of two component systems, and for any quantity on the first component and any eigenspace of it, there is a unique *associated state* of the second called the *relative state*. The idea of *associated state* can be developed from the orthodox interpretation that the state prescribes probability distributions for measurement results, though the idea does not rest on this interpretation. In these terms, 'associated' means that immediately after measuring the given quantity on the first component and obtaining the result corresponding to the given eigenspace, a measurement of any quantity on the second component will have the probability distribution prescribed by the relative state. In short, after measuring the first component the second component "looks as if" it is in the relative state.

I should clarify four points (in ascending order of importance).

1. The first/second contrast here is obviously immaterial: we can similarly define the relative state of the first component for a given eigenspace of the second.

2. For readers familiar with quantum theory, it may be helpful to put the idea of relative states in terms of the formalism. For simplicity's sake, I shall take the given eigenspace as one-dimensional, i.e., choose a given state on the first component. Recall that any pure state of the composite can be written as a superposition of product states in arbitrary bases on the components, i.e., $\sum_{ij} c_{ij} |\phi_i\rangle |\psi_j\rangle$. Given this initial state, we define for the given state $|\phi_i\rangle$ of system 1, the relative state of system 2 to be $\sum_j c_{ij} |\psi_j\rangle$ (up to normalization).

3. I have defined the relative state in terms of the state of the composite system *at a time*, and to that extent the idea of a relative state is not relativistically invariant. But relativity requires only that there should be a suitable meshing of the state-evolutions associated with two different foliations of spacetime. As I will mention in section 8, one can give a foliation-independent generalization of relative states using the local algebras approach to quantum theory, in which quantities are associated directly with spacetime regions—an approach advocated by Saunders.

4. Because the notion of relative state assumes a composite system with two components, one might think that this decomposition has to be in some way given *ab initio*—independent of the dynamics. But this is *not* the case. No matter how one

---

[23] The detailed physics of decoherence is also an active research area, so the first point above, that vagueness can be a matter of working definition, also applies.

defines the composite system and its decomposition into components, the notion of relative state applies. This is important for a reason often ignored in discussions about interpreting elementary quantum theory: according to quantum field theory—our best theory of matter—particles are excitations of underlying fields. Thus, the very existence of these discussions' systems depends on what the state is. Two morals ensue: the notion of relative states nevertheless applies; and irrespective of this notion, we must accept state-dependent (and so dynamics-dependent) definitions of physical systems.

Let us now see how Everettians use the notion of relative state to define their notion of 'world' (or 'branch' or 'macrorealm'). They will differ over how to split up (*factorize*) the Hilbert space for the universe into Hilbert spaces of component systems and over which quantity to select on a given component. As I mentioned in the previous section, there are two main approaches. One is naturally called "many minds," since it specifies the preferred components and quantities in broadly mental terms (like "mind" and "seeing the pointer to be in a definite position").[24] The other is naturally called "many worlds," since it uses broadly physical terms (like "apparatus" and "pointer-position") to address the questions of factorization and quantity selection.

Within the second approach one finds two views of *macrosystem*: the many worlds-apparatus view and the many worlds-decoherence view. The more traditional many worlds-apparatus approach assumes a primitive notion of apparatus (a distinguished set of subsystems of the universe) and a distinguished quantity on such an apparatus (e.g., position of the apparatus's pointer). This approach was developed in the 1960s and 1970s, before work on decoherence took off in the 1980s. More recently, the many worlds-decoherence approach has appealed to decoherence in order to avoid assuming these primitive notions. Again, this means that it assumes the notions of macrosystem (perhaps dynamically defined) and of a dynamically-induced preferred quantity (typically, a quantity "close to" position). The relativism of Saunders and Wallace is a version of this third approach.[25]

In order to explain the third approach, I first need to sketch the more traditional many worlds-apparatus approach, which will face objections regarding vagueness and the lack of a definite macrorealm. The many worlds-decoherence approach will claim to avoid these objections, in part by appealing to anthropocentric notions.

In sketching the many worlds-apparatus approach, it will be clearest to focus first on the idealized case of a single apparatus, together with the rest of the universe. The Everettian proposes that a branch is specified by an eigenspace of pointer-position on the apparatus, together with the relative state of the rest of the universe (defined by Ψ) for this eigenspace. Within the branch, the rest of the universe *is* in the relative state and, using the eigenvalue-eigenstate link within a branch, has values for exactly those quantities of which the relative state is an eigenstate.

---

[24] I have discussed this approach in "Worlds, Minds and Quanta," secs. 8–10, and "Quantum Curiosities of Psychophysics."

[25] This classification should of course be qualified. Although Everett's essay is regarded as initiating many-worlds interpretations—"many minds" being a much more recent phrase—several commentators maintain that he may well have intended to express a many-minds view. Also, some versions of the many worlds-decoherence approach, including Saunders' and Wallace's version, appeal to anthropocentric considerations in their accounts of macrosystems, bringing them closer to the many-minds approach.

However, this proposed definition of a branch seems inadequate, for it seems unlikely to secure definiteness (in a branch) across the whole macrorealm for those quantities that appear definite to us. Why should the relative state for my pointer being here be an eigenstate of my table being there and of the Moon being way over there? The Everettian, of course, takes the case of a single apparatus to be an unrealistic simplification. Supposing there are N apparatuses, she will define a branch as follows: (i) the state-space of the universe is to be factorized into N + 1 factors, one factor for each apparatus plus one for the rest of the universe; (ii) a branch is specified by an N-tuple of eigenspaces of pointer-position on the N apparatuses, together with the relative state of the rest of the universe for this N-tuple. One simply treats the N apparatuses collectively as the first component system in the definition of relative state.[26]

This approach faces various objections,[27] but here I will focus on two which the many worlds-decoherence approach claims to avoid. The first concerns vagueness. It seems wrong to base one's interpretation of our fundamental physical theory on vague notions. Indeed, commentators have criticized the appeal to "apparatus" and "pointer position" as vague and have even called this appeal question-begging with regard to the measurement problem.[28] One might also object that apparatuses are still too few and far between to secure a definite macrorealm (on a branch). Why should the collective relative state for all these pointers being here and there be an eigenstate of the Moon being way over there? In defense, the many worlds-apparatus Everettian might allow such indefiniteness even within a branch, and this would bring her closer to the many-minds approach to defining branches.[29]

In the simplest version of the many worlds-decoherence approach, all systems that decohere in interaction with their environment (*macrosystems* in the sense of section 4) constitute preferred subsystems. "Macrosystem" is vague, since "decoherence kind of interaction" is vague. But it certainly includes all familiar macroscopic objects—from tiny dust particles to larger objects like apparatus pointers, tables, cats, and people. The approach takes as the preferred quantity on each macrosystem that given by the components of the macrosystem's improper mixture after the decoherence interaction. That is, the quantity has as its eigenspaces the spaces in terms of which the improper mixture (the macrosystem's density matrix) is very nearly block-diagonalized as a result of the decoherence interaction. This orthogonal family of spaces is often called the *decoherence basis*, with *basis*

[26] More precisely, not every N-tuple need define a branch—only those that have a non-zero coefficient in the expansion of Ψ in terms of these N + 1-fold products.

[27] For example, this definition of a branch seems to assume a division of the world into N apparatuses and 'the rest', independent of the state and of dynamics—contradicting point (4) above. However, it might well be able to accommodate (4) by giving a more sophisticated account of 'apparatus'.

[28] See, for example, Howard Stein, "The Everett Interpretation of Quantum Mechanics: Many Worlds or None?" *Nous* 18 (1984): 635–52, esp. 646–7; John Bell, *Speakable and Unspeakable in Quantum Mechanics* (Cambridge: Cambridge University Press, 1987), 96–7, 124–26; Adrian Kent, "Against Many-Worlds Interpretations," *International Journal of Modern Physics A* 5 (1990): 1745–62, esp. 1754.

[29] Cf. my comment on *Indefiniteness* near the end of sec. 6. This approach would also be similar to the way in which the Wigner-Stapp version of the *Dynamics* and *DefApp* strategy envisages indefinite macrorealms; cf. sec. 5.

now being understood more loosely to allow multi-dimensional subspaces as well as rays.[30]

The many worlds-decoherence Everettian now proposes that a branch is specified by taking for every macrosystem a component of the improper mixture that is its state, together with the relative state of the rest of the universe (defined by Ψ). The proposal is, therefore, similar to the many worlds-apparatus approach's invocation of N apparatuses to define a branch. The difference is that N is much greater because now all macrosystems, not just all apparatuses, are subsystems.

What about the two objections against the many worlds-apparatus approach mentioned above? Regarding vagueness, the amazing efficiency and speed of decoherence mean that the preferred quantity is far more exactly specified. Regarding the second objection, designating all macrosystems, not just all apparatuses ("N being much greater"), as preferred subsystems eliminates the risk of the collective relative state representing an indefinite macrorealm.

Returning to the idea of anthropocentrism, I should again note that some proponents of the many worlds-decoherence approach—Saunders and Wallace among them—appeal to anthropocentric notions in specifying preferred systems and the preferred quantity on them. Thus, my simple version of the many worlds-decoherence approach—decoherence applied to all macrosystems—is set aside. Any such anthropocentric appeal will of course lessen the gap between this approach and the many-minds approach.[31]

7.3 Saunders and Wallace

At long last, I turn fully to "my" authors, Saunders and Wallace. I shall first briefly discuss their appeal to anthropocentric notions. I shall then present their arguments against making a once-and-for-all precise definition of 'branch' or similar words, i.e., of preferred subsystem and/or preferred quantity.[32]

---

[30] This specification of the decoherence basis is vague in three related ways: regarding the definition of *decoherence kind of interaction*, regarding when this interaction (which of course never "turns off") has "done its stuff" by making off-block-diagonal terms small enough, and regarding which of the many different quantities that nearly block-diagonalize the density matrix to choose. Vagueness apart, though, the quantity thus specified is typically "very close to" a position-like quantity, such as the position of the macrosystem's center of mass.

[31] Perhaps the best-known example of such an appeal is the work of Gell-Mann and Hartle, who adopt the consistent-histories approach—which I have eschewed not least because it builds in time from the beginning (see sec. 8). I cannot here explain their definition of 'consistent-history space' (or its special case, 'quasiclassical domain'), which is the analogue of 'branch' (or 'world' or 'macrorealm'). As I have mentioned, they relate decoherence to the production of stable records, and thereby to organisms or other macrosystems that exploit such records to survive (they call such systems IGUSes—information gathering and utilizing systems); see, for example, Gell-Mann and Hartle, "Quantum Mechanics in the Light of Quantum Cosmology."

[32] As I mentioned previously, Saunders casts his views in terms of consistent histories. Indeed, on the topic of defining 'branch'—or better, the consistent-histories analogue—his views are close to Gell-Mann's and Hartle's, which explains at least part of his resistance to talk of 'branches' and 'worlds'; Saunders, "Time, Quantum Mechanics, and Decoherence," 240–44, 251–53; idem, "Relativism," 128–32. He also casts his views in terms of the local algebras approach to quantum theory, which brings to the fore the idea of spatiotemporally bounded events, together with the correlations between them—topics that I will return to briefly in sec. 8.2. Wallace, on the other hand, does not adopt the consistent-histories

According to Wallace's terminology, 'branch' stands for the sort of approximate anthropocentric notion he and Saunders agree is needed to secure *DefApp*. He roots this notion in decoherence theory and perhaps also, in sympathy with Saunders, in the theory of consistent histories. He also requires that branches, so understood, should be temporally extended rather than defined only at an instant and should evolve stably. But exactly how branches relate to decoherence, consistency, and stability is complicated and controversial—not least because the physics of decoherence is still being explored.

Wallace also agrees that it is perfectly legitimate to resolve the universal quantum state Ψ into any basis (and to factorize the universe's Hilbert space arbitrarily). He emphasizes that although Ψ is specified (and so the universe completely described) by any such resolution, almost all such resolutions are useless. That is, they do not clarify any empirical facts, whether microscopic or macroscopic. Physicists would call such resolutions "unphysical" or "a bad choice of variables." (In spite of Saunders' preference for the consistent-histories approach, he would no doubt agree with these points.) But since all such resolutions of Ψ are perfectly legitimate (though usually useless), it is convenient to have a word for the corresponding 'aspects of the quantum universe (multiverse)'. Wallace proposes 'world'. So in his terminology, worlds, unlike the branches, are precise, defined arbitrarily, and in general defined only at an instant.

I cannot here explore the physics of decoherence, but only discuss Saunders' and Wallace's general doctrines about how the branches secure definite appearances. Generally, they claim that (i) we probably will never get a once-and-for-all precise definition of 'branch' or similar words, i.e., precise definitions of 'preferred subsystem' and/or 'preferred quantity' that are once-and-for-all best for securing definite appearances, but that (ii) we do not need such definitions! In what follows, I will assume with Wallace that the notion of 'world' is best taken as precise, arbitrary, and in general instantaneous. The debate, then, is about whether 'branch' can or should be defined as a precise, but also nonarbitrary and temporally extended, notion.

First, one should *not* take position as the precise preferred quantity. At first sight, position is the obvious candidate, since the measurement problem originally threatened to prevent macroscopic objects from being in localized states. However, there are technical and conceptual reasons against it. As to technical reasons, I think the lack of exact eigenstates, which arises even in elementary quantum theory, is not so important. More important, the concept of position in relativistic quantum theories, including quantum field theory, is problematic.[33] One cannot reply to these difficulties simply by saying that for quantum field theory the preferred basis is the basis of definite field configurations (which is indeed a natural analogue of elementary quantum theory's notion of particle position). This basis is also unsuitable for describing macroscopic objects because the particles that they comprise are really superpositions of these basis states. (Recall from section 7.2 that the very existence of such "basic" systems is a state-dependent matter.)

---

approach. His views are generally closer to Zurek's and Deutsch's than are Saunders'.

[33] See Gordon Fleming and J. Butterfield, "Strange Positions," in *From Physics to Philosophy*, J. Butterfield and C. Pagonis, eds. (Cambridge: Cambridge University Press, 2000), 108–65, for a review of the issues—both for relativistic quantum particles and quantum field theory—in flat spacetime. In more general spacetimes, the problems are worse.

From a conceptual point of view, choosing position prevents any object of our familiar ontology, such as dust-particles, apparatus pointers, and people, from being an 'inhabitant' in a branch. It suggests instead that any such object is a collective feature of many branches. If we define branches in terms of the position of each and every microscopic constituent, not just a coarse-grained notion like the position of the center of mass, then the state of any familiar object resolves into simultaneous position eigenstates of all its constituents. But because the chemical bonds that hold molecules together involve delocalized electrons, the object's electrons for any such position eigenstate are in a superposition of binding and not binding pairs of atoms. The eigenstate would represent an unstable "atomic soup" rather than a state of the familiar object. Conversely, any state of the familiar object depends on interference between many different branches. In light of this, it seems wrong to say that the object nevertheless exists in each such branch. Instead, we should say that the object, like its state, is some kind of collective feature of the branches. (Here again, the idea that the existence of what we normally consider a system is in fact state-dependent.)

One can generalize this conceptual reason against choosing position. Other apparently natural, precise specifications of the preferred quantity (e.g., energy) equally prevent ordinary objects being in a branch and make them instead collective features of many branches because interference between branches thus defined is important to the processes that maintain the object in existence. Presumably, the only way to avoid this problem is to make the preferred quantity some exact *decoherence basis*, i.e., some member of the vaguely delimited class of quantities selected by decoherence (and perhaps also satisfying some other constraints). But as I mentioned in section 4, the physics of decoherence is much more subtle and varied than my summaries suggest. Even the vaguely delimited class varies considerably across different kinds of physical system (e.g., dust-particle in air or outer space) or regimes for a system (e.g., high or low temperature). Thus, it might be unduly restrictive to specify an exact decoherence basis in terms of coherent states or localization in phase space, important though these notions are. On the other hand, to say only that the preferred quantity is "some exact decoherence basis" is in effect a vague promissory note about future research on decoherence. (Of course, this vagueness is intended to be temporary in the sense of section 7.1.)

Saunders and Wallace conclude from these difficulties that we should liberally accept resolutions of the universal quantum state Ψ into any arbitrary basis—or at least an arbitrary basis that is a fine-graining of "the" decoherence basis. In Wallace's terminology of 'worlds', they consider continuously many bases (even if they restrict themselves to bases that fine-grain "the" decoherence basis), and so commit themselves to continuously many worlds.

This is certainly a dizzying ontology. After all, each of these continuously many worlds is "inhabited." Each world is not just a component of a state (where states represent reality) but also has a "system" (albeit not an ordinary object!) actually in it. They would still have continuously many worlds even if for each basis they said—as they do not—that only one world is "inhabited." But this dizzying ontology—this dizzying pill for curing us of the measurement problem—is sugared by three other considerations.

The first is related to the interpretation of probability and is due to Saunders rather than Wallace (though I believe they agree on the point). Some Everettians propose that each component of Ψ in a preferred quantity is inhabited by a whole population of systems. Only with a population, they say, can we make sense of various different probabilities for future alternatives. Given a population, more of them can transit to future alternative A than to B—which makes sense of A being more probable. This proposal is largely independent of whether one takes the traditional view that there is just one preferred quantity or Saunders' and Wallace's "liberal" view that there are many.[34] This proposal is also independent of whether any such preferred quantity is specified in physical terms (many worlds) or mental terms (many minds).[35]

But Saunders and I (and, I believe, Wallace) reject this proposal as a naïve attempt to ground probability in counting measures, or in some version of the principle of indifference. These are both dead horses in the philosophy of probability—which I will not flog.[36] Suffice it to say here that I believe instead that the Everettian can maintain that there is only one system in each world, and that systems indeed "split" when the number of worlds increases (cf. section 8 for further discussion). The probability proposal clearly does *not* sit well with the Everettian idea that all the worlds are actual, "inhabited." If, according to this proposal, there is supposedly no splitting, but only a stochastic process continually re-partitioning an overall population of persisting systems, it is hard to see why the Everettian should postulate more than one system, i.e., more than one "inhabited" world.[37]

The second bit of sugar comes from what I identified earlier as the conceptual reason against specifying position as the preferred quantity, namely, that any familiar object is a collective feature of many worlds. Putting this the other way around, the "system" inhabiting a world is not a familiar object, unless the world is defined by one of the (vaguely delimited!) class of decoherence bases (in Wallace's terminology: unless the world is a branch).

The third bit of sugar comes from an analogy between worlds and the spacelike slices of relativistic spacetime, conceived on the *tenseless* or *block universe* conception of time. This analogy builds on the analogy between worlds and times I gave in section 6 (*Branching*) in response to the charge of contradiction. Just as someone who accepts the tenseless conception of time can readily accept instants (spacelike slices of spacetime) as useful and even indispensable for describing phenomena, and yet not involving any substantive ontological commitment beyond a commitment to spacetime, so also an Everettian can readily accept worlds as useful

---

[34] I say "largely," rather than "entirely," because of a point stressed by Meir Hemmo and I. Pitowsky, "Probability and Nonlocality in Many-Minds Interpretations of Quantum Mechanics," in preparation. They point out that if there is more than one preferred quantity, and these quantities do not commute, then one will have to allow a sort of contextuality about how systems are assigned to the various populations in order to prevent commitment to a violated Bell inequality.

[35] This proposal is made by advocates of Worlds and Minds approaches; for example, respectively, David Deutsch, *The Fabric of Reality* (London: Penguin, 1996), and Michael Lockwood, " 'Many-Minds' Interpretation of Quantum Mechanics," *British Journal for the Philosophy of Science* 47 (1996): 159–88.

[36] Saunders, "Time, Quantum Mechanics, and Tense," sec. 6; idem, "Time, Quantum Mechanics, and Probability"; Butterfield, "Whither the Minds?", sec. 5.

[37] This, I believe, lies behind the suggestion that Everettians can and should avoid this postulate, opting instead for the pilot-wave interpretation and its postulate of just one actually possessed value of the preferred quantity; cf. Bell, *Speakable and Unspeakable in Quantum Mechanics*, 97, 133.

and even indispensable, and yet not involving any substantive commitment beyond the commitment to describing actuality by the universal state.[38]

Recall that a classical relativistic universe can be described as a collection of instants (spacelike slices) and their temporal relations to each other. There is great arbitrariness in specifying these slices (i.e, how spacetime is foliated). Any foliation breaks some symmetries of the spacetime. But any foliation, together with its slices' temporal relations, gives a complete description of the universe. In everyday circumstances, or in treating some specific problem in physics, there is an approximately defined best choice of foliation—though the details are always arbitrary, especially regarding how to foliate spacetime far away (in space or in time) from the region containing the events one aims to describe.

In the Everettian analogue, a quantum universe *at a time* can be described as a collection of worlds with amplitudes c_i. There is great arbitrariness in specifying these worlds (i.e., in specifying a basis, or more generally, a quantity). Any specification breaks some symmetries of the quantum universe at the time. But any choice of basis, together with its elements' amplitudes, gives a complete description of the quantum universe at the time. In treating some specific problem in quantum physics (and in everyday circumstances, to the extent that quantum theory is needed for them!), there is an approximately defined best choice of basis, or more generally, quantity—though the details are always arbitrary, especially regarding how to specify the quantity for events other than those one aims to describe.

As the parallel wordings show, this is a telling analogy. It suggests that Everettian worlds should be as innocuous to someone who "takes quantum theory seriously" as instants are according to the tenseless conception of time.

The analogy has other facets as well. I shall briefly mention two here. First, just as the tenseless conception of time must reconcile its account with the intuitive but problematic "flow of time," so also the Everettian must reconcile her account of a deterministically evolving Ψ with the intuitive but problematic notion of probability. Second, the treatment of time by classical physics and metaphysics provides an analogue to the idea that any familiar object is a collective feature of many worlds (which militated against specifying position as the preferred quantity and suggested that the inhabitants of worlds were "unfamiliar systems"). The analogue is clear: the states of objects, as we conceive them both in everyday life and in science, are usually not instantaneous. We say that the gun fires, the vial breaks, the cat falls—each requires the object to persist for some (vaguely defined!) period of time. Nor are mental states instantaneous. To see the color yellow, an organism must be conscious for approximately a fifth of a second. Instantaneous states are a minority, and this in spite of the supreme success of the concept of instantaneous state in classical physics.[39] Most of what we call *states of objects* make no sense from the perspective of an instant. These two points raise the issue of time and thus bring us to our final topic, the Everettian treatment of time.

---

[38] This rich and multi-faceted analogy is developed in detail in Saunders, "Time, Quantum Mechanics, and Decoherence," secs. 1, 7; idem, "Time, Quantum Mechanics, and Tense"; Wallace, "Worlds in the Everett Interpretation," *Studies in History and Philosophy of Modern Physics* (2001): forthcoming. As Saunders notes, it goes back at least to Robert Geroch, "The Everett Interpretation," *Nous* 18 (1984): 617–33.

[39] Indeed, it has been so successful that nowadays physicalism is often combined with my point about mental states to argue against mental states supervening on instantaneous physical states.

8 Time

The Everettian view of time involves both philosophical and technical issues.[40] In section 8.1, I state another facet of section 7.3's closing analogy, which leads to related philosophical issues about identity through time. I conclude in section 8.2 with a few remarks about dynamics.

8.1 Identity through Time

From the classical spacetime side of the analogy, we have no exact criterion of identity over time for familiar objects like dust particles, tables, cats, and people. Nevertheless, we can make do both in everyday and scientific life—everywhere, that is, but in philosophy's puzzle-cases of teleportation, fission, brain-transplant, etc.—with vague and approximate criteria of identity. So also, says the analogy, the Everettian can admit to having no exact criterion of identity over time for systems that are inhabitants of worlds, the vast majority of which, as we have seen, are not familiar objects. But the Everettian can make do with vague and approximate criteria of identity.

To explain this, recall first that ever since section 6, I have defined Everettian worlds only at an instant. Although 'world' (and equally 'branch', 'system' and 'inhabitant') suggests persistence over time, the discussion hitherto has not needed any Everettian account about what such persistence involves—in philosophers' jargon, what the criteria of identity for worlds are. The present suggestion, endorsed equally by Saunders and Wallace, is that the Everettian need not give a uniform account for all worlds (or branches, etc.). Nor need she give a precise but non-uniform account that first divides worlds into different kinds, and then gives different exact accounts for different kinds.

Indeed, for Saunders and Wallace, she *should not* do so. For as we have seen, they allow arbitrary resolutions of Ψ to define worlds, almost all of which are "unphysical." These worlds will be useful for describing macroscopic facts only if they are a decoherence basis, i.e., some member of the vaguely delimited class of quantities selected by decoherence (they could perhaps also be moderate fine-grainings or coarse-grainings of a decoherence basis). Therefore, we should not expect a useful notion of persistence to be definable for arbitrary worlds.

On the other hand, a few of these decoherence bases do describe familiar macroscopic objects, from dust particles to people. So we *should* expect to be able to define persistence for the worlds given by these resolutions—or rather, we should be able to do this vaguely and approximately, just as we do for dust particles and people. In short, persistence makes no sense for arbitrary worlds, but it does make vague and approximate sense for macroscopic objects.

All this I find plausible. But the topic of criteria of identity for worlds over time also raises issues having to do with the "splitting" of worlds over time, which are not part of section 7.3's analogy. Although the Everettian defines worlds only relative to an instant, she must accept different numbers of worlds at different times because the number of worlds changes whenever an eigenstate unitarily evolves into a superposition. For example, in my toy-model of measurement (section 3), there was one world before the measurement interaction and two after. From this understand-

[40] The following discussion complements my "Worlds, Minds and Quanta," secs. 6, 7, and "Whither the Minds?", sec. 4.

ing of measurement comes the widespread use of the word 'branch', based on the image of a tree trunk splitting into branches. But how exactly should we understand such splitting?

I agree that Everettians who hold that a whole population of systems inhabits each component of Ψ in a preferred basis can avoid the number of worlds changing in time. For they can say that different members of such a population evolve to different later components. Members simply "go different ways" during a so-called "splitting." But in section 7.3, I joined Saunders and Wallace in rejecting the idea that each component is inhabited by a whole population because it seems motivated only by a bad reason—the attempt to ground probabilities in counting, or in some version of the principle of indifference. Let us assume, then, that at any one time there is only one inhabitant in each world. How should we understand splitting on this view?

Saunders has plausibly addressed this question, partly by comparing it with the question—much studied in philosophical discussions of identity over time (especially personal identity)—of how to understand cases of fission of ordinary objects (including people), and partly by connecting it to his discussion of probability.[41] Yet there is one aspect of the question I must address since, although it is independent of the discussion about vague criteria of identity, it is so central to the topic of identity over time.[42] Essentially, it turns on an analogy between Everettian worlds and a well-known philosophical dispute about the identity of objects over time.

Recall that philosophers treat change of properties over time in two rival ways, depending on whether they conceive objects that persist over time as strictly self-identical across time, or as having stages (also known as temporal parts). Let us adopt a recent jargon used in the dispute over temporal parts[43] and say that an object that persists by having temporal parts at various times 'perdures', while an object that is wholly present at different times 'endures'. The neutral word, 'persistence', covers perdurance and endurance.

These distinctions suggest similar ones for worlds. Should we say that one object (the pointer) has contrary properties (two non-overlapping positions) relative to distinct worlds or branches? Or should we say that there are multiple objects, similar to one another (hence often called *copies*) but with different positions?

Most Everettians, including Saunders and Wallace, opt for the copies view. I agree that there are good reasons to do so, reasons that hold even if one resists the analogous idea of temporal parts. But it is important to take note of the one-object view as well.[44] In particular, we should beware of two *bad* reasons for adopting the copies view. First, this view is for many Everettians simply a corollary of their view (which Saunders, Wallace, and I reject) that each component of Ψ in a preferred basis is inhabited by a whole population of systems. Note that the converse fails: the copies view does not require their view. This leaves Saunders and Wallace free to combine the copies view and genuine splitting. There is no conflict between this

---

[41] Saunders, "Time, Quantum Mechanics and Probability," secs. 4–6.

[42] In fact, it is more closely related to the basic analogy between worlds and times from section 6 (*Branching*) that it is to the relativistic analogy from section 7.3.

[43] David Lewis, *On the Plurality of Worlds* (Oxford: Blackwell, 1986), 202.

[44] Paul Tappenden, "Identity and Probability in Everett's Multiverse," *British Journal for the Philosophy of Science* 51 (2000): 99–114, suggests that we can think of an object relative to a world as a new kind of part of the object, which he calls a 'superslice'.

combination and conservation laws, such as the conservation of mass or charge. We should *not* say that a pointer's mass is doubled when it splits during a measurement-process into two pointers with non-overlapping positions. After all, we do not say in the temporal analogue that the advocate of perdurance is committed to the existence of a two kilogram mass just because she believes that a one kilogram pointer has two temporal parts.

Second, most discussions ignore the point that on any exact definition of 'world', ordinary objects are not features of one world, but collective features of many worlds (section 7.3). Indeed for the sake of clarity, I just ignored this point myself when I presented the choice between the one-object and copies views as a choice between one or two pointers. Once we notice this point, the one-object view becomes more plausible. If an object is "extraordinary," why not let it have contrary properties relative to different worlds (or even corresponding parts as Paul Tappenden advocates)?

What is a good reason for adopting the copies view? It seems to me that the contrary properties involved in two alternative outcomes of a measurement process might well be vastly more disparate than two non-overlapping positions—so much so that it seems wrong to say a single object is involved. I have in mind not just such pairs of properties as being alive or dead—as in the case of Schrödinger's proverbial cat—but more extreme cases such as when one of two measurement outcomes vaporizes the pointer. What is the common object then?

You might reply that we should identify an object with its microscopic constituents, so that in the case of vaporization we could trace these in space and time. But this suggestion faces objections from both philosophy and physics. From a philosophical perspective, the suggestion wrongly implies that the pointer survives vaporization. From the physical point of view, because quantum field theory treats particles as excitations of fields, our attempts to make 'microscopic constituents' precise must face the fact that the very existence of the particles we take as constituting objects depends on the quantum state (of the field-system)—and this state can vary between Everettian worlds in such a way that a particle exists in one but not the other of two worlds after a split.

8.2 Dynamics

I turn finally to what an Everettian should say about dynamics. I will just make brief remarks, first about determinism, and then about relativistic covariance.

Let us distinguish between the dynamics of the universe as a whole, and the dynamics of its subsystems. Of course the detailed physics of the universe as a whole is impossibly complex and will surely forever remain beyond our ken. But the fact that the universe is an isolated system (strictly speaking, the only one) means that according to orthodox quantum theory—and the Everettian—it, and it alone, obeys the deterministic unitary evolution of the Schrödinger equation. On the other hand, its subsystems are not isolated. They are "open," able to be influenced by their environment, so that their evolution is nonunitary. In general, the theoretical description of their evolution is much more complex than that of isolated systems.

As is often remarked, Erwin Schrödinger and the other pioneers of quantum theory were lucky that nature supplied subsystems of the universe (namely, atoms) that were both sufficiently isolated to make the Schrödinger equation an excellent approximation, and sufficiently simple to allow at least some aspects of their behavior to be understood (at least by a pioneer of genius!). Isaac Newton and his contemporaries were likewise lucky that nature supplied sufficiently isolated and simple subsystems of the universe—the ball on the inclined plane, the planets, etc.

Thus we return to our opening theme of determinism versus indeterminism. For the Everettian, deterministic evolution of the whole universe is compatible with indeterministic evolution of generic subsystems—which is only to be expected since the subsystems are open—and with almost deterministic evolution of subsystems that are almost isolated (e.g., a molecule in a gas). How we are to recover these two types of evolution from an over-arching unitary evolution is of course an ongoing research area in physics, in which decoherence will undoubtedly play an important role.

I turn finally to relativistic covariance. It is of course intended as a principal merit of Everettian interpretations that they obey relativistic covariance fundamentally rather than solely at a phenomenological level (unlike the pilot-wave interpretation, and perhaps also the modal interpretation and theories with a physical process of wavepacket collapse; cf. section 5). But again, it is important to distinguish the dynamics of the universe as a whole from the dynamics of its subsystems. Everettian discussions almost always focus on the relativistic covariance of the former. As I shall spell out in a moment, I see no problems of principle lurking there. But since the relativistically covariant theories we in fact confirm are theories of subsystems of the universe (and very simple and tiny ones at that!), the Everettian also owes us an account of how her postulated relativistically covariant evolution of the universe yields relativistically covariant evolution for appropriate subsystems—an important lacuna, albeit perhaps one easily filled. Let me end more positively, though, by outlining how Everettian dynamics of the universe as a whole can be fundamentally relativistically covariant, thereby returning to Saunders.

Relativistic covariance is obscured by the traditional way of presenting the Everettian strategy—which I adopted in sections 6 and 7—in terms of relative states defined by the instantaneous state Ψ. But for three reasons (in increasing order of technicality) there is no real problem here. The first two in effect give us different notions of relativistic covariance. First, a fundamentally relativistic quantum theory has no problem using foliations of spacetime with states Ψ associated with slices, provided it allows arbitrary spacelike foliations, such that slices of different foliations are suitably related (in particular by a unitary transformation). Second, the local algebras approach to relativistic quantum theories, which associates an algebra of quantities with each bounded spacetime region, lends itself easily to foliation-independent notions and techniques. Third, the consistent-histories approach to all quantum theories (including relativistic ones) affords a generalized and foliation-independent notion of relative state.

Much more can be, and has been, said about these three reasons, and in particular about the relations between them. Here I can note only that Saunders combines these last two approaches—local algebras and consistent histories—so that the projectors defining the histories become elements of local algebras. In particular, he defines a relation of *relative definiteness* between histories H_1 and H_2: roughly, "H_1 has probability 1, given that H_2 is realized." He then discusses the relation of the consistency condition on histories to properties like the symmetry and transitivity of relative definiteness.[45] For now, this summary must suffice as a way of suggesting that there are good prospects for formulating the Everettian strategy in relativistic quantum theories, and in particular in the local algebras approach.

---

[45] Saunders, "Time, Quantum Mechanics and Decoherence," secs. 2, 3; idem, "Relativism," sec. 3.

Acknowledgment. Many thanks to the editors and conference participants, and to Meir Hemmo, Adrian Kent, Paul Tappenden, and Lev Vaidman—and especially to Simon Saunders and David Wallace—for discussion and comments on previous versions. If only I could have consistently acted on all suggestions!

A more detailed version of this essay is available on the Internet at: http://philsci-archive.pitt.edu and http://xxx.lanl.gov.

THE TANGLED STORY OF NONLOCALITY IN QUANTUM MECHANICS[1]

Michael Redhead

1 Introduction

There are many purported proofs of nonlocality in quantum mechanics. A representative selection of such proofs will be critically examined so as to expose as precisely as possible the assumptions on which they are based. In order to keep the discussion to a reasonable compass the detailed treatment will be restricted to phenomena involving so-called entangled states, i.e. many-particle states that cannot be expressed as a simple product of one-particle states, but only as a superposition of such products. (In this sense this essay is less wide-ranging than Raymond Chiao's contribution to this volume.)

My essay is predicated on the assumption that nonlocality in the sense of instantaneous causal action-at-a-distance is to be avoided if at all possible, on pain of violating at any rate the spirit of special relativity. It is, of course, controversial as to what exactly special relativity does prohibit. I take the view that relativity is more than just a phenomenological invariance principle, and is essentially grounded in the causal structure of spacetime.[2] Thus there are essentially three principles at issue in discussing possible conflicts with special relativity.

Firstly, there is the First Signal Principle (FSP), which asserts that we cannot transmit information in the form of a signal faster than the speed of light in vacuo. Then there is the Invariance Principle (IP), which claims the invariance of physical theories under Lorentz transformations, i.e., if any physical process is allowed then so is any Lorentz-boost of that process. Finally, there is what I will call the Philosophically Grounded Invariance Principle (PIP), which asserts that causal influences cannot operate outside the light-cone and thus provides a causal underpinning for the characteristic light-cone structure of Minkowski spacetime.

Note the following logical relationships between these principles:

$$\text{IP} \nrightarrow \text{FSP} \quad (1)$$

and

$$\text{PIP} \rightarrow \text{FSP}, \quad (2)$$

but

$$\sim\text{PIP} \nrightarrow \sim\text{FSP}. \quad (3)$$

In my view PIP is the basis of special relativity. ~ PIP means that causal influences operate at space-like separation, i.e. propagate faster than light, but in view of (3) this is perfectly consistent with the FSP, i.e. no signal can be sent faster than light. The reason for this is that to constitute a signal the causal process must be suitably controllable, and this may be inherently impossible in the quantum situation. Some

---

[1] This essay is an updated and re-evaluated version of material originally presented in Michael Redhead, "Nonlocality and Quantum Mechanics," *Proceedings of the Aristotelian Society*, supp. vol. *65* (1991):119–40 and Michael Redhead and Patrick La Rivière, "The Relativistic EPR Argument," in *Potentiality, Entanglement, and Passion-at-a-Distance: Quantum Mechanical Studies for Abner Shimony*, vol. 2, R. Cohen, M. Horne, and J. Stachel, eds. (Dordrecht: Kluwer,1997), 207–15.

[2] This view is defended, for example, in Michael Redhead "Relativity, Causality and the Einstein-Podolsky-Rosen Paradox: Nonlocality and Peaceful Coexistence," in *Space, Time and Causality*, R. Swinburne, ed. (Dordrecht: Reidel, 1983):151–89.

authors such as David Bohm have advocated interpretations of quantum mechanics that do exactly that, namely, allow superluminal causal processes but not superluminal signaling. If special relativity entails FSP but not PIP, then Bohm works. But if—as I believe—special relativity entails the stronger PIP as well, then Bohm is definitely inconsistent with special relativity.

The Bohm interpretation is also deterministic. It can therefore be objected to on two counts:

1. the *scientific* count of fudging the relativity issue;
2. the *theological* count of not allowing "room" for divine action.

Point 2 seems prima facie right, but is of course disputed by philosophers and theologians defending a so-called compatibilist position between determinism and the possibility of divine action, free will, etc. I shall not go into these arguments here, but simply assume for purposes of exposition an incompatibilist stance.

I shall therefore be looking particularly at indeterministic settings and examining the question of whether proofs of objectionable forms of nonlocality can be blocked, thus remedying the two defects in Bohm-type interpretations noted above (despite their undoubted attraction in terms of picturability—trajectories—and the absence of a measurement problem). I also examine recent demonstrations of nonlocality in the context of relativistic quantum mechanics. I conclude that an indeterministic framework in which measurements actualize potentialities (as advocated by Abner Shimony, for example) offers the best prospect for avoiding the twin objections to the Bohm approach.[3]

2 The EPR and Bell Arguments

It is often maintained that quantum mechanics is obviously "nonlocal," in that it treats of extended wavefunctions, so an electron in a hydrogen atom, for example, is in some sense everywhere at once![4] But things are not that simple. There are three issues involved: (a) Are electrons "spread out"? (b) If they are, do they nevertheless have local properties, such as propensities to be at particular locations, the "spreadingoutedness" reflecting the fact that a whole spectrum of such properties can be possessed by the electron at the *same* time? (c) Can these local properties "act" on one another instantaneously at a spatial separation (speaking nonrelativistically for the time being)?

The puzzle starts with (a). When you measure the position of an electron you find it somewhere, so it is tempting to think that the sense of "nonlocal" in (a) reflects an ignorance of where the electron is, rather than an "ontological" spreading out. But if you answer "yes" to (a) then (b) also gets a "yes," so the properties themselves are not "spread out." (c) trivially gets the answer "no" if you answer "no" to (a), but "yes" if you go along with "yes" to (a) and (b). This is really the famous issue of the projection postulate. If I change the propensity "here" to one, the propensity everywhere else goes to zero, and that looks like action-at-a-distance.

Then, what about the famous two-slit experiment? If I answer "yes" to (a), there does not seem to be any *extra* nonlocality involved in the effect of opening and

[3] However, cosmological considerations tell against such an approach and may favor the realist-holistic treatment, as I advocate in *From Physics to Metaphysics* (Cambridge: Cambridge University Press, 1995). This will be discussed further in section 6 below.

[4] Indeed, the "range" of a wavefunction may be much longer than the "range" of interaction, as in the case of so-called Efimov states in nuclear physics.

shutting the second slit. If I answer "no" to (a) then it is not clear whether any instantaneous effects can be propagated—it depends on the detailed "hidden-variable" dynamics of the experiment.

So, it is all very confusing. Clarification began in 1935 with the Einstein-Podolsky-Rosen (EPR) argument[5] for the incompleteness of the quantum formalism. The EPR argument deals with the quantum mechanics of *two-body* systems. This opens up the possibility of a much more precise discussion, but also new sorts of issues arise: (A) If you have two distinguishable particles in quantum mechanics, like an electron and a proton, does each of them have individual properties?[6] If the answer to that is "yes" then we can go on to ask the question: (B) Can these properties be affected nonlocally by events occurring even at space-like separation? EPR assumed the answer to (A) was "yes" and proceeded to pose a dilemma: the formalism of quantum mechanics implies either that:

(B) gets answered "yes"

or

(a) gets answered "no."

EPR assumed (B) is answered "no," and hence argued that (a) gets the answer "no."

The crunch came in 1964 when John Bell showed[7] that, if the answer to (a) is "no," then the answer to (B) is "yes"! So the conclusion of the two arguments put together is: the answer to (B) is "yes." But since the argument is predicated on assuming that the answer to (A) is "yes," the conclusion should really be rendered as: the answer to (B) is "yes," *or* the answer to (A) is "no." Here, the first disjunct (B) offers us action-at-a-distance between individual properties. The second disjunct (A) offers us "nonseparability," a holistic involvement of the two particles, in which not all their properties can be attached individually to each particle. But unfortunately that is not the end of the story, because Bell's argument brought in some other assumptions, call them collectively P, so the conclusion is really: either P is false, or we have nonlocality (in the precise sense that we have either action-at-a-distance *or* nonseparability). So if we want to avoid the unpalatable conclusion of nonlocality then we can always escape by denying P. So we had better delve into P. We must be a *little* more precise about the details of Bell's argument.

3 Presuppositions of Bell's Argument

We have introduced (a) in the version: Are electrons spread out? But the question is much more general. Do superposed states "spread" properties generally among the properties appropriate to the component states? So far we have talked about position, but for a general observable a, in a superposition of eigenstates of the associated operator A, what can be said about the value of a? The original EPR argument did apply to the positions of two particles and purported to show that in an appropriately chosen superposed state, the positions of the particles were sharp, i.e.,

---

[5] Albert Einstein, Boris Podolsky, and Nathan Rosen, "Can Quantum-Mechanical Descriptions of Reality be Considered Complete?", *Physical Review* 47 (1935): 777–80.

[6] In the case of indistinguishable particles question (A) becomes more controversial since the observables of quantum mechanics are restricted to symmetric functions of the particle labels. See Steven French and Michael Redhead, "Quantum Physics and the Identity of Indiscernibles," *The British Journal for the Philosophy of Science 39* (1988): 233–46.

[7] John Bell, "On the Einstein-Podolsky-Rosen Paradox," *Physics* 1 (1964): 195–200.

not in any sense spread out. But in 1951 Bohm[8] developed a version of the EPR argument that used the spin components of two spin-½ particles projected along a particular direction, say the Z-direction, in a state, the so-called singlet state, involving a superposition of products of eigenstates of the spin-projections along the Z-axis, and demonstrated that, even in such a superposition, the spin-projections of each particle had a definite sharp value.

Bell employed the Bohm version of the EPR argument, but now considered spin-projections for the two particles along *different* directions. Assuming that correlations between spin-projections predicted by quantum mechanics could be expressed as a weighted average over the products of the spin-projections for *given* values of the hidden variables and the values of the spin-projections were *determined* by the "hidden" variables describing the "complete" state of the source, he derived an inequality, the so-called Bell inequality between correlation coefficients, which was violated by the predictions of quantum mechanics, and also, as it turned out, by experiment.

So one can now identify two elements in the additional assumptions P, used by Bell in his original paper to arrive at his conclusion.[9] First, a probability structure was laid out, exhibiting the values of the spin-projections as random variables over a classical probability space equipped with a probability measure expressing our ignorance as to the exact hidden state of the source. Secondly there was an assumption of determinism. In particular, the first assumption identified a joint probability distribution for the values of spin-projections along different directions for the *same* particle although, according to the usual interpretation of the uncertainty principle, such components for the same particle could never be measured simultaneously. We shall call this particular assumption JD (for joint distribution).

Two major controversies in the literature have arisen out of Bell's argument:

1. Can one get the argument off the ground if one denies JD?
2. Can one get the argument off the ground if one denies determinism?

With regard to JD, things looked promising when a proof of the Bell inequality was developed by Henry Stapp and Phillipe Eberhard,[10] which did not use the machinery of formal probability spaces at all, but expressed the relevant correlations directly in terms of limiting frequencies in a long run of repetitions of the experiment. Making appropriate randomness assumptions about these sequences, it appeared that one could derive the Bell inequality without committing to JD.[11]

---

[8] See David Bohm, *Quantum Theory* (Englewood Cliffs, N.J.: Prentice-Hall, 1951), 614–9.

[9] Bell, "On the Einstein-Podolsky-Rosen Paradox."

[10] Henry Stapp, "S-Matrix Interpretation of Quantum Theory," *Physical Review D* 3 (1971): 1303–20, and Phillipe Eberhard, "Bell's Theorem without Hidden Variables," *Il Nuovo Cimento* 38B (1997): 75–80.

[11] This point was particularly stressed by Redhead, "Relativity, Causality and the Einstein-Podolsky-Rosen Paradox"; idem, *Incompleteness, Nonlocality and Realism: A Prolegomenon to the Philosophy of Quantum Mechanics*, 2nd revised impression (Oxford: Clarendon Press, 1989). But meanwhile Arthur Fine, "Hidden Variables, Joint Probability and the Bell Inequalities," *Physical Review Letters* 48 (1982): 291–5, had claimed to show that satisfaction of the Bell inequality was a *sufficient* condition for JD to obtain. Fine's result has been generalized to consider the conditions under which any set of experimental probability distributions can be returned as marginals on a common probability space. For a complete discussion of what has been achieved in this direction, see Itamar Pitowsky, *Quantum*

With regard to the second controversy concerning determinism, I argued against Stapp that determinism was a concealed assumption in the Stapp-Eberhard proof. The issue here related to the necessity of using counterfactual formulations in the Stapp-Eberhard proof. This turned on the question, What can one say about the possible result of a measurement that one could have performed but did not? Could these counterfactual results for one particle vary according to what measurement was carried out on the remote particle? If the measurement results were generated indeterministically, I claimed, such variation could be routinely expected in the absence of any mysterious nonlocal influence of measurement procedures on one particle affecting the state of the remote particle.[12]

So, can one prove the Bell inequality assuming indeterminism? The answer is "yes," and such a proof was explicitly produced by John Clauser and Michael Horne[13] (who developed a general setting for ideas presented more sketchily by John Bell[14]). This proof involves giving up the idea that the measurement results are determined by the hidden variables, but allows these results to be linked stochas-

---

Probability—Quantum Logic (Berlin: Springer, 1989). Fine had claimed, contra Redhead, that the Stapp-Eberhard "proof" must have a concealed premiss equivalent to JD, so the violation of the Bell inequality could be imputed to a failure of JD, rather than any unacceptable violation of locality. The situation here was clarified in the work of George Svetlichny, Michael Redhead, Harvey Brown, and Jeremy Butterfield, "Do the Bell Inequalities Require the Existence of Joint Probability Distributions?", *Philosophy of Science* 55 (1988): 387–401, who pointed out that Fine's mathematics showed only that satisfaction of the Bell inequality was sufficient for the existence of a classical probability space returning the measured experimental joint distributions as marginals. This space would have relative frequency models satisfying JD, but there was no argument to show that the relative frequencies generated in the real world should conform to these purely mathematical models. So it was perfectly consistent to claim both the denial of JD and the satisfaction of the Bell inequality, and Fine's maneuver failed to vitiate the Stapp-Eberhard proof. Another telling objection against giving up JD is that, for states involving strict correlation or anticorrelation, joint distributions for incompatible observables on the *same* particle is a direct consequence of the correlation (anticorrelation), so a general prohibition on JD as applied to these important special cases, is not a viable option. Furthermore, Fine seems committed to the view that altering the experimental set-up for one particle from a Bell-conforming to a Bell-violating setting can mysteriously cause JD to obtain or not obtain on the distant particle. For full discussion see Svetlichny et al., "Do the Bell Inequalities Require the Existence of Joint Probability Distributions?"

[12] Stapp attempted to modify his proof to meet my objection, but it has been claimed by Rob Clifton, Jeremy Butterfield and Michael Redhead, "Nonlocal Influences and Possible Worlds—A Stapp in the Wrong Direction," *The British Journal for the Philosophy of Science* 41 (1990): 5–58, that none of Stapp's later proofs can be made to work, and that the success of the Stapp-Eberhard approach in avoiding JD was offset by its involvement with determinism. See Redhead, *Incompleteness, Nonlocality and Realism*, 90ff., for a detailed discussion. This whole topic of evaluating counterfactuals under an assumption of indeterminism has been the subject of a long-running debate between Stapp and myself. The most up-to-date list of references on the topic can be found in Rob Clifton and Michael Dickson, "Stapp's Algebraic Argument for Nonlocality," *Physical Review A* 49 (1994): 4251–6; and Henry Stapp, "Reply to 'Stapp's Algebraic Argument for Nonlocality'," *Physical Review A* 49 (1994): 4257–60.

[13] John Clauser and Michael Horne, "Experimental Consequences of Objective Local Theories," *Physical Review D* 10 (1974): 526–35.

[14] John Bell, "Introduction to the Hidden Variable Question," in *Foundation of Quantum Mechanics: Proceedings of the International School of Physics 'Enrico Fermi', Course 49*, Bernard d'Espagnat, ed. (New York: Academic Press, 1971), 171–81.

tically to the hidden variables describing the complete state of the source and/or the measuring apparatus. Since these measurement results are no longer to be thought of as random variables on a single probability space, there is no commitment to JD (further reinforcing the argument of Svetlichny, Redhead, Brown and Butterfield).[15] But in this proof a new formulation of locality has to be provided, including a potentially controversial assumption of conditional stochastic independence of measurement outcomes for the two particles—conditional, that is, on the precise specification of the hidden variables.[16]

4 The Algebraic Proofs of Nonlocality

The proofs of nonlocality discussed so far have all depended on proving a Bell inequality relating certain statistical correlations and then using the fact that this inequality is violated experimentally, and also by the predictions of quantum mechanics. A quite different approach has tried to demonstrate that local hidden-variable theories are actually self-contradictory, so we would not need to do experiments, involving inevitable interpretation, with auxiliary assumptions about the working of the measuring apparatus that can always be called in question, to avoid impugning locality. This approach is generally referred to as the algebraic approach to demonstrating nonlocality, since the contradiction demonstrated amounts to showing that two unequal numbers, such as zero and one, are equal.

An algebraic contradiction for the hidden-variable description of a single system described by a state-space of dimension greater than two had been demonstrated by Simon Kochen and Ernst Specker in 1967,[17] assuming that value assignments to

---

[15] I here correct the statement made in Michael Redhead, "Undressing Baby Bell," in *Harré and his Critics*, R. Bhaskar, ed. (Oxford: Blackwell, 1990): 122–8. See also Jeremy Butterfield, "Bell's Theorem: What It Takes," *The British Journal for the Philosophy of Science* 43 (1992): 41–83.

[16] The discussion of the locality assumptions required in the derivation of Bell inequalities in the stochastic hidden-variable framework was given a very careful formulation by Jon Jarrett, "On the Physical Significance of the Locality Conditions in the Bell Arguments," *Nous* 18 (1984): 569–89. Two locality conditions were introduced, which are now usually referred to as parameter independence and outcome independence (following Abner Shimony), These will be defined and discussed more fully below. A further twist to the argument was given in 1976 by Patrick Suppes and M. Zanotti, "On the Determination of Hidden-Variable Theories with Strict Correlation and Conditional Stochastic Independence of Observables," in *Logic and Probability in Quantum Mechanics*. P. Suppes, ed. (Dordrecht: Reidel, 1976), 445–55, who showed that conditional stochastic independence was anyway incompatible with the strict mirror-image correlations built into the singlet spin-state used in the original 1964 proof, except in the degenerate case when determinism was restored. In point of fact it can be argued that we do not need to start with the stochastic hidden-variable framework, in order to show that for the case of strict correlation or anticorrelation, locality implies determinism. We can use the EPR argument to justify the fact that the measurement results for the distant particle cannot evolve in time stochastically, since the predicted results for the distant measurement hold good for all times subsequent to the first measurement, so stochasticity fails for the distant spins after the first measurement, but, by locality, this lack of stochasticity cannot be induced by the first measurement, so there is never any stochasticity, i.e., determinism must be true. The argument was first given in Redhead, "Undressing Baby Bell" (which was written and widely circulated in 1981). See also Geoffrey Hellman, "EPR, Bell and Collapse: A Route Around 'Stochastic' Hidden-Variables," *Philosophy of Science* 54 (1987): 558–76.

[17] Simon Kochen and Ernst Specker, "The Problem of Hidden Variables in Quantum Mechanics," *Journal of Mathematics and Mechanics* 17 (1967): 59–87.

physical magnitudes obeyed the same functional relationships as the associated quantum-mechanical operators. The resolution of this paradox for a single system lay in recognizing that the values assigned to degenerate quantities (for which there may be many states associated with a given eigenvalue) depended on the context provided by different incompatible nondegenerate quantities, of which the degenerate quantity could be regarded as a function in a well-defined mathematical sense.[18]

The route to an algebraic proof of nonlocality consisted in applying these ideas to the quantum mechanics of two separated systems. Even if a quantity, like a spin-projection along a particular direction, is nondegenerate (or maximal, as it is often referred to) when referred to the state space of a single particle, it becomes degenerate (nonmaximal) when considered with reference to the joint state space of the two particles. So could it be that giving value assignments to uniquely prescribed locally maximal quantities could lead to a Kochen-Specker paradox, which would force the conclusion that locally maximal quantities depend in their value assignments on a holistic context provided by incompatible quantities maximal in the joint state-space? If all this turned out to be the case we would have a demonstration of nonseparability in a well-defined mathematical sense. The question was posed by Jeffrey Bub in 1976,[19] but the answer that finally emerged in 1980 was that no such contradiction could be derived.[20] Nonseparability was not forced on pain of algebraic inconsistency.

In 1983 Peter Heywood and I took up the challenge and proved the following result:[21]

$$\text{VR} \wedge \text{FUNC*} \wedge \text{O-Loc} \wedge \text{E-Loc} \wedge \text{Determinism} \rightarrow \text{Contradiction.} \quad (4)$$

The principles VR, FUNC*, O-Loc and E-Loc express the following ideas:

- VR (Value Rule): Values shall not be assigned to quantities that have a zero probability of turning up as measurement results.
- FUNC*: Functional relationships between maximal observables shall be preserved by value assignments.
- O-Loc (Ontological locality): Locally maximal quantities are uniquely specified, independently of a holistic context.
- E-Loc (Environmental locality): Value assignments to locally maximal quantities for one particle are not affected by any physical manipulation of the environment of the remote particle.

---

[18] Cf. Bas van Fraassen, "Semantic Analysis of Quantum Logic," in *Contemporary Research in the Foundations and Philosophy of Quantum Theory*, C.A. Hooker, ed. (Dordrecht: Reidel, 1973), 80–113.

[19] Jeffrey Bub, "Hidden Variables and Locality," *Foundations of Physics* 6 (1976): 511–25.

[20] See William Demopoulos, "Locality and the Algebraic Structure of Quantum Mechanics," 119–44, Jeffrey Bub, "Comment on 'Locality and the Algebraic Structure of Quantum Mechanics' by William Demopoulos," 149–53, Paul Humphreys, "A Note on Demopoulos's Paper 'Locality and the Algebraic Structure of Quantum Mechanics'," 145–7, all in *Studies in the Foundations of Quantum Mechanics*, Patrick Suppes, ed. (East Lansing, Mich.: P.S.A., 1980).

[21] See Peter Heywood and Michael Redhead, "Nonlocality and the Kochen-Specker Paradox," *Foundations of Physics* 13 (1983): 481–99. A similar approach was arrived at independently by Simon Kochen, although never published (private communication).

Heywood and I then showed that, modulo the other assumptions in the proof, O-Loc and/or E-Loc must be violated. Essentially, Bub's idea had been to demonstrate a violation of O-Loc on its own. This, as we have seen, could not be done, and the weaker conclusion was all that could be established. Probability enters the proof via VR, but there is no mention of statistics or correlation functions, so this first "quasi-algebraic" nonlocality proof seemed a good deal more direct than the original Bell-type proofs.[22]

A completely new turn to the developing story of nonlocality proofs was introduced by Dan Greenberger, Michael Horne, and Anton Zeilinger in 1989[23]—henceforth referred to as GHZ—who produced a new algebraic proof of nonlocality, *apparently* quite unrelated to the Kochen-Specker paradox.[24] They considered a four-body decay problem in which an initial spin-one system decayed into two spin-one systems, each of which in turn decayed into two spin-½ systems. Denoting the four emerging particles by A, B, C, and D, GHZ considered the spin-projections on directions at right-angles to a common line of flight of the decaying particles. Let $A(\theta_A)$ be the spin-projection for particle A along direction θ_A, similarly for $B(\theta_B)$, $C(\theta_C)$, and $D(\theta_D)$, then GHZ chose the quantum state so that the expectation value of the product of the four spin-projections had the following property:

$$\text{if} \quad \theta_A + \theta_B - \theta_C - \theta_D = 0\,, \; < A(\theta_A) \cdot B(\theta_B) \cdot C(\theta_C) \cdot D(\theta_D) > = -1, \tag{5}$$

$$\text{but if} \quad \theta_A + \theta_B - \theta_C - \theta_D = \pi\,, \; < A(\theta_A) \cdot B(\theta_B) \cdot C(\theta_C) \cdot D(\theta_D) > = +1. \tag{6}$$

GHZ then gave an argument to show that no local hidden-variable reconstruction of these expectation value results was possible, on pain of contradiction. The GHZ proof suffered from two defects: (1) it assumed determinism; (2) the proof, as sketched by GHZ was actually fallacious![25] (It effectively assumed that the measure of an uncountable infinity of intersections of measure-one sets was itself one!)

Both these problems were set aside in the work of Rob Clifton, Michael Redhead and Jeremy Butterfield[26]—hereafter CRB—who showed how to avoid the illegitimate assumption in the GHZ proof, and also generalized it to the framework

---

[22] The Heywood-Redhead proof has since been improved in a number of directions. Firstly, it should be noted that we found it necessary to introduce value assignments for locally nonmaximal quantities to make the proof work. Allen Stairs, "Quantum Logic, Realism, and Value Definiteness," *Philosophy of Science* 50 (1983): 578–602, showed that a more elegant proof could be given under weaker conditions, requiring only consideration of value assignments to locally maximal quantities. The Stairs proof was further examined and more carefully formulated in respect of all relevant presuppositions by Harvey Brown and George Svetlichny, "Nonlocality and Gleason's Lemma. Part 1. Deterministic Theories," *Foundations of Physics* 20 (1990): 1379–88. The work of Brown and Svetlichny brought the quasi-algebraic proofs of nonlocality to the same degree of sophistication as the Bell proofs, but, as we have stressed, with the added advantage of not requiring analysis of potentially complex correlation experiments, as in the case of the Bell approach.

[23] Dan Greenberger, Michael Horne, and Anton Zeilinger, "Going Beyond Bell's Theorem," in *Bell's Theorem, Quantum Theorem and Conceptions of the Universe*, M. Kafatos, ed. (Dordrecht: Kluwer, 1989), 69–72.

[24] But see David Mermin, "Simple Unified Form for the Major No-Hidden-Variables Theorems," *Physical Review Letters* 65 (1990): 3373–6.

[25] I stress that the GHZ proof, as published, is only a descriptive sketch, but in my opinion no proof of the sort described is rigorously possible.

[26] Rob Clifton, Michael Redhead, and Jeremy Butterfield, "Generalisation of the Greenberger-Horne-Zeilinger Algebraic Proof of Nonlocality," *Foundations of Physics* 21 (1991): 149–84.

of stochastic hidden-variable theories. The proof proceeded in two stages. First CRB showed that, subject to appropriate locality assumptions, the satisfaction of (5)-(6) required the theory to be deterministic "almost everywhere," as the mathematicians put it.[27] Secondly, CRB showed how to produce a contradiction using a finite (indeed rather small) number of settings for θ_A, θ_B, θ_C, θ_D.

One may note that the CRB contradiction involves a *fixed* setting for the fourth particle. So one suspects that a similar type of argument could be produced using a three-body rather than a four-body decay. That this is so has been shown in explicit detail by David Mermin.[28] Further discussion of experimentally realizable examples for the three-body and the four-body arrangements was given by Dan Greenberger, Michael Horne, Abner Shimony, and Anton Zeilinger.[29] These authors also independently derived the CRB contradiction.

It is also interesting to note that Mermin's proof can be generalized to the general case of N particles, and the question was posed, Does the proof go through even in the case of an infinite number of particles? This problem has been examined by Constantine Pagonis, Michael Redhead, and Rob Clifton[30] who find that the proof of nonlocality actually breaks down in the classical limit of an infinite number of particles. This confirms the intuition that the nonlocality under discussion is a peculiarly quantum mechanical phenomenon.

5 The Relativistic EPR Argument

At first glance, the realist interpretations of quantum mechanics such as Bohm's offer many advantages over standard interpretations of the theory. In particular, they give a clear, intuitive picture of many potentially paradoxical physical situations, such as the two-slit experiment and the phenomenon of barrier penetration. At the same time, their chief drawback—a form of nonlocality that seems, as we have claimed, to conflict with the constraints of relativity theory—is apparently shared by the standard, "anti-realist" interpretations that reject hidden variables and assume completeness, as was demonstrated by the original Einstein-Podolsky-Rosen argument.

However, while the Bell argument that establishes nonlocality for realistic interpretations such as Bohm's has been formulated in a relativistic context,[31] there is no well-established relativistic formulation of the EPR argument. In the absence of such a formulation, it seems hasty to conclude that the tension between the standard interpretations and relativity theory is just as great as that between Bohmian

---

[27] Compare the Suppes-Zanotti result referred to above in fn. 16 for the two spin-½ correlations.

[28] David Mermin, "Quantum Mysteries Revisited," *American Journal of Physics* 58 (1990): 731–4; idem, "Simple Unified Form for the Major No-Hidden-Variables Theorems." Mermin's proof has been reworked with more careful attention to all concealed assumptions by Adam Stocks and Michael Redhead, "A Value Rule for Non-Maximal Observables," *Foundations of Physics Letters* 9 (1996): 109–19.

[29] Dan Greenberger, Michael Horne, Abner Shimony, and Anton Zeilinger, "Bell's Theorem without Inequalities, *American Journal of Physics* 58 (1990): 1131–43.

[30] Constantine Pagonis, Michael Redhead, and Rob Clifton, "The Breakdown of Quantum Non-Locality in the Classical Limit," *Physics Letters A* 155 (1991): 441–4.

[31] S.J. Summers and R. Werner, "The Vacuum Violates Bell's Inequalities," *Physics Letters A* 110 (1985): 257–9; L.J. Landau, "On the Non-Classical Structure of the Vacuum," *Physics Letters A* 123 (1987): 115–8.

interpretations and relativity. Clearly, if a relativistic formulation of EPR could be given that did *not* entail nonlocality, anti-realist interpretations would have an advantage over the Bohmian interpretation.

Let us now investigate the possibility of a relativistic formulation of the EPR argument. First, I review the standard nonrelativistic version of the EPR argument and consider the problematics of translating it into a relativistic context, paying particular attention to the need for a reformulation of the so-called reality criterion. Then, I introduce one such reformulated reality criterion, due to GianCarlo Ghirardi and Renata Grassi,[32] and show how it is applied to the nonrelativistic EPR argument. Next, I discuss the application of the new reality criterion in a relativistic context and point out a flaw in Ghirardi and Grassi's argument that appears to undermine their conclusion of peaceful co-existence between quantum mechanics and special relativity. Finally, I engage issues related to the evaluation of counterfactuals that reveal a hidden assumption of determinism in Ghirardi and Grassi's proof, while offering a way of salvaging their conclusion.

A relativistic version of the EPR argument must differ from the nonrelativistic version in two principal ways. First, the particle states must be described by a relativistic wavefunction. The details don't concern us here; we need only require that the wavefunction preserve the maximal, mirror-image correlations of the nonrelativistic singlet state. And indeed, I have recently demonstrated the existence of maximal correlations in the vacuum state of relativistic algebraic quantum field theory.[33] Second, the argument must not depend on the existence of absolute time ordering between the measurement events on the left and right wings of the system, for in the relativistic argument these may be space-like separated. As it turns out, the nonrelativistic version of the argument *does* invoke absolute time ordering. To see how to get around this problem, we must briefly review the standard formulation of the incompleteness argument.

For EPR, a necessary condition for the completeness of a theory is that every element of physical reality must have a counterpart in the theory. To demonstrate that quantum mechanics is incomplete, EPR need simply point to an element of physical reality that does not have a counterpart in the theory. In this vein, they consider measurements on a pair of scattered particles with correlated position and momentum, but the formulation of the argument in Bohm,[34] in terms of a pair of oppositely moving, singlet-state, spin-½ decay products of a spin-0 particle, is conceptually simpler. In this case, the formalism of quantum mechanics demands a strict correlation between the spin components of the two spatially separated particles, such that a measurement of, say, the Z-component of spin of one particle allows one to predict with certainty the outcome of the same measurement on the

---

[32] GianCarlo Ghirardi and Renata Grassi, "Outcome Predictions and Property Attribution: the EPR Argument Reconsidered," *Studies in History and Philosophy of Science* 25 (1994): 397–423.

[33] Michael Redhead, "More Ado About Nothing," *Foundations of Physics* 25 (1995): 123–37. The argument actually applies more generally to any state of bounded energy; see also idem, "The Vacuum in Relativistic Quantum Field Theory," in *PSA 1994*, vol.2, D. Hull, M. Forbes and K. Okruhlik, eds. (East Lansing, Philosophy of Science Association, 1995), 77–87. For a unified treatment of the relativistic EPR and Bell arguments, see Michael Redhead and Fabian Wagner, "Unified Treatment of EPR and Bell Arguments in Algebraic Quantum Field Theory," *Foundations of Physics Letters* 11 (1998): 111–25.

[34] Bohm, *Quantum Theory*.

distant particle. This ability to predict with certainty, or at least probability one, the outcome of a measurement is precisely the EPR criterion for the existence of an element of reality at the as-yet-unmeasured particle. By invoking one final assumption, a locality assumption stating that elements of reality pertaining to one system cannot be affected by measurements performed "at a distance" on another system, EPR can establish that the element of reality at the unmeasured particle must have existed even *before* the measurement was performed at the distant particle. But the quantum formalism describes the particles at this point with the singlet state, and thus has no counterpart for the element of reality at the unmeasured particle. It follows that the quantum description was incomplete.[35] Schematically,

$$\text{Quantum Formalism} \wedge \text{Locality} \rightarrow \sim \text{Completeness.} \quad (7)$$

Alternatively, if one assumes completeness, the argument may be rearranged as a proof of nonlocality:

$$\text{Quantum Formalism} \wedge \text{Completeness} \rightarrow \sim \text{Locality.} \quad (8)$$

The problematic assumption of absolute time ordering entered the argument in the reality criterion, which turns on the possibility of predicting with certainty the outcome of a measurement along one wing *subsequent* to having obtained the result of a measurement along the other. Of course, for space-like separated events, notions like precedence and subsequence are reference-frame dependent, not absolute. So to translate the EPR argument to a relativistic context requires a modified criterion for the attribution of elements of reality that is not contingent on the time ordering of the measurement events. Ghirardi and Grassi have undertaken to formulate just such a criterion, and thus to salvage the EPR argument in a relativistic framework. For the sake of clarity, we shall first describe how this criterion applies to the nonrelativistic version of the argument.

Ghirardi and Grassi's criterion rests on the truth of certain classes of counterfactual statements[36]—statements of the form "if φ were true, then ψ would be true," where the antecedent φ is in general known to be false. In particular, they wish to "link... the attribution at a time t of the property corresponding to [observable a having value a] to the truth of the counterfactual assertion: if a measurement of [a] were performed at time t, then the outcome would be [a]."[37]

[35] I follow here a streamlined version of the EPR argument, as introduced in Redhead, "Relativity, Causality and the Einstein-Podolsky-Rosen Paradox," and Geoffrey Hellman, "EPR, Bell and Collapse: A Route Around 'Stochastic' Hidden-Variables," *Philosophy of Science* 54 (1987): 558–76. Historically this version of the argument seems to have been known to Einstein. See, for example, Arthur Fine, *The Shaky Game* (Chicago: University of Chicago Press, 1986). This is also the version used in the paper under discussion, Ghirardi and Grassi, "Outcome Predictions and Property Attribution."

[36] An analysis of the EPR argument using counterfactuals (as distinct from purely conditional reasoning), though not specifically in the context of a relativistic reformulation, has been undertaken by Linda Wessels, "The EPR Argument: A Post-Mortem," *Philosophical Studies* 40 (1981): 3–30. She seeks to uncover the full logical structure of the EPR argument by formulating the original paper's somewhat ambiguous reality criterion in precise modal terms. Among four possible modal readings of the EPR reality criterion, she lists a counterfactual reading similar to Ghirardi and Grassi's, which is the one I also adopt in the present work.

[37] In order to evaluate the truth of such statements, they call on David Lewis, *Counterfactuals* (Cambridge: Harvard University Press, 1973): Let us denote the counterfactual "if φ were true, then ψ would be true" as "$\varphi \Box\!\!\rightarrow \psi$." Then Lewis proposes the

With this criterion in hand, Ghirardi and Grassi can now run the *nonrelativistic* EPR argument essentially as before. They assume a measurement of property *a* is performed on the right-hand particle at time t_R, yielding a specific result a. To ascertain whether an element of reality corresponding to property $a = a'$ exists at the left-hand particle, they must assess the truth of the counterfactual assertion: "if I were to perform a measurement of property *a* at the left-hand particle at time t_L, I would obtain the result a′." In the nonrelativistic case, the truth of this counterfactual assertion follows naturally from the presence of absolute time ordering. For if $t_R < t_L$, then the outcome of the right-hand measurement can be assumed to be the same in all of the "accessible" (most similar) worlds used to evaluate the counterfactual, because it is strictly in the past of the counterfactual's antecedent. The strict correlation laws of quantum mechanics, also assumed to hold in all accessible worlds, then demand that the result of a measurement on the left wing also be fixed in all possible worlds (specifically, the laws require that a′ = −a). Thus the counterfactual is true, and an element of reality can be said to exist at the left-hand particle. From here, the argument unrolls in the usual way, and by supplementing this reality criterion with a locality assumption (they call it G-Loc, after Galileo), Ghirardi and Grassi can deduce that quantum mechanics is incomplete. Once again, we can represent their argument schematically by

$$\text{Quantum Formalism} \wedge \text{G-Loc} \rightarrow \sim \text{Completeness} \tag{9}$$

or

$$\text{Quantum Formalism} \wedge \text{Completeness} \rightarrow \sim \text{G-Loc.} \tag{10}$$

While these conclusions seem sound, the locality principle, G-Loc, bears further investigation. It reads: "A system cannot be affected by actions on a system from which it is isolated. In particular, elements of physical reality of a system cannot be influenced by actions on systems from which it is isolated." An examination of the structure of Ghirardi and Grassi's argument reveals that they make use not of the general principle stated but of a special case of this general principle, namely that elements of reality cannot be brought into existence "at a distance." It is this special case of G-Loc, call it ER-Loc (for elements of reality) that enters toward the end of the argument to establish that the measurement at the right wing could not have created an element of reality at the left wing and thus it must have existed prior to the measurement at the right wing, when the quantum formalism said the particles were in the singlet state. Thus they conclude that quantum mechanics is incomplete. All is well so far, but when one turns the argument around, assuming completeness and dispensing with locality, one must ask, can one be more precise as to which locality principle should be given up: the principle they label G-Loc, or the special case ER-Loc? Indeed it is the latter, for only it entered into the argument. As it turns out the distinction between G-Loc and ER-Loc does not affect their conclusions in the nonrelativistic case, because the conclusion they choose to highlight—the creation of elements of reality at a distance—is precisely one that does follow from dispensing only with ER-Loc.

---

following truth condition: $\varphi \Box\!\!\rightarrow \psi$ is true at world w if either (i) there are no possible worlds at which φ is true or (ii) some world where both φ and ψ are true is more similar ("closer") to w than any world in which φ is true and ψ is false. Obviously one has to specify the possible worlds one is taking into account; this is done by assigning to each world w a set of worlds S_w called the sphere of accessibility around w.

In the relativistic case,[38] however, greater care must be taken with the statement of the locality principle, this time called L-Loc (after Lorentz by Ghirardi and Grassi), because a locality principle must enter at the very beginning of the argument as well as in the usual way at the end. The argument begins in the same way as in the nonrelativistic case, with the occurrence of a measurement on the right-hand side, but now the absence of absolute time ordering means the result of this measurement can no longer tacitly be assumed to be the same in all the accessible worlds used to evaluate the element-of-reality counterfactual at the left-hand side. Locality must be invoked to establish the independence of the outcome of the right-hand measurement from the occurrence of the measurement at the left. This done, Ghirardi and Grassi then demonstrate the existence of an element of reality at the left-hand side following the same reasoning as above. From here, the argument unrolls once again in the usual way and locality makes a second appearance in its familiar place at the end of the argument. In this way, Ghirardi and Grassi can again prove that standard quantum mechanics plus "locality" implies incompleteness.

But there are two quite distinct cases of L-Loc that are actually being employed, one used in getting the argument started and the other appearing in the conclusion. Ghirardi and Grassi define L-Loc as the following: "An event cannot be influenced by events in space-like separated regions. In particular, the outcome obtained in a measurement cannot be influenced by measurements performed in space-like separated regions; and analogously, possessed elements of physical reality referring to a system cannot be changed by actions taking place in space-like separated regions." As in the nonrelativistic case, it is not the general principle but rather the two special cases, call them OM-Loc (for outcome of measurement) and ER-Loc (again for element of reality), that are doing the logical work in their argument. OM-Loc affirms that the outcome of a measurement cannot be influenced by performing another measurement at a space-like separation, while ER-Loc affirms that elements of reality cannot be created by performing a measurement at space-like separation. Ghirardi and Grassi invoke OM-Loc at the beginning of the argument while applying the counterfactual reality criterion, as discussed above, and they invoke ER-Loc at the end of the argument, as they did in the nonrelativistic case. So if we write L-Loc = OM-Loc $\wedge$ ER-Loc, then, schematically, their argument looks like this:

$$\text{Quantum Formalism} \wedge \text{OM-Loc} \wedge \text{ER-Loc} \rightarrow \sim \text{Completeness} \tag{11}$$

or

$$\text{Quantum Formalism} \wedge \text{Completeness} \rightarrow \sim \text{OM-Loc} \vee \sim \text{ER-Loc}. \tag{12}$$

Ghirardi and Grassi now argue, in effect, as follows. Assuming OM-Loc we can again demonstrate from Completeness a violation of ER-Loc, i.e. Einstein's *spooky* action-at-a-distance *creating* elements of reality at a distance. But if we don't assume OM-Loc, then we cannot deduce a violation of ER-Loc. All this is quite correct, but the price we have to pay for *not* being able to demonstrate a violation of ER-Loc is precisely that we have to accept a violation of OM-Loc!

[38] A discussion of the EPR set-up in a relativistic context has also been provided by G.J. Smith and R. Weingard, "A Relativistic Formulation of the Einstein-Podolsky-Rosen Paradox," *Foundations of Physics* 17 (1987): 149–72. They argue that any relativistic formulation of EPR should employ a relativistic correlated state. They derive such a state and demonstrate the relativistic invariance of the correlations. However they fail to pursue the analysis beyond the existence of the correlations, i.e. to develop the full EPR argument.

In other words, the relativistic formulation of the EPR argument does not help with the thesis of peaceful coexistence between quantum mechanics and special relativity, unless one argues that violating ER-Loc is more serious than violating OM-Loc from a relativistic point of view. This is hard to maintain since violating OM-Loc involves a case-by-case version of what Shimony refers to as violating parameter independence,[39] i.e. the independence of the probability of the outcome on one wing of the EPR experiment with respect to the parameters controlling the type of experiment being performed on the other wing. By analogy, violating ER-Loc is also a form of parameter dependence. Thus we find ourselves unable to agree with Ghirardi and Grassi's claim that in the relativistic context "the conclusion that quantum mechanics implies... effects of parameter dependence, is not justified."

To justify peaceful coexistence, we need to identify an additional assumption omitted from (12), which, if challenged, could undermine the inference. Recall that to run the argument in either the nonrelativistic or relativistic case, Ghirardi and Grassi must establish that the outcome of, say, the right-hand measurement is the same in all accessible worlds. With this established, the correlation laws of quantum mechanics imply that the outcome of the left-hand measurement is the same in all accessible worlds, and hence establish the truth of the counterfactual assertion about the left-hand measurement result that permits the attribution of an element of reality to the left-hand particle. In the nonrelativistic case, the constancy of the right-hand result is a natural consequence of the absolute time ordering as discussed above; in the relativistic case, it's not so simple. A premise akin to one that I, following Stapp, label the Principle of Local Counterfactual Definiteness (PLCD) is needed to do this sort of work.[40]

In the present case, PLCD may be taken to assert that the result of an experiment that *could* be performed on a microscopic system has a definite value that does not depend on the occurrence of a measurement at a distant apparatus. Ghirardi and Grassi implicitly assume that PLCD is licensed by their locality principle, for they invoke only OM-Loc to establish the constancy of the right-hand outcome in all accessible worlds. But I argue that PLCD does *not* follow directly from any typical locality principle, certainly not from one like OM-Loc, which asserts that the outcome obtained in a measurement cannot be influenced by measurements in space-like separated regions. The reason is quite simple: while invoking locality may prevent measurements on the left-hand particle from influencing the result at the right and from breaking the constancy of the accessible worlds as far as the right-hand result is concerned, it does not prevent *indeterminism* from wreaking that sort of havoc. Intuitively, we can imagine that we run the world over again, this time performing the measurement on the left-hand particle. If we consider this left-hand measurement schematically as a point event with a backward light cone identical to that in the actual world, we are concerned with what will happen in the complement of the forward and backward light cones. Under indeterminism the events in this complement (the absolute elsewhere) simply *cannot* be assumed to remain the same.[41]

---

[39] See Abner Shimony, *Search for a Naturalistic World View*, vol. II (Cambridge: Cambridge University Press, 1993), 138, for his preferred terminology in this matter. For more discussion of this terminology see below.

[40] Redhead, *Incompleteness, Nonlocality and Realism*, 92.

[41] This claim is not uncontroversial, however, for Lewis himself has argued that the events in the complement can be assumed to be fixed; thus for Lewis, OM-Loc *does* licence PLCD;

I thus maintain that Ghirardi and Grassi need both OM-Loc and an assumption of determinism to get their argument off the ground. Schematically, (12) is replaced by

$$\text{Quantum Formalism} \wedge \text{Completeness} \wedge \text{Determinism} \rightarrow \sim \text{OM-Loc} \vee \sim \text{ER-Loc} \quad (13)$$

It seems then, that Ghirardi and Grassi's reformulation of the EPR argument in a relativistic context may be less general than they would have us believe, for its scope is limited to deterministic systems.

6 Conclusion

So where does the discussion of nonlocality in quantum mechanics rest in the light of the various proofs whose ideas we have been sketching?

Taking first the realist option, in which all observables have sharp values at all times, if we assume determinism or restrict the discussion to cases of strict correlation or anticorrelation where we can derive determinism on plausible assumptions, there seems very little scope for avoiding the conclusion of nonlocality for any "realist" reconstruction of quantum mechanics. But experimentally we never observe absolutely *strict* correlations, so it may be argued that discussions of nonlocality should not deal with this ideal case, but should be based upon the experimentally realistic nonideal case. I am not convinced by this argument, since idealization is an essential aspect of any scientific theorizing. Be that as it may, the assumption of determinism for the nonideal case where correlations are less than perfect, may well be suspect. In such cases one is forced, as we have seen, to go to the stochastic hidden-variable framework, and much of the recent discussed has focused on the significance of violating the locality assumptions involved in this framework. Following Jon Jarrett[42] these break down into two classes.

1. Independence of the probability for a particular outcome of measurement on one particle conditionalized on the collective hidden variables, from the *type* of measurement being carried out on a remote particle. This is Shimony's parameter independence, which I have already referred to above.
2. Independence of the probability for a particular outcome of measurement on one particle, conditionalized on the collective hidden variables, from the *outcome* of measurement carried out on a remote particle. This is what Shimony calls outcome independence.

Violation of (1) has been generally held to mark probabilistic causal dependence between the outcome of a local measurement and the "setting" of a remote piece of apparatus, in prima facie conflict with standard interpretations of special relativity

---

David Lewis, private communication. See also David Lewis, *Philosophical Papers*, vol. II (Oxford: Oxford University Press, 1986). His argument turns on a dual reading of the "might" counterfactual implicit in our description of "re-running" the world: if I were to run the world over again and perform the left-hand measurement, the right-hand outcome *might* be different than it was in the actual world. This "might," he argues, could be read either as "would be possible" or as "not would not," but that the first reading does not contradict the negation of the second reading. I am not enamored of this slippery semantic solution to the problem that is forced on Lewis by his insistence on including events in the absolute elsewhere in assessing the similarity relation between worlds.

[42] Jarrett, "On the Physical Significance of the Locality Conditions in the Bell Arguments."

as prohibiting causal links between space-like separated events. Violation of (2) has led to more controversy. It seems to demonstrate that the precise specification of the state of the source, or more generally of the particles before the measurements are undertaken, cannot be regarded as a common cause of the measurement outcomes. And this in turn suggests that such a common cause must be overlaid by a direct causal connection between the events consisting of the measurement results on distant particles obtaining in the way they actually do. I have challenged this view by proposing to interpret violation of (2), not in terms of a *combination* of common cause and direct cause, in the way proposed, but in terms of a noncausal direct dependence between measurement outcomes that lacks the necessary "robustness" in respect of how the measurement results are brought about, to merit description as a causal connection.[43] This approach gives up the stochastic hidden-variable framework in favor of trying to understand the correlations between distant events in terms of a harmony-at-a-distance, or, as Shimony describes it,[44] a passion-at-a-distance.

This idea ties in with, but is really quite distinct from, the fact that violation of (2) cannot be used to transmit signals between distant locations,[45] since the necessity to recover quantum probability distributions by averaging over the hidden variables, means that we have no way of controlling the local marginal distributions for the results of measurement on one particle, by varying the type of measurement performed on distant particles. The no-signaling result is often cited to defuse the tension between nonlocality in quantum mechanics and the constraints of special relativity, but, at a deeper level, my "nonrobustness" argument may be preferable.

Turning to the anti-realist option, I have examined Ghirardi and Grassi's attempt to reformulate the EPR argument in a relativistic context and argued that it is flawed by an ambiguously stated locality principle and a hidden assumption of determinism. By making explicit the logical structure of their argument, I have undermined the conclusion that in the relativistic case the existence of action-at-a-distance is not a valid deduction from the EPR argument.[46]

This conclusion can, however, be rescued if an additional hidden assumption of determinism is exposed. Assuming indeterminism then, I claim to avoid the EPR inference to action-at-a-distance, and the concomitant challenge to peaceful co-

---

[43] Michael Redhead, "Relativity and Quantum Mechanics—Conflict or Peaceful Coexistence," *Annals of the New York Academy of Sciences* 480 (1986): 14–20; idem, *Incompleteness, Nonlocality and Realism*; idem, "Nonfactorizability, Stochastic Causality, and Passion-at-a-Distance," in *Philosophical Consequences of Quantum Theory*, J.T. Cushing and E. McMullin, eds. (Notre Dame: Notre Dame University Press, 1989): 145–53; idem, "Propensities, Correlations, and Metaphysics," *Foundations of Physics* 22 (1992): 381–94. For a careful discussion of these issues, see also Jeremy Butterfield, "David Lewis Meets John Bell," *Philosophy of Science* 59 (1992): 26–43.

[44] Abner Shimony, "Controllable and Uncontrollable Nonlocality," in *Proceedings of the International Symposium: Foundations of Quantum Mechanics in the Light of New Technology*, S. Kamefuchi et al., eds. (Tokyo: Physical Society of Japan, 1984), 225–30.

[45] For further discussion of the no-signaling results see Redhead, *Incompleteness, Nonlocality and Realism*, 113ff, and other references detailed there, and also Phillipe Eberhard and R. Ross, "Quantum Field Theory Cannot Provide Faster-Than-Light Communication," *Foundations of Physics Letters* 2 (1988): 127–49.

[46] Note in particular that my version of the relativistic EPR argument makes no reference to the contentious issue of relativistic state-vector collapse. In this sense my discussion transcends any proposed resolution of that notorious problem.

existence between quantum mechanics and special relativity. Thus I end up agreeing with Ghirardi and Grassi, but for different reasons from the ones they present in their paper.

I should stress that I have been concerned in this discussion with violations of locality principles such as OM-Loc and ER-Loc, which actually figure explicitly or implicitly in the original EPR argument and its extension by Ghirardi and Grassi. Even if, as I have argued, violations of either ER-Loc or OM-Loc cannot be derived under an assumption of indeterminism, there remains of course the question of how to interpret the violation of outcome independence. Assuming completeness, outcome dependence famously follows. In the EPR set-up this means that when measurements are performed at space-like separation on the two wings of the experiment, the results are mirror-image correlated. As one potentiality gets actualized on the left, say, how does this happen exactly in tandem with the opposite result on the right? Are we faced with a causal effect, namely result-to-result causation, so that peaceful co-existence with relativity is still challenged even if OM-Loc and ER-Loc are not violated? Similar answers involving no-signaling or nonrobustness may be provided as in the discussion above of the "realist" option.

If these arguments are accepted, then our results concerning the EPR argument may be seen as closing *additional* gaps in the peaceful co-existence argument arising from the possibility of violating OM-Loc and ER-Loc over and above the fact of outcome dependence. But it must be stressed that the mysterious harmony of the result-to-result correlations remains arguably "spooky" even if it does not involve *causal* dependence.

Another distinguishing feature of this harmony is its symmetrical character, quite unlike the asymmetry that one would normally want to ascribe to a causal connection. Shimony's phrase "passion-at-a-distance" seems exactly the right one to capture what is going on, even if one concedes that the mystery of the EPR correlations is not eliminated merely by introducing an apt nomenclature.

For the anti-realist the role of measurement is to actualize potentialities. If there are no measurements, then there are no actualities! This observation is particularly relevant to cosmological applications of quantum mechanics, where there is nothing "outside" the universe to serve as a measuring device! So, from the cosmological perspective, one can argue that the realist option is after all to be preferred.

And here one can argue that the nonseparability approach, as exemplified in blocking the nonlocality proof by denying the O-Loc principle, may in the end be the best way of understanding the peculiar features of entangled states in quantum mechanics.[47]

Many people find it difficult to see the distinction between violating O-Loc (nonseparability) and violating E-Loc (action-at-distance). A simple example due to David Lewis[48] may help here. Consider someone with what she calls a bilocal hand. There is in reality just one hand that is manifested in two different places. If you shake hands with this curiously disabled person, her *other* hand will move in synchrony with the one you are shaking, not because there is an interaction between the two hands, but just because, in reality, there is only *one* hand, bilocally located! According to Lewis, this is an example of nonseparability as distinct from action-at-distance.

[47] This is the approach I defend, for example, in *From Physics to Metaphysics*.

[48] Private communication.

The aim of this essay has been to produce *arguments* for invoking either indeterminism or holistic nonseparability in the interpretation of quantum mechanics. The theological implications of such interpretation should be clear. Indeterminism, as I have claimed, allows "room" for divine action on particular occasions, while allowing overall statistical laws to remain inviolate. Holism is an anti-reductionistic thesis that shows how every element of the universe has for its ground of being the totality of the whole, which pantheists would want to identify with God. I venture these remarks merely as a basis for the theologians to enlarge on. But too often the *scientific* arguments for these sort of claims are vague and woolly. I have tried to show that this need not necessarily be the case, although, as with much conceptual discussion, the arguments I have presented are involved and intricate. The important point is that rigorous arguments are in fact possible on these important questions.

Acknowledgment. I have benefitted greatly from fruitful discussions on these topics with members past and present of the Philosophy of Physics Group at Cambridge. I would particularly like to mention Guido Bacciagaluppi, Jossi Berkovitz, Thomas Breuer, Jeremy Butterfield, Rob Clifton, Andrew Elby, Meir Hemmo, Patrick La Rivière, Constantine Pagonis, Adam Stocks, and Fabian Wagner—all of whom have patiently educated me in so many ways.

Appendix—Glossary of Locality Principles Used in this Essay[49]

Realist Interpretations

E-Loc: Local elements of reality cannot be affected by changes in a distant, space-like separated environment.

O-Loc: Local elements of reality can be specified independently of a holistic context.

Anti-realist Interpretations

G-Loc: Events cannot be affected by a distant measurement performed simultaneously.

L-Loc: Same as for G-Loc but with "simultaneously" replaced by "at space-like separation."

ER-Loc: Elements of reality cannot be *created* by distant measurements performed simultaneously/at space-like separation—a special case of G-Loc/L-Loc.

OM-Loc: The outcome of a measurement at one location cannot be affected by a distant measurement performed at space-like separation—another special case of L-Loc.

PLCD: The outcome of a measurement that *could* be performed has a definite value independent of whether or not another measurement takes place at a distant space-like separated location. OM-Loc implies PLCD under an assumption of determinism, but not of indeterminism.

Parameter Independence: The *statistics* of measurement results at one location is independent of the parameters defining a measurement procedure at a distant space-like separated location. OM-Loc is a case-by-case version of Parameter Independence. It is thus a logically stronger principle than the latter.

Outcome Independence: The *statistics* of measurement results at one location is independent of the outcome of measurements performed at a distant space-like separated location.

---

[49] The formulation of the Locality Principles in this appendix is adapted to the precise role they play in the arguments related to realist interpretations on the one hand and anti-realist interpretations on the other, as presented in this essay.

THE HISTORIES INTERPRETATION OF QUANTUM THEORY AND THE PROBLEM OF HUMAN/DIVINE ACTION

Chris Clarke

1 The Context of This Discussion

A volume such as this is characterized by the diversity of the authors' approaches, and so it seems particularly necessary here to assist the reader at the start by explaining—only in the briefest way—the general background to my own approach.

I see our aim as that of helping people, individually and as social groups, to live *good lives*. Of course, finding the meaning of this last phrase is a large part of the religious quest. In pursuit of this aim we tell stories about the world we live in, because this is part of the essence of human life, and the stories are effective when they are well grounded in experience. Some stories derive from science, in which case there is a spectrum between mathematical *models* that are testable experimentally and *pictures* in the sense of contributions to a general worldview supported by science. Other stories are drawn from the great myths of religion. There is nothing to stop us weaving together these stories to create a greater whole in the light of our deepening experience; but Immanuel Kant (and, in a way more applicable to us today, Maurice Merleau-Ponty) has taught us that we must guard against thinking that any part of such a story represents a reality that stands independently of ourselves as a thing in itself. Nor are all parts of such a composite story equally supported by experimental tests.

To say that a part of a story (religious or scientific) is not independent of ourselves does not imply, however, that it is arbitrary or entirely subjective. I have developed my personal position in a recent essay,[1] of which a very rudimentary sketch can be given here. I start from the phenomenological position of Martin Heidegger and, more especially, Merleau-Ponty, who accept that the everyday world is a construction, but that its ground is a pre-conceptual experience of Being-in-the-world, of presence. Unlike the Cartesian picture, the ground of experience is not sensations (the elements revealed by neurophysiology, though this was unknown to these authors) but this encounter with a totality. What we call the "objective world" arises from the overlaps between the total worlds of different beings. This overlapping is more than just a passive intersection, however, but a dynamic dialogue, characterized by Merleau-Ponty's description of perception, when "suddenly the sensible takes possession of my ear or gaze, and I surrender a part of my body, even my whole body, to this particular manner of vibrating and filling space."[2] So-called objective reality is thus primarily located neither in the subject nor the object, but in the second-person relationship between the two. The overlapping of the domains of perspectives on the universe, which arises in the relativistic version of the histories approach described below, is a static and physicalist version of this conception of reality. When viewed in its fully dynamic form, however, it

[1] Chris Clarke, "Construction and Reality: Reflections on Philosophy and Spiritual/Psychotic Experience" in *Psychosis and Spirituality: Exploring the New Frontier*, Isabel Clarke, ed. (London: Whurr, 2000).

[2] Maurice Merleau-Ponty, *Phenomenology of Perception*, Colin Smith, trans. (London: Routledge, 1962), 212.

provides an articulation of the participatory approach to reality of John Heron and Peter Reason.[3] According to them, any dividing line between the subjective self and the Other must remain indeterminate, the witness of philosophy since Kant indicating to me that the project of determining such a line is based on a misunderstanding of the human condition. My aim is not to discover what part of my view coincides with ultimate reality, but to grasp the concreteness of the world as it in fact emerges in the interplay of self, society, and Other. Quantum theory plays an essential role in understanding this interplay, but it does not thereby impose an unnecessarily realist metaphysics on the basic openness of our experience.

2 Quantum Theory as a Conceptual Framework

Quantum theory emerged historically as a way of handling the physics of the very small. This is reflected in the role within the theory of a fundamental constant, Planck's constant h, whose value when expressed in conventional laboratory units is a very small number. But the formalism of the theory happily extends to arbitrarily large systems. If we knew nothing about the behavior of small systems we could have used a quantum formalism to describe statistical Newtonian mechanics, from which the mechanics of single Newtonian systems can be derived, though not (as Michael Berry shows in this volume) simply by letting Planck's constant tend to zero. Indeed Hamilton's approach to mechanics was almost exactly this. From this point of view, which focuses on the formalism, quantum mechanics is not an *alternative* to classical mechanics, but a *generalization* of classical mechanics. Should we not, therefore, resolve to use a quantum mechanical description for everything, so as to encompass all our physics in a unified way? At this point we encounter the central problem of the theory: it has appeared that, if we do this, then the theory predicts macroscopic phenomena that seem flagrantly at variance with our experience—the standard case being the Schrödinger's cat "experiment," involving the appearance of a superposition of a live and a dead cat.[4]

Several other papers in this volume (e.g., Jeremy Butterfield) give careful analyses of the options that have been tried to avoid this problem. The question used to indicate the nature of the problem is often expressed as follows: How does the classical world emerge from quantum mechanics? From the point of view of the histories formalism to be examined here, this problem can be rephrased, Why does the classical world not contain any states corresponding to the superpositions that appear in the quantum world? We can reformulate this question, not quite equivalently, in terms of three sub-problems pertaining to the *logic* of the classical world (that is, the rules for combining yes/no propositions about states using "or," "and," and so forth). Thus we can ask:

---

[3] John Heron and Peter Reason, "A Participatory Inquiry Paradigm," *Qualitative Inquiry* 3 (1997): 274–94.

[4] For the reader who has not encountered this: Schrödinger imagined a hypothetical closed box containing a radioactive atom, whose decay triggered a lethal device that killed a cat also in the box. According to quantum mechanics, the entire combined system should evolve into a state that involved the absurd superposition of a live cat and a dead cat. Does the observer collapse the state into one or the other when she opens the box and sees only one of these possibilities? Why doesn't the cat collapse the state in the same way? (There is a need for a more humane version of this paradox.)

1. Why does the classical world obey Boolean logic (a distributive lattice of propositions) rather than quantum logic (an orthocomplemented lattice)?[5]
2. Where the classical behavior is random as a result of quantum superpositions that are not manifested at the classical level (the cat is either "live" or "dead," but not "live + dead"), then:
   (a) What determines the particular distributive lattice (i.e., Boolean logic) of quantum states that have corresponding classical states? (This is known as the "basis selection problem.")
   (b) How is the statistical mechanics of these probabilistically chosen classical states derived from the dynamics of quantum theory?
   (c) On each individual occurrence, what determines which one of the classically allowed states is in fact actualized?
3. When the classical behavior is (at least to a high accuracy) deterministic, how is the classical dynamics derived from the quantum dynamics?

There is no formalism that solves all of (1)–(3). Michael Berry (in this volume) surveys some of the subtle issues involving (2b) and (3). I shall examine the histories formalism's attempt to solve (1) and (2a) and offer some speculations on (2c).

As James Cushing carefully explains in this volume, Bohmian mechanics remains an option for understanding what is happening here in a way that in principle avoids the above problem. In the early days of quantum physics, when our underlying beliefs favored a particle ontology for the universe, this mechanics was probably the front runner for an interpretation, and only failed to achieve that status because it arrived late on the scene. The main difference that adopting Bohmian mechanics makes is that, because of its particle-ontology—more generally, its position-space-ontology—the answer it gives to question (2a) about what Boolean logic to use is: the logic generated by pure states in which the particles have definite positions. At first sight this seems very reasonable. After all, the problem with Schrödinger's cat is precisely that it involves superpositions of particles in different positions. On the other hand, the criterion is less obvious when applied to, say, electromagnetic fields. And when it comes to, say, deciding whether or not quantum theory has anything to do with consciousness, then do we really want to give the positions of microscopic particles precedence over the global state of all the fields in the brain? For the purposes of this essay, in any case, I will stick to orthodoxy.

The traditional way of avoiding paradoxes such as Schrödinger's cat had been through the approach of Niels Bohr, later rigidified by John von Neumann, which rejected the idea of bringing all of the physical world within quantum theory. The Bohr/von Neumann approach works well in the laboratory, but it depends on: (i) a division of the world into a (microscopic, quantum) system and a (macroscopic, classical) environment; and (ii) the introduction of two separate processes of time development in the quantum part of the world, namely, Hamiltonian deterministic evolution and stochastic state-reduction. Von Neumann's demonstration[6] that it did not matter (subject to conditions that could be, and have been, stipulated) where the division in (i) was placed ensured a consistent and workable scheme, but I would take the position (based on a sort of principle of sufficient reason) that a basic

[5] For a discussion of quantum logic, see Enrico G. Beltrametti and Gianni Cassinelli, *The Logic of Quantum Mechanics* (Reading, Mass.: Addison Wesley, 1981); their formalism coincides with mine, though their use of the word "logic" is slightly different.

[6] John von Neumann, *Mathematical Foundations of Quantum Mechanics*, Robert T. Beyer, trans. (Princeton, N.J.: Princeton University Press, 1955).

ontological division between two categories of reality, macroscopic and microscopic, cannot be entirely arbitrary—it needs to rest on some definable criterion.

If we are looking for a story that includes divine action, and if we want to enlist some of the basic features of quantum theory (e.g., indeterminism, connectivity) to support such a story, then an overall picture of the world is precisely what is wanted. The story cannot be confined, moreover, to a laboratory setting, since divine action extends to the whole arena of life. The story needs to include physics within it, in such a way that a consistent physical formalism applies to all of the physical world, placing the observer/apparatus and the observed system on the same footing.

The essential difficulty with the von Neumann approach lies in the enigmatic nature of the process of state-reduction, and so the theories in which we are interested can be classified, following Butterfield's *Dynamics* versus *ExtraValue* dichotomy (in this volume) roughly into: (1) those that introduce a specific physical mechanism to carry out the process of state reduction; and (2) those that try to reformulate quantum theory in such a way that it can be claimed that the idea of state reduction is unnecessary. Many of those who hold to type 2 also hold to the "Definite Appearance (*DefApp*)" (but indefinite macroscopic reality) pole of Butterfield's second, independent, dichotomy, leading to theories that seem contrary to some of our intuitions. For example, Wojciech Zurek has remarked that "the only 'failure' of quantum theory is its inability to provide a natural framework for our prejudices about the workings of the universe."[7] I shall nail my own colors firmly to the mast of recognizing that it is only appearances (at best) that are definite.

Typically, a dynamics that allows state reduction (type 1) cannot be added onto quantum theory without some consequent modification of the theory itself, which might be minimal[8] or might involve the adoption of a whole new theory (or putative theory, if the hoped-for one does not yet exist) as in the work of Roger Penrose.[9] Modification in the dynamics can usually be tested experimentally in principle. Their details vary widely and in some cases, including that of Penrose, a great many essential details remain to be worked out.[10] In this paper I shall, like Butterfield, not consider these type 1 modifications in detail, but will focus mainly on one particular example of type 2 only briefly covered by Butterfield in this volume—the consistent-histories approach—though it will be helpful to make comparison with the many-worlds and many-minds views as we progress.

3 Quantum Histories[11]

A *history* is a time-sequence of asserted propositions (later I will generalize this somewhat to arrays that are not strictly sequences). The times of the propositions

[7] Quoted in Michael Lockwood, "Many-Minds Interpretation of Quantum Mechanics," *Brit. J. Phil. Sci.* 47 (1996): 159–88.

[8] GianCarlo Ghirardi, A. Rimini, and T. Weber, "Unified Dynamics for Microscopic and Macroscopic Systems" *Phys. Rev. D* 34 (1986): 470.

[9] Roger Penrose, *Shadows of the Mind* (Oxford: Oxford University Press, 1994).

[10] However, this division is not without question since Adrian Kent, "Quantum Histories," *Physical Scripta* 76 (1998): 78–84, has argued that Ghirardi's interpretation (here, type 1), could be reinterpreted as a kind of a histories formalism (type 2).

[11] Fay Dowker and A. Kent, "On the Consistent-Histories Approach to Quantum Mechanics," *J. Stat. Phys.* 82 (1996): 1575; Roland Omnès, *Understanding Quantum Mechanics* (Princeton, N.J.: Princeton University Press, 1999).

might be assigned as part of the specification of the history, or it might be that only the temporal order of the propositions is significant. A history might be a formalization of the life-experiences of a person, of a society, or of the whole sequence of events in the course of the universe. It might also be a formalization of the trajectory of a fundamental particle. In the next section I give examples of several three-event histories in the case of a particle going through three Stern-Gerlach-type observations. For simplicity, but in contradiction with the tenets of relativity theory, I assume in the main text that it is possible to refer all events to a chosen time coordinate, and I will not discuss the question of the choice of this coordinate.[12] Also for simplicity I assume that each history involves only a finite sequence of propositions.

A *quantum history* is one where "proposition" is understood with the flavor that this word has in quantum theory. Typically, this will involve the idea that, at any given time, there is a much wider range of propositions available than in the classical case, because in addition to the assertions of the form "particle P is in the region of space R" there are also assertions of the form "particle P is in a superposition of being in regions R_1 and R_2." Moreover, not all such propositions will be simultaneously decidable: because of the uncertainty principle it may not, for example, be possible to determine both that the particle is in a space-region R and also that its momentum is in a mathematical region Q of momentum-space. This has consequences for the nature of the logic that must be employed if we want to handle the totality of quantum propositions available at any one time.[13]

The role of a *physical theory* is to assign probabilities to histories. These can either be absolute probabilities, or probabilities conditional on a given initial state. We will work with mixed states in the quantum case, or probability distributions in the classical case. Thus we will usually be concerned with the quantities $\boldsymbol{P}(\rho, H)$, the probability of history H conditional on the initial state ρ.[14] There is also the possibility of including the specification of a final state, giving an interesting time-symmetric concept of causation. We choose the Heisenberg formalism for quantum theory, in which time dependence is built into the operators (including the propositions that make up histories) and the states stay fixed in time.

As an example, consider a deterministic theory of single-particle motion, using classical histories and mechanics. Suppose the history H consists of two propositions $H = (P_1, P_2)$, where the first specifies the position and momentum of the particle at some initial time to be (x_1, p_1) and the second asserts that at some later time the particle is in some region R of space, and suppose that the initial state ρ is an atomic probability distribution concentrated at (x_0, p_0). Then the probability $\boldsymbol{P}(\rho, H)$ is 0 unless we have both that $(x_0, p_0) = (x_1, p_1)$ and also that, on the given theory, the particle arrives in the given region of space at the later time, when the probability is

---

[12] This restriction is not essential. A fully relativistic version of the concept of history has been developed in Christopher J. Isham, "Quantum Logic and the Histories Approach to Quantum Theory" *J. Math. Phys.* 35 (1994): 2157–85; Christopher J. Isham, N. Linden, K. Savvida and S. Schreckenberg, "Continuous Time and Consistent Histories," *J. Math. Phys.* 39 (1999): 1818–34; and is also implicit in Matthew J. Donald, "A Mathematical Characterisation of the Physical Structure of Observers," *Foundations of Physics* 25 (1995): 529–71.

[13] See Beltrametti and Cassinelli, *The Logic of Quantum Mechanics*.

[14] Quantum mechanically, this probability is given in the Heisenberg picture by $\boldsymbol{P}(\rho, H) = \mathrm{Tr}(P_n, P_{n-1}, \ldots, P_1 \rho P_1, \ldots, P_{n-1}, P_n)$, where $P_1, P_2, \ldots$ are the quantum projection operators corresponding to the history $H = (P_1, P_2, \ldots, P_n)$.

1. In a classical theory probabilities can often be 0 or 1, whereas in quantum theories probabilities are almost inevitably strictly between 0 and 1. As another example (developed in the next section), in quantum theory a history containing a succession of propositions about the results of measurements might be used to describe a particle passing in turn through a sequence of measuring devices, each one preparing the particle for the next.

The "histories approach" to quantum mechanics can be thought of as nothing more than *the linking together of a sequence of Bohr preparation/measurement pairs, in which each measurement becomes the preparation for the next.* The only difference from the Bohr approach is that the entire sequence gives a context for each "measurement," and that this context offers the hope of avoiding the von Neumann "cut" between macro and micro. If we think of it in this way, however, then we must realize that the term 'measurement' is hardly appropriate when we are considering an arbitrary sequence of manifestations.

The language of histories is a flexible tool covering many approaches. It can be used, for example, in the context of David Albert and Barry Loewer's[15] version of the many-minds approach, in which minds migrate along the branching pathways provided by brain-states, which periodically split into numbers of different quantum possibilities.[16] The experience of one of these minds can conveniently be described in terms of a history, with one proposition corresponding to each of the splitting-points of the brain-state.

The hope is, then, that the context of the entire history can provide a unified scheme (free of notions such as "many worlds," which might be seen as unnecessary metaphysical impedimenta) for discussing the whole range of physical phenomena, quantum and classical. Such a hope is, however, soon challenged. The formalism has to overcome the problems, enumerated earlier, that face any nondeterministic version of quantum theory. Schrödinger's cat strikes again with the realization that, whenever a measurement takes place on a state that is a superposition of two states giving definite values for the measurement (i.e., two eigenstates), then in the histories formalism a nonzero probability would be assigned not only to the histories consisting of the preparation followed by one or other of the definite outcomes, but also to the history in which the measurement is followed by a superposition of the two outcomes of the measuring apparatus. Even when we have solved this problem, the formalism only gives us probabilities for the different histories, it does not tell us which one is actually instantiated. So far, it appears that nothing has been achieved by stacking together the Bohr "measurements" to obtain histories.

4 Consistent (or Decoherent) Histories

The *consistent*-histories approach tries to rule out superpositions of macroscopically distinguishable states by demanding at the outset that all the histories being considered are chosen from a set that satisfies a mathematical condition (see footnote 20) that ensures probabilities emerging from the formalism will have the form that they should have in classical logic. I will describe the motivation for this and then explain some of the formalism for handling it. The theory (also known, rather confusingly, as the *decoherent histories* approach because of its intimate connection

---

[15] David Albert and B. Loewer, "Interpreting the Many-Worlds Interpretation," *Synthese* 77 (1988): 195–213.

[16] Lockwood, "Many-Minds Interpretation of Quantum Mechanics."

with the phenomenon of decoherence) was first introduced by Robert Griffiths in 1984 and later by Murray Gell-Mann and James Hartle in the context of cosmology. The theory received a severe blow through the work of Dowker and Kent,[17] who showed that it could not deliver all that was claimed for it. Subsequent work[18] has been devoted to (largely unsuccessful) attempts to refine the consistency condition in order to avoid the problems highlighted by Dowker and Kent, or to shift the philosophical rules so as to make their conclusion acceptable. My approach here will be based on a replacement of the condition that removes many of the problems and enables us to make contact with the themes of human and divine action.

The problem with the approach as I have presented it so far is that there are too many histories. Something is needed to weed out all those that we do not, in fact, observe when the history refers to the experience of an individual—namely, those involving *unacceptable superpositions* of states that are quite different at a macroscopic level. On the face of it, Schrödinger's cat seems to demonstrate that the "something" is not already contained implicitly in the dynamics, and so we require a selection criterion that needs to be imposed as an additional ingredient. (We have already cited the way in which Bohm's mechanics would give such an extra ingredient, on a physical basis.)

Essentially all interpretations face this selection problem in some form or other. One can ask, What exactly constitutes a measurement, a macroscopic state, a branching of worlds, a branching of minds, or an acceptable history? The attraction of the histories approach is that it highlights one particularly natural way of weeding out many of the unacceptable histories. I shall describe how this works and then go on to examine the implications of the fact that this still does not eliminate enough of the unobserved histories.

If we consider one history in isolation then there is nothing about it that can characterize it as acceptable or unacceptable purely from the point of view of the formalism of quantum mechanics. This is because of the *unitary invariance* of the rules for forming probabilities: the probability of a history remains unchanged if we apply the same unitary transformation (the transformation in Hilbert space that is analogous to a rotation in physical space) to all the propositions and to the initial state. But a unitary transformation can be chosen so that it turns any given state into any other given state. Therefore, given any history with a nonzero probability, we can always obtain from it another history that has a nonzero probability for states that are unacceptable superpositions. As a result, any criterion must either (i) refer to a particular representation of the quantum formalism in terms of, for instance, wavefunctions in physical space; or (ii) it needs to be applied to whole sets of histories and their interrelations; or (iii) we need to ask more sharply, How do we restrict the possibilities of *future* histories, given a *fixed* (and "acceptable") history up to time t_0?

Option (i) could be involved if, for instance, we decreed as unacceptable any wavefunction that represented a superposition of two states in which a given macroscopic object was in two positions widely separated in space. However, unless one adopts the *ontological* position of Bohm's approach, then ruling out things like this by *fiat* seems arbitrary. Where does one draw the line? How big is "macroscopic"? How wide is "widely"? To my mind the only solution to such questions lies

[17] Dowker and Kent, "On the Consistent-Histories Approach to Quantum Mechanics."

[18] A recent survey is Angelo Bassi and G. Ghirardi, "Decoherent Histories and Realism," to appear in *Statistical Physics*; available at http://www.arXiv.org/quant-ph/abs/9912031.

in a modification of the dynamics such as has been proposed by Penrose. Here I shall consider option (iii), looking at an unmodified dynamics, but within *sets* of histories. I will concentrate on the simplest such sets, those consisting of all histories of the form $(P_1, P_2, \ldots, P_n)$ where each P_i is required to come from a given fixed set of possible propositions σ_i.

If we take this approach then a simple and strong limitation on the possible sets of histories can be derived from a particular definition of *consistency*. It is this that defines the standard *consistent-histories approach*. In the next section I will examine whether or not this condition is in fact adequate to the task of narrowing the range of histories to those that are in fact observed. I will build up this restriction in three stages: first, examining the relation between the propositions in one particular σ_i, then moving to a minimal condition, and finally generalizing this to the condition conventionally imposed.

The first restriction (which I shall from now on take as a standard restriction on all sets of histories, consistent or otherwise) is essentially logical. We choose the propositions $\{P_i^{(1)}, P_i^{(2)}, \ldots, P_i^{(ni)}\}$, which make up a given σ_i, to be exclusive in the sense that if $P_i^{(j)}$ is found to be true, then an immediately following examination of $P_i^{(k)}$ for k different from j will prove to be false.[19] In addition we require their sum to be the identity operator—which is not really a restriction, since it can always be achieved by adding a further proposition that takes in everything that the others miss out.

Following our selection of exclusive propositions, we require that the probabilities $\boldsymbol{P}(\rho, H)$ satisfy the appropriate laws of probability, namely, that if (for fixed index i) we consider histories

$$H' = (P_1, P_2, \ldots, P_i', \ldots, P_n)$$
$$H'' = (P_1, P_2, \ldots, P_i'', \ldots, P_n)$$
$$H = (P_1, P_2, \ldots, P_i' + P_i'', \ldots, P_n)$$

(where P_i' and P_i'' are exclusive members of σ_i, and $P_i' + P_i''$ denotes the disjunction "or" of the two propositions) then

$$\boldsymbol{P}(\rho, H) = \boldsymbol{P}(\rho, H') + \boldsymbol{P}(\rho, H''). \qquad (1)$$

At this stage we can give the standard example of where this does *not* hold, which illustrates the role of this condition (and hence of the consistency condition that implies it). Suppose we are considering an uncharged particle with minimum nonzero magnetic moment made to pass through three successive magnetic fields, having intensity gradients such that the deflection of the particle in the two fields measures first its spin (which I here take to be equivalent to magnetic moment) in the x-direction and then in the y-direction and finally in the x-direction again (the Stern-Gerlach experiment). The results of such measurements are quantized, yielding values that are either $+1/2$ or $-1/2$ in standard atomic units. We can then consider the histories:

$$H' = (P_1, P_2', P_3) \qquad \text{where } P_1 = \{x\text{-spin is } +1/2\},\ P_2' = \{y\text{-spin is } +1/2\}$$
$$H'' = (P_1, P_2'', P_3) \qquad P_2'' = \{y\text{-spin is } -1/2\},\ P_3 = \{x\text{-spin is } +1/2\}$$
$$H = (P_1, P_2' + P_2'', P_3)$$

[19] The reader is reminded that, in the quantum area, the logic becomes slightly generalized (technically, it is no longer distributive). This demands caution in handling conditions such as this. For instance, classically, A and B are exclusive if there is no third proposition that implies both A and B; but this is not true in the quantum case.

with an initial state of $\rho = \frac{1}{2}I$ (no initial information). Then the quantum mechanical probabilities are

$$\boldsymbol{P}(\rho, H') = 1/8$$
$$\boldsymbol{P}(\rho, H'') = 1/8$$
$$\boldsymbol{P}(\rho, H) = 1/2.$$

The failure of equality between $\boldsymbol{P}(\rho, H') + \boldsymbol{P}(\rho, H'')$ and $\boldsymbol{P}(\rho, H)$ arises because the last of these involves $P_2' + P_2''$ ("spin +1/2 or spin -1/2") expressed as a single measurement, which is equivalent to no measurement at all being performed; whereas the first two involve a measurement being performed, and then the probabilities of both possible outcomes being added together. In quantum mechanics the very act of measurement has an effect, even if the result is not taken into account in the final analysis, and so the two quantities being compared represent quite different physical sequences. In logical terms, the failure arises because

$$P_1 \text{ and } (P_2' \text{ or } P_2'') \neq (P_1 \text{ and } P_2') \text{ or } (P_1 \text{ and } P_2''), \qquad (2)$$

the definition of a logic that is *nondistributive*.

This is as far as the restrictions on histories can be taken on the basis of purely logical considerations. In practice the restriction is strengthened in a third step by generalizing the expression that arises when the condition expressed in equation 1 is evaluated using quantum mechanical rules.[20] Several authors give a full discussion of this generalization.[21] The strengthening enables the condition to be expressed in a more generally useful form. Sets satisfying this strengthened condition are called *consistent*.

The significance of this restriction for ruling out unacceptable superpositions now comes from the phenomenon of decoherence, described by Butterfield. If the system interacts with the environment in between the determination of propositions, then random phases are introduced into its quantum state. When these are taken into account, it turns out that the sets of "ordinary" histories—i.e., histories that do *not* have propositions that select superpositions of macroscopically distinct states—are consistent to a very high degree of approximation. Moreover, it can plausibly argued that close to any very-nearly-consistent set there is a consistent set that is observationally indistinguishable from it, so that nothing is lost by demanding as a principle that all sets involved are consistent.

5 Adequacy of the Consistent-Histories Approach

The question now is: Is the imposition of consistency enough to restrict the histories to those we actually observe, or is some further structure required? First, it helps to clarify the whole question of what structure we are talking about when we refer to quantum theory. If by "quantum theory" we follow a purist textbook and mean only the specification of an abstract Hilbert space, a Hamiltonian (the operator that gives the time evolution) and a vector giving the state at some initial or fiducial time, then it is clear that this is hopelessly inadequate. This is because such a structure is completely characterized by the spectrum of the Hamiltonian (essentially a measure on the real line) and the state by a function on this spectrum, so that there is almost

[20] Explicitly, the condition is:

$$\mathrm{Tr}(P_n^{(i_n)}, \ldots, P_1^{(i_1)} \rho P_1^{(j_1)}, \ldots, P_n^{(j_n)}) = \delta^{i_1 j_1} \ldots \delta^{i_n j_n} \boldsymbol{P}(\rho, (P_1, P_2, \ldots, P_n)).$$

[21] See, for example, Dowker and Kent, "On the Consistent-Histories Approach to Quantum Mechanics"; Omnès, *Understanding Quantum Mechanics*.

no scope here for distinguishing features such as space, particles, etc.—as is needed for interpretation. On the other hand, the quantum theory that is used in practice involves a much richer structure. It is actually a hierarchy of theories, going down through quantum field theory, with various successive layers of fundamentality and unification, and moving out of sight of experiment in speculative attempts at quantum gravity that go beyond space and time; and then up through various phenomenal theories, solid state theories (e.g., lasers, superconductivity), and perhaps jumping to quantum cosmology. More precisely, it manifests, in a rather compressed form, the epistemic hierarchy discussed by Ellis and Murphy.[22] The links between the levels are subtle and often not formalized. Each level uses not an abstract Hilbert space, but a particular representation in terms of structures on space and time. The states are not bare vectors but already have interpretations in terms of a wealth of special operators imported from other levels (definitions of what constitutes a particle and so on). Granted, we never start from a blank slate, but how much of this richness is part of the ongoing development of the details of scientific theory, and how much is absolutely essential for quantum theory to make any sense at all?

Returning to our initial question as to whether extra structure is needed in order to single out acceptable histories, we can ask more precisely, Though all the acceptable histories are consistent (because of decoherence), is the converse true? The answer is, No. More strongly, Dowker and Kent make a convincing case that if a history is acceptable (i.e., "classical") up to a certain point, then there is nothing to stop it from becoming, with a reasonably high probability, unacceptably non-classical thereafter and still remaining within a consistent family. On this approach, although in the past live cats and dead cats have never entered into superposition, yet it could be that tomorrow this may just be possible.

Yet we need to be cautious about this way of speaking, because it marks a transition from the language of histories, which talks about the probabilities of different eventualities, to the language of scientific realism, which talks about what "actually happens." As a first step, let us go back to Schrödinger's hypothetical cat. The first proposition of the history might be the specification that the apparatus has been set up appropriately, and the second proposition the ascertainment of whether the cat is alive. To illustrate the idea with a cat having only two degrees of freedom (!), we can denote the corresponding quantum states for the second proposition by $|\text{live}\rangle$ and $|\text{dead}\rangle$. Then the problem seems to be that, while we can derive probabilities for each of these states from the consistent-histories approach, and interpret them without difficulty, we can equally well derive probabilities for $(|\text{live}\rangle + |\text{dead}\rangle)$ and $(|\text{live}\rangle - |\text{dead}\rangle)$, or for any other of infinitely many possible superpositions. The idea obviously generalizes to many degrees of freedom. The temptation is to ask, But what really happens? This well illustrates Zurek's remark (above) about our prejudices on the universe.

The key point here is that there is no way in which we can practically test for the occurrence of some particular set of superpositions of live and dead states,[23] whereas

[22] See the essay by George Ellis in this volume, and Nancy Murphy, "Supervenience and the Nonreducibility of Ethics to Biology," in *Evolutionary and Molecular Biology: Scientific Perspectives on Divine Action*, Robert J. Russell, William R. Stoeger, S.J.; Francisco J. Ayala, eds. (Vatican City State/Berkeley, Calif.: Vatican Observatory/Center for Theology and the Natural Sciences, 1998), 463–90, 466.

[23] Note also that we cannot test for the fact of there being a superposition: the sentence

we can very easily test for the occurrence of the live, or dead states. The consistent-histories approach does not set out to tell us "what really happens." Rather, it gives us probabilities for those propositions that we can actually measure (or observe with our senses), within the historical context within which we measure them. It enables us to understand the past and to (probabilistically) predict the future. Prediction, however, involves two steps: first the determination of what set of propositions (Boolean logic) is being considered out of the continuously infinite range of mutually inconsistent logics allowed by the theory; and second, the determination of probabilities within that chosen logic. This prediction is reached not just from the abstract formalism of the interpretation, because, as Dowker and Kent have shown, this is inadequate for the first step in such a prediction. It requires the formalism together with our knowledge of the contingent circumstances that determine what is measurable for us. The results of the propositions so far determine the set from which the next proposition can be drawn. A consequence of this (representing a variation of the picture described so far) is that in principle the set of consistent histories within which we operate is not defined at the outset, but progressively constructs itself—though in practice (with exceptions to be indicated later) it can be stipulated in advance for laboratory purposes.

I propose, therefore, that the "additional structure" we were seeking earlier is to be found not in the general laws of the theory, but in the contingent circumstances within which we operate the theory, and which are expressed through the past history up to the present moment, from which we desire to predict the future history.[24] Within these circumstances, I have introduced[25] a physically based condition, namely, a form of stability in relation to the repetition of propositions, defined in terms of the time-scale characteristic of the present circumstances.

The view that the past history determines criteria for the subsequent propositions, has some similarities with the views of Bohr with which we started—that the present state determines the nature of any subsequent measurement—bearing in mind that the past history and the present state are equivalent. Both views recognize the contingent role of the actual macroscopic situation. But my proposal differs from Bohr's in important respects. First, there is no suggestion of a division into classical and quantum domains. On this way of construing reality, the whole universe is described in the language of quantum theory, and all sets of histories satisfying the stability condition are on an equal footing in terms of prediction. Second, there is no collapse of the wavefunction—indeed, there is no wavefunction! There is no world[26] (in the realist sense of an entity that is supposed to be independent of perceiving beings and yet rationally discoverable) other than the initial (fixed, in the Heisenberg picture) state.

I should stress that much needs to be done in this formalism to find the most satisfactory form for the consistency condition, which still has some *ad hoc* features.

"the state is a superposition of some live and dead states" does not convert into a valid quantum proposition.

[24] The combination of a past history with an initial state is equivalent to giving the present state, so that we have essentially a state-dependent interpretative principle.

[25] Chris J.S. Clarke, "The Histories Interpretation: Stability Instead of Consistency?", *Found. Phys. Lett.* 14 (2001): 179–86, available at http://arXiv.org/abs/quant-ph/0008060.

[26] Some may object to this view, which I have held in some form for many years. See Chris J.S. Clarke "Reply to Stanley Kerr," *Philosophy of Science* 43 (1976): 583–4, and references therein.

The next aim of the project would be to reduce "stability" to a more fundamental physical basis. I am claiming, however, that in this formalism the "measurement problem" is freed from the disguise it carries in other systems, which throws into clearer relief just what more is needed. Though this formalism could be applied to the state of the universe as a whole, it seems to me much more natural, and more in keeping with the realities of our human experience, to apply it at the local level by examining what it means for the selection of my set of histories, and then to extend this to examine how this process, operating through a variety of organisms, builds up an overall dynamics. This is in keeping with the spirit, being developed here, of not adopting a realist view based on a particular physical ground that is declared to be fundamental, but rather recognizing that science is concerned with a hierarchy of explanatory levels meshed together. If, like Dowker and Kent, we ask why, given that the universe has been classical so far, it will continue to be so, then we now give an answer not in terms of the logic and dynamics of the formalism, but in terms of the stability built into a construction of histories in which the history so far determines each successive proposition.

To end this section, I want to return to the idea of decoherence in order to analyze more closely the issue of the persistence of the classical world. The core idea of Zurek, Omnès, and others is that decoherence is a well-analyzed physical process that takes place in the world, the result of which is that histories constructed out of laboratory-measurable propositions are in fact (almost) consistent. This results in probabilities (including as a special case certainties) that (almost) obey a classical "logic."[27] I have argued that this can be made a strict deduction, provided that one replaces consistency with stability, and provided that one is, given the past history, in the sort of world we in fact live in where decoherence operates as it does.

This gives a different emphasis from that placed on the subject by Dowker and Kent. For them, consistency is a regulative principle placed on the system in order to obtain a classical world, which makes its relation to coherence rather problematic. On my approach stability is a regulative principle (but one that I believe is reducible to a more fundamental one with fewer *ad hoc* features) and decoherence, and hence (approximate) classicality, is contingent on the past history. We find that the universe is approximately classical because we are verifying a histories approach from the vantage point of our own personal histories, which are, as a contingent fact, governed by decoherence.

6 Decoherence and Life

Our aim is to connect quantum theory with divine action, which I am here viewing as an analogical extension of human action. To do this I need to examine what light the quantum mechanics developed so far sheds on human action; that is, on the histories of the organism that is myself. If I wanted to specify just what propositions were determined as the next step in my personal history, then I would need an analysis of the sort of system I am. At this stage, therefore, we need to address the arguments of Mae-Wan Ho[28] that living organisms exhibit coherence. The word is used in a rather generalized sense, but it includes the claim that living organisms can

[27] Omnès, it should be noted, regards logic in a fundamental sense as a basic principle of the world, and not as a contingent matter. We are dealing here with two senses of the word; see Roland Omnès, *Quantum Philosophy*, (Princeton, N.J. Princeton University Press, 1999).

[28] Mae-Wan Ho, *The Rainbow and the Worm*, 2nd ed. (London: World Scientific, 1998).

maintain the phase relations between the quantum states of their constituents over considerable distances and times. If this is the case, then (as I shall discuss shortly) this would significantly alter the picture given above, and would open the possibility that our experiences might not obey a classical logic. The evidence for there being sufficient control over quantum phases for this to be possible falls into two classes. First, there is specifically quantum-related evidence—controversial in both data and interpretation—from observations of anti-bunching in biophoton emission.[29] Anti-bunching is a strictly quantum effect indicative of second-order coherence in the light emission,[30] and it is hard to see how this could be achieved without a corresponding coherence of the light-generation mechanism, which is thought to extend over the whole organism. Second, there is evidence for large-scale coordination in organisms, which could be explained by underlying large-scale quantum effects. Examples are the alignment of bi-refringent intracellular structures over large distances, and the global coordination of muscular contraction.

Possible mechanisms for such quantum effects have been discussed for some time. I have reviewed[31] the most promising of these, involving the formation of a Bose pseudo-condensation[32] arising from a generalization of the Fröhlich mechanism. In the particular case of the application of these ideas to consciousness, Max Tegmark[33] has claimed that decoherence invalidates mechanisms for large scale quantum coherence. However, he addresses almost entirely one particular model, due to Stuart Hameroff, and his calculations have been shown[34] to be in error by some four orders of magnitude. He fails entirely to address the Fröhlich mechanism, referring only to popularized accounts of it. Unlike other quantum effects, such as superconductivity, it increases in strength with increasing temperature.

In the light of this we can now return to the issues raised at the end of the last section, where I argued that the choice of a set of histories was determined by the contingent situation in which it was possible to verify predictions. That was an implicitly first-person perspective: I have certain experiences that I want to explain, which are predicated on my situation. The histories appropriate to me are to be determined by my structure and its state at each stage. We now need to make the decisive transition to a perspective in which we recognize that other people, and other organisms, are also to be described in the same way. I observe them, and at the same time they carry out their own processes as described by the histories formalism. At the philosophical level we start to engage with the "many-minds" problem; at the

---

[29] Fritz A. Popp and K.H. Li, "Hyperbolic Relaxation as a Sufficient Condition of a Fully Coherent Ergodic Field," *Int. J. Theoret. Phys.* 32 (1993): 1573–83.

[30] Rodney Loudon, *The Quantum Theory of Light*, 2nd ed. (Oxford: Oxford University Press, 1983).

[31] Chris J.S. Clarke, "Distributed Pumped Dipole Systems Do Not Admit True Bose Condensations," *J. Phys A* 27 (1994): 5495–500.

[32] This refers to the behavior of a system as the number of degrees of freedom tends to infinity. A Bose condensation occurs if, in this limit, a finite proportion of the energy remains in the ground state. A pseudo-condensation resembles this for practical purposes, but while, in the limit, the relative energy per mode tends to infinity as the ground-state is approached, the proportion of energy in the ground state tends to zero in the limit.

[33] Max Tegmark, "The Importance of Quantum Decoherence in Brain Processes," *Phys. Rev.* 61 (2000): 4194–206.

[34] S. Hagen, S. Hameroff, and J.A. Tuszynski, available at http://xxx.lanl.gov/abs/quant-ph/0005025.

scientific level we encounter the "Wigner's friend" variant of the Schrödinger's cat paradox: What is the quantum mechanical description of an organism that is both observer and observed? If I observe another organism then our histories become intermeshed (in ways indicated below); I will be exploring those aspects of its responses that I can interact with, which will on the whole involve gross movements rather than subtle quantum information. On the other hand these movements may arise from that organism's detection of quantum information. In particular, if this organism can respond to quantum phase information then I can potentially observe behavior that arises from histories *pertaining to that organism* that are nonclassical and "inconsistent." Returning to the first person perspective, if I am such an organism, then my own experience can be nonclassical. I conclude, therefore, that if one accepts the histories interpretation, with a stability criterion but without the arbitrary imposition of consistency, then there may be non-Boolean aspects of an organism's own history (its own consciousness, if that language is applicable) that are observationally detectable by an external observer.

The interplay, and the distinguishing, of different "persons" is crucial to these ideas. As a first-person subject, I see the world displayed through my history, over which, as I discuss in the next section, I have some creative control. Within this I discern other entities as external objects, treating them through a third-person perspective, even, at times, when they may have their own first-person aspect. Mediating the two, however, is the second-person perspective, the I-thou relationship that can be established through empathy with another sentient being, in which we share a subjectivity. Quantum mechanically, we share histories, at least momentarily. It is the possibility of this that is the key to the "other minds" problem. I see this as the essential component in linking the phenomenology of Heidegger with the concept of the objective world through the interrelation of beings-in-the-world discussed in section 1. This second-person process goes beyond Heidegger into what Levinas[35] called the "face-to-face." It allows the consciousness of others to play a role in my world, even when at the classical third-person level consciousness is merely epiphenomenal (as I have discussed elsewhere[36]).

Whether or not we accept the arguments of Ho, and their consequences just sketched, makes a crucial difference to the theory. If they are not accepted, then all the histories of organisms are consistent as a result of decoherence. To see a particular consequence of this, suppose that we have two organisms with histories $H = (P_1, \dots, P_i, P_{i+1} \dots, P_n)$ and $K = (Q_1, \dots, Q_j, \dots, Q_m)$ respectively, and with Q_j chronologically between[37] P_i and P_{i+1}. Then we might ask whether we should really be including Q_j in this position in the history H. But (as we discussed in detail above) the consistency condition implies that if we include Q_j without specifying its value (that is, we include the mere fact that some proposition is in that place), then this does not make any difference to the probabilities for H. Probabilities for histories can be computed without knowing whether or not other organisms are in the neighborhood, and without knowing what sets of histories they might be

[35] Emmanuel Levinas, *Totality and Infinity* (Pittsburgh: Duquesne University Press, 1969). See also the discussion of Levinas's work in Arthur Still, "Immanence and Transcendence in Psychotherapy," *Changes* 18.3 (2000): 215–25.

[36] Chris J.S. Clarke, "Is Psychology Possible?", available at http://www.scispirit.com/Psych-laws.htm.

[37] I am here using a chronology relation $\rightarrow$ between sets defined by $A \rightarrow B$ iff $(\forall x \in A, y \in B)(x \in I(y))$.

determining. I shall suggest later that this determination of histories is connected with free will. Hence, in this situation the exercise of free will has no external effect—it is purely an epiphenomenon.

On the other hand, if Ho's arguments are accepted the situation changes dramatically. All the probabilities become interlinked; we cannot discount the free interactions between organisms. The universe changes from an orderly collection of histories, each of which is defined and computable in isolation from the others, to an interconnected holarchy in which any computation is conditional on the actions of other parts. There is considerable leeway available as to how we generalize the formulae for probabilities in this case, which will require further exploration.

These considerations have taken us from the rather logic-dominated realm of the history formalism to a perspective that starts to take on many of the characteristics of a many-minds approach involving a theory of mind. The differences are perhaps differences of style rather than content, the style being that of avoiding metaphysical accoutrements. Probabilities are not derived from randomly migrating minds, but inserted "by hand" from the formalism of the consistent-histories approach. Ontological assumptions about entities that I am not observing (other "branches" of a many-minds view) are avoided.

I have proposed that the choice of sets of histories is determined by the current situation, but *how* is it to be determined? For I have not demonstrated that the stability condition singles out a unique set of histories, but only that those that it does single out have the properties that we require of them. The arguments I have presented so far suggest that this determination requires a theory of *life* as much as a theory of mind—perhaps the key factor is sentience. Moreover the emphasis as far as the first-person perspective is concerned would be on a theory of external awareness rather than, for instance, Donald's theory[38] of switching networks whose relation to awareness is unclear.

7 History Selection and Organismic (Including Human) Action

If we now move from the rather Spartan realm of quantum mechanics to the sorts of stories that can include the core issues of our living in the world, it might be appropriate to elaborate the account just given of the interaction of first-person and third-person perspectives.[39] I necessarily attribute reality to my own being-in-the-world, a reality that is shared by those other beings that show themselves within my world.[40] This reality is, as I have described, progressively constructed through the unfolding of a probabilistic history and meshes with the similar histories of other beings that I observe. To anticipate later discussion in this section, a part of this unfolding of histories involves activity that corresponds to the subjective experience

---

[38] Donald, "A Mathematical Characterisation of the Physical Structure of Observers."

[39] The second-person perspective, Martin Buber's "I-thou," is even more important, and relates to the discussions by Ho of the extent to which coherence can extend between organisms, thus enlarging the boundaries of the self. This would, however, take us beyond the scope of this paper; see also footnote 40.

[40] Thus I am not inductively imputing autonomous being to other organisms, but following Heidegger's definition of the phenomenological approach as that which "let[s] that which shows itself be seen from itself in the very way in which it shows itself from itself"; Martin Heidegger, *Sein und Zeit*, in English *Being and Time*, John Macquarrie and Edward Robinson, trans. (Oxford: Blackwell, 1962). See footnote 1 and the discussion in section 1 for the phenomenological background.

of free will and that has a creative aspect to it. It is appropriate to describe this whole process in terms of the co-creation of the universe through a set of intermeshing histories. I suggest that it is this co-creative picture that will provide a framework for an understanding of human action, and hence for the analogical extension of this understanding to stories about divine action. There is little point in including divine action, however, before we understand human action; and if the latter is creative then the former can shed light on the larger creative processes in the universe. This brings us to a consideration of free will.

Many of the issues surrounding free will are well examined philosophical problems that are independent of quantum considerations.[41] Some of the problems appear to be pseudo-problems arising from the imposition of a Cartesian ontology onto experiences that are in themselves not problematic. We might be puzzled, for example, about how it is that I can exercise my will on my body in such a way that my arm lifts: What is the causative action involved here? Of the mind on the body? Why should this causative action of will be restricted to the body? The apparent problems here arise entirely from a curious, though by now well established, conceptual splitting between "I" and "my body." The sentence "I raise my arm" does not, in actual experience, imply the existence of two quite distinct entities, "I" and "my arm." It is only a reflexive idiom used to describe my arm-raising in an emphatic manner suggestive of deliberation. If we avoid Cartesian and related dualities then many of these pseudo-problems are eliminated.

A second class of problems having to do with free will involve a basic antinomy, which runs as follows. If we have sufficient grounds for making one particular choice, then it could be argued that our choice is determined by those grounds and is not free. On the other hand, if there are insufficient grounds for the choice then it would appear to be a purely random event and not a true choice or a rational act of will. Free will appears to be "free" only if it is not "will." On the histories approach there certainly are nondetermined aspects to my future behavior whose probabilities are computed to be strictly between 0 and 1. The emergence of such an action may be a conscious process, and this may be experienced as making a "choice." It is not clear, however, either from physics or introspection, what there is about these actions that distinguishes them from mere randomness. We are tempted to say that our consciousness is the agent of the choice; yet this is an illegitimate deduction. The evidence of introspection[42] is that at this stage we participate passively, not actively: we observe and report a decision process whose actual origin is nonconscious. The point is discussed in some detail by Max Velmans.[43] His main evidence comes from the work of Benjamin Libet,[44] whose experiments are usually interpreted to imply that we make our decisions subconsciously[45] a short time before we are consciously aware of them. Velmans goes on to point out that consciousness does seem to be

---

[41] See, for example, the survey in Storrs McCall, *A Model of the Universe* (Oxford: Clarendon Press, 1994).

[42] Julian Jaynes, *The Origin of Consciousness in the Breakdown of the Bicameral Mind* (New York: Penguin, 1993).

[43] Max Velmans, *Understanding Consciousness* (London: Routledge, 2000), 211–13.

[44] Benjamin Libet, "Are the Mental Experiences of Will and Self-Control Significant for the Performance of a Voluntary Act?", *Behavioral and Brain Sciences* 10 (1987): 783–86.

[45] Of course a subconscious decision is still "my" decision, in the sense that it is arrived at within the total organism that I refer to as "me." But the issue here is whether or not it belongs to that particular subsystem endowed with the interesting property of subjectivity.

involved in the process of *attention*, in which a certain range of propositions is assembled for the purpose of scrutiny and processing, and this act of attention and selection continues until a decision emerges. But the decision itself is not an act of consciousness. The purely logico-deductive part of decision-making (which for Descartes was the only part of consciousness that he regarded as unquestionably nonmechanical) is something that can be, and increasingly is, fully automated and so has little necessary connection with the conscious process.

There is, however, a class of decisions that is suggestive, introspectively, of a category that is neither determined nor random, and that matches well with a characteristic feature of the histories approach, namely the need to determine the choice of a set of histories out of the range of possibilities. This class is involved with changes of meaning, or changes of the "reference frame"—the whole fabric of meaningful connections. When we make decisions, we start off in practice by trying a variety of (usually incompatible) rules of thumb, combined with a process of imagining possible outcomes of decisions and evaluating the nature of our emotional reactions to these outcomes. So far this is essentially the same as the construction of a chess-playing computer (except that the evaluation process is not called "emotions"). If a decision does not then emerge, we do not resort to a randomizer; rather, *we alter the rules of the game*. We re-construe, re-categorize the situation.

As Farhad Delal[46] has pointed out, this involves radical disruptions to classical logic, alternating between classical logic and a wider context in which there is a local application of "rules of deduction" that would be destructively inconsistent if applied globally. Associated entities are identified, even (indeed, especially) to the extent of identifying propositions with their own contraries. It is this, I suggest, that enables a shift from one consistent Boolean logic to another to take place, and thereby constitutes the essence of conscious human decision making.[47] From the point of view of an external description employing a fixed classical logic, it is an a-causal maneuver that, by default, can only be described as "random." From the person's point of view, however, it is an act of creative meaning-making: the situation comes to make sense, and the change of cognition itself makes sense, but only retrospectively, from the perspective of the new frame of reference once achieved.

There is a parallel, amounting to an identity of logical structure, between this change of reference frame and the determination of one out of a range of possible sets of disjoint projections for the next such set in the histories available to an organism that utilizes coherence. Each set of histories is governed by classical logic, but the collection of all available propositions requires quantum logic (as noted above), so that shifting projection-sets requires moving from classical logic to quantum logic and back again, just as suggested by Delal in the case of radical human thinking. In both cases we are dealing with a change that is neither a random choice within a given set of possibilities (as happens with the outcome of a quantum observation within a given set of exclusive possibilities) nor a determined choice (which is really a special case of a random choice with probability 1), but a third possibility, a shift of underlying meaning, in the case of free will. On the basis of

---

[46] Farhad Delal, *Taking the Group Seriously: Towards a Post-Foulkesian Group Analytic Theory* (London: Jessica Kingsley, 1998).

[47] Chris J.S. Clarke, "Consciousness and Non-Hierarchical Physics," to appear in *The Physical Nature of Consciousness*, Philip van Looke, ed. (Philadelphia, Penn.: Jon Benjamins Publishing), available at http://www.scispirit.com/consc2.htm.

these parallels, and supposing that Ho's thesis is valid, I propose that we are capable of acts of free will whose physical correlates are selections of projection sets.[48]

This sort of act of will, which I call *creative will*, is characterized by the simultaneous creation of a volition and of the framework of meaning within which that volition is justified. It is this simultaneity of volition and meaning that distinguishes this act from, on the one hand, an act that is determined by prior psychological and physical factors and, on the other hand, an act that is random in that one of a number of pre-existing alternatives is selected with no perceptible rationale. (These last two cases can, or course, be combined in various proportions.) The existence of this category of creative will is the solution to the basic antinomy of free will noted above. The approach here is quite separate from that of Donald MacKay. He recognizes the potential in the indeterminism of quantum theory for a separate determinative factor, which could be God or human will,[49] and he recognizes[50] the role of meaning in the human capacity for information processing, which goes beyond the sort of information theory that was available at the time, but he fails to unite the two in the change of frame of meaning. It is this connection that both gives a new understanding of the uniquely creative nature of sentient action and links it with the most distinctive feature of quantum theory.

8 Divine Action

It is only once we have a picture that contains human will together with the physical universe that we can usefully extend our story to include divine will, on the basis of a metaphorical extension from human will. I want to begin by drawing attention here to a particular feature of quantum theory that I shall call *entrainment*. It relates to the idea of "top-down causation" in the context of the action of will, whether human or, if the metaphor is to mean anything, divine.

Let us imagine a chaotic physical system such as one vessel of water draining turbulently into another, initiated in a state that is a superposition of two quantum states that are indistinguishable by macroscopic measurements. If we suppose an essentially quantum universe whose state is occasionally constrained by macroscopic information, then such superpositions will be the normal state of affairs. The evolution of this system will then, in a time depending on the Liapounov exponents of the chaotic modes, evolve to a superposition of states that are macroscopically distinguishable. If we now ascertain which one of these distinguishable states is in fact realized, and retrodict from this what was happening in the past, we will retrodict precisely that sequence of detailed microscopic fluctuations that led from the initial state to the final state now realized.

---

[48] The ability to choose which set of projections is to be measured, and to do it repeatedly, can be used actually to alter a state. Suppose, for example, that we have a particle with spin initially in an eigenstate of the z-angular momentum component, and that we then measure its spin n times in directions in the (z,x) plane making angles of $\pi/n, 2\pi/n, \ldots$ with the z-axis. Then for $n>2$, at the end the particle is more likely to have its spin reversed, and this probability tends to 1 as n increases. Thus the ability to choose projections implies the ability to modify quantum states at will.

[49] Donald M. MacKay, *Science, Chance and Providence* (Oxford: Oxford University Press, 1978).

[50] Donald M. MacKay, *Information, Mechanism and Meaning* (Cambridge: MIT Press, 1969).

For example, suppose that we had also filmed the system from the start through a microscope. Then (following the usual quantum mechanical argument) the film would be in an entangled quantum state such that, when we ascertained the final state, if we then looked at the film it would show a sequence of events that led precisely to the final state that we had ascertained.

Expressed in terms of histories, if ρ is the initial state and the two histories terminating at the propositions that distinguish the final superposed states are $H = (P_1, P_2, \dots, P_n)$ and $H' = (P_1', P_2', \dots, P_n')$, then the probability of $(P_1, P_2, \dots, P_{n-1})$ conditional on P_n becomes much higher than the probability of $(P_1', P_2', \dots, P_{n-1}')$. On the other hand we are supposing that the organism can have some say in the choice of which pair (P_n, P_n') appears in the history, and some say as to which one is exercised at the time when these propositions are added to the history. Indeed, as indicated previously (footnote 48), this may even extend to a determination of which proposition out of (P_n, P_n') is in fact realized. Thus the organism appears also to be "influencing" the previous events $(P_1', P_2', \dots, P_{n-1}')$ leading up to P_n, even though these were at an earlier time.

A more accurate account, however, would seem to be that it is *the fact of these events having occurred* that is realized at the time of P_n, rather than the occurrences themselves. I shall describe this situation as the *entrainment* of these preceding events into the final determination of P_n. The organism in question will in general have no interaction with the details of these preceding events—it will merely be observing, or even influencing, the final event. But in the course of doing this the smaller scale predecessors are automatically entrained into the process.

It seems to me that this account of the action of human will and choice within a quantum mechanical setting provides the most natural metaphor for divine action. I find myself part of a nested sequence of organisms. I contain cells, organelles, etc.; and I am contained in a family, a community, in Gaia, etc. Each of these has its proper activity, which includes, in all probability, freedom of the sort that I have been discussing. We can note that the freedom of the parts is not compromised by being included in the freedom of the whole, a situation that is even more true in quantum theory than in classical theory because of the greater number of real degrees of freedom that are contained in the phase relationships between the parts that make up the whole. This makes sense of the way that I experience the intermeshing of my own freedom with the choices of the organisms of which I am a part. My experience of the divine is of a guidance that is immanent, in being part of the concrete flow of events around me, and transcendent in the sense of not being contained within any given contextual framework, of being always greater than my current horizons. It thus makes sense for me to think of the divine as analogous to an ideal outermost system, beyond any imaginable context, and of divine action as being the top-down action of will through entrainment that coordinates and informs all the individual acts of will that it contains.[51]

For me this top-down description of divine action accords more with experience than does the idea of a god who influences each atomic process individually so as to build up the whole. Top-down action is a story that builds on the metaphor of my own acts of will, which are also delivered at the macroscopic level. The picture that this produces goes somewhat beyond pantheism, in that the entire universe is strictly

---

[51] I am assuming in this, despite the fact that "a thousand ages in Thy sight are like an evening gone," that the characteristic timescale relevant to the action of god in the universe is short enough to allow this possibility within the stability condition.

within god; in this sense it is panentheistic. That word tends, however, also to imply that god is in some sense also embodied in the world, and this is not obvious in the present system. I would regard such embodiment, the complementation of transcendence with immanence, as necessary for a fully religious perspective. This I define in terms of the second *mahavakya*[52] of the Upanishads: "This atman is Brahman." Religion is living one's life in the light of the experience that the transcendent ultimate within is identical to that without. Our current physical picture cannot close up the nested sequence of inclusions into the circle that this implies, in which god is both the ground of being and the context of meaning.

[52] *Ayam atman Brahamasti*. See Bede Griffiths, "The Vision of Non-Duality in World Religions," in *The Spirit of Science: From Experiment to Experience*, David Lorimer, ed. (Edinburgh: Floris Books, 1998).

III. THEOLOGICAL ISSUES

John Polkinghorne
Michael Heller
Philip Clayton
Thomas F. Tracy
George F.R. Ellis
Robert John Russell

PHYSICAL PROCESS, QUANTUM EVENTS, AND DIVINE AGENCY

John Polkinghorne

1 The Problem

Quantum theory has been spectacularly successful during the seventy-five years of its intensive exploitation. It affords us predictive powers apparently as reliable for quarks as for atoms, and its picture of microscopic process has enabled us to understand many physical phenomena that otherwise would have been opaque to us, ranging from the stability of atoms to the existence of superconductivity. Yet our pleasure in these achievements is qualified by a deep perplexity, for we do not properly understand the theory itself. The central unresolved difficulty concerns how the cloudy quantum domain is related to the clear world of our everyday experience. How does it come about that an electron, which quantum mechanically has a certain probability of being found "here" and a certain probability of being found "there," actually manifests a perfectly definite experimental answer whenever we actually measure its position (though not the same answer on all such occasions of measurement)? This conundrum is the celebrated "measurement problem"[1]: how do specifically determinate results emerge, on each particular instance of measurement, from the cloudiness of apparent quantum indeterminacy?

One highly popular, but by no means universally accepted, way of stating this problem is to talk about "the collapse of wavepacket." In much conventional thinking about quantum theory, the wavefunction, ψ, generates a probability distribution for the possible outcomes of measurement. From this point of view, when a specific outcome has actually been realized, the probability has then collapsed onto that single actuality, which has now become a certainty. This implies that ψ has been subjected to a discontinuous change. Yet this "jump" is in complete contradiction to the basic dynamics of quantum theory, for the Schrödinger equation prescribes a completely smooth variation for ψ. The hypothesis of collapse has, therefore, to be imposed upon the formalism from the outside. Brutal though this extra prescription may be, it certainly yields a successful technique for doing the sums, for it is found to give precisely correct experimental predictions. Empirically there is no problem, but in terms of our fundamental understanding of the theory there is obviously very great difficulty. At present this difficulty has not been resolved to universal satisfaction, and the interpretation of the measurement process has been the subject of many contending and contentious proposals.

Putting it another way, our problem may be said to be, How does the fitful quantum world interlock with the reliable world of macroscopic measurement? In fact, this is rather too crude a way of posing the dilemma, but essentially it reproduces the manner in which the pioneers of quantum theory, led by their philosopher-king Niels Bohr, began to address the issue. The unsatisfactoriness of this formulation of the problem lies in the way in which it seems to speak dualistically, as if there were two physical worlds, one quantum, the other classical, that have to intermesh. In reality, of course, there is a single physical reality, for the reliable classical measuring apparatus, on which the pioneers laid such stress, is itself made up of quantum constituents. Michael Berry's contribution to this volume

[1] See my *The Quantum World* (Princeton, N.J.: Princeton Univ. Press, 1984), chap. 6.

makes it clear that, in fact, there is a subtle interweaving of the quantum and classical domains of behavior, with no possibility of simply invoking a scale factor ("large is classical") to define a boundary between the two.

2 Proposals

In this section, I shall outline and classify a number of different proposals and considerations that have been brought to bear on the measurement problem. The reader will detect overlaps and connections with Jeremy Butterfield's discussion, but here the details of presentation and organization are along lines that seem natural to a physicist, rather than to a philosopher. Combining the two chapters should give a "binocular vision" that may be of assistance to the reader.

First, let me say that I do not regard it as a satisfactory strategy simply to settle for a level of understanding that enables consistent calculations to be done, but which leaves the deeper interpretative problems unresolved. I therefore reject a purely statistical interpretation, in which it is held that quantum theory refers only to ensembles and does not seek to speak of individual occurrences. So restricted an understanding is hardly acceptable for a fundamental theory. The consistent-histories approach to quantum theory also seems to be in this class, for it represents a more detailed, but still insufficient, account of how to generate calculational schemes that yield a consistent classical account of their results. Approaches of this kind represent a kind of refined positivism. It may turn out that this is the best that can be done—that quantum physics is ultimately apophatic physics—but this is a conclusion only to be reached as a last resort. Let us, therefore, consider other possibilities.

1. *Known physics*. It has long been recognized that quantum probabilities seem to have a different character from that of the classical probabilities that arise from ignorance of the details of the state of affairs of a system that, in itself, is actually perfectly deterministic (as in the tossing of a coin). This difference stems from the way in which quantum probabilities are calculated (as the square of the modulus of ψ) and from an extremely important contrast between the kinds of particle states that are allowed in classical physics and in quantum physics. In the classical way of thinking, an electron is either "here" or "there," and there is no possibility of any mixing between these two mutually exclusive possibilities. Quantum mechanically, things are different, for an electron can be in a state that is an (unpicturable) mixture of "here" and "there." This adding together of what classical physics, and common sense, keep rigidly apart is called "the superposition principle."[2] It leads to the existence of quantum "interference effects," which arise from overlaps between different states in the superposition. (The canonical example is given by the wavelike patterns generated in the "double slit experiment," corresponding to adding together the two states "passed through the top slit" and "passed through the bottom slit.")

One way of expressing the difference between quantum and classical probabilities (albeit a way that is rather technical in detail) is through what is called "the density matrix," relating to the possible outcomes of some process. In classical statistics, this matrix is diagonal; in quantum physics, it has off-diagonal elements as well, corresponding to interference overlaps between different states in the superposition.

A very important recent discovery has been the existence of the phenomenon of *decoherence*. This results from the interaction between the quantum process and the general environment in which it takes place. Radiative interaction with the

[2] Ibid., chap. 3.

environment has the effect (through phase randomization) of almost completely washing out the off-diagonal elements of the density matrix, and doing so with quite amazing rapidity. One important distinction between quantum and classical statistics has thereby been rendered practically almost negligible.

Significant though this discovery certainly is, it by no means solves the measurement problem. It explains why probabilities rapidly become quasi-classical (technically, how a pure state quickly comes to approximate to a mixed state), but it does nothing to explain how, on a particular occasion of measurement, a particular outcome, corresponding to a specific one of the diagonal elements, actually occurs. Classically, this process is unproblematic; it simply corresponds to the actual detailed state of the system that was always there but of which we were ignorant. Quantum mechanically, the realization of a particular final state is highly problematic. According to conventional understanding, one of the states that was potentially available because it was a component in the superposition has become actual (the wavepacket has collapsed onto it). We therefore need to go on to consider suggestions about how this collapse might be brought about or, alternatively, how the necessity of thinking in this way might be circumvented.

2. *Hoped-for physics*. This is the category under which we may classify the most widely held point of view among physicists in general. It may be termed the *neo-Copenhagen interpretation*, the prefix being justified by the more refined and cautious attitude taken toward the role of the measuring apparatus than was adopted by the founding fathers of the subject in the late 1920s. The central, hoped-for feature is that interaction with "large" systems brings about the collapse of the wavefunction. Because quantum theory contains within itself a natural scale—Planck's constant enables one to define parameters such as the Compton wavelength—it can give a meaning to "large" (but note the caution expressed above about the interweaving of the quantum and classical domains). Yet, the neo-Copenhagen proposal remains an aspiration rather than an achievement, for no full explanation has yet been given of how a determinative property of large systems could emerge from the indeterminate quantum substrate of their constituents.

One possibly hopeful feature, however, is the coexistence of another property of large systems that, though also not well-understood, might correlate with the capacity for measurement: the irreversibility of such systems' behavior. The fundamental constituent interactions of physics are reversible; that is to say, they do not define a natural arrow of time pointing from "before" to "after." Yet the macroscopic world is full of irreversible processes (before, the egg was whole; after, it is shattered beyond restoration). There is clearly a connection between this macroscopic arrow of time and the second law of thermodynamics, for the increase of entropy (disorder) in isolated systems also defines an arrow pointing from past to future. Since measurement involves an irreversible macroscopic registration of a microscopic state of affairs, it also defines a similar arrow (before, we were ignorant; after, we know). It may be that all these mysterious arrows are related to each other and that the infant science of complexity theory, studying the detailed behavior of complicated systems, will one day help us to understand these matters better. At present, however, this can amount to no more than a pious hope.

3. *Unknown new physics*. A prime example of this approach is the proposal of GianCarlo Ghirardi, Alberto Rimini, and Tullio Weber (GRW). They suggest that wavepacket collapse onto specific position states is a universal stochastic physical process, but that its rate is strongly correlated with the amount of matter involved. In consequence, macroscopic states collapse almost instantaneously, while micro-

scopic particles virtually never do so on their own. Although this proposal has been subject to various criticisms (mostly concerned with what is felt to be its air of *ad hoc* contrivance), it does suggest the existence of a novel physical process that could, in principle, be the subject of experimental investigation. (Study of systems of a size intermediate between particles and the macroscopic might reveal the occurrence of the process of collapse.) Obviously, the proposal's fate depends upon whether this investigation is actually made and what its outcome proves to be. A less well-articulated proposal of a similar kind is Roger Penrose's suggestion that quantum gravity, when properly formulated, will also solve the measurement problem as well.

4. *Hidden new physics.* The highly instructive ideas of David Bohm fall into this category (see James Cushing in this volume). Here, I consider principally their formulation appropriate to the quantum mechanics of particles.

For Bohm, the wavefunction is a guiding wave field. Probabilities arise in a classically unproblematic way from uncertainty in knowledge of the positions of the particles, which are separate entities in this theory, totally distinct from the wave. These particles have their motion influenced by the wave (which encapsulates instantaneous information about the total environment), but they are themselves as classical and picturable in their nature as even Sir Isaac Newton himself could have desired. There is, of course, no collapse of the wavepacket in Bohm's theory. Measurement is just the establishment of what is unproblematically the case. It is the additional influence of the guiding wave that yields results simulating conventional quantum-type probabilities, despite their arising simply from ignorance of fine detail.

There are two notable special features in Bohm's theory. One is the preferred role assigned to position, as compared to other dynamical variables such as particle momenta. (In the version of the theory appropriate for fields, this preferred role is played by certain ontological entities that, following John Bell, Bohm called "beables.") We shall return later to the issue of whether all states and dynamical variables are to be considered on an equal footing, or whether there is a "preferred basis" that selects out certain states or variables to play a uniquely significant role. The other property is the remarkable one that the empirical predictions of Bohm's (determinate) theory are identical with those of conventional (indeterminate) quantum mechanics.[3] In this sense, the hidden physics it postulates is unknowable because no scientific test could establish its exclusive validity. Of course, a similar objection can be made against conventional quantum physics.

It follows that a choice between Bohm and Bohr has to be made on the basis of extra-scientific metaphysical principles, rather than by appeal to empirical adequacy. On reflection, this may cause us no great surprise, for questions of causality are ultimately, and legitimately, metaphysical questions, and their answers are consequently metaphysical responses. Therefore if, for example, theologians believe that they have their own proper reasons to believe that God's creation should be indeterministic in the character of its physical process, they are fully entitled to reject Bohm and to state a preference for conventional quantum mechanics. However, they must not pretend that it is science that has enabled them to do so.

5. *Unknown but "nearby" metaphysics.* The word in scare quotes is intended to indicate that the metaphysics being appealed to is not general in character but it has about it something that might cause its proponents to see it as a "natural"

---

[3] Technically this requires the assumption that a state of quantum stability has been attained in which the initial probabilities of the particles coincide with those defined by ψ. This would then be maintained for all subsequent times.

extension of scientific understanding, taken beyond its actual bounds. (This would contrast with a metaphysics that acknowledged that some of its motivation came from a totally different discipline, such as theology.) Two kinds of current proposals may be held to fall within this classification.

One is an appeal to the role of the *consciousness* of the observer in bringing about the collapse of the wavepacket. Although proponents of this point of view have sometimes appealed to toy models that they have constructed (technically, using a nonlinear modification of the Schrödinger equation to decompose the linear superposition of states), so little is understood of the physical basis for consciousness that invoking its effect must, presently at least, be classified as a metaphysical move. There are many well-rehearsed difficulties with this proposal. Whose "consciousness" can do it (just humans, or also Schrödinger's cat, or even a worm?). What is being asserted about physical process in all that long cosmic time, and in all those many cosmic places, in which no consciousness has been present?

One might think that process philosophy, with its pan-experiential attitude that declines to see a fundamental difference of principle between a person and a proton, might have something to say here. However, there does not seem to have been much serious engagement by process thinkers with the detailed problems of quantum theory.[4] Moreover, process philosophy's episodic, event-dominated metaphysics is not easily reconcilable with the continuity of the Schrödinger equation.

The second approach falling within this category has been to suggest that physical reality is much larger than the world present to our direct experimental and observational investigation. The universe of our experience is held to be but a component in a vast multiverse. This is the strategy of the many-worlds and many-minds approaches to the interpretation of quantum theory. At the heart of both lies a belief that we should take the basic formalism of quantum mechanics absolutely seriously and not tamper with it by imposing the alien requirement of wavepacket collapse. It is claimed, therefore, that all the states in a quantum superposition endure. It is only a trick of perceptual perspective that makes us think that measurement produces a single outcome. The other possibilities are also there, "somewhere else," either in parallel universes or perceived by parallel minds. Once again, there are many well-rehearsed criticisms of this approach, mostly centering on its breathtaking ontological prodigality. It is interesting to note that, from the overall point of view of the multiverse, these approaches treat quantum theory as deterministic (the Schrödinger equation is all). It is only from the perspective of my world or my mind that physical process appears to be indeterminate.

3 Preferred Bases

The proposals surveyed above display a great diversity of approach to the interpretation of quantum theory. Yet many of them also exhibit a common feature,

---

[4] Only a few articles on the topic of quantum mechanics have appeared in *Process Studies*. See Henry Stapp, "Quantum Mechanics, Local Causality, and Process Philosophy," *Process Studies* 7.4 (Winter 1977): 173–82; Charles Hartshorne, "Bell's Theorem and Stapp's Revised View of Space-Time," *Process Studies* 7.4 (Winter 1977): 183–91; William B. Jones, "Bell's Theorem, H.P. Stapp, and Process Theism," *Process Studies* 8.1 (Spring 1978): 250–61; Henry J. Folse, Jr., "Complementarity, Bell's Theorem, and the Framework of Process Metaphysics," *Process Studies* 11.4 (Winter 1981): 259–73. See also the two recent issues of *Process Studies*, vols. 26.3–4 (1997), guest edited by Timothy Eastman and devoted to the question of the relation between process thought and physics.

in that they suppose either that the collapse onto a single realized possibility, or the perspectival illusion that such a collapse has occurred, depends upon an effect that gives a preferred role to a certain set of states (or to the dynamical variables that are involved in defining these states). This feature can be summarized as follows:

1. *(Neo-)Copenhagen*. Although Bohr and his companions simply appealed in their initial formulations to the role of the measuring apparatus in general—and they emphasized that different choices of apparatus by the observer would lead to differing and complementary effects and descriptions—the suspicion lingers that, in fact, spatial position plays a preferred role in actual measurement procedures. Talk of "positions of a pointer on a scale," and of marks on photograph plates or tracks in bubble chambers, are all expressed in spatial terms. It may be that this fact will need more explicit recognition if this kind of approach is to progress further.

2. *Many worlds*. Here divergence between the different branching worlds of the multiverse is correlated with an act of measurement, so that the remarks made under (1) would seem also to apply to this case.

3. *Consciousness*. If consciousness has a role to play, and if conscious states are correlated in some way with brain states, presumably the latter would form a preferred basis, linked with the effect of inducing the perception of specific outcomes.

4. *Many minds*. Here the branching of the multiverse (or of the observer) is related to specific brain states, much as in (3).

5. *GRW*. This theory postulates a collapse of the wavefunction in position space.

Although Bohm's theory does not involve any notion of collapse, we have already noted that in its particle version it too has a preferred spatial basis, assigning a particular role to particle positions. We may note also that decoherence, the effect of the environment on the quantum entity, is a position space phenomenon.

If the recognition of the role of a preferred basis is indeed an important requirement for progress with the measurement problem, this would imply a significant modification in our thinking about physics. From the first formulation of classical Hamiltonian dynamics onwards, and certainly extending to conventional quantum theory as set out, for instance, in Paul Dirac's classic *The Principles of Quantum Mechanics*, it has always been assumed that all dynamical variables, as well as their associated states of motion, are to be treated in an equal way.

4 Process

All the proposals set out above, with the exceptions of Bohm's theory and GRW, attach special significance to the act of measurement. "Measurement" is not necessarily meant to refer to the intervention of a conscious observer (except, of course, in those approaches that assign a special role to consciousness), but rather to the irreversible macroscopic registration of a property that is assigned to the microscopic quantum entity involved. This kind of discussion is well adapted to physicists' preoccupation with making predictions about the results of laboratory experiments. Such investigations have a definite beginning (preparing the state to be considered) and a definite end (making the measurement). However, we are concerned not only with such artificial arrangements, but also with the general flux of process in the physical world. It is far from clear that a theory of measurement, even if successfully formulated, would answer all our questions. The physical world involves processes taking place on all scales, with the truly microscopic and the truly macroscopic being endpoints of a continuum. Since there is a single physical world, and not two disjoint worlds of classical and quantum physics, we need a single integrated account of the universe's process. If we are to talk properly of what is happening, the discourse must be able to refer to all levels of complex physical reality. We do not currently possess such an ability.

Even the account of truly macroscopic physical phenomena may need to be more complicated and subtle than a straightforward appeal to conventional classical physics would suggest. Studies of computer models of complex systems have revealed that they possess astonishing powers spontaneously to generate patterns of large-scale order. This property encourages the idea that some kind of pattern-forming causality of a holistic kind ("active information") may be needed to complement the energetic constituent causality of the customary physical way of thinking. Proposals based on an ontological reinterpretation of the unpredictabilities of chaos theory have led to the suggestion that the equations of classical physics themselves might be no more than "downward emergent" approximations of a more supple account of macroscopic physics.[5] (They are also, presumably, "upward emergent" from quantum physics, illustrating the potentially complex and interleaved nature of a true account of physical process.)

These considerations suggest that, if we are looking for a physical basis on which to ground the metaphysical construction of a worldview, the specific, episodic instances represented by quantum measurements may only be part of the necessary input into the account of its causal nexus. An immense task confronts us if these issues are to be carried further.

We cannot bypass these difficulties by appealing to a quantum/classical separation based simply on some scale defined in relation to Planck's constant. Collaborative effects involving the phenomenon of Bose condensation, in which many particles are all in the same quantum state, manifest quantum properties on macroscopic scales (e.g., superconductivity and superfluidity). Chaos theory, with its discovery of the exquisite sensitivity of many classical systems to the finest details of their circumstances, shows that classical mechanics is not impervious to effects located at apparently quantum microscopic scales. The two regimes are inextricably intertwined and the eventual true theory of physical process will have to reconcile and incorporate these interlinkings.

The presently unresolved perplexities about how to understand quantum chaology indicate what a demanding task this is likely to be. Chaotic dynamics characteristically has a fractal structure, whose unending chain of self-similarity implies that no intrinsic scale is present. Quantum theory characteristically sets scales through relating phenomena to Planck's constant (as in the uncertainty principle, for example, which sets limits to the degree of fine detail that is accessible). The energy spectra of classical mechanics are characteristically continuous; in quantum theory they are frequently discrete. Chaotic dynamics characteristically involves nonlinearity, while quantum theory's linearity lies at the heart of the superposition principle. Resolving these contradictions in a consistent synthesis is an urgent matter whose successful treatment still lies in the future.

Our perplexity stems in part from not knowing what meaning to attach to "quantum event." The complex many-layered character of the flux of the physical world does not encourage the belief either that the phrase should refer simply to those occasional happenings that could be classified as measurements, or that it is

---

[5] John Polkinghorne, *Reason and Reality* (London: SPCK/Trinity Press International, 1991), chap. 3; idem, *Belief in God in an Age of Science* (New Haven: Yale University Press, 1998), chap. 3; see also Ilya Prigogine, *The End of Certainty* (New York: Free Press, 1997).

to be reduced to something like the rather banal unfolding of the consequences of the Schrödinger equation.

5 Prospects for Theology

A vigorous and adequate theology will wish to speak in terms of agency freely exercised, both in the acts of moral beings and by the providential action of the Creator. (I reject a compatibilist account of human agency since I do not believe that the mere impression of choice can be the foundation of moral being.) This has led to the concern with questions of divine action that has characterized this sequence of CTNS/Vatican Observatory conferences. In addressing these issues, theology will want to exercise its own proper autonomy. But in constructing the metaphysical backing for its discourse, theology will wish to take careful account of what science can suggest about the process of the world and the character of its causal nexus. Physics does not determine metaphysics, but it can certainly constrain it. It can never be the case that "anything goes."

Many contemporary theologians wish to try to speak of divine special providence as God's interaction with creation taking place within the grain of natural physical process. To suppose that God habitually acts against the grain of nature would seem to suggest the theological absurdity of God acting against God, since theologically the laws of nature are to be understood as expressing the faithful will of the Creator who ordains them. If a general concept of divine action of this kind is to prove feasible, it must surely be the case that the grain of nature is sufficiently open to afford causal space for the operation of that divine agency. (Equally, one might add, such a degree of openness would also seem necessary to accommodate the acts of moral agents.) From these considerations come the widespread tendencies, in one way or another, to look to the intrinsic unpredictabilities known to be present in nature (quantum mechanically and chaotically) and to seek to interpret some or all of them as indicating not mere epistemological limitation, but actual ontological openness. All schemes have in common the supposition that, properly metaphysically interpreted, the reductionist account of physics does not exhaust the description of the causal powers actually at work in bringing about the world's future.

As admirable as this idea may be, our discussion has made it clear that many unresolved problems beset all attempts of this kind. One can summarize some of the issues by reference to the two strategies that have principally been pursued by the theologically motivated:

1. Those who have appealed to (an assumed indeterministic) quantum theory[6] have to recognize how ill-defined and problematic is the concept of "indeterminate quantum events." The only events that we might feel reasonably free to invoke with some degree of confidence (assuming we have espoused the appropriate interpreta-

---

[6] See, George F.R Ellis, "Ordinary and Extraordinary Divine Action: The Nexus of Interaction," 359–96, Nancey C. Murphy, "Divine Action in the Natural Order: Buridan's Ass and Schrödinger's Cat," 325–58, Thomas F. Tracy, "Particular Providence and the God of the Gaps," 289–324, all in *Chaos and Complexity: Scientific Perspectives on Divine Action*, Robert J. Russell, Nancey C. Murphy, and Arthur Peacocke, eds. (Vatican City State/Berkeley, Calif.: Vatican Observatory/Center for Theology and the Natural Sciences, 1995); Robert J. Russell, "Special Providence and Genetic Mutation: A New Defense of Theistic Evolution," in *Evolutionary and Molecular Biology: Scientific Perspectives on Divine Action*, Robert J. Russell, William R. Stoeger, S.J. and Francisco J. Ayala, eds. (Vatican City State/Berkeley, Calif.: Vatican Observatory/Center for Theology and the Natural Sciences, 1998), 191–224.

tive proposal) are measurement events. Their strictly episodic nature does not obviously fit them to describe agency, which must surely be assumed to have a more free-flowing character. (The very short time-scales associated with decoherence phenomena show that recurrence times could be extremely short, but these events would still form a discrete punctilinear series.)

Of course, there are specific processes (such as gene mutation) in which the outcomes of atomic events generate macroscopic consequences, as well as ingenious contrivances (such as that to which Schrödinger's cat was subjected) that may operate similarly. However, it does not seem to me that these particular instances, and their discrete consequences, manifest much promise of modeling the flexible actions of agents. A vastly greater range of openness than is suggested here would appear to be required. This would seem to necessitate an attempt to generate much wider macroscopic consequences by appealing to some much more generalized notion of "the amplification of quantum uncertainties." However, this attempt may then be expected to run foul of the presently unresolved problems of micro/macro relationships that were considered in the brief discussion of quantum chaology.

2. Those who have appealed to chaos theory, ontologically reinterpreted as reflecting openness and not mere ignorance, and potentially affording a desirably continuous account of active process,[7] do so partly because the exploration of a strange attractor seems to have a much wider and more flowing character than is afforded by a discrete series of measurement events. Chaotic regimes occur in many physical contexts (self-gravitating systems, fluid motions, physiological states, etc.) that are of potential significance for the unfolding of history.

However, this approach is not without its own problems, though they do not arise from the claim that chaos theory is intrinsically deterministic,[8] since this depends upon taking the Newtonian equations as given and unquestionable—a decision that simply preempts the metaphysical issues from the start. (We have seen that, for those who take the ontological option, these classical equations are considered to be downward emergent approximations of a more supple ontology.) The actual difficulty is that macroscopic dynamics cannot possibly be the whole story, since quantum theory clearly operates at microscopic levels. The fractal geometry assigned to strange attractors must therefore be modified on some scale. This implies that the zero energy difference between paths through the attractor, appealed to in support of introducing the notion of pure "active information," cannot hold absolutely,[9] but must be modified, at least at the level of Heisenberg energy uncertainties. Once again, we run afoul of the unresolved problems of quantum chaology in any attempt to appeal to a "hybrid scheme."[10]

It is interesting to note that William Pollard, who pioneered the consideration of how quantum mechanics might be held to relate to the issue of divine action, clearly believed that a many-level approach would be necessary. He wrote:

> When we survey the whole field of natural phenomena from electrons to man, giving equal weight to everything that happens, it is surprising to discover what a rare and

[7] See John C. Polkinghorne, "The Metaphysics of Divine Action," in *Chaos and Complexity*, Russell, Murphy, and Peacocke, eds., 147–56, particularly footnote 3.

[8] This claim was made by several contributors to *Chaos and Complexity.*

[9] I am grateful to Nicholas Saunders for drawing my attention to this point.

[10] John Polkinghorne, *Scientists as Theologians* (London: SPCK, 1996), 37.

> severely limited class is formed by phenomena of the kind studied in classical physics, in which only one alternative exists.[11]

He gives turbulence in fluids (a chaotic phenomenon) as an example. A little earlier he had written

> Elsewhere [above the atomic level] in these sciences where variability, alternatives and probability are found, we must suppose that they arise out of as yet undefined principles or sources of indeterminacy proper to biological organisms or man as such.[12]

Clearly we must await further scientific advance before more adequate metaphysical conjectures can be made. This does not mean that either of these primitive strategies has been a waste of time. Metaphysical proposals have often to be defended by a degree of appeal to general appropriateness, since detailed and convincing arguments are frequently not available, reluctant though some of the graver proponents may be to acknowledge that this is so. An element of hand-waving cannot be avoided, but the key thing is to try to wave one's hands in a suitably promising direction. Either of the suggested strategies is sufficient to indicate that it is possible to take physical science absolutely seriously without being condemned automatically to denying the possibility of the acts of agents, either human or divine. There is room for metaphysical maneuver.

In our present state of knowledge, no proposal relating to the conceivable causal joint of divine providential interaction can be more than what the physicists would call a "zero-order approximation," a crude starting point from which one may hope that better developments might eventually spring. Science itself often starts (and sometimes gets stuck) with zero-order approximations. The art is to try to choose the most satisfactory starting point, the one that will be the best approximation to the essential character of the more complicated reality one is trying to understand.

Those who appeal to quantum theory are supposing that its widely held (but poorly understood) indeterminism is the appropriate place at which to begin. Those of us who appeal to chaos theory are supposing that its direct connection with macroscopic phenomena, where human and divine agency are both expected to manifest themselves, makes it the more promising zero-order approximation. Both groups surely know that there is much more work to be done.

This might seem to be a disappointingly modest conclusion to the discussion, but problems of the exercise of agency are of immense difficulty and are unlikely to yield, after centuries of effort, to a contemporary quick fix. We can nevertheless be affirmed in the value of continuing the enterprise by our belief in our direct experience of human agency, and by our religious intuition of divine providence, both of which assure us that these fundamental experiences must be capable of being accommodated within an adequate worldview.

There are two basic strategies that might be pursued in the hope of fostering further progress. One is that of bold metaphysical speculation, which takes science into account but is relatively uninhibited in pressing on to grander designs. The alternative, more cautious, approach recognizes that science alone will never provide the key to the nature of agency, yet seeks patient clarification of the complex character of physical process as being the best basis for the elaboration of further metaphysical ideas. I am of this second school. My immediate hope is for the further progress and clarification that the insights of quantum chaology may have to offer.

---

[11] William Pollard, *Chance and Providence* (London: Faber and Faber, 1958), 56.

[12] Ibid.

GENERALIZATIONS: FROM QUANTUM MECHANICS TO GOD

Michael Heller

1 Introduction: Evolution within the Conceptual Framework

The evolution of concepts is a driving force of scientific progress. A new concept contains in itself the heritage of its predecessors and, at the same time, is open to future generalizations. However, concepts never evolve "by themselves"; they participate in the struggle to solve problems. There is always a problem to be solved that challenges the way a concept is currently used. If an attempt to put new content into an old concept produces paradoxes and inconsistencies, a crisis ensues. When many such critical situations arise, they make the concept less resistant to change and in this way initiate a "conceptual revolution."

In fact, there are no isolated concepts in science. Every change in the content of one concept results in shifts the meaning of many others. It is, therefore, more correct to speak about evolution within the conceptual framework rather than about the evolution of particular concepts. There are various mechanisms of this kind of evolution: from gradual modifications to sudden jumps. To give detailed descriptions of them in particular cases is the task of historians of science; to reconstruct different patterns of this evolution is the challenge for philosophers of science.

The evolution of concepts is reflected in the evolution of language. Some philosophers are inclined to reduce the evolution of science to the evolution of scientific language. It is usually easier to analyze terms as they appear in various linguistic contexts than to explore the meaning of concepts as they unfold in the historical development of science. However, people engaged in creating science think rather than speak, or first think and only then speak to each other; and thinking happens with concepts, whereas speaking happens with terms. This is why I prefer to focus on concepts rather than on terms. Nevertheless, the only way to concepts is through their linguistic counterparts, that is, through terms.

Conceptual evolution is especially clearly seen at the extremities of the history of science, in its beginnings and in the current frontier of research. Let us consider an example.

The "great miracle," the origin of rational discourse concerning the world, happened around the sixth century BCE in the Greek colonies on the coast of Asia Minor. Central to this process was the formation of the concept of necessity. This concept gradually replaced the former idea of the world as a stage of passion and a "free game" between gods or primitive elements. In order to express the idea that necessity produces the natural order by linking different phenomena into chains of causes and effects, Greek thinkers used the term *ananke*, which in everyday language "literally meant the various means, from persuasion to torture, by which a criminal was made to confess."[1] In this way, the everyday concept changed its meaning to become a predecessor of such important concepts as law of nature, determinism and causality. In the subsequent history of science these concepts were never static, but underwent many changes and transformations.

[1] Olaf Pedersen, *The Book of Nature* (Vatican City State: Vatican Observatory Publications, 1992), 8.

Now, let us go to the other extremity of history, to current research in quantum mechanics. The very same concepts—determinism, causality, law and chance—are at the focus of the contemporary debate. Although these concepts are the products of long evolutionary chains with various adaptations and mutations, their contemporary meanings are shaped by problems in which they are involved and which remain unsolved. Problems over decoherence, entanglement, nonlocality (to name only the most widely discussed) permeate these concepts and change their meanings, as it were, from the inside.

There are many patterns according to which concepts evolve, but in all these "adventures of ideas" one can quite easily discern a certain common feature: as concepts evolve, they become more and more general. The history of physics clearly shows that although the tendency to generalization has seldom been a leading motive for change, new concepts have usually been more general than the old ones. This process is by no means linear. It has many side branches and ramifications. In particular, it may well happen that as time passes some concepts become more refined and more specific. After all, the goal of science is to explain each specific phenomenon. But to explain a phenomenon always means to place it in a more general pattern. This seems to be consonant with the fact that in the process of generalization old concepts are not eliminated, but instead restricted to a smaller domain of validity than originally envisaged. It is precisely the tendency towards more and more general horizons that constitutes the central current in the evolution of science. One might even risk the statement that the generalization of concepts and theories determines an "arrow of time" within the process of scientific transformation.

The history of science could be looked upon as a great attempt by humanity to catch reality in the net of concepts and theories. Every time we think the endeavor has succeeded, it turns out that reality is richer than the concepts and theories we have elaborated so far. The situation is not unlike the one in theology where one speaks about God even if one believes that God is infinite and, consequently, transcends all human conceptual and linguistic possibilities. Some theologians go so far as to say that negation (*via negativa*) is the only road open to us for asserting things about God. For example, if we say that God is infinite, we are in fact denying God the property of being finite—the property that we know from our own experience, and about which we are able to speak sensibly. In the view of many theologians, when other concepts (whose origin is not so manifestly by negation as the concept of infinity) are used to refer to God, "though partly true, [they are] ultimately false and must be transcended."[2] The concept of Being itself belongs to this class—it was Duns Scotus who claimed that if we say "God exists," we have spoken more falsehood than truth.

Does this mean that we should remain silent about God? Traditional theology never surrendered to this temptation. It is better to say something about God rather than nothing, even at the cost of transcending the syntactic and semantic rules of ordinary language. Syntactic and semantic rules are indeed at the center of interest in modern analytic philosophy. After the first period of enthusiasm, it soon turned out that the purely formal approach to language is not enough. No language can fulfill its role without an interpretation, i.e., without a world of "objects" or "states of affairs" to which it *refers*. The problem of *reference* became the central problem of the philosophy of language. Ludwig Wittgenstein in his *Tractatus Logico-*

---

[2] Robert Cummings Neville, *A Theology Primer* (New York: State University of New York Press, 1991), 179.

philosophicus went so far as to ask the question: What structure should be ascribed to the world to ensure the meaningfulness of language? He tried to answer this question in a purely formal way. However, he soon realized that the formalistic standpoint is not enough. In his *Philosophical Investigations* he claims that the meaning of a term is determined by the rules of its usage in a given context within a certain "linguistic game." If so, can we meaningfully speak about what goes beyond the domain of our "linguistic games," i.e., beyond the domain of our everyday activity? And what about the language of science and philosophy? In both of these fields we try not only to describe what we see, but also (or even first of all) to understand what cannot be seen (for instance the subatomic world). Jean Ladrière regards this as a challenge for the "Great European Rationalism."[3]

In contemporary physics we are faced with insistent interpretative problems regarding the correlation of our language with what we have deciphered of the quantum world. The issue of interpretation is even more insistent in theology. While it is true that we can speak about God only "on the ruins of our semantics," I think the breakdown of language, as well as our attempts to understand how it breaks down, can lead to insights about God deeper than those gained simply by adding a negation functor to our utterances. When language breaks down in modern physics this is an unmistakable sign that underlying concepts should be generalized. The assumption underlying the present study is that theology can learn something in this respect (by analogy rather than in detail) from physics.

Of course, there is one great difference between physics and theology that should constantly be kept in mind. A generalized concept always transcends the concept that it generalizes. In physics, we can compare the two concepts with each other and clearly see the semantic mechanisms that are at work in the generalization process. If this strategy is applied to language about God, we know that our everyday concepts must be transcended, but we can only speculate how this should be, and we shall never know to what extent the newly elaborated concepts are adequate. In contemporary physical theories the degree of generalization is immense. Concepts that are now standard in physical theories are very distant from those we use in everyday life. In theology the degree of transcendence goes much further.

The goal of the present essay is to look, in this respect, at contemporary quantum theory and to derive from it a lesson for theology. To do so more effectively, I shall use not quite standard (but perfectly legitimate) analytical tools. My starting point is the fact that the main distinguishing feature of quantum mechanics is its noncommutativity. To show the degree of generalization already present in this physical theory, I shall rely upon the recent discovery of noncommutative geometry. It not only clearly shows the generalization mechanisms underlying present quantum theory, but also points towards further possible generalizations.

2 Algebraic Formulation of Quantum Mechanics

How can we gain access to the quantum world, which, in itself, is closed to our sensory perceptions? We can only hope that some quantum phenomena, after being magnified to the classical level (spontaneously or through our artful inventions), will be detectable by some of our measuring devices. Perhaps the fact that this sometimes does happen is not entirely surprising. After all, our macroscopic world has somehow emerged from the more fundamental quantum level. These properties of

[3] Jean Ladrière, *L'articulation du sens*, vol. 2 (Paris: Éd. CERF, 1984), 109.

the quantum world, which can, at least in principle, be measured by our macroscopic devices, are called *observables*. The question arises, Can we construct a full theory of the quantum world based only on observables? The question itself and the affirmative answer to it, favored by some adherents of the so-called Copenhagen interpretation, are evidently philosophical in character. They were prompted by empiricist presuppositions that, at the time, seemed entirely reasonable. Historically, of course, these presuppositions owed much to the philosophy of logical positivism.

Today we know that quantum theory can be expressed mathematically in terms of observables alone (i.e., operators—see below), in terms of the states of a quantum system alone (i.e., vectors of a Hilbert space), or in ingenious combinations of the two (i.e, expectation values or transition matrix elements). We can choose as primitive elements of the theory either states or observables. In fact, the empiricist formulation of quantum mechanics based on observables came after the formulation in terms of unobservable states (represented as vectors in a Hilbert space) was already well known and thoroughly explored. Of course, what one needs is a certain mathematical representation of observables; but once one has it, the entire range of quantum mechanics can be expressed in terms of this representation. From the works of Pascual Jordan, John von Neumann, Eugene Wigner, Irving E. Segal, Rudolf Haag, and Daniel Kastler, the formulation of quantum mechanics emerged that is now known as its *algebraic formulation*.[4] If one first defines an abstract C^*-algebra (read: "C-star algebra"), it turns out that its (self-adjoint) elements can be regarded as mathematical representatives of observables (as it happens, these elements of C^*-algebras are also called *observables*). From the C^*-algebraic formulation of quantum mechanics, one can recover its original formulation in terms of Hilbert spaces.[5] In fact, the C^*-algebraic formulation of quantum mechanics is more general than its Hilbert space formulation. Quantum statistical mechanics and quantum systems with an infinite number of degrees of freedom (field theories) are somewhat beyond the reach of the Hilbert space formulation, whereas the C^*-algebraic formulation works well in these areas. It came as a nice surprise to physicists that C^*-algebras had been known to mathematicians for some time. Such algebras naturally appear in the important mathematical theory of so-called Banach spaces.

Physicists invented the C^*-algebraic formulation of quantum mechanics by isolating some properties connected with measuring procedures from the usual Hilbert space formulation of the theory. In the standard formulation, vectors in a Hilbert space represent states of the quantum system under consideration. When a measurement is carried out on a quantum system in a given state, the measuring device interacts with this system, changing its state. Therefore, the measurement consists of a transition from one state to another—the act of measurement is what causes this transition. Since states are vectors in the Hilbert space, the measurement can be described formally as something that acts on a vector in the Hilbert space to change it into another vector in the same Hilbert space. Something that does this is known in mathematics as an operator. We can depict this in the following way:

---

[4] See, for example, Eugene Wigner, "On the Algebraic Generalization of the Quantum Mechanical Formalism," *Ann. Math.* 35 (1934): 29; Irving E. Segal, "Postulates for General Quantum Mechanics," *Ann. Math.* 48 (1947): 930–48; idem, "Irreducible Representations of Operator Algebras," *Bull. Am. Math. Soc.* 53 (1947): 73–88; Rudolf Haag and D. Kastler, "An Algebraic Approach to Quantum Field Theory," *J. Math. Phys.* 5 (1964): 848–61.

[5] See, for example, Walter Thirring, *Lehrbuch der mathematischen Physik* (Wien: Springer, 1979).

operator: vector 1 → vector 2

Since various devices measure various physical properties (position, momentum, etc.), various operators can be identified with various physical properties of quantum systems. This is why operators on a Hilbert space are also called simply *observables*, the name often being used interchangeably for physical properties and operators corresponding to them.

Once we are in the realm of mathematics, we can profit from its rich possibilities. For instance, we can use the mathematical concept of representation to understand the connection between the theory of C^*-algebras and the theory of Hilbert spaces. To say that an abstract mathematical structure has a representation in a concrete mathematical structure means that the concrete mathematical structure has all the formal properties of the abstract structure—it is, as it were, a concrete incarnation of the abstract structure. This is exactly what happens in our case: every C^*-algebra has a representation in a subset of (bounded) operators in a Hilbert space. The fact that the set of observables in quantum mechanics has the formal properties of a C^*-algebra is not just a happy coincidence; it is rooted in deep mathematical results.

Now, we come to our main concern. C^*-algebras that find their application in quantum mechanics, in particular in algebras of (bounded) operators on a Hilbert space, are *noncommutative* algebras. We say that a certain operation, e.g., multiplication, is commutative if the order of factors does not affect the result. For instance $3 \times 7 = 7 \times 3$, which means that the multiplication of numbers is commutative. If this property does not hold, we say that the operation is *noncommutative*, and in fact the multiplication of operators in a Hilbert space is, in general, noncommutative. As we can see, the concept of commutativity is a very simple concept, but one with far-reaching consequences. I would risk the statement that (besides linearity) it is precisely the noncommutativity of observables in quantum mechanics that is responsible for all the peculiarities of the theory. Let us consider an example.

For a newcomer to the field of quantum mechanics, the Heisenberg relations are always a great surprise. It turns out that they are the direct consequence of noncommutativity. For instance, let x be the position of a particle and p its momentum. If x and p multiply in the commutative way (as in the case of classical physics), we have

$$xp = px$$

or

$$xp - px = 0.$$

But in quantum mechanics we have

$$xp - px \neq 0,$$

and indeed this difference cannot be less than the Planck constant $\hbar$, i.e.,

$$xp - px \geq \hbar .$$

This is exactly the Heisenberg relation for position and momentum, which can be interpreted in the following way. If we first measure the momentum of a particle and then its position (first p then x, but we write the reverse, xp), the measurement of momentum perturbs the position of the particle, and consequently the result is different from when we first measure the position of a particle and then its momentum (px), since then the measurement of the position disturbs the momentum (consequently, $xp \neq px$). The Planck constant $\hbar$ determines the lower bound of this perturbation.

It is customary to abbreviate the expression $xp - px$ to $[x,p]$ and to call it the *commutator* of x and p. Therefore, if x and p are the position and momentum of a classical particle, e.g., a billiard ball, we can write

$$[x,p] = 0;$$

and if x and p are the position and momentum of a quantum particle, we have

$$[x,p] \geq \hbar .$$

Let A be a C^*-algebra. If it is a commutative algebra, the commutators of all its elements vanish, i.e.,

$$[a,b] = 0$$

where a and b are any elements of A. If A is a noncommutative algebra, the commutators of at least some of its elements, for instance of a and b, do not vanish, i.e.,

$$[a,b] \geq k$$

where k is a deformation parameter. Every commutative algebra can be made noncommutative by suitably perturbing it. In light of this we can say that, from the mathematical point of view, quantum mechanics is but a noncommutative C^*-algebra with the Planck constant playing the role of a deformation parameter.

In the next section we shall explore some important consequences of noncommutativity more deeply.

3 One Step Further

One of the main trends in modern physics is the tendency to geometrization. This can be traced back to Descartes, who claimed that extension (which is a purely geometric property) is the fundamental property of material bodies and, consequently, that physics should be carried on *more geometrico*. This tendency reached its peak with Einstein's general relativity, which is *par excellence* a geometric theory; its main idea is to present gravity as the curvature of spacetime. On the other hand, quantum mechanics could hardly be called a geometric theory. Its strongly probabilistic features take it rather far away from anything we would be inclined to associate with geometry. This is certainly one of the reasons why it is so difficult to unify quantum mechanics with general relativity. However, we could at least try to look at quantum mechanics with a "geometric eye," and to geometrize at least some of its aspects. It is rather evident that in this process the very concept of geometry will have to be generalized. It often happens that if a mathematical or physical theory admits more than one formulation, only one of these formulations can serve as a suitable starting point for the chain of subsequent generalizations. It turns out that in the case of quantum mechanics this role is fulfilled by its algebraic formulation (briefly presented in the preceding section) rather than by its usual formulation in terms of Hilbert space. To show this we must first turn to the concept of geometric space.

In modern differential geometry, the concept of geometric space has been rigorously elaborated and is known under the technical term of 'differential manifold' or simply 'manifold'. The standard way of defining this concept is in terms of coordinate systems with which this "space" can be equipped. This way of defining a manifold closely follows the practice of geometers, who usually do their calculations in terms of coordinates of various geometric objects (points, vectors, curves...) on a given manifold. It turns out, however, that this approach is rather rigid and thus not suitable for further generalizations. Happily enough, there is another equivalent method of defining a manifold that proves to be more flexible in this respect. Instead of considering coordinate systems on a given manifold we can consider a family of all smooth functions on this manifold. It can be shown that all relevant information about the manifold is contained in this family of functions. Here is the essential point: this family of functions forms an algebra. In principle, then, one can forget about coordinates and construct the geometry of the manifold entirely in terms of the algebra of smooth functions. To deal with functions is usually more difficult than to compute with the help of coordinates but, as I noted above, the "functional" approach turns out to be more suitable for further generalization.

First, we can take any algebra of functions, and treat it—*ex definitione*—as consisting of smooth functions on a certain space. Spaces that are introduced in this way are called *differential spaces* (or *structured spaces*), and they are more general than differential manifolds: each manifold is a differential space, but not every differential space is a manifold.[6]

Since functions are multiplied in a commutative way, all functional algebras (i.e., algebras that have functions as elements) are commutative algebras. But we can go even further and claim that any algebra, not necessarily a commutative one, defines a certain space. In such a case, we speak about a *noncommutative space*. This is the central idea of the new branch of mathematics nowadays known under the name of *noncommutative geometry*. It has recently been elaborated in great detail by many mathematicians and theoretical physicists.[7]

4 Pointless Spaces

Noncommutative geometry is indeed a vast generalization of standard geometry. "Sets" or "objects" that had previously resisted usual geometric methods are now perfectly workable noncommutative spaces. One of the chief peculiarities of noncommutative spaces is their totally global character. In principle, no local concept can be given any meaning in the context of noncommutative geometry. Typical examples of local concepts are those of point and its neighborhood. It turns out that, in general, we cannot say that a noncommutative space consists of points. Noncommutative spaces are indeed pointless spaces.

If we want to reach the very roots of the concept of "point" we need to think in algebraic terms, since noncommutative geometry is based on algebraic concepts. In the standard approach, points of any differential manifold (which is a commutative space) are identified with the help of their coordinates, but how can one identify a point with the algebra of smooth functions? The answer is surprisingly straightforward: every point is uniquely determined by all smooth functions that vanish at it. The family of such functions has certain properties that—in mathematical jargon—qualify them as *maximal ideals* of the algebra of smooth functions on the manifold under consideration. We can say that in the algebraic approach, the role of points is

---

[6] See Michael Heller and W. Sasin, "Structured Spaces and Their Application to Relativistic Physics," *J. Math. Phys.* 36 (1995): 3644–62. Structured spaces are somewhat more general than differential spaces.

[7] See, for example, Alain Connes, *Noncommutative Geometry* (New York: Academic Press, 1994); Giovanni Landi, *An Introduction to Noncommutative Spaces and Their Geometries* (Berlin; Springer, 1997); John Madore, *An Introduction to Noncommutative Differential Geometry and Its Physical Applications*, 2nd ed. (Cambridge: Cambridge University Press, 1999); José M. Gracia-Bondia, Joseph C. Várilly, and Héctor Figueroa, *Elements of Noncommutative Geometry* (Boston-Basel-Berlin: Birkhäuser, 2001).

played by maximal ideals of a given algebra. In other words, in the standard (commutative) geometry a point can be equivalently described either by specifying its coordinates or as a maximal ideal of the algebra of smooth functions on the manifold (space) under consideration. And here we touch on an important property of noncommutative algebras: in general, they have no maximal ideals. Consequently it is impossible to define points in noncommutative spaces. Similar reasoning shows that for noncommutative spaces the concept of a neighborhood (of a given point) is meaningless.

To see the vastness of the generalization encountered here, let us look at this problem from yet another point of view. It is sometime claimed that set theory can serve as a logical basis for all other mathematical theories. Of course, the crucial concept in set theory is that of "belonging to a set." For this concept to have meaning, one must be able to identify elements of the collection under consideration by means of an at most denumerable family of properties. In noncommutative spaces, in general, there is no such possibility; one cannot distinguish the elements of such a collection from each other by means of a denumerable family of properties. (Each such property would define a measurable subset of the collection being considered).[8] Alain Connes concludes: "The noncommutative sets are thus characterized by the effective indiscernability of their elements."[9]

The concept that is closest in this context to the concept of point is that of state, which is well known in both classical and quantum physics. Let us note, however, that it is a global concept. For instance, in classical mechanics the state of a system composed of *n* particles, where *n* can be as large as one wishes, is characterized by providing information about the positions and momenta of *all* particles. In quantum mechanics, states are represented by vectors in a Hilbert space and play the crucial role in the structure of the theory. The nonlocal character of this concept can be clearly seen, for instance, in the Einstein-Podolsky-Rosen (EPR) type experiments, in which two quantum particles that once interacted are described by the same state (or by a vector in a Hilbert space) even if they are now at opposite edges of the galaxy.

Clearly, there is a certain affinity between the main characteristic of quantum mechanics and that of noncommutative geometry. It looks as if a noncommutative space were underlying quantum phenomena. The Heisenberg relations, nonlocal effects, the importance of the state concept—these are doubtlessly results of some underlying noncommutativity. Recall from section 2 that quantum mechanics can be expressed in terms of a noncommutative C^*-algebra. It is a straightforward task to associate a noncommutative space with such an algebra. However, physicists' ambitions go far beyond obtaining yet another formulation of quantum mechanics. The main goal of current research in theoretical physics is to unify quantum mechanics with general relativity, and it seems that noncommutative geometry can offer new possibilities here. In fact, there have been many attempts to formulate a noncommutative version of general relativity with the larger goal of developing a theory of quantum gravity.[10] In a series of papers I and my coworkers, Wiesław

[8] This is valid provided one decides to use only measurable maps between spaces; see Alain Connes, *Noncommutative Geometry*, 74.

[9] Ibid.

[10] See, for example, Ali H. Chamseddine, G. Felder, and J. Frölich, "Gravity in Non-Commutative Geometry," *Commun. Math. Phys.* 155 (1993): 205–17; Ali H. Chamseddine and A. Connes, "Universal Formula for Noncommutative Geometry Actions: Unification of

Sasin, Dominique Lambert, and Zdzisław Odrzygóźdź, have presented a working model of how could this be done.[11] This is certainly not yet a complete physical theory, but rather an attempt to find a new pathway that could lead to the unification of quantum mechanics and general relativity. The first results are encouraging, but there is still a long way to go. In the present study, I shall not go into details of this approach, but rather sketch the philosophy underlying the project. In this essay, I am interested not so much in physical results, as in the "adventures of concepts" and in what can be learned from physical models about the limits of language and imagination.

5 Noncommutative Regime

The main idea of the new approach is to suppose that the physics ruling the universe on the fundamental level is based on a noncommutative geometry, and that at this level there is no distinction between physical processes and the (spatio-temporal) stage on which they develop. There is instead a nonlocal (noncommutative) "pre-geometry" that encompasses everything. Only when we go from the fundamental level to the upper layers of the world's structure does the distinction between the spatio-temporal arena (governed by the ordinary, commutative geometry) and physical processes emerge. By the fundamental level I mean everything that happens below the Planck threshold. This threshold is encountered in two directions: first, traveling *backwards in time* to the moment at which the typical dimension of the universe was on the order of 10^{-33} cm (the so-called *Planck length*); second, traveling *now* to smaller and smaller distances until one reaches the dimension 10^{-33} cm. In fact, these two directions coincide, since below the Planck threshold there is no space and time in the usual meaning of these terms.[12]

This idea did not appear from nothing; it was suggested by our work on classical singularities.[13] Roughly speaking, the (initial) singularity is a geometric counterpart of the Big Bang, and "classical" means that when investigating it we do not take into account quantum effects. To decipher the geometric structure of the Big Bang, we have used the methods of noncommutative geometry, which turn out to be very effective in this respect. Our result was that at the level described by noncommutative geometry there is no spacetime and no distinction between singular and non-singular states. Both spacetime and singularities appear as soon as we change to the usual, commutative description. Moreover, some mathematical techniques (operators on a Hilbert space) typical for quantum mechanics appear almost automatically. It was quite natural to take one further step and speculate that these features are not

---

Gravity and the Standard Model," *Phys. Rev. Lett.* 24 (1996): 4868–71; John Madore and J. Mourad, "Quantum Space-Time and Classical Gravity," *J. Math. Phys.* 39 (1998): 4423–42.

[11] Michael Heller, W. Sasin and D. Lambert, "Groupoid Approach to Noncommutative Quantization of Gravity," *J. Math. Phys.* 38 (1997): 5840–53; Michael Heller and W. Sasin, "Emergence of Time," *Phys. Lett. A* 250 (1998): 48–54, idem, "Noncommutative Unification of General Relativity and Quantum Mechanics," *Int. J. Theor. Phys.* 38 (1999): 1619–42; Michael Heller, W. Sasin, and Z. Odrzygóźdź, "State Vector Reduction as a Shadow of Noncommutative Dynamics," *J. Math. Phys.* 41(2000): 5168–79.

[12] The Planck threshold is also characterized by other magnitudes: the *Planck time* equals 10^{-44} s and the *Planck density* equals 10^{95} g/cm^3.

[13] Michael Heller and W. Sasin, "Noncommutative Structure of Singularities in General Relativity," *J. Math. Phys.* 37 (1996): 5665–71; idem, "Origin of Classical Singularities," *Gen. Relat. Grav.* 31 (1999): 555–70.

just interesting coincidences, but rather indicate that the fundamental level is indeed noncommutative.

Let me describe in more detail what the noncommutative regime of the fundamental level might look like. The most striking feature of this regime is its entirely nonlocal character. This also creates the main difficulty we encounter in thinking about it, since all our thinking takes place in local terms (I am a localized individual at a concrete place and in concrete instant of time). Happily enough, the language of mathematics transcends the reach of imagination, and gives us some insights into the realms of noncommutativity.

In the noncommutative regime, there is no time and no space with their accompanying concepts of point and time-instant. Does this mean that there is no motion and dynamics? Without the usual concepts of space and time one cannot have the usual concept of motion. But this does not mean that there can be no dynamics in a more generalized sense. Usually, dynamical magnitudes (such as velocity and momentum) are described by vectors that, as local concepts, do not appear in a noncommutative setting. But "vector field" evidently has a global aspect (e.g., a vector field can be defined on an entire space), and as such it has its counterpart in noncommutative geometry. This counterpart is constituted by the so-called *derivation* of a given algebra. Such a derivation transforms one element of the algebra into another element of the same algebra, modeling a certain change in the system, and satisfies axioms typical for a vector field.[14] It turns out that it is possible to write down dynamical equations in terms of derivations. Although there is no space and no time, there can be an authentic (albeit generalized) dynamics (see below, section 7.2). Interestingly, this would appear to falsify the claim of some philosophers and theologians that, in the absence of time, one is necessarily confronted with a purely inactive, static situation.[15]

As I have mentioned above, one can meaningfully speak about states of the noncommutative universe. And, as our model shows, all such states are on an equal footing; there is no distinction between singular and nonsingular states. Therefore, the very idea of the beginning of the universe will have to undergo a radical transformation. If there is no time, one cannot speak about a temporal beginning. This does not necessarily mean, however, that the idea of a "quantum tunneling out of nothing" could not be adapted to the noncommutative conceptual framework.

One of the central concepts of Western philosophy is that of the individual. This concept preserves its importance in classical physics, but already in quantum mechanics, and especially in quantum field theories, very serious difficulties connected with this concept appear.[16] If our hypothetical model is at least approximately true, these difficulties are but the remnants of the totally global character of the noncommutative stage.

I should mention yet another concept, which in the noncommutative regime is highly generalized: the concept of probability. Standard probabilistic concepts come into play only when there are many individuals of a certain type. It seems that in quantum mechanics one can apply certain probabilistic concepts even to a single particle. In the noncommutative setting, there are no individuals, but the concept of

---

[14] Linearity and the Leibniz rule.

[15] See, for example, Paul Tillich, *Systematic Theology*, vol. 3 (London: SCM Press: 1978), 305.

[16] See, for example, Paul Teller, *An Interpretative Introduction to Quantum Field Theory* (Princeton: Princeton University Press, 1995), chaps. 2 and 3.

probability survives in a generalized sense.[17] This concept has very little in common with the idea of "counting the ratio of the frequency of favorable events to all possible events," although it reproduces this idea in the correct (commutative) limit. It preserves more abstract properties of probability (such as that the outcome of probabilistic calculations must be always positive and not greater than one). The probabilistic character of quantum mechanics, together with all its probabilistic peculiarities, seems to be but a consequence of the deeper and more general concept of probability permeating the entire noncommutative regime. I shall return to this issue in section 7.2.

6 Emergence of Standard Physics

Of course, every scientifically valuable model of the pre-Planck era must reproduce the standard physics of the post-Planck universe. This can be achieved quite naturally in our approach. A center of a noncommutative algebra is the set of all its elements that commute with all other elements of this algebra. If we restrict the original noncommutative algebra to its center (or to a subset of the center), we recover an ordinary commutative algebra. This happens in our model. In this way, we recover the usual physics with general relativity presenting gravity as the space-time curvature, and quantum mechanics as a probabilistic theory. This restriction of the original algebra to its subset can be interpreted as the first "phase transition" in the history of the universe, which gave birth to space, time and multiplicity.

It is interesting that the detailed analysis of this "phase transition" shows that, together with spacetime, classical singularities are born.[18] The situation is as follows. From the point of view of an observer situated on the fundamental level (if such there could be), the universe has no temporal beginning (all its states are on an equal footing), but a macroscopic observer can truly say that the universe had a temporal beginning in its finite past (the initial singularity) and will possibly have an end in its finite future (the final singularity, if the universe is spatially closed). This is a new possibility, one that has so far never been considered in cosmology. Previously people had envisaged two mutually exclusive possibilities: either the future quantum gravity theory will remove singularities from our picture of the world, or not. Now, a third possibility appears: singularities are but a part of our macroscopic perspective. From the fundamental level, the question of the existence or non-existence of singularities is meaningless.

By restricting the original noncommutative algebra to its center, we obtain ordinary physics, but also something more. In "ordinary physics" there are certain phenomena of typically nonlocal character that either remain unexplained or require some additional (and very often artificial) hypotheses to explain them. Let us mention the EPR-type experiments in quantum mechanics, in which space-like separated particles that once interacted "know about" each other although there is no physical signal by which they could communicate. To explain such phenomena, some exotic interpretations of quantum mechanics have been proposed. Another typically nonlocal phenomenon appears in cosmology as the so-called horizon problem: How can parts of the universe now far distant from one another that have never been in causal contact with one another be characterized by exactly the same values of

---

[17] The noncommutative counterpart of the concept of probability is found in the so-called von Neumann algebras.

[18] See Heller and Sasin, "Origin of Classical Singularities."

certain parameters (e.g., temperature corresponding to the microwave background radiation)? This phenomenon is usually explained by the so-called inflationary model, which is an extra cosmological hypothesis (one, by the way, to which some cosmologists strongly object).

All these nonlocal phenomena find their natural explanation within the noncommutative approach. Since the fundamental level is entirely nonlocal, it is no wonder that some quantum phenomena (such as EPR experiments), which are rooted in that level, exhibit some nonlocal effects; they are but the tip of the iceberg of the "fundamental noncommutativity" that somehow survived the "phase transition" to the usual physics.[19] To explain the horizon problem, we should adopt another perspective and regard the "fundamental noncommutativity" as situated "in the beginning," in the pre-Planck epoch. According to our hypothesis, this epoch was totally global; no wonder, then, that when the universe was passing through the Planck threshold, it preserved some global characteristics even at those places which never (after the Planck threshold) causally interacted with each other.

As I have mentioned, the noncommutative regime allows for a generalized "global dynamics" even though it is timeless. In one of our works,[20] we have proposed an equation describing such a dynamics. Our model has (as it should) two "limiting cases": general relativity and quantum mechanics. If one goes to quantum mechanics, noncommutative dynamics leads to the (unitary) evolution described by the Schrödinger equation. But if one goes to general relativity, noncommutative dynamics "projects down" to a process occurring in spacetime that can be interpreted as the act of quantum measurement in which the so-called reduction of the state vector takes place.[21] This supports the idea, long defended by Roger Penrose,[22] that quantum gravity is responsible for the phenomenon of the vector state reduction.

Although all the above effects seem to corroborate the idea of a fundamental noncommutative level, I should stress once more that in the present essay I am not so much interested in developing a noncommutative "final theory" as I am in illustrating the "adventures of concepts" in modern physics.

7 Adventures of Concepts

Quite independently of whether the hypothesis of a noncommutative origin of the universe is true or not, we can learn a lesson from it. Many important concepts

---

[19] To see why certain nonlocal effects survive the Planck threshold and can be observed, one should look at some mathematical properties of the noncommutative regime. Generally speaking, the transition from the noncommutative level to the level of ordinary physics in our model has the character of a projection. Such a projection, being always "onto," switches off all possible "short-distance correlations," but can leave "long-distance correlations." See, for example, Michael Heller and W. Sasin, "Einstein-Podolski-Rosen Experiment from Noncommutative Quantum Gravity," in *Particles, Fields and Gravitation*, Jakub Rembieliński, ed. (New York: American Institute of Physics, 1998), 234–41; and Heller and Sasin, "Noncommutative Unification of General Relativity and Quantum Mechanics," secs. 7 and 8.

[20] See, Heller and Sasin,"Emergence of Time."

[21] Heller, Sasin, and Odrzygóźdź, "State Vector Reduction as a Shadow of a Noncommutative Dynamics."

[22] See Roger Penrose, *The Emperor's New Mind* (Oxford: Oxford University Press, 1989).

involved in current scientific (and philosophical) research appear in the noncommutative context in a strongly generalized form. I do not claim that when we refer our everyday concepts to God, we should generalize them in a similar manner. My intention is to examine the way in which a concept can be generalized, and to treat this as a warning for our theological discourse. Roughly speaking, if such drastic generalization of meanings is possible in physics, how much (infinitely!) more this can occur when we are speaking or thinking about God.

In the following, I shall take a few concepts that are of great importance for philosophy and theology (such as causality, probability, and chance) and apply them to the environment of the "noncommutative world." This world is powerfully shaped by its two principal (and mutually interconnected) properties: timelessness and nonlocality. Therefore, if we want to obtain "noncommutative counterparts" of these concepts, we must, first of all, strip them of their usual involvement with time and locality. Happily enough, in doing so we are not condemned to work only with the help of our intuition; we can also be guided by strict mathematical methods that lead securely from our well-behaved commutative world into the regions of noncommutativity.

7.1 Causality

The problem of causality is difficult even in the commutative world. I do not intend here to go into all its subtleties; I shall confine myself rather to its most commonsense aspects. It seems obvious that causation presupposes distinct events. According to John-Michael Kuczynski:

> If *a* causes *b*, then *a* and *b* are distinct—that is, they are not identical and neither is a part of the other. The lightning and the thunder are distinct, and this is why it is possible for the lightning to cause the thunder.[23]

How is one to imagine causation in a totally global setting where no well-localized entities can exist?

It also seems that causality presupposes a temporal order: causes usually precede their effects. As is well known, there were attempts to reduce causality to temporal sequences. According to Hume (and many of his followers), we cannot know that "*b propter a*," but only that "*b post a*." What remains of causality if we strip it of all temporal aspects?

We can, at least partially, answer these questions if we remain in the domain of mathematical models. Spacetime in general relativity is equipped with what is called its *causal structure*. This structure is defined by the so-called "Lorentz metric," which at each point of spacetime introduces a light-cone determining how physical signals (or causal influences) can propagate throughout spacetime. Every such cone is generated by light-rays emanating from a given point in spacetime. No physical signal can propagate outside the light-cone, which is equivalent to the fact that the velocity of light is the maximal physical velocity in nature. In this sense, the light-cone structure in spacetime determines a "net of channels" through which causal influences can propagate.

When we change to a noncommutative setting, the causal structure changes drastically. First of all spacetime, as a set of well-localized points and their neighborhoods, disappears. We are left only with a "global" entity that can be in

[23] John-Michael Kuczynski, "A Solution of the Paradox of Causation," *Philosophy in Science* 8 (1999): 81.

various states. In Connes' approach,[24] one cannot define a Lorentz metric, only a Riemann metric. The latter opens up light cones in such a way that there is no limiting velocity imposed on physical signal propagation: everything in spacetime can be causally connected with everything else. However, by using the Riemann metric, one can define the distance between various states. It seems, therefore, that there are states that can be causally connected. In our approach, one can define a Lorentz metric that, however, is totally nonlocal, in the sense that it does not define local light-cones, but specifies which vector fields (regarded as global objects) can influence each other. Remember that derivations of a given algebra are noncommutative counterparts of vector fields and, in our model, are responsible for the dynamics of the system. Therefore, we are entitled to say that, in this model, causality is incorporated into its noncommutative dynamics.

The general conclusion of this discussion is that it is possible to generalize the concept of causality so as to free it from involvement with time and locality. What remains still deserves the name of (generalized) causality, since it gives rise to the dynamics of the system. It seems that the essence of causality is a dynamical nexus rather than the distinctness of the cause and its effect, or their temporal order.[25]

7.2 Probability and Dynamics

The standard concept of probability also requires well-defined individual events, the probability of whose occurrence is to be computed. On the set of all such events, the so-called *distribution function* ascribes to every event a real number from the interval 0 to 1. The *probability* of a given event is the value of the distribution function at this event. Additionally, one assumes that the sum of the probabilities of all events is equal to 1 (one of the events must occur). The latter assumption is often expressed by saying that probability is "normed to 1." In the standard example of throwing dice we have six (elementary) events, and the distribution function for each of them ascribes the real number one-sixth. The sum of these numbers, for all events, is evidently one. The concept of probability in the above sense is a special instance of a more general concept, namely that of measure; and the theory of probability is but a subchapter of mathematical measure theory. Measure is a function on a family of (measurable) subsets of a certain (measure) space; one assumes that values of this function must be positive numbers (including zero and plus infinity) without necessarily assuming that this function is normed to 1.

How can the idea of probability be transferred to the noncommutative environment where all concepts presupposing locality and individuality are in principle meaningless? It cannot be simply transferred, but it can be generalized. A hint of how to do this is contained in the fact that the concept of probability is strictly connected with the concept of function (the distribution function). I cannot here go into details; let me mention only that the concept of measure finds its generalized counterpart in the so-called "von Neumann algebras." A von Neumann algebra is, roughly speaking, a C^*-algebra (see section 2) with a certain positive "functional" that is normed to 1 (such a functional is called a *state*).[26] The concept of "functional"

---

[24] Alain Connes, *Noncommutative Geometry*, chap. 6.

[25] This conclusion is consonant with the traditional pre-Humean notion of causality as linked primarily to agency and not to time-order.

[26] For a precise definition see V.S. Sunder, *An Invitation to von Neumann Algebras* (Berlin: Springer, 1986).

generalizes the concept of "function,"[27] and the fact that it is positive and normed to 1 is a trace of an affinity of this concept with the usual concept of probability.

Happily, the concept of generalized probability in the noncommutative regime is strictly connected with another important property: von Neumann algebras are "dynamical objects." In section 5, I said that in the noncommutative regime, a generalized dynamics can be constructed in terms of derivations of a given algebra, but if the algebra in question is a von Neumann algebra the dynamics significantly improves. On the strength of the famous Tomita-Takesaki theorem,[28] von Neumann algebras guarantee the existence of a certain parameter[29] which "imitates time" and in terms of which we can write generalized dynamical equations.[30] It is not yet the "true time," since it depends on the state of a physical system. In each state we have an essentially different "flow of time" with no possibility of synchronizing them! But we can write truly dynamical equations with the above parameter playing the role of the independent variable.

It is remarkable that in von Neumann algebras the concepts of generalized probability and generalized dynamics are unified. We know from quantum mechanics that quantum systems evolve in a probabilistic manner. The dynamical equation, which in the case of quantum mechanics is the Schrödinger equation, does not describe the evolution of an elementary particle such as an electron, but rather the evolution of probabilities with which we ascribe certain properties (e.g., position or spin) to something we call an elementary particle. Let us notice that these evolving probabilities have a nonlocal character. Because of this nonlocal character, the act of measurement of a certain quantum property performed at a particular location can affect an act of measurement separated from that location by a space-like interval, i.e., an interval that could be covered only by a superluminal signal. We could regard this peculiarity as a trace or a remnant of the entirely nonlocal, noncommutative, fundamental level.

Probability, as we know it from standard mathematical measure theory or from its quantum mechanical version, could be but a shadow of the noncommutative properties of the fundamental level encoded in a von Neumann algebra.

Finally, let us go back to the issue of time. Associated with a von Neumann algebra A there is a group called a *unitary group*. In terms of this group we can define a certain condition (called *inner equivalence*), and if the algebra A satisfies this condition (which glues together some elements of the algebra A, making it "coarser"), all state-dependent times can be synchronized to obtain the single "unitary" flow of time. We recognize in it the same time that measures the unitary evolution of observables in the usual quantum mechanics. However, only if the algebra A is restricted to its center does one recover the familiar macroscopic time. It seems, therefore, that the "emergence of time" is a gradual process that can be mathematically modeled.[31]

---

[27] Roughly speaking, functional is a "function on functions."

[28] The Tomita-Takesaki theorem essentially says that if A is a von Neumann algebra then there exist mappings $a_t: A \to A$ forming one-parameter groups, called *modular groups*, which are dependent on the state on the algebra A (see Sunder, *An Invitation to von Neumann Algebras*, chap. 2). These one-parameter groups can be used to parametrize the dynamics.

[29] Strictly speaking, the existence of one-parameter groups.

[30] See Heller and Sasin, "Emergence of Time."

[31] Details can be found in Heller and Sasin, "Emergence of Time"; see also Alain Connes

8 Theological Consequences

8.1 The Primary Cause

As is well known, traditional philosophy and theology identified God with the Prime Cause. Although philosophical and theological texts were full of declarations that divine causality should be understood in an analogous manner, it seems that in practice this concept was treated much more univocally than other concepts referring to God. The very distinction between the Primary Cause and secondary causes is based on the lesson of everyday experience "that individual things have their own operations, through which they are proximate causes of things, not of all things but only of some."[32] Our own operations are so overwhelmingly causality-laden that we subconsciously attribute the same to God.

With the advent of modern science, the tendency arose to narrow the concept of causality only to efficient causality, and the great influence of Cartesian philosophy added to this concept an intuition of a direct contact between the cause and its effect. These ideas concerning causality were strengthened and consolidated by the progress brought about by Newtonian mechanics, although this branch of physics, with its *actio in distans* and its principle of extremal action (discovered soon thereafter), was rather far from such simplistic ideas.[33] In our time, quantum mechanics has virtually put an end to mechanistic and deterministic interpretations of the causal nexus, but some other forms of causality are in circulation.[34] So-called "bottom-up" causation exhibits a strong reductionist flavor, at least if the reductionism is understood in Steven Weinberg's sense of "petty reductionism," in which a whole system—its structure and functioning—is regarded as arising from the sum of its constituent units, their properties and their behavior.[35] This type of causation is often correlated with (not necessarily opposed to) "top-down" causation. The latter was first recognized in nonlinear dissipative systems, in which

> the changes at the microlevel, that of the constituent units, are what they are because of their incorporation into the system as a whole, which is exerting specific constraints on its units, making them behave otherwise than they would in isolation.[36]

---

and C. Rovelli, "Von Neumann Algebra Automorphisms and Time-Thermodynamics Relation in Generally Covariant Quantum Theories," *Class. Quantum Grav.* 11 (1994): 2899–917.

[32] Thomas Aquinas, *Scriptum super libros Sententiarum Petri Lombardi*, bk. 2, dist. 1., q. 1, a. 4; English translation: *Aquinas on Creation*, Steven E. Baldner and William E. Carroll, trans. and eds. (Toronto: Pontifical Institute of Medieval Studies, 1997), 83.

[33] Newton himself was much subtler in his understanding of causality than many of his followers. At the end of his *Principia* (vol. II, translated by F. Cajori (Berkeley: University of California Press, 1962), 546) he wrote:

> Hitherto we have explained the phenomena of the heavens and of our sea by the power of gravity, but have not yet assigned the cause of this power. This is certain, that it must proceed from a cause that penetrates to the very centres of the sun and planets, without suffering the least diminution of its force; that operates not according to the quantity of the surfaces of the particles upon which it acts (as mechanical causes used to do), but according to the quantity of the solid matter which they contain, and propagates its virtue on all sides to immense distances, decreasing always as the inverse square of the distances.

[34] In some of these forms of causality "the same cause is always followed by the same effect" does not necessarily hold.

[35] Steven Weinberg, "Reductionist Redux," *The New York Review* 5 (October, 1995): 39–42.

[36] Arthur Peacocke, "God's Interaction with the World—The Implications of

It was soon extended beyond its original domain to denote something "more open and more nonlocal than that."[37]

It might seem that the causality encountered in noncommutative models (section 7.1) is a kind of "top-down" causality. However, if we take into account the fact that, according to these models, this type of causality operates at the most fundamental level, it could equally well be labeled "bottom-up." I think its distinctive feature is not its specific "vertical" mode, but rather its a-temporal and nonlocal behavior. If we could learn from it something about God's possible ways of acting in the world, it would reveal first of all the drastic difference between this type of causality and what would traditionally qualify as causality. I think, however, that in spite of this difference, the former deserves the name 'causality'. This terminological decision is justified by the fact that noncommutative causal structure (in the mathematical sense) is a legitimate generalization of the usual causal structure of spacetime studied in general relativity. By no means do I want to say that when speaking about God's causality we must regard it as a sort of noncommutative causality; I only want to emphasize that we must treat the doctrine of the analogous or metaphoric character of theological language very seriously.

We think of God as the Prime Cause because we regard the world (both in its entirety and in its details) as an effect of God's action. This causal nexus between God and the world is called *creation*. And it is precisely at this point where we must revise our theological concepts. In light of my analysis, we should take into account the possibility that the concept of causality does not necessarily presuppose local interaction between the cause and its effect (causality's global aspect), and that it does not necessarily include a temporal order (causality's a-temporal aspect). Let us notice that the time-honored metaphysical question,"Why is there something rather than nothing?" (so persistently asked by Leibniz) has a strong globally causal flavor. By asking this question we do not restrict the world's existence (which we want to justify) to any particular place or time-instant. The possibility that the world could exist from eternity and nevertheless be created by God (considered very seriously by St. Thomas Aquinas) emphasizes an a-temporal aspect of God's causality. It is truly worthwhile to read the old masters! However, one should not repeat their doctrines blindly but look at them with an eye sharpened by the enlargements of imagination prompted by the achievements of modern science.[38]

Ernan McMullin has recently examined the traditional doctrine of creation. Alluding to St. Augustine's view, he contemplates the "Creator in the fullest sense,"

Deterministic 'Chaos' and of Interconnected and Interdependent Complexity," in *Chaos and Complexity*, R.J. Russell, N.C. Murphy, and A. Peacocke, eds. (Vatican City State/Berkeley: Vatican Observatory/Center for Theology and the Natural Sciences, 1995), 273. In this context Bob Russell favors the term "whole-part causality."

[37] John Polkinghorne, "The Metaphysics of Divine Action," in *Chaos and Complexity*, 151. In reference to God's action in the world, the problem of the "top-down" causation was discussed extensively in the volume *Chaos and Complexity* of the CTNS/Vatican Observatory series; see papers by Arthur Peacocke, John Polkinghorne, Willem Drees, and William Stoeger.

[38] For more about theological lessons that can be learned from noncommutative models see my "Cosmological Singularity and the Creation of the Universe," *Zygon* 35 (2000): 665–85.

a "Being from whom the existence of all things derives."[39] Such a Being cannot be subject to any constraints, and temporality is certainly a severe constraint.

> Time is a condition of the creature, a sign of dependence. ... The act of creation is a single one, in which what is past, present or future from the perspective of the creature issues as a single whole from the Creator.[40]

In this sense, the Creator is "outside" the transient passage of time. Of course, it is only a metaphor but an important one. If we enrich it with Boethius's famous definition of eternity as "the whole, simultaneous and perfect possession of boundless life,"[41] it conveys not so much an image of timelessness, as of fullness of time.

The concept of existence is crucial for the traditional doctrine of creation. However, it is notoriously vague and fuzzy.[42] In contrast with the concept of a-temporality, one can hardly find in contemporary physics anything that would help one to understand better the meaning of "existence." Perhaps the only hint that the process of physicalization might one day embrace the idea of existence is the dispute in the conceptual foundations of quantum field theories concerning the identity of elementary particles.[43] The discovery that they are lacking what could be termed "primitive thisness"[44] is at least a signal that the problem of *modus existendi* could have a physical component.

8.2 Chance and Purpose

We usually think about chance and purpose in terms of probabilities. For instance, if an event of very small probability happens, we say that it has either happened by chance or has been purposefully chosen by an intelligent agent. We claim that, given long enough span of time, even events of the slightest probability will be realized. If this type of argument is applied to the universe as a whole, it often assumes the form of various teleological or anti-teleological disputes. All of them tacitly assume that the "classical concept of probability" is valid on all levels of reality. As we have seen above (section 7.2), such a view is simply naïve. There are strong reasons to believe that "the classical concept of probability" is valid—as are many other concepts—only within the "classical domain" (i.e., within the domain of linear dimensions comparable to those of our body). Such a possibility has a powerful impact on the philosophical and theological polemics in this context.

To claim that God uses a noncommutative probabilistic measure in designing the universe would display an equally enormous naïveté. But the very existence of noncommutative dynamical models shows that the idea of an a-temporal existence is not contradictory in itself (as has been claimed by many thinkers, especially those

---

[39] Ernan McMullin, "Evolutionary Contingency and Cosmic Purpose," *Studies in Science and Theology* 5 (1997): 91–112, 104.

[40] Ibid., 105.

[41] Boethius, *The Consolation of Philosophy* (Oxford: Clarendon, 1999), 5.6.

[42] The term "existence" has been analyzed by Gottlob Frege, Alfred Tarski, Stanislaw Leśniewski and others. However, their analyses are valid only in the context of a given formal logic, and it is by no means clear which kind of logic one would have to use with regard to the sought-after fundamental theory of physics.

[43] See, for example, Teller, *An Interpretative Introduction to Quantum Field Theory.*

[44] This term has been used by Paul Teller (ibid., 17) as a modern counterpart of the medieval *heacceitas*.

belonging to the Whiteheadian school of process philosophy). It would be interesting, then, to look at processes that happen by chance (in the macroworld) from the perspective of the hypothetical Creator who exists "outside" the macrocosmic temporal order and creates the world in an a-temporal manner (see section 8.1).

> Such a Creator knows the cosmic past, present and future in a single unmediated grasp. God knows the past and the future of each creature, not by memory or by foretelling, then, as another creature might, but in the same direct way that God knows the creature's present.[45]

Our concepts of design and teleology are heavily laden with the all-pervading idea of temporality: pursuing a purpose is a temporal process, and knowledge of its outcome is possible only if the process is strictly deterministic and if its unfolding does not too sensitively depend on the initial conditions. However, the a-temporal Creator's knowledge "is not discursive"[46]: God does not infer or compute future states of the universe from knowledge of its previous states. He knows what we call the past and the future by inspection, and consequently in his planning there is no aspect of expectation. "For God to plan is for the outcome to occur. There is no [time] interval between decision and completion."[47]

The shift of meanings of such interconnected concepts as probability, chance, and purpose has obvious consequences as far as philosophical and theological disputes on the "design argument" are concerned. Let us once more hear what McMullin has to say on this topic:

> It makes no difference, therefore, whether the appearance of *Homo sapiens* is the inevitable result of a steady process of complexification stretching over billions of years, or whether on the contrary it comes about through a series of coincidences that would have made it entirely unpredictable from the (causal) human standpoint. Either way, the outcome is of God's making, and from the Biblical standpoint may appear as part of God's plan.[48]

There still remains the problem of creaturly freedom in such an a-temporally created temporal world. Let us only notice that if God does not compute the future but rather sees it from the divine a-temporal perspective, it is immaterial whether the process itself is strictly deterministic or not. For God the outcome is simply present.

9 Concluding Story

In his commentary on the *Sentences* of Peter Lombard, St. Thomas Aquinas deals with the following problems:

> 1. whether there is only one principle; 2. whether from that principle things come forth by way of creation; 3. whether things are created only by that one principle, or whether they are also created by secondary principles; 4. whether one thing is able to be the cause of another in some way other than by way of creation; 5. whether things have been created from eternity; 6. on the supposition that things have not been created from eternity, in what way God is said to have created the heavens and the earth "in the beginning."[49]

---

[45] McMullin, "Evolutionary Contingency and Cosmic Purpose," 105.

[46] Ibid., 106.

[47] Ibid.

[48] Ibid.

[49] *Aquinas on Creation*, 63 (*Scriptum super libros Sententiarum Petri Lombardi*, bk. 2, dist. 1, q. 1).

It is interesting to notice that in the first four questions there is no direct mention of time. The temporal aspect of the creation problem appears only in article 5, "Whether things have been created from eternity," in which St. Thomas presents his famous, and for many surprising,[50] doctrine that the concept of the world existing from eternity but nevertheless created by God is not contradictory. In his view, we know by revelation that the world had a temporal beginning, but this cannot be demonstrated by reason alone. After pondering all arguments for and against the temporal beginning of the world and finding that neither side can convince the other, St. Thomas anticipates the disappointment of his readers and switches to a more rhetorical style (unusual in his writings):

> ...If someone should argue on the basis of the full-grown man what must be true of the man in an incomplete state in the womb of his mother, he would be deceived. Accordingly, Rabbi Moses, *The Guide of the Perplexed*, tells the story of a certain boy whose mother died in his infancy, who was raised on a solitary island, and who, at the age of reason, asked someone whether and how men were made. When the facts of human generation were explained to him, he objected that such was impossible, because a man could not live without breathing, eating, and expelling wastes, so that it would be impossible for a man to live for even one day in his mother's womb, let alone nine months. Like this boy are those who, from the way that things happen in the world in its complete state, wish to show either the necessity or the impossibility of the beginning of the world.[51]

Today we ask questions such as: How old is the universe? Did it initiate in a Big Bang? Will the future theory of quantum gravity remove the initial singularity appearing in the standard cosmological model? Is the fundamental level of the world a-temporal and nonlocal?—and many other similar queries. All these questions are purely scientific, and we hope that, with the continual progress of our theoretical and empirical tools, we shall sooner or later find answers to some of them. I do believe that this will greatly contribute to our better posing philosophical and theological questions, and more cautiously formulating tentative answers to them. The main lesson we should learn from science in this respect is that we must always be open to broader and broader horizons. St. Thomas drew a similar conclusion from Rabbi Moses' story:

> What now begins to be begins through motion; hence what causes motion must always precede [the motion] in duration and in nature, and there must be contraries; but none of these are necessary in the making of the universe by God.[52]

Acknowledgment. I express my gratitude to Ernan McMullin for valuable suggestions and for polishing my English.

---

[50] To another of St. Thomas's works on this topic, posterity gave a telling title: *De aeternitate mundi contra murmurantes* (*On the Eternity of the World against Those Who Murmur*), see *Aquinas on Creation*, 114–22.

[51] *Aquinas on Creation*, 97.

[52] Ibid., 97.

TRACING THE LINES: CONSTRAINT AND FREEDOM IN THE MOVEMENT FROM QUANTUM PHYSICS TO THEOLOGY

Philip Clayton

1 Introduction—Should One Apologize for Trying to Do Theology?

Fools rush in where angels fear to tread. Many of the essays in this collection—essays by theologians as well as by professional scientists—describe the difficulties of our shared project. In fact, this book is replete with warnings about how hard it is to determine the viable options for interpreting the quantum mechanical results, then to interpret quantum theory, and finally to specify which theories of action (human or divine) are or are not consistent with the physics of the very small. Perhaps more wisely than I, the majority of the authors have remained at the descriptive level with regard not only to the physics but also to the theology (if any) about which they write.

I am about to dive into the morass of constructive metaphysics, and perhaps even constructive theology, in a rather less cautious manner. Preparing to do so causes one to worry about what flaws of mind (or character?) might be responsible for this lack of reticence. Do I not realize—as James Cushing has shown in many fine publications—that the empirical data underdetermine the interpretive position that one takes? Even if this underdetermination should be contingent on the current state of science rather than expressing some necessary limitation, am I not aware that experts are deeply divided on interpretive ("ontological") questions regarding quantum physics, making forays into this field by non-experts all the more precarious? Finally, do I not realize that theology as a discipline is itself highly suspect, if not downright unsavory, to many scientists; that profoundly difficult questions are raised by *any* attempt to draw connections between technical science and theology; and that—even if scientists did agree on the interpretation of quantum mechanics—additional premises and arguments would be required to reach even the most rudimentary and tenuous of theological conclusions? And if I am aware of these things, why would I still wish to do constructive theology in dialogue with quantum physics?

Two reasons come to mind. First, if there *is* to be any contemporary theology, it cannot be carried out in ignorance of natural scientific results, even the most difficult ones. The one thing worse than a theology that attempts to draw connections between physics and God is a theology that believes it has no need of any such connections, a theology that believes it can concoct the divine out of metaphysical whole cloth. An intellectually responsible theology has no choice, I suggest: theologians either wrestle with the best physical knowledge available or condemn themselves to the subjectivity that *sola fides* has come to represent in the modern world.

The second reason for proceeding is that I am willing to countenance a type of theology (and metaphysics, for that matter) that is much more hypothetical, fallible, open to revision, and provisional than was traditionally allowed. Call it *eine Theologie von unten*: theology from below, theology in the trenches. In various publications over the last decade[1] I have defended a view of theology that allows for

---

[1] Philip Clayton, *Explanation from Physics to Theology: An Essay in Rationality and*

multiple models, open-ended discussion, and the underdetermination of theological theory by data. Now admittedly it is significantly harder to specify exactly which parts of such a pluralistic theology are *true*, since one acknowledges both the viability of competing truth claims and the malleability of one's own claims. But to acknowledge that one does not yet know which of the competing claims will finally turn out to be true does not mean that one must then become a relativist about all arguments. Sometimes a stronger case can be made for one theological option than another, so that—given the state of the data and the discussion at some particular time—one is more justified in assenting to one position than to its competitors. We won't know until we try.

One way to ensure that theological discussions do not amount to "tennis without a net" is to tie them where possible to existing fields of empirical study. In order to do this, one first considers the array of interpretive models in a specific scientific field, looking for areas of compatibility (or incompatibility!) with existing theological models *and reevaluating one's theology, or perhaps even one's science, as a result.* One then repeats this activity across a variety of scientific (and, if the questions are theological, also nonscientific) disciplines, attempting to discern which theological options become more and which become less credible in light of the overlap between the various fields. A theology subjected to these sorts of rigors becomes a sort of third-order discipline: reflections on the "data set" of theories from multiple disciplines, which are in turn responses to the data within those disciplines. Like a complicated Venn diagram that contains many more than three circles, this synthetic task entails looking for common points and exclusion relationships between rather diverse fields of data and theoretical options. In the present essay I will be satisfied if I have been able to illustrate how this process might work when the starting point is the difficulties associated with the interpretation of quantum physics.

2 Why Physics Might Provide Constraints on How God Might Act

Physicists regularly comment on the ways that quantum mechanics has transformed our view of the nature of reality. Authors often use classical or Newtonian physics as the background in order to dramatize the transformation brought about by the twentieth-century understanding of quantum processes. The technique is effective: even the lay reader is struck by what a radically different world it is that she encounters in the realm of the very small.[2] It is as if one were confronting a new level of reality, a new metaphysical space. (This fact alone demands theological response!)

The conceptual space defined by the quantum mathematical formalism and the associated empirical observations neither proves nor disproves the existence of any divine being. Nor will it, by itself, establish—or rule out—any claims about divine action. But it *may* tell us something about how a being (human or divine) must act if it acts in the physical world and in conformity with physical law. That is, the conceptual space of quantum physics may constrain the ways such a being could be manifested and the sorts of actions a human observer could in principle detect. All this assumes, of course, that any inklings we might have of a divine being would

---

Religion (New Haven, Conn.: Yale University Press, 1989); idem, *God and Contemporary Science* (Edinburgh: Edinburgh University Press; Grand Rapids: Eerdmans, 1997); idem, *The Problem of God in Modern Thought* (Grand Rapids: Eerdmans, 2000).

[2] See e.g., James Cushing, *Philosophical Concepts in Physics: The Historical Relation between Philosophy and Scientific Theories* (Cambridge: Cambridge University Press, 1998); John Polkinghorne, *The Quantum World* (Princeton, N.J.: Princeton University Press, 1985).

have to be drawn from its creative actions (as reflected, say, in the structure of the physical world or the evolutionary development of the cosmos) or from its interactions with us (as reflected in human experience, including claims regarding religious experience and revelation in the various world religions).[3] Let me use the word 'God' to designate whatever might be the actual nature of the divine and 'divine action' to designate the manifestations of God (if any) in the cosmos and its history. If God exists and has not "acted" at all, or if these actions fail to indicate anything about the divine nature—and especially if the actions lead us to infer things about the divine nature that are false—then we are completely sunk, epistemically speaking (and perhaps in other ways as well!). In such cases our best reflection will yield only false conclusions about the divine.

On the other hand, it is possible that the constraints of physics do represent the context within which God chooses to act; thus it is possible that the constraints imposed by the physical order *themselves* tell us something interesting about the nature of that order's Creator. Of course, when one reads "possible" rather than "probable" or "necessary," one realizes that this is theology in a hypothetical mode—rather unlike the old certainties of the faith. If theologians must proceed with this sort of tenuousness, one is always justified in choosing not to play. The same holds, by the way, for any theology of divine action in light of contemporary science. Whether one plays will depend on how "expensive" one judges the wager to be (do I lose credibility by even considering the God-hypothesis seriously?) and how valuable one thinks the possible outcome is. Clearly one's decision on *these* questions depends on subjective factors that go far beyond the context of this (or any) academic essay. The most I can show here is that the wager is not irrational. There is no point in wagering on an impossible option, making bets one can only lose. But it is not irrational to wager on a possible outcome.

What, precisely, is the wager? Formulated negatively, the wager is that none of the three "if's" in the sentence two paragraphs above is true. It is a wager because no empirical result can determine the answer, at least at present. Nonetheless, it is the *positive* side of the wager that sets the agenda for the present essay. Put positively, one can wager that the structure of the physical world sets parameters on—and thus gives us some knowledge of—the manner in which God *could* act. The physical world would thus provide us some epistemic access to divine action (if God acts); it would be conducive to knowledge of the source of these actions and of the nature of that hypothetical source. Indeed, if there *is* a God who creates, the bet's not a bad one, for wouldn't one expect the nature of the Creator to be represented in some way in the structures of what has been created?[4] As Owen Thomas has pointed

[3] This is a more philosophical formulation of what theologians call "Rahner's Rule." See Wolfhart Pannenberg, *Systematic Theology*, vol. 1, Geoffrey Bromiley, trans. (Grand Rapids, Mich.: Eerdmans, 1991).

[4] James Cushing refers to the problem of evil at this point. My approach, rather than dodging this difficult set of issues, puts them right at the center. The history of evolution in general, and human history in particular, includes incredible waste and suffering, and the wager suggests that the biological and psychological structures that cause this suffering are somehow indicative of the nature of the underlying divine cause (if any). One pursuing this method must therefore introduce the categories of evil and good, ask whether God can consistently be called good, and examine the reasons that a divine being might have had for allowing biological and psychological structures of this sort. Such debates belong to the field of *theodicy*; see e.g., John Hick, *Evil and the God of Love* (New York: Harper & Row, 1966); John Feinberg, *The Many Faces of Evil: Theological Systems and the Problem of Evil*

out in conversation, my approach also amounts to the wager that there is some analogy between human and divine action, for *our* actions are certainly constrained by the physical world. Hypothesizing some similarity between the human and divine agent gives us some basis for understanding divine action (if it exists), whereas hypothesizing that divine action is utterly *sui generis* would rule out any general knowledge of it.[5]

But why treat quantum physics separately? If we had a unified science in which the interrelationships between the various special sciences were fully understood, as in the third-order form of theology proposed above, we could use the shared structures common to all scientific fields as the starting point for asking about the nature and action of the divine. The task would be difficult, of course, even with an agreed upon empirical and theoretical basis on which to draw. At present, however, we are far from unified science; and major areas of physics, such as quantum physics and gravitational theory, remain theoretically distinct. There is no other option, then, but to consider the various scientific fields *seriatim*, asking what divine action would mean in that context, how it might occur (if it occurs) and, given the laws and structures in question, what the nature of the divine source might be. Whether or not the conclusions that one reaches within the various fields—for example physics, evolutionary biology or genetics, the neurosciences, and the social sciences—fit together into a single picture is a separate question.[6]

I recognize that these proposals are controversial; there are opponents on all sides ("Cannon to the left of them, cannon to the right of them/Volleyed and thundered…") and the debates are often heated. If one uses not only the present book but also other recent publications as data, one finds at least five alternative positions[7]:

1. No reasons can be given, other than purely subjective ones, for *any* theological position (Cushing);
2. Serious theological positions can be defended in light of science *in some cases*, but quantum physics is too unclear, and subject to too much difficulty, to give rise to helpful theological conjectures (Polkinghorne);
3. Some constructive theology can be written on the topic of divine action and quantum physics, even if our conjectures remain highly speculative (the present essay, but also, *inter alia*, those by Chiao, Russell, Stoeger and Tracy);

(Grand Rapids, Mich.: Zondervan, 1994); Thomas F. Tracy, ed., *The God Who Acts: Philosophical and Theological Explorations* (University Park, Penn.: Pennsylvania State University Press, 1994); David Ray Griffin, *Evil Revisited: Responses and Reconsiderations* (Albany, N.Y.: SUNY Press, 1991); Jane Mary Trau, *The Co-existence of God and Evil* (New York: P. Lang, 1991); Richard Worsley, *Human Freedom and the Logic of Evil: Prolegomenon to a Christian Theology of Evil* (Houndmills, Basingstoke: Macmillan Press; New York: St. Martin's Press, 1996). Needless to say, I cannot resolve the debate in this essay, though the successful outcome of my argument requires that I eventually address it.

[5] This is the opposition of faith and reason first formulated by Tertullian and associated in the twentieth century with the "No!" of Karl Barth in his debate with Emil Brunner.

[6] In *The Emergence of Spirit* (forthcoming) I argue for the affirmative, but obviously that case cannot be made here.

[7] Predictably, each of these schools view those to their "left" as unnecessarily empiricist and positivist and those to their "right" as insufficiently aware of the power and rigor of scientific thought.

4. Rather strong theological conclusions can be reached on the basis of modern physics, presumably including quantum physics. Thus, for example, the physicist Cyril Domb finds clear evidence of the Creator in the world, and the "intelligent design" theorists (William Dembski, Michael Behe, et al.) argue that evolution requires a prior intention and in-built design on God's part.
5. The convergence between scientific conclusions and the teachings of the religious traditions is so great that they should no longer be viewed as separate realms that need to be connected but rather as one integrated whole. The "Mystics and Scientists" conferences have produced a variety of calls for their unification[8]; Fritjof Capra has long been famous for touting the role of intuition and holism in quantum physics; and much popular and "New Age" thinking presupposes that the science-religion separation is now defunct. For many of these individuals quantum physics actually serves as the central argument for their position.

Those of us who write in the neighborhood of (3) make the plea to advocates of (1) and (2) that they would at least consider our proposals with an open mind, recognizing that they are not excluded by sound empirical science and that they are different from many of the dogmatic and unquestioning theologies of the past. Likewise, we caution advocates of (4) and (5) to be aware of the hypothetical and contingent nature of all such theological reflection, as well as to observe the continuing distinctions between theology and the sciences.

3 Some Quantum-Mechanical Constraints on the Possibility of Divine Action

In the following paragraphs I explore three examples of issues that have arisen in debates concerning the interpretation of quantum mechanics: the role of subjectivity, the "many-worlds" interpretations, and the debate over indeterminacy and free will. Note that these classic debates have arisen independently of theological concerns and need not presuppose belief in God or divine action. Nevertheless, it is my hypothesis that debates such as these are deeply relevant to the question of divine action. These are the sorts of "level-two" discussions that naturally emerge out of the field of quantum physics and that the theologian of type (3) must master before she proceeds to any constructive theological work.

3.1 The Role of the Observer

One fundamental issue in the interpretation of quantum physics is the role of subjectivity. Since the early days of quantum theory, many leading physicists have held—to put it briefly—that the observer's choice regarding what to measure determines which quantum propensity becomes actual. In opposition to the subjectivity theories, other interpretations dispense with any real reduction of the wavepacket, or argue that the reduction occurs automatically as quantum systems interact with macrosystems, or appeal instead to "consistent histories" or "many worlds" in the effort to avoid giving any indispensable place to conscious intentions in explaining quantum phenomena.

Those who do posit an irreducible role for the observer may do so either as minimalists or as maximalists. Minimalists introduce the smallest amount of meta-

---

[8] See David Lorimer, ed., *The Spirit of Science: From Experiment to Experience* (New York: Continuum, 1999).

physics necessary to explain the reduction of the wavepacket through the act of observation. They dispute the need for a theory of the conscious observer, arguing that it is enough for there to be some macrophysical act of measurement or recording by an observer. No lofty metaphysic is at work here, they assert, since it is trivially true that it takes human agents for there to be science in the first place. All we need to note is that the observer is never *within* the quantum mechanical system being studied. For there to be an observation there must be an observer; hence, according to minimalists, the notion of observation is irreducible in quantum mechanics.

For maximalists, by contrast, these answers stop half-way; there is no way around a metaphysics of the observer. Concepts such as the subject, subjectivity, consciousness, free will, and spirit must be introduced in order to explain what happens when a scientist constructs an experiment and makes an intentional measurement. Now it is true that many leading quantum physicists have taken maximalist views, and we will return to these below. Still, it is important to ask why many others are committed to finding an interpretation that avoids the need for any theory of subjectivity. What are the intuitions that underlie their effort? At the deepest level, as Jeremy Butterfield has argued in the context of the CTNS/Vatican conference, physics has an inherent resistance to invoking subjectivity. Subjectivity is certainly not part of that family of physical properties (such as mass, charge, location, time, and entropy) that makes up standard physical explanations.

It does seem true that one does not immediately need to invoke a full metaphysics in order to interpret the transition from quantum propensities to actual measurements. Some level of analysis lies between the straight physical report and the robust metaphysics of subjectivity or holism that some interpreters favor. In this middle level of analysis one can formulate a more minimalist account of the transition. What occasions the move from quantum coherence to decoherence? Is it a sheer result of size, of the number of particles in a system, or does the act of measurement, or even the *intent* to measure, play a crucial role in this occurrence? The minimalist wants to know only what is entailed by the physics—or, to put it differently, whether anything is presupposed in doing physics and formulating physical theories that (presently, or perhaps necessarily) lies outside the scope of physics. Some minimalists thus argue that an observer is presupposed by quantum theory, and that there is no place for the observer within that theory as currently formulated.

It is important to ask how strong a role is played in this debate by another assumption that has been a part of the history of physics in the modern era. We might call it the "ladder of disciplines" or "ladder of the sciences" assumption. That is, the success of science is based on the explanatory reduction of one discipline to another. If chemistry were a unique domain of its own, not connected via physical chemistry to the fundamental laws of physics, then (it is argued) we would have a situation very similar to the age of alchemy: chemistry would be a completely separate discipline, governed by its own rules, laws, and principles. But (they argue) such isolation of explanatory fields would cast questions on the unity of science and thus on the prospect of the completion of science, or even of genuine scientific advancement. Likewise, if some unique principle of "life" characterized all the biological sciences, such as the striving for perfection or self-development (entelechy), then biochemistry would not be sufficient to explain the functioning of living beings, and again the ladder of the sciences would fail. Isn't some such concern, at any rate, at the root of the resistance to allowing the interpretation of quantum mechanics to rest on a "higher-order" principle such as conscious observation rather than on a principle proper to physical theory itself?

Now consider an alternative metaphysical framework—if only as a *Gedanken* experiment.[9] On this imagined view one expects interconnections between scientific disciplines but not relationships of reduction between all of them. More complicated natural systems are sometimes genuinely or "strongly" emergent from underlying physical systems. New emergent levels presuppose the law-like relations that we find at the lower levels, but they are not fully explainable in terms of those laws. Indeed, phenomena occurring at the "higher" levels are sometimes actually constitutive of the lower-level processes. Systemic patterns describable only at higher levels of analysis affect what occurs at the lower levels—whether it be epigenesis affecting the way a cell's genetic code is actually expressed, or qualities of an ecosystem influencing the behavior of particular organisms in the ecosystem, or the observations of a "subject" affecting which of the quantum mechanical probabilities are in fact observed and become a part of the macrophysical world.

The choice between these two models, which we might call the reductionist and the emergentist models respectively, is a difficult one, and there is much to be said for both and against both. At the most cautious level, it may suffice to note that the interpretation of the quantum mechanical formalism is deeply affected by one's metaphysical inclinations on this matter. But there is also a less cautious response to the question—one that a number of well-known quantum physicists have pursued. An additional set of conclusions can be drawn about the measurement problem by those who are inclined to postulate that observers are a basic part of the furniture of the universe. In its strong form (cf. the strong anthropic principle) this view holds that subjectivity in certain important respects *makes* the physical world to be what it is. Perhaps as a result of being more strongly dualist in its theory of the human person than the emergentist view sketched above, the "subjectivity is basic" view opens the door more readily to a theory of God and God's actions. If subjectivity has this sort of foundational role in the becoming of the physical world, and if a divine being exists, then it would not be unreasonable to imagine that the divine subject plays a role in influencing the outcome of all (or at least a significant number) of the quantum mechanical events in the world.

Like many physicists, I shy away from an overtly robust metaphysics of subjectivity as a tool for interpreting quantum phenomena; and the stronger claims about divine action at the quantum level raise some perplexing problems.[10] But I must at least grant the inherent interest of a metaphysic that, were it successful, would solve the measurement problem and offer a synthesizing perspective on both quantum physics and theology.

3.2 Everett and DeWitt's Many-Worlds Interpretation Versus Subject-Based Interpretations

A particularly interesting set of methodological issues is raised by the famous many-worlds interpretations of quantum mechanics. Many-worlds theories hold that there is no need for a shift from an indeterminate to a determinate state of affairs, no need for a transition from quantum superposition to definite states by means of the so-called collapse of the wavepacket. Instead, the set of possible measurements of a

---

[9] I have developed these ideas in more detail in the above-cited works; I list them here only to illustrate some of the effects of alternative metaphysical or theological frameworks.

[10] See the forthcoming book by Nicholas Saunders from Cambridge University Press, which explores some of the difficulties associated with this view.

quantum mechanical event constitutes a (large) number of branches, each of which represents a different actual universe. The observer, when she makes the measurement, finds herself to inhabit one of these worlds rather than the others. The other universes continue to exist, even though we can have no further causal contact with them.

Note that there are some philosophically important differences among many-worlds theories. Hugh Everett's original formulation emphasized memory: the different memory traces in one or another observer separate the universe as remembered one way or another.[11] In the original 1957 article in *Modern Physics* Everett does not appeal explicitly to multiple (actual) universes. By contrast, Bryce DeWitt appealed to multiple actual universes from the start. When the observer observes the precise location (or momentum or spin) of a subatomic particle, she causes a branching among the various universes, which were identical up to that point; after the observation she inhabits one of these actual universes and not the others. In either case, no actual reduction of the wavepacket occurs. It's not that the physical world changes from indeterminate to determinate; it's rather that a branching of universes occurs and the observer subsequently finds herself in only one of them.

Everett was explicit that his interpretation was designed to avoid the consequence that some mysterious subject should cause an ontological change in the physical world, namely the collapse of the wavepacket. He was thus reacting against the Copenhagen interpretation of quantum mechanics, which held this sort of view. Consider, for example, the position of Werner Heisenberg, who explained the Copenhagen interpretation by taking a fundamentally Aristotelian view of quantum mechanics. Heisenberg believed that quantum indeterminacy was like the world of Aristotle's metaphysics, in which (actual existing) potentials strive to become actual. In this theory the subject acts as a sort of final cause, pulling a certain (real) potential into actual existence. Note that this view reverses the stance of classical (Newtonian) physics, which requires that the subject ultimately be explained in terms of physical laws. For the Copenhagen theorists, by contrast, when a definite measurement is made at the subatomic level, the resulting macrophysical state is a *combination* of a quantum-physical probability distribution and the scientist's decision of what, when, and how to measure. Indeed, on this view the subject's role is in one sense the primary one: the "world" is merely potential until the moment of observation, when the conscious observer resolves it into an actual state. In its most extreme form, the form propounded for instance by John Wheeler, the entire universe may have existed in a state of quantum potentiality until the first observer emerged, at which point it was *retroactively* resolved into macrophysical structures such as stars, planets, and the like. Wheeler even applied this view backwards to the creation of the universe:

> Is the very mechanism for the universe to come into being meaningless or unworkable or both unless the universe is guaranteed to produce life, consciousness and observership somewhere and for some little time in its history-to-be? The quantum principle shows that there is a sense in which what the observer will do in the future defines what happens in the past—even in a past so remote that life did not then exist, and shows even more that "observership" is a prerequisite for any useful version of "reality."[12]

[11] See David Bohm and Basil J. Hiley, *The Undivided Universe: An Ontological Interpretation of Quantum Theory* (New York: Routledge, 1993), 296ff.

[12] John Wheeler, quoted in Paul C.W. Davies, *Other Worlds: A Portrait of Nature in Rebellion, Space, Superspace, and the Quantum Universe* (New York: Simon and Schuster, 1980), 126.

The debate between the many-worlds and the subject-centered interpretation cannot yet be physically resolved. (It may be resolved in the future if many worlds is an entailment of string theory or hyperspace or a cosmology of universes birthing universes, as some claim—and *if* these theories are in fact empirically checkable. Of course, it would also be resolved if one could produce a physical theory that explained the collapse of the wavepacket.) It is therefore at present a philosophical debate, and one that, as I hope to show, is deeply influenced by metaphysical intuitions or assumptions. At the risk of oversimplification, we might state the basic opposition in this way: if you take it to be crucial that the explanation of the world be given ultimately in physical terms, then you will be justified in rejecting explanations that are essentially subject-based—even at the cost of an incredible loss of parsimony. For it certainly *seems* like ontological exuberance (or over-kill) of the worst sort to assert, with DeWitt, that "our universe must be viewed as constantly splitting into a stupendous number of branches" and that "every quantum transition taking place on every star, in every galaxy, in every remote corner of the universe is splitting our local world into myriads of copies of itself. Here is schizophrenia with a vengeance!"[13] But if one holds, as many many-worlds theorists have, that this is the only viable scientific interpretation that interprets quantum mechanics in a purely physical fashion, and if one has a strong enough commitment to avoiding any reference to conscious observers, then it may be a cost one is willing to pay.[14]

But what if you believe that subjects are irreducible parts (inhabitants) of the one universe? In this case your metaphysical belief will incline you to see quantum mechanics as *evidence* for a metaphysics of the subject—as a number of its leading theorists have in fact maintained (Eugene Wigner, John von Neumann, Henry Stapp). Instead of multiplying worlds unnecessarily, you'll argue, one should see quantum mechanics as a (the?) point at which the physical and the mental connect. Thus the quantum physicist Carl Friedrich von Weizsäcker argued in the 1950s that quantum physics was the vindication of Kant's dualism, his sharp separation between the kingdom of causes and the kingdom of means and ends.[15] This was also the view taken by Eugene Wigner and his followers. Wigner used the quantum revolution to argue that "the minds of sentient beings occupy a central role in the laws of nature and in the organization of the universe, for it is precisely when the information about an observation enters the consciousness of an observer that the superposition of waves actually collapses into reality."[16] Interestingly, one of Roger Penrose's arguments against many-worlds theories also appeals to subject-based considerations. He calls them "zombie theories of the world" because "the continual branching of the world and the threading of my own consciousness through it would seem to result in my becoming separated from the tracks of consciousness of all my friends."[17] Penrose insists that one needs an adequate theory of consciousness *before*

[13] Ibid., 136.

[14] Put more strongly, it sometimes seems that the major motivation for many-worlds theorists is that Copenhagen or subjectivity-based views would stand in the way of a strong, unambiguous reduction of all sciences, including the sciences of human subjectivity, to physical objects, forces, and laws.

[15] Carl Friedrich von Weizsäcker, *Zum Weltbild der Physik*, 4th ed., revised (Stuttgart: S. Hirzel, 1949).

[16] Eugene Wigner quoted in Davies, *Other Worlds*, 132f.

[17] Roger Penrose, "Singularities and Time-Asymmetry," in *General Relativity – An Einstein Centenary Survey*, S.W. Hawking and W. Israel, eds. (Cambridge: Cambridge

one can make sense of the many-worlds view as an interpretation of quantum mechanics.

Now there are also serious objections to the subjective interpretation, objections that emphasize its counterintuitive features. Every text on the philosophy of quantum physics includes diagrams of the counterexamples of Schrödinger's Cat and Wigner's Friend. Another form of the objection imagines that a meter is set up to permanently register whether the radioactive particle has decayed at the end of a minute (assuming an experimental set-up in which there is a 50% probability of this occurring). Two photographs are then automatically taken of the meter reading, first photo A and then photo B. The photographs are developed but no one looks at them. Imagine that ten years are allowed to pass during which no subject observes either the meter or the photos. At the end of that time a subject looks at photo B, and suppose that she observes the meter to register a radioactive decay. On Wigner's view—according to the critic—at that moment, but not before, the superposition of states will be collapsed, the particle will (retroactively) have decayed, the meter will (retroactively) register its decay, and photo A (which no one has yet looked at) will suddenly show a picture of the meter in its "on" position. Before that moment photo A was still indeterminate; the observation of photo B makes A determinate—despite the fact that A was taken *before* B.

At this point you may be inclined to dismiss *both* options, if not the entire field of subatomic physics, as hopelessly counterintuitive. Yet counterintuitiveness seems to go with the quantum territory; after all, our intuitions have been nurtured on millennia of macrophysical experiences. The formalism is extremely well supported; it's the *interpretation* that is raising the difficulties. Can theological intuitions (or, for the critic, the intuition that theological intuitions are mistaken) help us at all here?

3.3 Indeterminacy and Free Will

On the subject of quantum physics and free will, at least, it is not only the philosophers who tend toward speculative exuberance; the early founders of quantum mechanics and the initial architects of the Copenhagen interpretation were already quick to draw connections.[18] According to defenders, the free choice of an observer to decide where, when, and how to measure plays an irreducible role in the outcome of quantum physical experiments. The observer thus helps to determine whether the world will be this way or that way, and he does so in a manner that (thanks to Heisenberg's uncertainty principle) is completely consistent with the laws of physics. Since quantum indeterminacy is an inherent feature of the physical world and not a hole that further advances could plug, these physicists argued, physics and genuine human freedom are fully compatible. Indeed, quantum physics opens the door again to free will for the first time since Newton.

Of course, the quantum argument for free will has been widely criticized as well. Indeterminacy at the quantum level may well be cancelled out by the time one reaches the macrophysical level. Thus it might be that the microphysical world is indeterminate but that this indeterminacy disappears when one reaches systems as large as the ones encountered in biology and psychology. If so, quantum physics

---

University Press, 1979).

[18] Werner Heisenberg, *Physics and Philosophy: The Revolution in Modern Science* (New York: Harper and Row, 1958); Carl Friedrich von Weizsäcker, *Zum Weltbild der Physik* (Stuttgart: S. Hirzel, 1949).

cannot support the doctrine of freedom at the level of human actions. Also, physical indeterminacy—the claim that our epistemic limitations in knowing the quantum physical world reflect genuine indeterminacy in the way things actually are at that level—may not be sufficient for human freedom. After all, human freedom is normally parsed in terms of choice, conscious decision, rational selection among multiple alternatives, and similar notions. None of these factors are, as far as we know, operative at the quantum physical level. And, of course, the Copenhagen defense of strong ontological indeterminacy, especially if it is to be transformed by the "choice" of a particular subjective agent, is only one of multiple viable interpretations allowed for by the empirical data.[19] Even if it is the majority view in the field, this fact cannot be taken as decisive support for this interpretation.

Nonetheless, there are various respects in which quantum indeterminacy (assuming it exists) is significant for metaphysical discussions of human and divine action. First of all, indeterminacy seems to be a necessary condition for free will. This debate, which philosophers know as the debate about *compatibilism*, has been a subject of attention for most major modern philosophers and an absolutely central question in twentieth-century philosophy. Compatibilists hold that a fully deterministic universe is compatible with one type of freedom (the freedom necessary for moral responsibility), even if it rules out another traditional type of freedom ("counterfactual" freedom: the freedom that one might have done otherwise, even in precisely the same situation). Thus if a causal chain of events results in some action *A*, such that, given the first event in the chain, *A* would necessarily result, compatibilists would still call *A* a free action as long as the penultimate link in the causal chain ($A - 1$) happens to be the will of the agent in question. On this compatibilist definition an act is free as long as it is the result of the agent's will, whether or not the agent could have willed differently (counterfactual freedom). Conversely, incompatibilists reject this argument. An agent's action is only free if the agent could have done otherwise, all previous moments of her life (and of the universe's history, for that matter) remaining the same. This notion of counterfactual freedom continues to have strong intuitive appeal, even though it has been subjected to some pretty vigorous criticism and, at least until recently, was clearly the minority view among philosophers.[20]

There is no danger that we will resolve this debate here, though some progress has been made in recent years.[21] In this context we can ask only about the effects of the debate on the interpretation of quantum mechanics and theology, and vice versa. If one accepts compatibilism, then the whole issue of physical determinacy or inde-

---

[19] James Cushing, *Quantum Mechanics: Historical Contingency and the Copenhagen Hegemony* (Chicago: University of Chicago Press, 1994).

[20] See, e.g., the numerous publications by Donald Davidson; cf. also Richard Double, *The Non-Reality of Free Will* (Oxford: Oxford University Press, 1991). Neuroscientists have also weighed in on this side, e.g., Richard M. Restak, *The Modular Brain: How New Discoveries in Neuroscience are Answering Age-old Questions about Memory, Free Will, Consciousness, and Personal Identity* (New York: Scribner's, 1994). But the incompatibilist side has, if anything, become stronger in recent years. Among many examples see especially Peter Van Inwagen, *An Essay on Free Will* (Oxford: Oxford University Press, 1983); Timothy O'Connor, ed., *Agents, Causes, and Events: Essays on Indeterminism and Free Will* (Oxford: Oxford University Press, 1995); John Martin Fischer, *The Metaphysics of Free Will: An Essay on Control* (Oxford: Blackwell, 1994); Derk Pereboom, ed., *Free Will* (Indianapolis, Ind.: Hackett Pub. Co., 1997); Robert Kane, *The Significance of Free Will* (Oxford: Oxford University Press, 1996).

[21] Van Inwagen, *An Essay on Free Will* represents a particularly strong example.

terminacy is clearly irrelevant to the question of freedom and deserves no further mention here. By contrast, what happens if one shares the incompatibilist intuition *and* holds that humans are, at least on some occasions, genuinely free (as I do)? One answer is to side with a strongly dualist view of the physical and mental realms. For the dualist (of the Cartesian as well as the Kantian variety), it doesn't matter if the physical order is deterministic, since the action of the mental agent[22] is by itself sufficient to guarantee that the action is free, *whatever* the state of the physical world.

But many of us do not find such dualistic views credible as a theory of human nature and action. For nondualists who are incompatibilists, there must be some place or places in the physical order where an outcome in the natural world is *not* determined by the set of antecedent conditions and states. Call it the Nondeterminism Postulate. As Robert Russell has written, an ontological indeterminacy of this type seems necessary if human beings are to enact their own choices in the world.

We might look, for example, to see whether brain functioning allows for an openness of outcome that is sufficient for counterfactual freedom. Could the same complex brain state result in more than one subsequent outcome (assuming that we had the knowledge to establish that it *was* the same brain state that was correlated with two different outcomes in two different cases)? Those of us who accept the Nondeterminism Postulate argue that there must be this sort of openness at least somewhere in the hierarchy of natural phenomena. And, given what we know of the microphysical world, it's at least plausible that the required openness of outcome has its first (and perhaps only) source at the quantum level. As long as the openness could be amplified up through the causal chain so that it remained relevant to the description of some of your actions—e.g., to the complex physical state underlying your choice to commit a crime or not—then you could be said to be free and thus responsible for your actions. Only in this sense could quantum indeterminacy (if it exists) be said to be the necessary condition for human free will. Incidentally, note that nothing in this account makes indeterminacy *sufficient* to establish robust free will in humans; it is only a prerequisite, a first step in showing how genuine freedom might arise at the level of complex organisms like ourselves.

In this third example there is again room for one to engage in some serious metaphysical reflection, though not all will wish to do so. Imagine that you accept incompatibilism as defined above, as well as some version of the Copenhagen interpretation of quantum physics: you hold that the quantum world is genuinely indeterministic. Suppose also that you do not think that indeterminacy arises at any other or higher physical level than quantum mechanics. The *minimal* formulation of your view is that quantum indeterminacy is a necessary condition for human freedom. But you might also postulate and look for other kinds of openness as well. You might hold, for instance, that the hierarchical structure of the physical world, rather than eliminating the indeterminacy, actually augments or amplifies it. You might look for expressions of indeterminacy at multiple levels of the physical hierarchy, from the macrophysical level of measuring devices through genetic variation to indeterminacy in neuronal firing within the brain and the resulting behavioral plasticity. In your more philosophical moments you might argue that the existence of mentality in general, and free will in particular, are among the results of this openness of the world at whatever levels it occurs.[23]

---

[22] By "agent" I mean here Descartes's *res cogitans*, or a member of the "kingdom of means and ends," as in Kant's *Groundwork for the Metaphysics of Morals*.

[23] Having said this, I must add that I am not currently aware of any concrete results that

Your view would then commit you to giving some account of *how* quantum indeterminacy could find macrophysical expression. Could you suggest a physical mechanism for making this indeterminacy matter at the level of human decision makers? There are a variety of physical processes that augment quantum mechanical effects (see the essay by George Ellis elsewhere in this volume.) Or perhaps you would look (more speculatively) for the augmentation of quantum indeterminacy via chaotic systems to the point where they might have discernable macrophysical impacts. Or perhaps you would attempt to specify the mechanisms within the brain whereby quantum mechanical indeterminacy might directly contribute to the brain processes that underlie and give rise to conscious experience.[24] If you were convinced of the strength of these empirical accounts and were also interested in integrating the results with a theological position in the fashion suggested above, it would be natural to take an additional step. You might then postulate that God created a world with indeterminacy at the most fundamental level in order to allow the freedom required of human agents. In this case, the existence of these mechanisms would reveal something of the nature and intentions of the Creator of this physical order. For example, God would have to be such that God could intend to create conscious agents such as ourselves.

Clearly I have stopped short of a complete defense of quantum indeterminacy as the touchstone for human freedom. But I have tried to show the significance of the sorts of connections that might be drawn and to defend the *type* of reflection that is required for developing and assessing these various possibilities. It may not be the case that the different metaphysical alternatives would entail physically distinct observations. Yet from the perspective of a theory of human nature,[25] at any rate, they do represent significantly different options.

4 Taking Stock

In examining these three classic areas in the interpretation of quantum mechanics it has been my goal not so much to decisively establish one set of systematic conclusions as to defend the importance of this type of discussion. This task is necessary, since many scientists are constitutionally opposed to speaking of science and divine action in the same sentence, such as occurs in this book's subtitle. Such scientists pride themselves, in fact, on the distance of their own work from any theological subject whatsoever (whatever their own private religious stance might be).

4.1 The Continuum of Abstraction

Such outbreaks of animosity are not without precedent: for much of the twentieth century, and perhaps since Kant, a bipartite opposition was set up between physics and metaphysics. This contentious dichotomy unfortunately managed to obscure a third realm, wedged in between the two, in which interesting and productive philosophical interactions regularly take place. In its professional form this realm is known as the philosophy of physics and includes (among many other topics) the

---

suggest such openness at any other level than the quantum level.

24 Henry P. Stapp, *Mind, Matter, and Quantum Mechanics* (Berlin: Springer, 1993).

25 See e.g., Robert J. Russell, Nancey C. Murphy, Theo C. Meyering, and Michael A. Arbib, eds., *Neuroscience and the Person: Scientific Perspectives on Divine Action* (Vatican City State/Berkeley, Calif.: Vatican Observatory/Center for Theology and the Natural Sciences, 2000), hereafter *NAP*.

questions of the status of objects and entities in physics, the nature of physical law, the nature of inference and justification in physics, and the relations of physics to chemistry, biology, and mental events.[26] But this in-between realm has in fact a rather broader scope than the formal discipline of the philosophy of physics. Bench scientists engage in it on a regular basis—sometimes during the work day, and sometimes over a glass of wine afterwards. It is inevitable that both theorists and experimentalists spend time thinking about the specific conceptual problems raised by their field: quantum indeterminacy, observation, nonlocality, entanglement, and so forth. The preceding section provides a good example of questions raised by physics even though they are no longer physical questions in the direct sense.

I would like to call this type of reflection *meta-physical*, using the hyphen to indicate its borderline status. In one sense these questions come "after" (hence the Greek *meta-*) the physical work. Yet in another sense they are close cousins to physical questions, insofar as they arise directly out of experimental and theoretical work in mathematical physics. To grant their close kinship with physics is to acknowledge that physics and philosophy lie on a continuum. At the one endpoint—e.g., solving problem sets in your physics textbook—few, if any, philosophical questions are raised. At the other extreme, as one moves into the more lofty realms of metaphysical speculation, empirical questions and the controls of physics play a decreasingly significant role. But many of the most interesting questions (the ones I have called meta-physical) fall in the middle regions of this continuum.

Many factors influence how far along the continuum from problem sets an individual thinker is willing to go. These include taste, scientific training and field, intellectual influences, cognitive style (e.g., philosophical or anti-philosophical mind-set), and level of intellectual curiosity. When the topic is divine action, one's religious commitments become crucial. If one is a theist—and if one believes that her theism should be responsive to the results of the natural sciences—then she has a strong motivation, and perhaps obligation, to travel further along the continuum toward metaphysics and theology.

4.2 Taking the Next Step

One is often told that the problem is metaphysics: statements about divine action are metaphysical statements, and some physicists deny that there is anything interesting at all to say about any metaphysical topic. Indeed, there were some in this working group who took this position, maintaining that issues about the mathematical formalism and concrete experimental results represent the sum total of decidable questions; all else is subjective preference and hence beyond the scope of argument.

---

[26] Recent introductory texts to the philosophy of physics include James Cushing, *Philosophical Concepts in Physics: The Historical Relation between Philosophy and Scientific Theories* (Cambridge: Cambridge University Press, 1998); Peter Kosso, *Appearance and Reality: An Introduction to the Philosophy of Physics* (Oxford: Oxford University Press, 1998); Roger G. Newton, *Thinking about Physics* (Princeton: Princeton University Press, 2000). Significant works (among many others) in the field include Jarrett Leplin, ed., *The Creation of Ideas in Physics: Studies for a Methodology of Theory Construction* (Dordrecht: Kluwer, 1995); Bernard d'Espagnat, *Veiled Reality: An Analysis of Present-Day Quantum Mechanical Concepts* (Reading, Mass.: Addison-Wesley, 1995); Jeremy Butterfield and Constantine Pagonis, eds., *From Physics to Philosophy* (Cambridge: Cambridge University Press, 1999); Elena Castellani, ed., *Interpreting Bodies: Classical and Quantum Objects in Modern Physics* (Princeton: Princeton University Press, 1998).

But most who work on questions of quantum physics think that some interesting things *can* be said about at least some meta-physical issues raised by physics.[27]

We found that what might at first appear as a dichotomy (physics or philosophy) is in fact a continuum. There are interpretive statements one can make about quantum theory that, although they are not themselves part of physical theory, stand quite close to it; what I have called meta-physical debates are determined even less fully by the body of physical theory; and classic metaphysical debates are speculative in yet a stronger sense. Clearly statements about theology and the quantum world will fall in the latter category. Thus d'Espagnat's comments about quantum fields are only mildly interpretive; his postulation of a "deeper reality" begins to be speculative; and his advocacy of a strong Spinozistic monism, one phenomenal manifestation of which is the physical world, is a fully metaphysical position. Still, at no particular point does one encounter a definite point on the continuum, such that assertions made prior to that point reflect purely empirical knowledge and assertions made after it are purely speculative. I have attempted to show that there is value in reflecting one's way across the entire spectrum of this continuum.

It may be that there is no point of contact between those physicists whose various publications span the whole continuum (in this group, Heller, Polkinghorne, Russell, Stoeger, and Shimony) and those who have not engaged in theological or metaphysical reflection of this sort. But the idea of a continuum suggests that constructive contact is at least possible—especially if we avoid labeling the side closest to actual physical theory "good" and the side closest to classical metaphysical reflection or theology "bad." (The opposite value judgment must be avoided as well.) It is not intrinsically questionable to explore the more speculative reaches of the continuum, for example by engaging questions at the interface of theology and physics. Of course, claims about physics and theology *can* be poorly or dogmatically argued, and participants in the discussion (especially in more popular publications) sometimes substitute lofty global claims for serious argument. But there is no intrinsic reason why good argumentation cannot take place at *any* point along the continuum.

One protection against sloppy or dogmatic treatments of theology and physics is to take a pluralistic approach to the discussion, tracing the multiple interpretive possibilities of quantum mechanics rather than siding immediately with one or another. The reasonableness of a given position in speculative debates is not immediately obvious, as if one interpretation could be proven to have exclusive rights as the best interpretation of the data. Rather, like the owl of Minerva, judgments of reasonableness must come at the end of the day. Reasonableness in metaphysical debates lies in one's ability to tie multiple discussions together within a single interpretive framework. The more comprehensive the claim, the more domains will have to be part of the explanatory story.

Clearly, explanations that include the term "God" lie at a rather high point on the scale of comprehensiveness; they certainly cannot be *less* comprehensive than the

[27] Indeed, many have actually published on meta-physical questions raised by quantum physics, and some even have books in which the word 'metaphysics' appears in the title! So the problem cannot be that all metaphysical statements are strictly speaking meaningless and to be eschewed, as Professor Ayer famously held; A.J. Ayer, ***Language, Truth, and Logic***, 2nd ed. (New York: Dover, 1946). For positivists in Ayer's tradition, metaphysical questions are unacceptable whenever they introduce any entities or categories that cannot be directly justified by the mathematical formalism and the empirical data. On this view, debates about the foundations or interpretation of quantum mechanics might or might not be acceptable, depending on how they are pursued.

field of physics as a whole. Now the more comprehensive one's explanation becomes, the more difficult it becomes to decide on its truth, whereas, famously, claims about more limited data sets are easier to resolve. Strict and rapid decidability will not be the hallmark of debates about physics and theology. Thus a certain shyness about advancing any position at all is in order. But the shyness need not be stultifying. Science has often been advanced by this or that risky hypothesis; why shouldn't the same be true of the metaphysics of science as well? Hence, if we are to make any progress at all on research questions that involve moving further along the physics-philosophy continuum, it will be necessary to develop an even greater willingness to entertain speculative hypotheses and to look rigorously for reasons to select one and to abandon others.

5 Three Metaphysical/Theological Options

An entire genre of writings on the interpretation of quantum mechanics moves in this more speculative direction. These authors take the sorts of interpretive debates described in section 3 and attempt to provide a more systematic framework within which to address them. Here the goal is not only (for example) to solve the measurement problem, but also to construct an adequate systematic philosophical position. As test cases for this sort of discussion we might take the topics of philosophical monism, treatments of quantum physics in the context of Eastern philosophy, and theistic accounts of divine action.

5.1 Bernard d'Espagnat's Spinozistic Monism

Bernard d'Espagnat has developed an ontology based on quantum theory that represents a particularly interesting way of drawing lines between physics and metaphysics—and refusing to draw others. As he writes, in quantum field theory particles are viewed as quantized states of a field that extends over the whole of space. But quantum fields are not appropriately interpreted as things. The mathematics associates them with "operators" like the operators associated with observable properties in elementary quantum mechanics, properties such as position or momentum. To d'Espagnat a French example comes to mind: quantum fields are less like the Eiffel Tower than like some qualities that are in (or: that we observe in) the Eiffel Tower, such as its height, size, or shape. So we must ask: what is it that these qualities are qualities *of*? According to d'Espagnat, the only possible answer is that the state vector expresses properties of some deeper underlying reality. Since we know its manifestations to us—we know what it's like when measured—and since quantum physics forbids us to speak about what it's "really like" when not measured, d'Espagnat speaks of it as a "veiled reality."[28] His is a sort of realism at a distance: we can't say that reality is "just this way or that," since our observations and what we observe are intertwined; and yet we *can* say that the-world-as-observed is a manifestation of the real; reality really takes this or that form *in* our observations. D'Espagnat develops these insights into a strong form of philosophical monism.[29]

[28] See Bernard d'Espagnat, *In Search of Reality* (Berlin: Springer-Verlag, 1983). More recently, see his *Veiled Reality*. See also idem, *Realism and the Physicist: Knowledge, Duration, and the Quantum World* (Cambridge: Cambridge University Press, 1989), and his article in W. Schommers, ed., *Quantum Theory and Pictures of Reality: Foundations, Interpretations, and New Aspects* (Berlin: Springer-Verlag, 1989).

[29] Several critics are right to point out that d'Espagnat's monism is not a strict entailment

There is just the one reality, since quantum physics requires us to think of the world as interconnected and nonseparable. And yet this reality can take quite diverse forms. Take consciousness: it is not a separate kind of force, as in the "dualistic" or subject-based interpretations of quantum mechanics; rather, it is one "property" of that one reality—not a property of the particles in the brain. The brain particles and consciousness both represent properties of the veiled underlying reality. Thus, d'Espagnat writes, "the whole set of consciousness" and "the whole set of objects" are in fact "two complementary aspects of independent reality"; neither exists in itself, but "each one comes into existence through the other, somewhat in the same way in which the images of two mirrors facing one another give rise to one another."[30] With this parallelism (or "dual-aspect monism") d'Espagnat's allegiance to Spinoza comes to light. For Spinoza there was just one reality, which he called both God and nature: *deus siva natura*. Likewise for d'Espagnat, " 'God' means Being, and above all the unity of Being common to the indications of physics and to the most essential of Spinoza's intuitions."[31]

D'Espagnat is an ally to the divine-action theorist in several respects. He clearly makes the empirical world depend on a deeper order, and he allows that empirical science can at least hint at some of its qualities. Science cannot however provide definitive knowledge of this realm, since "the very nature of science is that its domain is limited to *empirical* reality."[32] Thus he does not try to derive subjectivity from (or reduce it to) the realm of the purely physical. On the other hand, d'Espagnat has also formulated the greatest competitor to divine action theories: Spinozistic monism. He reminds us that there are ultimately three major options for interpreting the physical world: there is no God, and physics ultimately defines reality; there is a God (and presumably also subjects), in which case Spirit is the more ultimate explanation of the physical universe; or there is the One that has both mental and physical qualities. Following the third option, d'Espagnat maintains that there is no God apart from the world; we don't need one, since the existing world admits both mental and physical properties as qualities or attributes or operators.

5.2 Eastern Mysticism

The best known exponent of the Eastern approach has been Fritjof Capra, whose book *The Tao of Physics* spawned a school of similar books, essays and disciples. Capra finds close parallels between modern physics and certain key tenets of Eastern religious thought. Kevin Sharpe provides a concise summary of the nine central parallels that Capra advocates:

> Eastern mysticism and modern physics perceive the cosmos as a unity, holding together opposites in a complementary yin and yang manner rather than in a conflicting dualism; they comprehend spacetime as [a] relative construct of the mind and not objective; the universe as dynamic and not static; that elements of matter cannot be understood as being isolated, but only in relation to the physical vacuum in which they are; that matter is performing a "dance" and not just quietly inert; he further sees a similarity between the

---

of his interpretation of quantum theory. The point of the continuum and the pluralistic model I am advocating is that broader metaphysical discussions are underdetermined by "formalism + empirical data," and even by the basic interpretive options, without thereby becoming purely arbitrary, "bad metaphysics."

[30] d'Espagnat, *In Search of Reality*, 96f.

[31] Ibid., 101.

[32] Ibid., 167.

paradoxical nature of quark symmetries and Zen koans; and between their both seeing no basic elements of matter, but only patterns of change and interpretation within the mutual interrelation and self-consistency of all phenomena (the "bootstrap" hypothesis).[33]

A host of authors have followed a similar tack. Dennis Postle, for example, finds a similar stress on the interconnectedness, interpenetration, and interdependence of all things in both Eastern philosophy and modern particle physics.[34] Both approaches, he believes, preserve a role for consciousness in affecting what the world becomes. In physics, Postle holds, an experimenter's attitude towards the experiment changes what is then looked at. But Eastern teaching goes further: "what we can know depends on our consciousness," and "knowledge is structured in consciousness." Physics (allegedly) points toward the Eastern unity of all things, though in the end it doesn't go quite far enough for these authors.

Bohm in his more mystical writings also seems to advocate a similar position:

One is led to a new notion of unbroken wholeness which denies the classical idea of analyzability of the world into separately and independently existing parts. . . . We have reversed the usual classical notion that the independent "elementary parts" of the world are the fundamental reality, and that the various systems are merely particular contingent forms and arrangements of these parts. Rather, we say that inseparable quantum interconnectedness of the whole universe is the fundamental reality, and that relatively independently behaving parts are merely particular and contingent forms within this whole.[35]

Bohm turned this interconnectedness into an ethical or religious view: "Indeed, the attempt to live according to the notion that the fragments are really separate is, in essence, what has led to the growing series of extremely urgent crises that is confronting us today."[36] Atomism leads to confused questions, which indicates the need for new forms of insight.[37] Based on the "quantum inter-connectedness of the whole universe," Bohm concludes, "one can no longer maintain the division between the observer and observed (which is implicit in the atomistic view that regards each of these as separate aggregates of atoms). Rather, both observer and observed are merging and interpenetrating aspects of one whole reality, which is indivisible and unanalyzable."[38] This leads to his main thesis: "So approaching the question in different ways, relativity and quantum theory agree, in that they both imply the need to look on the world as an *undivided whole*, in which all parts of the universe, including the observer and his instruments, merge and unite in one totality."[39]

Entanglement phenomena in quantum physics have often been cited as evidence for holistic conclusions. One finds mainline physicists who appeal to entanglement to defend an overarching interconnection of all things. Thus Henry Stapp defines the concept of local causality: "The principle of local causes asserts that what happens in one spacetime region is approximately independent of variables subject to the control of an experimenter in a far-away spacelike-separated region." Stapp shows

---

[33] See Kevin J. Sharpe, "Mysticism in Physics," in *Religion and Nature*, K.J. Sharpe and J.M. Ker, eds. (New Zealand: The University of Auckland Chaplaincy, 1984), 43f.

[34] Dennis Postle, *Fabric of the Universe* (London: Macmillan, 1976), 8f.

[35] Bohm and Hiley, *The Undivided Universe*, 96, 102.

[36] See David Bohm, *Fragmentation and Wholeness* (Jerusalem: The Van Leer Jerusalem Foundation, 1976), 1f.

[37] Ibid., 8.

[38] Ibid.

[39] Ibid., 10.

how Bell's theorem conflicts with the principle of local causes: "The statistical predictions from which this result follows... have been experimentally tested and confirmed.... Bell's theorem shows that no theory of reality compatible with quantum theory can allow the spatially separated parts of reality to be independent."[40]

Ken Wilber then uses Stapp's comments and the empirical tests of Bell's theorem to defend the holism of the Eastern traditions:

> It is common among the "new-paradigm" thinkers to claim that the basic problem with science is that, under the "Newtonian-Cartesian" worldview, the universe is viewed as atomistic, mechanistic, divided, and fragmented, whereas the new sciences (quantum/relativistic and systems/complexity theory) have shown that the world is not a collection of atomistic fragments but an inseparable web of relations. This "web-of-life" view, they claim, is compatible with traditional spiritual worldviews, and thus this "new paradigm" will usher in the new quantum self and quantum society, a holistic and healing worldview disclosed by science itself.[41]

The problem, in other words, was not that the scientific worldview was atomistic instead of holistic, since it was basically holistic from the start. No, the problem was that it was a thoroughly *flatland holism*. It was not a holism that actually included "all of the interior realms of the I and the WE (including the eye of contemplation)."[42]

When concepts such as these are fleshed out in full form by the more radical Eastern mystics, the results can be startling:

> ...in quantum physics the elements are not physical themselves; they do not exist as objects. Their very existence depends on the idea of their existence beforehand. They are treated as "tendencies to exist" rather than as already existing possibilities like the sides of a flipped coin. In the quantum world the quantum coin's sides do not appear unless someone calls for them to appear. ... Thus we conclude that the "new physics" introduces the element of consciousness into the material world. This consciousness will not arise from the molecule itself, as seen as a material unit, but will arise as a "risk-taking" psyche—that is, one that chooses. These choices cannot be made willy-nilly. "Reason" must begin to make its appearance, which surpasses the simple mechanism of cause and effect. We know that atoms do not follow the laws of cause and effect except statistically or on the average. To explain the evolution of learning, associative memory, and possibly even the more primitive forms of memory called habituation and sensitization, we must face the quantum. States of consciousness, feelings, emotional states, and psychology as a science may depend on the recognition that mind, the consciousness of the universe, arises through quantum physics.[43]

Now there may be inherent interest in the Eastern metaphysics that these thinkers are seeking to express. The problem arises when the authors couch their conclusions as if they were *implications* of the quantum physics. Physics needs metaphysical interpretation, and some quantum mechanical results do support a form of interconnectedness not unlike the metaphysics of holism. But physics does not directly support a particular metaphysic on the speculative end of the continuum. Of course, the founders of quantum mechanics were the first to stress that the resulting ontology, whatever it turns out to be, will be radically different from the everyday ontologies that we encounter in the macrophysical world and in traditional philosophical

---

[40] Henry P. Stapp, "Theory of Reality," *Foundations of Physics* 7 (1977): 313–23.

[41] Ken Wilber, *The Marriage of Sense and Soul: Integrating Science and Religion* (New York: Random House, 1998), 38.

[42] Ibid., 57.

[43] Fred Alan Wolf, *Star Wave: Mind, Consciousness, and Quantum Physics* (New York: Macmillan, 1994), 17–9.

theories. Clearly the physics requires some radical rethinking, though it is not clear that this particular direction is the best way to go, and it is certainly not the only one.

5.3 Theistic Metaphysics

In this final example I compare and contrast the previous two approaches with the tradition of theistic metaphysics, briefly considering both classical theism and panentheism. No pretense can be made that physics will determine the choice between them, any more than it can decide between theism and its competitors. Thus our only goal can be to show the coherence of one particular account of divine action with the physical constraints as we know them.

5.3.1 Classical Theism

Theism has generally been characterized by a two-fold assertion. On the one hand, the way the world appears to us is not an illusion: there are indeed multiple physical objects, law-like regularities, and so forth. On the other hand, this world has its origin in an ultimate principle characterized primarily as "spirit." This means, at least, that the divine is an active principle and that the actions of this principle are more like that of a person than like the operation of impersonal natural law.

There are certain advantages to theism's double assertion in comparison to some of the views encountered above. Surely it is some advantage to be able to incorporate the real existence of the world into one's philosophy rather than having to label it all illusion. But it is also an advantage to be able to grant the real *physicality* of the world—the existence of distinct physical objects that obey natural laws, change and develop, and have a beginning and an end in time. At the same time, postulating the existence of a God allows one to make sense of the existence of mental objects, or at least mental properties, in the world.[44] For theists, mental properties are not utterly different in kind from the deeper reality that underlies our everyday experience as creatures, since God, the creative source of all things, also manifests consciousness and agency. Indeed, from an explanatory perspective it would be a clear advantage if ultimate reality has (quasi-)personal features, since it would then have the resources to explain the higher-order features of human persons. By contrast, an impersonal ultimate principle (karma, say, or traditional materialism) must reduce mental or personal phenomena to the terms of its own ultimate principle(s).

On the other hand, classical philosophical theism (CPT) faces a few difficulties of its own.[45] Since God is understood purely as disembodied spirit, it is more difficult to specify in detail how God is to be presently related to the physical world. (I will assume for now that CPT can adequately answer questions about the initial creation of the world through the doctrine of *creatio ex nihilo*, though certain difficulties arise here as well.) Further, God's nature becomes perhaps *too* different from our own.

---

[44] Classically, Christian theology claimed that there were mental objects or "souls" that constituted the essence of (at least) each person. Recent dialogue with the neurosciences has led many theologians to think instead of mental properties rather than essentially mental things. See, e.g., the essays in *NAP*.

[45] I cannot do justice to the complicated criticisms in five sentences. In addition to other works cited here, see Philip Clayton, "The Case for Christian Panentheism," *Dialog* 37 (1998): 201–8; idem, "Panentheist Internalism: Living within the Presence of the Trinitarian God," *Dialog* 40 (2001), in press.

Human agents exist only as embodied,[46] whereas CPT understands the divine agent to be essentially *dis*embodied. At worst, one worries that no analogy will remain between human and divine agency, which would mean that one could no longer speak without equivocation of both humans and God as "agents."

But what CPT cannot provide in terms of a model, it *can* protect by means of a sharp separation between God and world. God dwells ultimately in mystery, and as the tradition noted, "his ways are not our ways." The theist may not be able to say how God is an agent, if he is utterly unlike us, nor how he makes a difference in the world. She may therefore be unable to supply a theory of how God can act in a purely physical world governed by natural laws. But she can always claim that God acts miraculously in the world, that God is not constrained by natural law, that God acts in time from outside of time, or that God is a co-cause of every event alongside inner-worldly causes. Surely the inaccessibility of such divine causes to explanation by human agents is disturbing. But it cannot be argued that it is incoherent.

5.3.2 Panentheism

As I have argued elsewhere,[47] a panentheistic form of theism avoids some of these difficulties. Panentheism is usually defined as the view that the world is within God, although God is also more than the world. If God includes the world with Godself—perhaps somewhat on analogy with the relationship between your mental properties and your body—then the question of divine action within the world is made less intractable than when God and the world are understood as fully ontologically distinct.

Specifically, we can imagine the regularities of the natural world as analogous to our bodies' autonomic functioning. Of course, a being that is omni-aware will know all the regular functions that are occurring in the universe and can be said to be in control of them to an extent far beyond a human's control over her autonomic bodily functions. This means that each physical event, no matter how law-like, can be understood as an expression of divine agency. In addition to such regular functions, panentheism also allows us to speak of focal divine actions, similar to the way that a human can carry out focal conscious actions through an act of attention and will. It is a matter of dispute among theists how many such focal actions God accomplishes in the world. But if they occur in a "top-down" manner, acting as a lure or partial motivation for individual human agents, then no natural laws need to be broken and hence no contradiction with the results of science needs to be introduced.

It is particularly fascinating to note the parallels between this sort of panentheism and some of the interpretations of quantum mechanics given above. One theme that has already been introduced in the reflections of several quantum physicists is the distinction between empirical appearances and an underlying reality. D'Espagnat distinguished between the manifest and the "veiled" or "hidden" reality, Bohm between the implicate and explicate (or implicit and explicit) order: "That is, there is a universal flux that cannot be defined explicitly, but which can be known only implicitly, as indicated by the explicitly definable forms and shapes, some stable and

---

[46] Even Christian theologians are now arguing that the notion of a (dualistically understood) soul does not make sense. See, e.g., Warren S. Brown, Nancey C. Murphy, and Newton Malony, *Whatever Happened to the Soul? Scientific and Theological Portraits of Human Nature* (Minneapolis, Minn.: Fortress Press, 1998).

[47] Clayton, *God and Contemporary Science*; idem, *The Problem of God*.

some unstable, that can be abstracted from the universal flux."[48] Some label this the fundamental metaphysical move: to distinguish between the world of appearances and something deeper. Think of Plato's distinction between the phenomena and the realm of the forms—or, for that matter, any of the other Greek attempts to specify the *archē* or ultimate principle. Note also that what the world ultimately turns out to be will depend on the nature of this deeper principle.

Consider some candidates for the nature of this underlying reality. In the Spinozistic tradition with which d'Espagnat aligns himself, the One is not an active principle; it is neither mind nor matter (though it manifests itself as both); it is unchanging, eternal, and in itself unitary and undivided. Contrast this position with the view of Bohm and the physicists who draw on the process philosophy of Alfred North Whitehead,[49] for whom the deeper reality is process or movement: "*What is* is a whole movement, in which each aspect flows into and merges with all other aspects. Atoms, electrons, protons, tables, chairs, human beings, planets, galaxies, etc. are then to be regarded as abstractions from the whole movement and are to be described in terms of order, structure, and form in movement."[50] One must then ask whether or not this reality-in-motion is conscious. The Hindu traditions, for example, have often understood it as a universal ground of consciousness. Thus the Hindu quantum physicist Amit Goswami solves the measurement problem by imagining all conscious observers to be manifestations of a universal, omnipresent ground of consciousness.[51] For Spinozists, by contrast, although mentality appears among the infinite attributes of the One, mind is no more basic to reality than matter. Panentheism can be seen to split the difference. It does not draw the sharp separation between this material world and its purely spiritual source that we found in CPT. On the other hand, it does not equate world and God, physical and mental, as Spinoza does. Rather, the relationship between them is analogous to our own relationship between our mental properties and our physical properties, which a number of thinkers have understood as a relationship of supervenience.[52]

Panentheistic theology tends to unify the world, understanding all its behaviors as expressions of the one unified divine will, whereas CPT conceives God as *present to* all parts of a world that is radically plural. Of the two, CPT is somewhat more dualistic in its understanding of the God-world relationship[53] and more pluralistic in its conception of the physical world. By contrast, panentheism shares with the Eastern traditions, with d'Espagnat, and with the later Bohm, a certain tug toward monism over atomism. This fact links it with those developments in contemporary physics that reveal a deeper level of interconnection within nature. Entanglement phenomena, for example, reveal this sort of interconnection in the quantum world,

---

[48] Bohm, *Fragmentation and Wholeness*, 10.

[49] See the two special issues of *Process Studies*—vols. 26.3–4 (1997)—edited by the physicist Tim Eastman and containing articles by (among others) the physicists David Finkelstein, Lawrence Fagg, and Eastman.

[50] Bohm, *Fragmentation and Wholeness*, 39.

[51] Amit Goswami, *The Self-Aware Universe: How Consciousness Creates the Material World*, with Richard E. Reed and Maggie Goswami (New York: Putnam's Sons, 1993).

[52] On the supervenience relation see several of the essays in *NAP*, e.g., Murphy and Clayton, and the literature cited therein.

[53] See Bede Griffiths, "The Vision of Non-Duality in World Religions," in *The Spirit of Science*, Lorimer, ed.

if indeed they show that several apparently discrete parts of the world are in fact best viewed as a single system or particle. The theory of quantum fields offers another framework in which apparently discrete objects (subatomic particles) are reconceived as manifestations of a unified field. The extent of such interconnections is of course contested. But if the best interpretation of quantum mechanics *should* turn out to be the one that emphasizes some degree of holism, interconnection, and interdependence, then panentheism becomes a level-three metaphysic that evidences a natural fit or coherence with quantum physics so understood.

6 Conclusion

In this essay I have attempted to trace some of the lines that connect quantum physics and theology. Admittedly, we have not found the sort of tight conceptual connections that sometimes arise within the philosophy of physics; in this sense there is certainly more freedom than constraint.

At least five tentative conclusions have emerged out of the discussion:

1. Strong connections or constraints would suggest a "tight" model of the science-theology relationship as involving strict entailments from one to the other. The much more tentative and inconclusive nature of the constraints defended here suggests a rather different model. Instead of working with a "physics *versus* metaphysics" dichotomy, we explored a three-level division of labor (physics, philosophy of physics interpretations, metaphysical interpretations) that, I believe, more accurately reflects the actual order in which the questions are raised. Interpreting the quantum results naturally gives rise to second-order philosophical debates; and these second-order options are in turn amenable to various theological interpretations, some of which may evidence a slightly better fit than others.

2. Such theological interpretations are admittedly speculative in nature. Whether this speculation is acceptable or unacceptable depends in part on what one's goals are. If your goal is to interpret, say, quantum physical entanglement with the absolute *minimum* metaphysical commitment, then something like Abner Shimony's seven features of potentiality will suffice. If you want a philosophically rich interpretation of quantum physics *but nothing more*, then one or more of the standard interpretations of quantum mechanics presented in this volume will do. By contrast, if you want an interpretation broad enough to also include (say) biological structures and the existence of creatures with psychological experience, then you will need a metaphysics at least as extensive as those developed in this volume by Chris Clarke and George Ellis. Finally, if you wish to reach a metaphysical dimension broad enough to include human religious experience, then you will have to go as far as the sorts of hypotheses covered in section 5 above. To characterize such a project in advance as "bad" or unacceptable is to use one's own disinterest as a criterion of epistemic adequacy—not a particularly powerful argument.

3. Imagine that one is looking for a level of theorizing that encompasses both quantum physics and theology. Of course, this project is not mandatory, yet neither have we found anything to suggest that it is incoherent. What we *have* discovered is that one cannot engage in such a project—one cannot do constructive work in science and theology—unless one includes the relevant intervening fields. Again, the essays by Clarke and Ellis in this volume[54] reveal that the best way of including these

---

[54] Cf. Nancey C. Murphy and George F.R. Ellis. *On the Moral Nature of the Universe: Theology, Cosmology, and Ethics* (Minneapolis: Fortress Press, 1996); Arthur Peacocke,

fields is by considering the pattern of emergence that characterizes the relation between them. In the case of quantum physics and theology, the relevant intervening data are (at least) biological structures and mental properties or experiences.

4. Even if conclusions 1–3 are granted, there remain multiple theological or metaphysical options. It would have been nicer for orthodoxy, perhaps, if a decision mechanism existed for making a clear rational decision between the theological options, but we have not found this to be the case. In a situation in which the metaphysical options are radically underdetermined by the theories and data of science, one must content oneself with affinities and compatibilities, with (at best) a fit or coherence between groups of concepts in radically disparate disciplines. Panentheism may represent an attractive middle ground between the complete holism of the Eastern mystical approaches and the stronger ontological separation of God and world that has often characterized CPT (section 5.3 above). But the looseness of the connection with physics—not to mention the deep ambiguities in the interpretation of quantum physics itself—proscribe any lofty claims on its part.

5. For those who begin to study the possible interrelationships between quantum physics and theism, and who are interested in the problem of divine action, panentheism offers a certain attraction. It provides a way of speaking of the law-like regularities of dynamics and mechanics as expressions of regularities within the being and character of the divine (autonomic divine action). At the same time, it allows (in principle) for focal divine actions that are consistent with physical law—as long as divine action is construed as a top-down (or whole-part) influence that lures the wills and mental dispositions of conscious agents. Now there may be some reason to wonder whether divine action actually occurs that is focused and specific in this sense. Certainly if the divine mental influence were understood as able to determine human thoughts and actions, it might begin to constrain human freedom and, by implication, to make God responsible for *non*-interventions that leave suffering and evil unrequited (the so-called problem of evil). For this reason, again, theologians are better advised to limit their claims to how God *could* act than to commit themselves to strong knowledge claims about how God *must* act in the physical world.

Quite a come-down from the old-style theology as "queen of the sciences"? Perhaps. The type of hypothetical theology that I have here advocated is keenly aware of its limitations and the difficulties of the task. These may not be awe-inspiring results. But they are accurate expressions of the difficulty of the task—but also of the opportunities—that face theists who wish to listen closely and respond to the scientific results. Our exploration does not, at any rate, represent a null result. And if I am right, it does provide at least the outlines for a continuing research program in the field of quantum physics and theology.

Acknowledgment. I am grateful to the entire workshop group for criticisms that have improved the argument of this essay. Once again, the CTNS/Vatican Observatory project has demonstrated the virtue of detailed and sustained critical interaction. Indeed, importing the ethos of scientific critique into theology may be the greatest long-term contribution of this fifteen-year project. I thank in particular John Polkinghorne, Owen Thomas, and Kirk Wegter-McNelly for their criticisms during the final writing phase.

Theology for a Scientific Age: Being and Becoming—Natural, Divine, and Human (Minneapolis: Fortress Press, 1993); Harold J. Morowitz, *Emergences: Twenty-Eight Steps from Matter to Spirit*, forthcoming.

CREATION, PROVIDENCE, AND QUANTUM CHANCE

Thomas F. Tracy

1 The God Who Acts in History

One of the central challenges of modern theology has been to explain what it means to speak of the "acts of God." The biblical texts at the foundation of Judaism and Christianity include a rich deposit of stories that depict God acting in history to advance the purposes for which the world was called into being. The extraordinary "main character" to whom we are introduced in these narratives is not a distant and impassive observer of the course of history. On the contrary, this God engages individuals and communities in a relationship that unfolds as a compelling drama full of tensions and surprises, reversals and renewals. The narratives present a vivid and often poignant account of a long history of divine initiative and human response, stretching from the covenant with Abraham to the liberation from bondage in Egypt and the giving of the law at Sinai to the establishment of an independent kingdom and the building of the temple and to the eventual bitter loss of both. Christians later take up this story, reading it in their own way and carrying it forward in the Gospels toward a "surprise ending" of stunning boldness.

These stories do not simply remain artifacts of ancient faiths. Rather, they continue to play a central role in shaping life within the religious traditions that preserved them as scripture. The liturgical practices of worshiping communities, for example, typically involve a movement back and forth between the biblical texts, in which God's "mighty acts" are depicted, and the contemporary context, in which God's ongoing presence is affirmed. In telling and retelling these stories, communities of faith renew their understanding of who God is and of what God is up to in the world. We might say that the narratives serve to delineate the character of God by showing the divine agent in action, not unlike the depiction of human character in a well-told story.[1] In light of this narrative understanding of the nature and purposes of God, the faithful seek to make sense of their own lives and of the world around them. This meaning-making process, as H. Richard Niebuhr has pointed out, involves not so much an incorporation of the story into one's life, as an incorporation of one's life into the story.[2] The biblical narratives provide patterns for discerning how God is at work in the world, for glimpsing this deepest of "plot-lines" in history. In understanding the world this way, the faithful come to see themselves as living out, even now in their own stories, a moment in the larger biblical drama.

1.1 Modern Challenges

Despite the central role that narratives of God's acts play in the biblical traditions, theologians in the modern era have persistently found themselves stumbling over the language of divine action, uncertain of what to make of it. To be sure, circumspect theologians throughout history have recognized that there is no direct and simple route from the biblical stories to a theology of divine action, and they have grappled

[1] Hans W. Frei, *The Eclipse of Biblical Narrative* (New Haven, Conn.: Yale University Press, 1974); idem, *The Identity of Jesus Christ* (Philadelphia: Fortress Press, 1975).

[2] H. Richard Niebuhr, *The Meaning of Revelation* (New York: Macmillan, 1941).

with questions of interpretation. Some degree of hermeneutical self-consciousness is virtually forced upon thoughtful readers by the biblical texts themselves, which do not all speak with a single voice, but rather reflect an internally diverse tradition; in order to construe these texts as a relatively unified story of God's acts, a series of important theological decisions must be made. This interpretive enterprise has become particularly problematic for modern theologians, however. There are several interrelated reasons for this, two of which are especially worth noting here.

First, the development of critical historical and literary techniques has deepened our understanding of the contingency and complexity of the biblical text. In particular, the application of historical criticism to biblical narratives has progressively loosened the connections between story and history. It now requires great determination to persist in taking biblical stories of God's acts at face value as descriptions of historical events. But if God does not, after all, perform just the actions described in the biblical stories, then what if anything does God do? If, for example, we are not prepared to say that God sent a series of plagues to Egypt, parted the waters of the Red Sea to let the Hebrew people pass, guided them with pillars of cloud by day and fire by night, and fed them heavenly manna in the desert, then in what way *is* God the agent of their liberation?[3] Having granted that the biblical narratives should not be read as direct reports of God's mighty deeds, modern theologians confront a host of difficult questions about how to interpret these stories and about what claims they warrant regarding divine action in the world.

Second, the rise of the natural sciences has profoundly changed the intellectual context of theological interpretation. Since the sixteenth century, the sciences have progressively disentangled themselves from the explicitly religious conceptions of the universe to which they initially were tied. For example, Newton's periodic divine corrections to the planetary orbits were replaced by the deterministic causal closure of Laplace; flood geology gave way to the uniformitarianism of Hutton and Lyell; the exquisite divine design of each creature for its place in nature (that Paley illustrated in his anatomical studies) was succeeded by Darwin's theory of natural selection. At every point the sciences have proven their ability to explain events in the world without appeal to a transcendent cause. Laplace spoke for the modern sciences generally in his famous remark, when asked about the role of God in his astronomical theories, that he had "no need for that hypothesis." The sciences, for their own explanatory purposes, not only get along perfectly well without God, they systemically exclude appeals to such an agent from their battery of explanatory strategies.

1.2 A Dilemma for Divine Action

Modern theologians have been acutely aware of these features of contemporary natural science and many have drawn the conclusion that we must give up, or at least

[3] Langdon Gilkey famously pressed this question with great effect against the biblical theology movement in his "Cosmology, Ontology, and the Travail of Biblical Language," *The Journal of Religion* 41 (1961): 194–205. The biblical theologians, e.g., G. Ernest Wright in *The God Who Acts: Biblical Theology as Recital* (London: SCM, 1952), argued that nineteenth-century liberal theology made a fatal error in identifying revelation with certain modifications of human religious consciousness. By contrast, Wright and others contended that we come to know God in response to God's self-revealing mighty acts in salvation history, as narrated in the biblical texts. Gilkey pointed out that the biblical theologians were unwilling to take these stories at face value and yet offered no alternative account of what they meant by an act of God. As a result, they were left in the embarrassing position of proclaiming God's self-revelation in action without being able to say what God has done.

profoundly qualify, talk of God acting in the world. This is often held to be a consequence of recognizing either that scientific *methods* in principle rule out divine action or that scientific *findings* in fact are inconsistent with the affirmation that God acts in the world. Neither claim will bear critical scrutiny. From the observation that "God" cannot appear as a term in scientific explanation, it does not follow that theologians, for their own distinctive purposes, cannot develop an account of divine action in the world. And while it is evident that scientific explanations sometimes refute *particular* religious claims about divine action (e.g., that God brought the universe into being in 4004 BCE), it is at least not obvious that anything the sciences have so far taught us about the world rules out the possibility of divine action within it. It is a mistake to conclude that modern human beings have adopted, in general, a "scientific way of knowing" and/or a "scientific worldview" that rules out talk of divine agency developed within a *theological* interpretation of the world.

The theologians who have made these claims, and used them as the basis for far-reaching theological revision, have almost always uncritically presupposed a deterministic picture of the natural world. We can see this in a long line of religious thinkers, from deists in the eighteenth century to Schleiermacher at the founding of liberal Protestant theology in the early nineteenth century to contemporary theologians like Rudolph Bultmann and Gordon Kaufman.[4] For these thinkers, a general metaphysical picture of the world as a closed causal continuum came to be invested with the authority of science by being treated either as a methodological given of scientific inquiry or as a well-established empirical result. This is one of the most important points at which theological appropriation of the sciences during the twentieth century lagged well behind the emerging openness and deep uncertainty of scientists themselves as they explored multiple ontological interpretations of their own results. Once universal causal determinism is assumed to be an essential feature of modern science, however, it is not difficult to see why one might think that scientific explanations are incompatible with theological claims about divine action in such a world. It appears that God will be able to affect the course of events in a deterministic world only by setting the initial conditions and laws of nature, which jointly determine each event in the world's history, and/or by interrupting this deterministic causal series to turn events in a new direction. This presents the theologian with a dilemma. The first option accepts the deliverances of science but does not give us divine action *within* nature and human history, only divine action at their foundation. The second alternative provides for divine action within the world, but does so by countenancing "violations" of the laws of nature, and so requires that we abandon a strictly deterministic worldview.

It now seems clear that the natural sciences do not require (on methodological grounds) or establish (on evidential grounds) an exceptionless causal determinism, though neither do they rule out such a metaphysical interpretation. This point alone is sufficient to disarm the dilemma that would force a mutually exclusive choice between modern science and the religious traditions that speak of a God who acts in history. But this by no means solves the problems that theologians face in interpreting this language in the contemporary world. A host of difficult questions remain

[4] See Friedrich Schleiermacher, *The Christian Faith*, H.R. Macintosh and J.S. Stewart, eds. (New York: Harper & Row, 1963), para. 46; Rudolph Bultmann, *Kerygma and Myth*, H.W. Bartsch, ed. (New York: Harper & Brothers, 1961); and Gordon Kaufman, *God the Problem* (Cambridge: Harvard University Press, 1972). Also see Gilkey, "Cosmology, Ontology, and the Travail of Biblical Language."

concerning the relation between theological talk about God's activity in the world and scientific descriptions of events as integrated within a lawful (even if not closed) causal structure. The scientific commitment to forming explanations strictly in terms an intelligible network of efficient, rather than final, causes has been enormously successful. This explanatory paradigm has come to carry tremendous authority, and it has had a deep impact on our expectations about how to make sense of particular events in the world around us. We look at the world very differently than did, say, the Irish monk who wrote the story of St. Brigid miraculously producing a vast quantity of good Easter ale from a small bag of malt.[5] Our initial skepticism about this story is based not so much on an assessment of the available evidence as on a general picture of the world and on habits of explanation that have been shaped by the natural sciences. Accordingly, miracles are very much out of favor, if by "miracles" we mean events that are brought about by God and depart from the laws of nature. Although nothing in the sciences entitles us to say that such events cannot occur, we know that there are important evidential hurdles facing any particular claim that one has occurred; on this point, critical approaches in historical analysis and our scientifically shaped understanding of nature reinforce each other. We have grown instinctively resistant to picturing the world as a place where God persistently breaks in with astonishing displays of divine power. So while the modern theologian's predicament is not as severely constrained as our initial dilemma suggested (namely, to a choice between the scientific enterprise as a whole or the God who acts in history), the options appear quite limited. We can speak of God as the creator who sets the terms of cosmic history, which then unfolds according to the natural laws God has established. But if we want to affirm that God acts *within* that history, then it appears that we must take up the epistemic burdens associated with miracles.

There are, I think, at least two ways to respond to this theological predicament. The first argues that God's relation to the world as creator, properly understood, provides the basis for an account of God's particular actions in history that is sufficiently theologically robust. This is to challenge the claim that constitutes the first horn of the dilemma we constructed; the strategy here is to show that traditional claims about God's special providence *in history* can be explicated by reference to God's activity as the creator *of history*. The second response addresses the other horn of the dilemma; it challenges the claim that if God acts to redirect the course of events in the world, then this must constitute an intervention that departs from the lawful structures of nature. *If* the structures of nature include events that are not fully determined by the past, and *if* these events have effects that sometimes are amplified in the causal sequences that flow from them, then God could shape the course of history by acting in these open interstices of creation without disrupting its immanent structures. It is at this point that quantum mechanical indeterminism *may* be relevant.

I want to explore some of the strengths and weaknesses of each of these approaches. Much of the discussion of the relevance of contemporary natural science to the theology of divine action has focused on variants of the second approach. In this paper, I would like to counterbalance this tendency by including extended consideration of the prospects for a position of the first type. There are two reasons for this. First, it is important not to underestimate the resources available in the classical theological tradition for giving an account of *special* divine action in history that does *not* appeal to causal openness in the structures of nature. Note that if a

---

[5] Anonymous, *Bethu Brigte*, Donncha O hAodha, ed. (Dublin: Dublin Institute for Advanced Studies, 1978), sec. 21.

position of this type could be sustained, there would no longer be as clear and pressing a theological need to develop a position of the second type; the theological stake in scientific debates about (for example) deterministic interpretations of quantum mechanics would be considerably reduced. Second, if we do go on to claim that God acts through indeterministic structures in the natural world, it is important to root this claim firmly in an account of the basic creative relation of God to nature.

2 God as Creator

God's fundamental action is the act of creating the world, i.e., the totality of non-divine things. As this idea developed in the West, it came to include three elements. First, creation is a free intentional action, rather than a necessity of the divine nature. Because God's being is complete, quite without the world of created things, creation is an act of gracious generosity. The effect of affirming the freedom of God's creative action is to emphasize the utter contingency of the existence of created things. This stands in contrast, for example, to Neo-Platonic conceptions of creation as a necessary and involuntary emanation of the super-abundant plentitude of the divine being. This classical understanding of creation also contrasts with the views of most process theologians. Whitehead's metaphysical scheme, for example, specifies that every individual entity must be a creative integration of relations to other entities. God is no exception to this scheme; God makes a uniquely pervasive contribution to the creative becoming of the world, but God and world are co-eternal.

Second, God's creative act cannot be understood on the familiar human model of refashioning materials already at hand. There is no prime matter, no chaotic primordial stuff, that is presupposed by and constrains God's creative work. Rather, God creates *ex nihilo*; apart from God's creative action, nothing but God would exist. Creation accounts for the very being of the creature, and not just for the way it is or for its properties over time. It follows that the divine creative act cannot be regarded as a species of change; in creating, God does not transform or modify the state of things, but rather brings it about that there are finite things at all.

Third, God's creative action includes the continuous "giving of being" to the created world in its entirety. Creation is not a particular event, completed at some time in the distant past, which leaves behind (as it were) a world that gets along perfectly well on its own (the understanding of eighteenth-century deism). But the mainstream of the theological tradition has held that created things do not possess a power of continuing in existence on their own; rather, the existence of the created world depends absolutely at every moment upon God's creative action. This has typically been expressed by saying that the act of creation includes the activity of sustaining, or "conserving," the existence of each creature.[6] If God were to cease this continuous creative action, finite things would cease to be.

---

[6] The idea of divine conservation of the existence of created things should not be confused with the scientific idea of conservation of mass and energy. The latter is concerned with physical interactions between entities, and it specifies that these interactions and the transitions they bring about cannot involve the creation or destruction of matter/energy. This does not conflict with the theological idea of continuous divine conservation of the being of finite things; on the contrary, the two ideas compliment each other, since both assert that interactions between created things involve changes of state but not the giving of being. For an extended discussion of this point, see J.L. Kvanvig and H.J. McCann, "Divine Conservation and the Persistence of the World," in *Divine and Human Action*, Thomas Morris, ed. (Ithaca: Cornell University Press, 1988), 13–49.

2.1 Direct and Indirect Divine Action

This understanding of God's relation to the world provides a rich context within which to interpret talk of divine action in history. God is universally and intimately present to every creature at every moment, and nothing takes place without God's agency. Thus we can speak not only of *creatio ex nihilo* but also of *creatio continua*, a continuous creative activity expressed in the unfolding history of the world. There is a sense in which everything that happens can properly be described as God's act, but we must be careful about just what sense this is. The theistic traditions have wanted to affirm that God gives to created things active and passive causal powers of their own, that is, the capacity to affect other things and to be affected by them. Aquinas held that this is part of God's providential governance of creation. "Divine Providence works through intermediaries. For God governs the lower through the higher, not from any impotence on his part, but from the abundance of his goodness imparting to creatures also the dignity of causing."[7] God could simply cause, say, a kettle of water to become increasingly warm until it begins to boil. But God instead grants to created things "the dignity of causing," so that the water is heated by the fire. This contrasts with the view of those "who have taken God's working in everything that acts to mean that no created power effects anything in the world, but that God alone does everything without intermediaries."[8] The position that Aquinas rejects here has come to be called "occasionalism," because it holds that the created entities (or events) identified as causes are merely occasions for God's own direct action. If we are to avoid occasionalism we must make a distinction between direct and indirect divine action. In causing the being of creatures *ex nihilo* God acts directly, without employing any subordinate agency as a means, since there are no such agents until God creates them. But in bringing about particular events in the world, God ordinarily acts through secondary causes, producing the result through the operation of created things.[9]

---

[7] Thomas Aquinas, *Summa Theologiae*, Thomas Gilby, O.P., ed. (Garden City, N.Y.: Image Books, 1969), Ia, 22, 3.

[8] Ibid., Ia, 105, 5.

[9] We can borrow a distinction from the philosophy of action to help explain the ideas of direct and indirect divine action. Philosophers of action have recognized that any instrumental intentional action—i.e. an action in which an agent does one thing as the means of doing another—must, on pain of infinite regress, have at its base an action that the agent intentionally undertakes without having to perform any prior intentional action as the means to it. This "basic action" is direct, in contrast to its intended result, which is indirect. There has been controversy about which element in an indirect human action should count as the "basic action," and we could, in a perverse mood, carry over this question into theology and speculate about whether there is some divine action that is intentionally prior to the act of creation. For my purposes it is enough to note that God's act of creating and conserving creatures *ex nihilo* obviously cannot have creaturely intermediaries, and so it is basic for all the indirect divine acts that flow from it.

It is also worth observing in this connection that there are two crucially different senses in which we may speak of "bringing about the existence of something." On the one hand, there is the act of creating/sustaining *ex nihilo,* which is unique to God alone. On the other hand, there is the bringing to be of a particular *arrangement* of matter/energy in the world. Finite agents create in this sense; we are able to bring about changes in things, and thereby cause complex individuals "to come into existence" or "to pass out of existence," as in birth and death. God can also be said to create in this second sense, by acting indirectly through

2.2 Indirect Divine Action through Created Causes

On this view, the changes of state from moment to moment that make up the history of the universe have as their proximate causes the interactions of creatures within the order of nature. These events can be regarded as acts of God, however, insofar as they result from a series of causal intermediaries that God has established. This pattern of attributing actions to an agent is familiar from human activity; we use a great variety of means to accomplish our ends indirectly, and we often describe such acts in terms of the outcome that the agent intends. If I make a mark on paper under the right circumstances, my action can be described not just as drawing an X but also as casting a vote. But not every description of an agent's intention is appropriate as a description of her action; the links between the act and the intended outcome may be circuitous or uncertain. We will hesitate to describe my vote for mayor as "ensuring adequate funding for public education," even if that is what I intend. Further, not every true description of the outcome of action (e.g, "causing distress to the ex-mayor's family") can be taken as a description of the agent's intention. Human agents are aware only of a narrow range of the vast number of true descriptions that can be given of our actions and their consequences.

These limitations do not apply to the indirect acts of the divine agent. God knows everything that it is possible to know about the causal history of the world, and God does not simply make use of causal structures that happen to be available in the world but rather establishes and sustains them. This vastly expands the range of events that may be attributed to God as indirect divine actions. It is easiest to see how this story might go if we work initially with a simple, deterministic picture of the natural world. In such a world, God will fix the entire history of created things by setting the laws of nature and the initial conditions under which those laws operate. It will be possible, therefore, to attribute every event to God as a divine act.[10] The number of causal intermediaries and the complexity of their interactions will be irrelevant; these causal chains may extend across the entire history of the created world, all of which will serve as the means to the ends now realized. When a strong east wind pushes back the water in the shallows of the Sea of Reeds and allows the fleeing Jews to escape from their pursuers, we can say quite straightforwardly that this wind was "sent by God." The wind, and the deliverance it makes possible, is no less God's act if it is a result of the lawful operations of the natural order than if it is the product of a divine intervention within that order. In either case, it is something that God intentionally brings about in accordance with God's overarching purposes for history. On this view, God's providential action *in* the world is principally a function of God's creative action *at the foundation of* the world.[11] The strong east wind is written into the course of history when God

secondary causes. Bearing this distinction in mind, we can say that all complex individuals (like ourselves), which are produced by the operation of secondary causes, are created by God both directly (in sustaining our being *ex nihilo*) and indirectly (by working through the order of nature).

[10] Also see William P. Alston, "God's Action in the World," in *Divine Nature and Human Language: Essays in Philosophical Theology* (Ithaca, N.Y.: Cornell University Press, 1989), 200–3.

[11] I add the qualifier, "principally," because it is possible to hold that God ordinarily acts through secondary causes, but sometimes intervenes directly to bring about effects outside the expected course of nature or beyond the natural powers of creatures (Aquinas's view).

establishes the laws of nature and the initial conditions of the created world, and the billions of years of cosmic history that follow are the means by which God carries out this action, along with an unimaginably vast range of other actions.

It is important to note that while every event in such a world will be God's act, our ability to describe these divine actions will depend upon our understanding of God's purposes. Jews, Christians, and Muslims might agree with the general principle that God as creator acts throughout the history of the created world, but the traditions disagree about some important aspects of the overarching "plot-line" that is being enacted and therefore about which intention-descriptions should be given of these actions. The differing stories they tell about God's acts have as their corollary diverging understandings of "who God is," i.e., of the identity of the divine agent.

2.3 Special Divine Action in a Deterministic World

If every event, taken under the right description, is an act of God, is there any sense in which we can single out some events as special, or particular, divine acts? God's action, on this account, is universal and uniform; God acts in the same way in every event, i.e., as the source of its being. So there is no basis for picking out some events as bearing a distinctive relation to God's agency or as being attributable to God in a way that other events are not. Nonetheless, there are at least two senses in which events may be singled out as special divine actions. First, events may play a special *epistemic role* if they become the occasion for our recognition of God's purposes, and thereby provide guidance in understanding other events as belonging to a wider pattern of divine action in the world. H. Richard Niebuhr remarks that "sometimes when we read a difficult book, seeking to follow a complicated argument, we come across a luminous sentence from which we can go forward and backward and so attain some understanding of the whole. Revelation is like that."[12] What makes the revelatory event special is that it enables us to see the world in a new way, namely, as caught up in a drama of divine action and therefore charged with a significance that we had not recognized before.

Second, events may play a special *causal role* in the developing course of the world's history. Even if every event is an indirect act of God brought about through created causes, some may play a particularly important role in advancing God's purposes, and this will be a fact about their function within the causal series and not just about our perception of them.[13] History may have turning points, and the special significance of these events is in no way diminished if they arise smoothly within the causal structures of the world. As a result, there can be *objectively special* divine acts even though they cannot be distinguished from other events with regard to the way in which God acts in them.[14]

---

[12] Niebuhr, *The Meaning of Revelation*, 68.

[13] Compare William Alston, "What we take to be special about them is simply that God has acted in such a way as to effect *this result*, that *this* is something that God *intended* to bring about. How God chose to do this is not the heart of the matter" ("God's Action in the World," 216). This is right, as far as it goes, but it does not yet give us a basis for marking out particular events as *special* acts of God, since *every* event (taken under an appropriate description) in a deterministic world will be a specific result that God intends.

[14] It is useful to map this idea onto the typology of divine action developed at earlier conferences and reprinted in sec. 2.1 of Russell's introduction to this volume. Both of the senses of special divine action that I have discussed are forms of "uniform divine action,"

2.4 Indirect Divine Action in an Indeterministic World

So far we have been considering a simple model of a thoroughly deterministic world; in such a world every event could in principle be deduced by applying the laws of nature to a complete description of the total state of things at any moment.[15] But how would this account of divine action be affected if the structure of the world were to include some events that have necessary but *not* sufficient conditions in the events that precede them? This might take either or both of two forms. First, there may be indeterministic chance, in which the most complete account of the transition from one state to another is probabilistic; in this case antecedent states of the system determine no more than a distribution of likely results for the next state. This is distinct from what we might call "epistemic chance," in which converging causal chains catch us by surprise and/or the causal series is too complex for us to unravel. Second, there may be indeterministically free intentional action, in which a rational agent's choices are informed but not determined by her physical and psychological history. The question of whether either of these forms of indeterminism occurs in our world is, of course, a matter of controversy. I will consider at length below (in section 3) the question of whether quantum mechanics can be understood to present a theologically relevant form of indeterministic chance. At this point we need only consider the hypothetical question of what impact such indeterminisms would have on our account of God's action in the world.

2.4.1 Chance

If the structures of nature in fact include a role for indeterministic chance, then one option for the theologian is to think of God as determining these events. In this case, chance events would be causally undetermined only in their "horizontal" relations to other finite events, but they would be fully determined by their "vertical" relationship to God. Note that in determining these finitely undetermined events, God would be acting *directly* in the world's history, rather than indirectly through secondary causes, but this direct action need not disrupt the causal structures of

---

within the terms of the typology. What I have called epistemically special action corresponds to what the typology calls subjectively special action. The second form of special divine action that I describe, however, cannot be located in the typology as it is currently formulated. I have suggested that an event may *both* be an expression of God's uniform action throughout creation *and* be objectively special by virtue of the role this event plays in realizing God's purposes in the world. What marks out the event is not that *God* plays a special causal role in producing it, but rather that the *event* plays a special causal role in the unfolding course of events. The escape of the Jews from Egypt may arise entirely through the ordinary interactions of natural causes and human agents, and yet it may also turn human history in a new direction and so be an objectively special, but *indirect*, divine act.

[15] Given the chaotic dynamics of some deterministic systems, however, no finite intelligence could specify the initial conditions with sufficient precision to make these calculations. Determinism asserts that the laws of nature and the initial conditions jointly entail every future state of the system, but determinism does not entail predictability for any knower other than God. See, for example, Wesley J. Wildman and Robert J. Russell, "Chaos: A Mathematical Introduction with Philosophical Reflections," in *Chaos and Complexity: Scientific Perspectives on Divine Action*, ed. Robert J. Russell, Nancey C. Murphy and Arthur Peacocke (Vatican City State; Berkeley, Calif.: Vatican Observatory; Center for Theology and the Natural Sciences, 1995), 49–92.

nature, since chance events, *ex hypothesi*, do not have sufficient secondary causes.[16] This is the second way of responding to the original dilemma we considered, and I will consider this possibility at greater length in section 3 below.

An alternative would be to say that God leaves some or all chance events undetermined, so that God really does play dice with the universe. To be sure, an extensive web of secondary causal conditions will be necessary for the occurrence of the chance event. But this causal nexus is not sufficient to produce the event, and if God does not determine it, then nothing does. This situation generates a conceptual puzzle. Is it coherent to say that God brings about a state of affairs in which an entity or system undergoes a change that has no sufficient cause, whether in creatures *or in God*? It is helpful here to recall the distinction between God's act of causing existence *ex nihilo* and the act of causing creatures to undergo various changes; the divine action of giving being to the entity does not cause the change of state that is the chance event; creation/conservation is not, we have said, a matter of working a change in the creature, but rather of positing the creature in existence. But in the special case of chance events, the creature that God creates/conserves undergoes a change that not even God determines. Perhaps God's creative act in such instances amounts to willing that one from among a set of possible states for the system shall be the one to which God gives being, without specifying which and without, of course, providing any means by which a selection is made.[17] This is a puzzling idea, but this or something like it appears to be required if we say that God is the creator of the world *ex nihilo*, that the world includes indeterministic chance, and that God does not determine chance events.

If it is a coherent possibility that God might build this kind of randomness into the structure of the world, how would this affect our account of divine action? The answer will depend on the role that chance plays within the world's unfolding history. If chance events at one level in the structures of nature are entirely subsumed within higher order deterministic regularities, then the account of God's indirect action through these structures will be unaffected. On the other hand, if indeterministic chance plays a significant role in shaping the direction of the world's unfolding history,[18] then the attribution of events to God as divine acts must be correspondingly qualified. In establishing the laws of nature, God determines how chance figures in the course of events, and sets the range of outcomes that are possible. But if God chooses not to determine these chance events, then at least some features of the world's future will be open, bounded but left unspecified in God's creative intention. The structures of nature will include within them a means for trying out novel possibilities not rigidly prescribed by the past; God would, in effect, make a world that must in some respects fill in the details of its own creation. If, for example, some of the genetic changes amplified by natural selection result from processes that involve not just epistemic chance but also indeterministic chance, then which living

---

[16] Again, see the typology in sec. 2.1 of Russell's introduction.

[17] Peter van Inwagen discusses this possibility with regard to God's creative choice between equally good alternative initial states of the world. God might, van Inwagen suggests, will that one from among a set of alternatives be actualized, without determining which it shall be. "It does not seem to me to be logically or metaphysically impossible that God should decree that *either* X or Y should be without decreeing that X should be and without decreeing that Y should be"; "The Place of Chance in a World Sustained by God," in *Divine and Human Action*, Morris, ed., 227.

[18] This is the question of "amplification," which I take up in section 3.3 below.

creatures appear over the course of cosmic history will not be written into the design of the world.[19] The natural order that God establishes may assure the emergence of diverse forms of life with a wide range of capacities, including eventually the ability to gain theoretical knowledge of the world and to wonder about its creator.[20] But on this view, God may not have provided specifically that personhood should be realized in a bipedal mammal; the particular identity of the rational agents that arise within the evolutionary process could be one of the accidents of biological history. God's agency would, of course, be at work throughout this history as the creator who sustains all of the secondary causes at work in it. And because God sets the boundaries within which chance operates, thereby designing the dice that are set rolling in cosmic history, the general result can certainly be attributed to God's action. But if, returning to our earlier example, the strong east wind at the Sea of Reeds happened to be the meteorological amplification of a chance event somewhere else in the structures of nature, then it seems more appropriate to view the wind as a stroke of good luck than as a particular act of God in history.[21]

2.4.2 Human Freedom

The second form of indeterministic transition that we noted above is a particular, and particularly controversial, form of free human action. One family of positions in the longstanding (and probably intractable) philosophical debate about freedom of the will holds that an action is free only if it is not determined by antecedent circumstances. On this view the past history of the universe and the laws of nature do not uniquely determine the agent's choice; under precisely these causal conditions the agent could do otherwise than she does. This is commonly referred as "incompatibilist" freedom because it holds that free action is incompatible with causal determinism.[22] Note that causal indeterminism is a necessary *but not sufficient* condition for incompatibilist free action; in order for a free act to be distinguished from a chance

---

[19] See section 3.3 below.

[20] Paul Davies, for example, suggests that "God selects very special laws that guarantee a trend towards greater richness, diversity, and complexity through spontaneous self-organization, but the final outcome in all its details is open and left to chance." See Davies, "Teleology Without Teleology," in *Evolutionary and Molecular Biology: Scientific Perspectives on Divine Action*, Robert John Russell, William R. Stoeger, S.J., and Francisco J. Ayala, eds. (Vatican City State/Berkeley, Calif.: Vatican Observatory/Center for Theology and the Natural Sciences, 1998). It is, of course, a matter of controversy as to whether the laws of nature and the conditions under which they operate make the emergence of intelligence to some degree probable in our universe. See, for example, Paul Davies, *The Accidental Universe* (Cambridge: Cambridge University Press, 1982), and John Barrow and Frank Tipler, *The Anthropic Cosmological Principle* (Oxford: Oxford University Press, 1986).

[21] The story here could be made more complex, however. If omniscience includes knowledge of how every random transition would in fact turn out if God were to permit it, then God could choose which total set of chance and determined events to permit (i.e., which world to create) with particular effects in mind. In this case, it seems to me, the east wind would be God's act by a different route but in just a strong a sense as if it were the deterministic outcome of a closed series of secondary causes.

[22] For some arguments that human freedom is incompatible with certain types of determinism see Peter van Inwagen, *An Essay on Free Will* (Oxford: Oxford University Press, 1983). For some representative compatibilist arguments see Daniel C. Dennett, *Elbow Room: The Varieties of Free-Will Worth Wanting* (Cambridge: MIT Press, 1984).

event, an account is needed of the agent's capacity for self-determination, and this account must not reduce to an explanation by appeal to the causal efficacy of antecedent events. This is the metaphysical burden carried by defenders of incompatibilist freedom, and it is important to remember, as we consider quantum mechanics, that searching out causal indeterminisms in nature (even if they are located in the brain) is not going to be *sufficient* to provide us with a theory of free action. My interest here, however, is simply to consider the impact that creaturely freedom of this sort would have, *if* the world were to include it, on the attribution of events in the world to God as divine acts.

Just as we saw in considering chance events, there are two ways of relating the divine agency to this second type of indeterministic transition. First, God might directly bring it about that the agent acts as she does. There are at least two ways to argue that this divine causal role in human action is compatible with the claim that the action is free. First, one might insist that because God acts directly as creator to constitute the finite agent and her act, God cannot be regarded as a determining cause that compromises the agent's freedom. Second, one might qualify the conditions for freedom of action so that indeterminism is required only on the horizontal level of relations within the world; created agents would possess indeterministic freedom in relation to other creatures, but not in relation to God. This second view combines a creaturely incompatibilism with divine determination, and so generates a distinctive theological compatibilism. This seems to have been John Calvin's position, and it has also been attributed to Aquinas, though some interpreters read him as taking a position of the first type, and the construal of Aquinas's view continues to be a matter of dispute.[23]

The alternative is to say that God empowers and permits human agents to make choices that are not determined by other creatures or by God. God's creative agency, of course, intimately and pervasively shapes the exercise of free human agency by establishing our powers of action, their limitations, and the circumstances under which they are exercised. In this respect, it is appropriate to say both that God always acts with the created agent, and that when free human actions conform to God's will, the human agent is the means by which God acts. But it is important not to miss the fundamental distinction between divine action by means of free human acts and divine action by means of secondary efficient causes. If God chooses to create finite agents who are free in this strong sense, then in establishing the laws of nature and the initial conditions of the world, God does not fix the whole course of history. Wherever a created agent faces a free choice, there will be branching alternatives for the world's future, and it will be up to the creature to determine

[23] For the first way of reading Aquinas see, e.g., David Burrell, C.S.C., *Aquinas: God and Action* (Notre Dame, Ind.: University of Notre Dame Press, 1979); idem, *Freedom and Creation in Three Traditions* (Notre Dame, Ind.: University of Notre Dame Press, 1993); and Kathryn Tanner, *God and Creation in Christian Theology* (Oxford: Basil Blackwell, 1988). For the second reading see, e.g., Thomas Flint, *Divine Providence: The Molinist Account* (Ithaca, N.Y.: Cornell University Press, 1998), and Thomas J. Loughran, "Aquinas: Compatibilist," in *Human and Divine Agency*, F. Michael McLain and W. Mark Richardson, eds. (Lanham, Maryland: University Press of America, 1999). The first approach faces important conceptual objections. See the discussion in my "Divine Action, Created Causes, and Human Freedom," and Kathryn Tanner, "Human Freedom, Human Sin, and God the Creator," both in *The God Who Acts: Philosophical and Theological Explorations*, Thomas F. Tracy, ed. (University Park, Penn.: The Pennsylvania State University Press, 1994). Also see David Burrell's reply to me, and William Hasker's reply to Tanner.

which of these alternative possibilities becomes actual. The agent's action will turn the course of events in a genuinely new direction, setting in motion a novel causal series; its consequences will spread outward in space and time like ripples in a pond. Both the free human act and its causal consequences are intentionally permitted by God, but it may be that they do not enact God's particular purposes.

It is apparent here that human freedom considerably complicates the account of divine action we have been considering. If we suppose that God acts in history exclusively by means of secondary causes, and if we also hold that God permits incompatibilist free action, then at least two interrelated theological concerns arise. First, as we have just seen, the attribution of particular events in the world to God as divine acts becomes more problematic. We no longer can say simply that the activity of creatures is the indirect action of God, since many events will have free human acts somewhere in their causal ancestry. For some theological purposes, this is a welcome conclusion. One of the most pressing problems with any form of theological determinism is that it makes God the cause of human moral wrongdoing, and this deepens the difficulty of offering a morally plausible response to the problem of evil.[24] Given the open future of a causally under-determined world, however, there will be many events that cannot be regarded as God's intentional actions, even though the divine agent acts in every event as its ontological ground. God gives creation some scope of freedom to go its own way, and while this freedom, along with all it makes possible, is embraced within God's purposes, some of its expressions can be at odds with the good that God intends for creatures.[25]

This leads to a second set of theological issues. The Christian tradition affirms that although history can and does go wrong through the misuse of human freedom, God's good purposes lie at its foundation and ultimately will be fulfilled. The freedom that God grants to creatures is a gift that expresses, rather than compromises, God's providential care for the created world. But how is this divine superintendence of history to be exercised if creatures have the capacity to stray from God's purposes? God is not only creator but also redeemer, and redemptive divine

---

[24] Although there are various strategies for blunting the force of this conclusion, they face important conceptual and moral objections. See, for example, Kathryn Tanner's careful discussion of this problem and William Hasker's reply in *The God Who Acts*, Tracy, ed.

[25] This idea lies at the heart of most modern responses to the problem of evil. God's good purposes in creation may require (as a logically necessary condition) that God permit various evils to occur. This can be argued with respect both to so-called "natural evils" (i.e., the harm that befalls creatures simply by virtue of the natural conditions of their lives) and moral evils (i.e., the misuse of moral freedom by rational agents). A full defense of God's goodness must identify the good for the sake of which evil is permitted, explain the relation between evils and this good, and argue that this good is worth having even at this price. I have argued elsewhere that there are important limits in principle on our ability to do this; we can make some helpful points about why, in general, a God of perfect goodness, power, and knowledge would create a world that includes the sorts of evils we see around us, but we cannot expect to give a full explanation of the magnitude and distribution of evils in the world. Rather than offering an explanation of evil, however, the central focus of Christian theology is on God's redemptive actions in response to it. See my "Evolution, Divine Action, and the Problem of Evil," in *Evolutionary and Molecular Biology*, Russell et al., eds., 511–30; idem, "Why Do the Innocent Suffer?" in *Why Are We Here: Everyday Questions and the Christian Life*, Ronald F. Thiemann and William C. Placher, eds. (Harrisburg, Penn.: Trinity Press International, 1988). Also see Russell's comments on the problem of evil in the context of evolution, in "Special Providence and Genetic Mutation: A New Defense of Theistic Evolution," in *Evolutionary and Molecular Biology*, Russell et al., eds., 220–3.

action would appear to require that God act *in response* to the actions of free creatures. If we insist, however, that God's action in history always takes the form of indirect action through the order of nature, then it is not clear that such responsive action is possible. The fundamental structures of the natural world are fixed and in place long before human agents appear on the scene and make the choices to which God responds. If human choices were determined by antecedent conditions, then both the human action and the divine response could be built into the causal program of the world. But indeterministic free human actions present problems for divine providence that cannot be addressed in this way. This provides a compelling theological reason to affirm that God not only acts *indirectly* through secondary causes but also acts *directly* among them. And this, in turn, motivates theological interest in points of under-determination in nature at which God could act directly and yet without a miraculous intervention.

2.5 Divine Response Without Direct Action in the World?

Before turning to the idea of direct divine action at points of causal openness in the world, however, it is important to note that the argument for moving in this direction is not as strong as it may at first appear. There are resources in the theological tradition for a fascinating and subtle reply to the problem we just noted, a reply that avoids relying on direct divine action in the world. The key to this view is found in a particular understanding of divine foreknowledge. In the midst of late sixteenth century disputes about divine sovereignty and human freedom, Luis de Molina argued that divine omniscience includes not only knowledge of all necessary truths and knowledge of all matters of fact, but also knowledge of what every possible free creature would freely choose to do in every circumstance in which it might exist.[26] This is not simply a matter of foreknowledge of the free actions of actual human beings. In addition to this it includes knowledge of what these created agents would freely choose to do in any conceivable set of circumstances, even though these circumstances never in fact arise. Further, it involves having this knowledge with regard to every possible free creature, including of course an infinite number who never actually exist. Molina called this third aspect of omniscience God's "middle knowledge," because it is neither logically necessary nor entirely dependent upon God's determining will, but rather is a knowledge of contingent matters of fact that are nonetheless independent of God's will, since their truth is fixed by the free choices of finite agents (i.e., the free choices these creatures would make if they were to exist in these circumstances). This idea has been controversial ever since Molina proposed it, and there is a lively contemporary discussion about whether there are any true propositions of the form Molina proposes (i.e., true propositions about what an actual or possible free agent would freely choose to do in circumstances that never actually exist).[27]

---

[26] Luis de Molina, *On Divine Foreknowledge: Part IV of the Concordia*, Alfred J. Freddoso, ed. (Ithaca, N.Y.: Cornell University Press, 1988). For helpful discussions of Molina and his dispute with Dominic Banez see Freddoso's introduction, and Kathryn Tanner, *God and Creation in Christian Theology*, chap. 4.

[27] These propositions have come to be called "counterfactuals of freedom," and a great deal has been written about them. For a small sampling of the contemporary controversy see, for example, Robert Adams, "Middle Knowledge and the Problem of Evil," *American Philosophical Quarterly* 14 (1977): 109–117; idem, "An Anti-Molinist Argument," *Philosophical Perspectives* 5 (1991): 343–53; Thomas Flint, *Divine Providence: The*

If there are such truths to be known, then omniscience will include them. This will put God in a position to respond to free human actions by acting indirectly through secondary causal chains built into the order of nature at the outset. It is, of course, extraordinarily difficult to imagine designing the causal laws and boundary conditions of the world in such a way that a particular set of free agents and a particular set of divine responses to the actions of those agents emerge entirely through the ordinary operation of secondary causes. But there is no reason to think that this is logically impossible; the fact that the design problem boggles *our* minds does not have much force as a refutation of the idea that *God's* providential intention works in this way.

It might be objected that an indirect divine action programed into the structure of nature from time immemorial is not what the faithful have in mind when they understand their lives to be lived as a responsive, interpersonal relationship with God. We need not, however, adopt a temporal picture of divine action that locates God's creative initiative at a moment in the distant past and imposes a temporal gap between our acts and God's response. As the creator of all things, including time, God has classically been understood to transcend time. One way to try to grasp this inevitably ungraspable idea is to imagine that the whole created world in its temporal extension is immediately present to God, so that God is simultaneous with every event in time even though these events are not simultaneous with each other. When God takes a free human action into account in the overall design of the created world, this "taking into account" does not occur either before or after the human action. The human action is explanatorily, but *not* temporally, prior to the divine act of taking it into account, and the events that constitute God's response take place at the time proper to them in the causal history of the world.[28]

There are, of course, important conceptual puzzles raised both by the idea of middle knowledge and by the notion of timeless eternity. These puzzles have kept theologians busy for centuries, and show promise of continuing to do so. Every theological position, however, brings with it various conceptual difficulties, and decisions between competing theological proposals inevitably involve judgments of art about which problems we want to cope with. So we do not need to settle these disputed questions about foreknowledge and eternity in order to see that we have here a powerful strategy for understanding particular divine action in history in terms of *indirect* action through the natural order. This considerably dampens the force of the theological argument I gave for supplementing indirect divine action with the claim that God also acts *directly* in the world. On the account I have been considering, events can be objectively special divine acts and particular divine responses to human acts, and yet be indirect acts brought about entirely through the working of created agencies without any direct divine action other than creation/conservation. If most of what theology needs to say about God's action in history can be provided in this way, then the theological motive for searching out openings in the causal structure of the world is undercut. This point applies, of

Molinist Account (Ithaca, N.Y.: Cornell University Press, 1998); William Hasker, "A Refutation of Middle Knowledge," *Nous* 20 (1986): 545–57.

[28] See, for example, William Hasker, *God, Time, and Knowledge* (Ithaca, N.Y.: Cornell University Press, 1989); Brian Leftow, *Time and Eternity* (Ithaca, N.Y.: Cornell University Press, 1991); Eleonore Stump and N. Kretsmann, "Eternity," *Journal of Philosophy* 79 (1981): 429–58; and Richard Swinburne, "God and Time," in *Reasoned Faith*, Eleonore Stump, ed. (Ithaca, N.Y.: Cornell University Press, 1993), 204–22.

course, not only to theological interest in quantum mechanics, but also to appeals to chaos theory or any other area of contemporary scientific work.

I argued in an earlier paper that *if* theologians want to say that God acts to alter the course of events once the world's history is underway, then there must be gaps (of the right sort) in the causal structures of nature.[29] That conclusion still seems to me to be correct. But in light of the theological options explored in this paper, it is less clear that there is a need to make claims of this sort about direct divine action in the world, except in a limited (but theologically crucial) set of instances (e.g., in explicating classical theological claims about Christ). In these special cases, however, traditional views seem to involve a mode of divine action more akin to "miraculous intervention" than merely to a redirection of events by means of a probabilistic flexibility built into the laws of nature.[30]

3 Direct Divine Action Through Open Structures in Nature

That being said, it is important nonetheless to consider whether we might think of God as acting directly at points of causal openness in the structures of nature. There are a number of reasons to explore this possibility. So far, we have been considering how rich a theology of divine action can be generated if we limit our account to direct action in creation and conservation and indirect action through secondary causes. If the idea of nonmiraculous direct divine action can be worked out satisfactorily, it could be conjoined with these modes of divine action in a combined approach that can more readily interpret traditional claims about God's active engagement with nature and history. Furthermore, if our best theories about the structures of nature support an indeterministic interpretation, then this is something that a theology of divine action will need to take into account. The creator of such a universe will be not only the Lord of natural law but also, and in a perfectly acceptable sense, the God of the gaps. We have already seen that the indirect action position, as I have sketched it, is able to accommodate indeterministic transitions of chance and of freedom. It is important to acknowledge the possibility that one of the ways God's providential care engages the world is through these open structures in nature. Finally, the theological approach I have so far been considering faces a variety of important objections, and so it is wise to consider alternatives. Of course, the idea of nonmiraculous direct divine action also faces a number of difficult challenges. Given the inevitably problematic nature of all theological constructions, there is good reason to explore a variety of possibilities. In developing the idea of divine action, we need not claim to know which of the possibilities comes closest to capturing God's ways with the world, but we do need to show that some coherent combination of these possibilities provides a means by which God could accomplish the purposes that we attribute to the divine agent.

We turn, then, to the suggestion that God might act *directly* at points of under-determination to shape the course of events without disrupting the structures of

[29] Thomas Tracy, "Particular Providence and the God of the Gaps," in *Chaos and Complexity*, Russell et al., eds., 289–324.

[30] It has often been noted that it is not possible to spell out very fully the action that is ascribed to God when Christianity affirms that God "raised Jesus from the dead." If we interpret this language as pointing to an eschatological transformation of the human creature, then the familiar notion of miraculous divine intervention in nature is not so much wrong as insufficiently radical. Certainly the new creation is not merely the disruption or violation of the old order, but rather its fulfillment.

nature. Any position of this type will require not only that the natural order be causally open rather than closed, but also that under-determined transitions at least sometimes make a significant difference in the development of the events that follow from them. If these conditions are met, then we can conceive of God bringing about particular effects in the world without displacing secondary causes. Note that this is *not* to say that God acts entirely without created causes. The effects God brings about will have an extensive network of causal antecedents in the world, but these will be necessary, rather than sufficient, conditions.[31] This general theological strategy can be deployed in a number of different ways, and the details will vary from case to case. I will focus here on the possibility of direct divine action through indeterministic events at the lowest levels in the structures of nature.[32] It is worth noting at the outset, however, that there may be causal incompleteness at other levels of the natural order; if the case can be made for the existence such open structures, then it may be possible to conceive of God acting directly through these structures as well.[33]

3.1 Multiple Interpretations of Quantum Mechanics

There are a number of challenges facing any attempt to make use of quantum physics in developing a proposal of this kind about divine action. Perhaps the first and most obvious is that quantum theory can be interpreted in a bewildering variety of different ways, not all of which are congenial to this theological project. The formalism of quantum mechanics is well established, but there has been a remarkable proliferation of different explanations of what that formalism might tell us about the world.[34] The behavior of quantum systems defies ready ontological

---

[31] Russell (in this volume) makes a distinction between mediated and immediate divine action. The former refers to divine action that presupposes secondary causal conditions and works together with them. The latter would be unilateral divine action. If an immediate divine action truly had *no* necessary causal conditions in the prior history of the world, however, it is not clear that it could be an action *in* the world at all. So all divine actions within nature and history will be mediated, whether those actions are performed indirectly by means of secondary causes or directly in the way we are now considering. God's direct act of creating/conserving the world, of course, will be unmediated.

[32] William Pollard is an early proponent of one version of this theological strategy. See his *Chance and Providence: God's Action in a World Governed by Scientific Law* (London: Faber and Faber, 1958). For contemporary varieties of this approach see Russell, "Special Providence and Genetic Mutation," and the articles by George Ellis, Nancey Murphy, and Thomas Tracy in *Chaos and Complexity*.

[33] John Polkinghorne, for example, argues that the unpredictability in principle of macroscopic chaotic systems suggests an underlying ontological openness. Although the nonlinear equations describing chaotic systems are deterministic, Polkinghorne suggests that this formalism is an abstract and approximate description of natural systems that are more flexible than the mathematics suggests. See his *Science and Providence: God's Interaction with the World* (London: SPCK Press, 1989); idem, "The Metaphysics of Divine Action," in *Chaos and Complexity*.

[34] A brief overview of competing interpretations of quantum mechanics can be found in John Polkinghorne, "The Quantum World," and Robert John Russell, "Quantum Physics in Philosophical and Theological Perspective," both in *Physics, Philosophy and Theology: The Common Quest for Understanding*, Robert John Russell, William R. Stoeger, S.J., George V. Coyne, S.J., eds. (Vatican City State: Vatican Observatory, 1988). Also see Shimony, Stoeger, Butterfield, and Polkinghorne in this volume. There are a number of good introductions to quantum mechanics written for the general reader. For example, see Nicholas Herbert, *Quantum Reality: Beyond the New Physics* (Garden City, N.J.: Anchor/Doubleday, 1985);

interpretation, and this leaves physicists grappling with the limits of our conceptual resources, with what is "speakable and unspeakable in quantum mechanics."[35] This predicament bears a striking resemblance to the classic struggles of theologians in attempting to speak of a reality that inevitably exceeds our grasp.

If quantum theory is going to be helpful for the theological purposes I have described, it obviously must be interpreted indeterministically. It is fair to say that some of the currently dominant interpretations of quantum mechanics meet this condition, but the question is by no means settled.[36] According to the Copenhagen view, the wavefunction of a quantum entity (e.g., an electron) describes a state that in certain respects is objectively indeterminate. Some of the properties of the entity have specific values, e.g., the mass, charge, and magnitude of spin of an electron. But other properties must be expressed as a sum of probabilities for these values; this is often the case, for example, with the electron's position, momentum, and spin orientation. The wavefunction describes the development of the entity in space and time, and is strictly deterministic. When a measurement is made, however, a particular value is obtained for the measured property, e.g., spin orientation. This "collapse" of the wavepacket to a single value for the measured property cannot be further explained beyond noting the probability of that particular outcome. On the Copenhagen interpretation, quantum theory is complete; there are no "hidden variables" that, if we knew them, would allow us to assign fully determinate properties to the entity at every moment and thereby explain the measured result as having been causally determined by antecedent conditions. It is at this point that we encounter the indeterministic character of quantum systems; the transition from the indeterminate superposition of possibilities to a particular determinate state represents a point of ontological chance and causal openness in the structure of the world.

This interpretation of quantum theory has not gone uncontested. Einstein was famously troubled by the idea that God would "play dice with the universe." The EPR experiment was designed to show that if quantum theory is complete, then it violates special relativity by requiring instantaneous action-at-a-distance when a measurement takes place. Einstein found this consequence of the completeness thesis too bizarre to be credible, and so concluded that quantum theory must be incomplete. Niels Bohr, on the other hand, held that the theory is complete, but that it places severe limits on a realist interpretation such as the kind Einstein hoped for (illustrating once again that one thinker's modus ponens is another thinker's modus tollens).[37] In the 1960s, John Bell broke through this impasse by showing that the theoretical predictions of quantum mechanics are incompatible with local hidden-variable theories.[38] This was not the end of hidden-variable theories, however; even

Peter Kosso, *Appearance and Reality: An Introduction to the Philosophy of Physics* (Oxford: Oxford University Press, 1998); John Polkinghorne, *The Quantum World* (London: Longman, 1984); Alastair I.M. Rae, *Quantum Physics: Illusion or Reality?* (Cambridge: Cambridge University Press, 1986).

[35] This is the title of John Bell's book (Cambridge: Cambridge University Press, 1987).

[36] Werner Heisenberg is well known for this indeterministic interpretation of quantum theory. See his *Physics and Philosophy: The Revolution in Modern Science* (New York: Harper & Row, 1958).

[37] Niels Bohr, *Atomic Physics and Human Knowledge* (New York: Wiley, 1958).

[38] Bell, *Speakable and Unspeakable in Quantum Mechanics*. "Local" stands for consistency with the "first signal" principle in special relativity and precludes action-at-a-distance. See the essays by Shimony and Redhead in this volume. Also see James T. Cushing

before Bell, David Bohm had put forward a nonlocal hidden-variable interpretation of quantum mechanics.[39] Bohm's version supposes that quantum systems are composed of classical particles with well-defined positions, and he accounts for the probabilistic character of our knowledge by postulating that these particles interact with a pilot wave, which is mathematically related to the wave equation of the quantum formalism. In order to explain the correlation of properties when measurement occurs on linked two-particle systems, the pilot wave must instantaneously incorporate information about the complete measurement situation. In this way Bohm constructs an interpretation of quantum theory according to which its probabilistic character is strictly an artifact of the limits of our knowledge. His interpretation does not reflect any indeterminateness in the properties of the quantum entities, nor any indeterminism in their causal histories.

Bohm's version of quantum theory has not been widely embraced. There are several reasons for this: e.g., worries about how well it coheres with special relativity, unease with its postulation of additional entities without experimental evidence, and its failure so far to suggest novel lines of empirical research.[40] But Bohm's account does save determinism and the principle of sufficient reason, and these are powerful considerations in its favor. James Cushing has argued that the current consensus in favor of the Copenhagen interpretation primarily reflects historical contingencies in the development of modern physics.[41] At this point in the development of quantum theory, the decision for or against a Bohm-like approach remains perhaps a matter more of metaphysics than of physics.

The alternative views I just sketched are by no means the only interpretative options that the theologian faces, nor is Bohm's account the only deterministic interpretation of quantum theory. In a rather different way, many-worlds interpretations are deterministic, insofar as they insist that when measurement takes place all the possibilities described by the wave equation are actualized. There is no indeterministic transition from superposed possibilities to a single actuality; the wave equation does not collapse. Rather, the world (read "universe"!) branches, and it does so in accordance with the deterministic evolution of the wavefunction.[42] The only uncertainty in this transition is epistemic; we know what outcomes are possible (i.e., what worlds will be spawned by our act of measurement) and we can precisely state the relative probability of each outcome (i.e., the likelihood of our world actualizing any one of these possibilities), but we cannot know which outcome will occur (i.e., which world we will find that we inhabit).

This interpretive pluralism creates both an opportunity and a hazard for theologians. On the one hand, it is perfectly legitimate under these circumstances for

---

and Ernan McMullin, eds. *Philosophical Consequences of Quantum Theory: Reflections on Bell's Theorem* (Notre Dame, Ind.: University of Notre Dame Press, 1989); Michael L.G. Redhead, *Incompleteness, Nonlocality, and Realism: A Prolegomenon to the Philosophy of Quantum Mechanics* (Oxford: Clarendon Press, 1987).

[39] David Bohm, "A Suggested Interpretation of Quantum Theory in Terms of Hidden Variables, I & II," *Physical Review* 85 (1952); idem, *Wholeness and the Implicate Order* (London: Routledge & Kegan Paul, 1980).

[40] See the essays by Cushing, Polkinghorne, and Redhead in this volume.

[41] James T. Cushing, *Quantum Mechanics: Historical Contingency and the Copenhagen Hegemony* (Chicago: University of Chicago Press, 1994).

[42] Bryce S. DeWitt and Neill Graham, eds. *The Many-Worlds Interpretation of Quantum Mechanics* (Princeton: Princeton University Press, 1973). See also Butterfield in this volume.

a thinker grappling with the theology of nature to prefer one interpretation to another on theological grounds. Indeed, there can be no theological appropriation of quantum mechanics that does not make use of one or another of the currently viable interpretations. On the other hand, in casting our theological lot with a particular interpretation, we take the risk that new developments in physics or in the philosophy of physics will significantly undercut our theological constructions. It is important to acknowledge this possibility in framing our discussion of these matters, and this suggests two caveats. First, the particular interpretive approach we favor should not be presented as "the" conclusion to be drawn from quantum mechanics. Second, proposals about the theological relevance of quantum theory should be regarded as tentative and provisional hypotheses reflecting the current uncertainty of the relevant science and the extraordinary difficulty of interpreting it.

3.2 The Measurement Problem

One of the considerations driving the proliferation of interpretations of quantum theory is the nest of puzzles generated by the role of "measurement" in the standard interpretation. As we have seen, when a measurement takes place, the superposed possibilities described by the wave equation collapse to a single determinate value for the measured property. The outcome of this transition is not determined by the prior state of the system; rather, one state is actualized from among a probabilistic-ally structured ensemble of possible states. Unless a measurement is made, the quantum system continues to evolve deterministically in accordance with the wave equation. Here we encounter one of the central puzzles of quantum theory. What is it about the act of measurement that induces the collapse of the wavefunction? Bohr was inclined to point out that the macroscopic apparatus in the laboratory registers determinate states that are distributed in conformity with the wavefunction, and leave it at that. But if we move beyond this instrumentalism and interpret the quantum formalism as representing an actual indeterminacy in the system studied, then a host of difficult questions arise about how and where the indeterminateness of quantum entities gives way to the definiteness of macroscopic objects.

The puzzles surrounding measurement, as it is understood by the standard account of quantum theory, have at least two kinds of consequences for theological uses of this interpretation. First, they provide a motive for adopting an interpretation that avoids the idea of wavefunction collapse, and this may well result in a view that is less congenial to theological use. In the perplexing enterprise of interpreting quantum mechanics, however, each approach engenders its own set of problems. We just noted, for example, that although David's Bohm's deterministic interpretation generates no measurement problem, it faces difficulties about the privileged role it gives to position, the postulation of the pilot wave, and how well it coheres with special relativity. Second, if we say that God acts through chance events at the quantum level, then it appears that this form of divine action is limited to occasions of measurement. John Polkinghorne has argued that this restricts God's action in a way that severely undercuts the usefulness of quantum indeterminism for a theology of divine action. If (at the quantum level) causal openness is found only in the collapse of the wavefunction, and if the wavefunction collapses only when there is an irreversible macroscopic registration of the state of the quantum system, then God's action appears to be discontinuous and episodic. "Occasions of measurement only occur from time to time and a God who acted through being their determinator would also only be acting from time to time. Such an episodic account of providential agency does not seem altogether satisfactory theologically."[43]

The measurement problem certainly raises an important knot of issues for theological appeals to quantum indeterminism. But questions about measurement are so basic and unsettled a part of quantum theory that it is unclear as yet how far-reaching a problem is posed by the apparently episodic character of measurement events. Two cautious observations may be pertinent here. First, it is important to note, as Robert Russell points out, that state reduction takes place throughout the natural world, and not only in the laboratory. "Such events occur *constantly* in the universe whenever elementary particles interact irreversibly with molecules, gases, solids, and plasmas."[44] Russell mentions a number of particular examples: e.g., Brownian motion, blackbody radiation, the photoelectric effect, fission and fusion. In radioactive decay an indeterministic quantum transition occurs that, at least on the customary interpretation, takes place whether or not a Geiger counter is present to record it. But these examples only point us back to the underlying puzzles about measurement. The radioactive material and our Geiger counters (and Schrödinger's infamous and unfortunate cat in the box) can all be described quantum mechanically, and yet we do not find macroscopic objects displaying superpositions of incompatible properties (e.g., we do not encounter cats that are both dead and alive). We are bought back to the question of when and under what circumstances the wave equation collapses, and this in turn prompts the second of my two points: it is not clear what constitutes "measurement." The indeterminate quantum world gives rise to the determinate world of observable objects; the two constitute one world, but as yet we cannot explain just how they do so. The conundrum about the collapse of the wave equation lies at the heart of this broader difficulty in the interpretation of quantum mechanics, and until some greater clarity is gained on these basic matters it will be difficult to assess the impact of this problem on theological efforts to enlist quantum mechanics in an account of divine action.[45]

3.3 Dampening and Amplification

Even if indeterministic transitions of the sort associated with measurement are a pervasive feature of the world, this alone would not provide a useable toehold for a theological proposal of the sort we are considering. As we have seen, a further condition must be met, namely, that quantum chance at least sometimes make a difference in the course of macroscopic events. There is a relatively straightforward sense, of course, in which the histories of quantum systems do make a macroscopic difference, namely, they jointly constitute macroscopic objects and are the underlying base upon which higher level properties supervene. But if indeterministic transitions are entirely dampened out by their accumulation in statistical patterns, and thus disappear into classical, deterministic regularities, then they will be largely irrelevant to the theologian's interest in special divine action in the world. It could contended that the probabilistic laws of quantum mechanics reflect the pattern of divine action

[43] Polkinghorne, "The Metaphysics of Divine Action." Also see Polkinghorne's remarks on this problem in this volume. The idea that "measurement" should be understood as the irreversible macroscopic registration of a quantum effect can be found both in Polkinghorne and Russell, in this volume, and in Russell, "Special Providence and Genetic Mutation," 212.

[44] Russell, "Special Providence and Genetic Mutation," 204.

[45] See Russell in this volume for an extended discussion of measurement.

in determining the outcome of every chance event in quantum systems.[46] But this is just to say that God establishes and sustains the structure of natural law; we get the same result if we say that God establishes the stochastic laws and leaves the particular transitions to chance. Nothing is gained (at least with regard to the question of special divine action) by the claim that God determines some or all of the otherwise undetermined events at the quantum level, unless those events sometimes set in motion particular causal chains with macroscopic consequences.

It is clear that indeterministic transitions in quantum systems *can* have macroscopic effects. On the standard interpretation, precisely this happens when physicists make measurements on quantum systems in the lab. The more controversial question is whether nature is arranged in such a way that this amplification of quantum effects can occur apart from human contrivance. This of course is a question of empirical fact, and it is an unsettled one. Theological proposals about special divine action through quantum transitions must be correspondingly cautious and tentative. There do appear to be structures in nature, however, that register and then amplify the results of chance events at the quantum level. Robert Russell and George Ellis have both noted, for example, that vision involves a dramatic biochemical augmentation of the interaction between photons and molecular structures in the retina.[47] The nervous system appears to rely extensively on amplification processes of this sort. Further, Russell and Ellis, together with a number of other authors, have pointed out that genetic mutation can be induced by a variety of quantum mechanical transitions. In discussing the measurement problem, Alastair Rae offers the following example.

> ... mutations can be caused by the passage of high-energy cosmic ray particles. But these cosmic rays are clearly subject to the laws of quantum physics and each cosmic ray particle has a range of possible paths to follow, only some of which give rise to the mutation. The mutation therefore fulfils the role of a measuring event, similar to the photon being detected by the polarizer.[48]

Mutation may in effect "record" the interaction with a quantum mechanical entity, and then the phenotypic expression amplifies this change, exposing it to the selective pressures of evolutionary processes that may in turn further amplify (or extinguish) it. Russell has offered a careful development of the idea that God might act in evolutionary processes by affecting quantum transitions that result in mutations in the germ-line of an organism.[49] Mutation, of course, is just one among a number of sources of variation in a species, but it clearly plays an important role and can occur at a wide variety of points in the processes by which gametes are produced.

We should also note, though even more hesitantly, the possibility that quantum transitions might serve as triggers for chaotic processes. Familiar deterministic but nonlinear macroscopic systems can be extraordinarily sensitive to their initial conditions, generating dramatically divergent results from infinitesimally different

---

[46] See Nancey Murphy, "Divine Action in the Natural Order: Buridan's Ass and Schrodinger's Cat," in *Chaos and Complexity*, especially sec. 4.4.

[47] George Ellis and Robert Russell, both in this volume. Also see Carl S. Helrich, "Measurement and Indeterminacy in the Quantum Mechanics of Dirac," *Zygon: Journal of Religion and Science* 35.4 (December 2000): 489–503.

[48] Rae, *Quantum Physics*, 61.

[49] Russell, "Special Providence and Genetic Mutation."

starting points.[50] This suggests the possibility that an interaction with, say, a single electron might be amplified by a chaotic system into significant macroscopic effects. The indeterministic quantum transition would provide the trigger for a deterministic development of wide scope. This is an elegant and intuitively appealing hypothesis, but it is accompanied by a number of fundamental scientific uncertainties. It is not clear, for example, whether chaotic processes really are pervasive within the structures of nature, how chaotic systems are related to nonchaotic systems, and how much the latter dampen out the effect of the former.[51] An even more basic set of issues concerns the relation of quantum mechanics and chaotic systems.[52] As has often been noted, the Schrödinger equation is linear, and the prospects are not promising at present for a nonlinear reformulation of the quantum formalism. So it is not clear how "deep" chaos goes in the structures of nature, or how chaotic behavior emerges at the macroscopic level from its quantum mechanical substrate. The idea of chaotic amplification of indeterministic quantum effects is an enticing possibility, but it remains to be seen whether it will become more than that.

4 Conclusion

A theological proposal tied to currently disputed scientific questions must, of course, be hedged about with qualifications and put forward with a significant degree of diffidence. But given the current state of knowledge, it remains a viable possibility to hold that God might act at points of indeterministic transition in quantum systems. In this way God could bring about particular effects in the world that were not built into history from the beginning, and yet do so without "intervening," if by intervention we mean that God interrupts the ordinary lawful operations of the natural order. Clearly, this conception of divine action depends upon a whole series of interpretive judgments and on unsettled questions of fact, and so it has more the character of a program for further research than of a thesis that can be confidently asserted.

How seriously we take this possibility will depend in part on how much we think a proposal of this kind is needed in contemporary theology. The key consideration is whether the idea of divine action in response to human actions requires that God act in ways that affect the course of events in the world once the world's history is underway. I have argued that responsive divine action does not require that God act directly to alter the course of events in the world, though some of the specific things Christians have traditionally said about *how* God responds to us (especially in Jesus Christ) do appear to require this. If this is right, then theologians have less at stake than it might first appear in the question of whether the science of quantum mechanics (or of chaos theory, or of higher-level emergent systems) provide openings in nature's causal structures through which God can act without intervening. Even if the natural order is deterministic, we can understand God to act responsively in

---

[50] James P. Crutchfield, J. Farmer, N.H. Packard, and R.S. Shaw, "Chaos," and Wesley J. Wildman and R.J. Russell, "Chaos: A Mathematical Introduction with Philosophical Reflections," both in *Chaos and Complexity.*

[51] See Jeffrey Koperski, "God, Chaos, and the Quantum Dice," *Zygon: Journal of Religion and Science*, 35.4 (December 2000): 545–59.

[52] This is the question of "quantum chaos." For helpful discussion of these issues, see the essays by Michael Berry and John Polkinghorne in this volume. Also see Abner Shimony "Conceptual Foundations of Quantum Mechanics," in *The New Physics*, Paul Davies, ed. (Cambridge: Cambridge University Press, 1989), 391–2.

history with particular intentions, bringing about events that reflect God's special providence and doing so in most instances without miraculous interventions.

We may find, however, that our best physical theories support (even if they do not require) an ontological interpretation that recognizes a significant role for chance within the structures of nature, so that chance and law are dynamically woven together in a way that allows creative new developments not rigidly prescribed by the past. This picture of the world would be consonant with theological understandings of God's good purposes in creation, and it invites theological interpretation. *If* what we think we know about the world suggests that the structures of nature are open in this way, then there is good reason for the theologian to consider the possibility that God might exercise providential care for creation in part by acting directly through these flexible structures without forcing or deforming them. It is important to bear in mind that this mode of divine action is limited and theologically secondary.[53] Clearly, it cannot by itself provide a full account of all that the theistic traditions have wanted to say about God's activity in the world. On the account I have given, God's foundational action directly establishes and sustains the existence of all finite things. Because this creative action gives creatures causal powers of their own, God also acts indirectly by means of created causes in an endless variety of ways. Now we tentatively add to this account the idea that God may also act directly at points of under-determination in these causal structures and thereby turn events in new directions that serve God's purposes in creation.

This last mode of divine action invites some familiar objections. We might worry, for example, that it returns us to the God of the gaps. The gaps in which God acts, however, are not merely points of incompleteness in our *knowledge* of the world; an appeal to divine agency to fill merely epistemic gaps is a clumsy and inevitably temporary expedient. Rather than preying upon what we do not yet understand about the natural world, this theological proposal would make use of what we claim to know, namely that there are (*ex hypothesi*) ontological gaps in the causal structures of nature.[54] It might be replied that this nonetheless treats God as one cause among others, on a par with secondary causes, busily pushing particles around the universe. This is a rhetorically vivid objection, but it does not carry much force unless we think of direct action at points of causal incompleteness as the only or the primary mode of divine action. God is never merely one agent among others. Rather, God is always the absolute source of the being of all finite things, acting continuously and universally as the primary cause. It *would* be an arbitrary limitation upon God's power if we denied that God could act among secondary causes, should God choose to do so. But this is a claim that Christians, in particular, should hesitate to make, given the radical affirmation in this tradition of God's freedom to enter fully into relationship with creatures without ceasing to be God.

---

[53] This limitation has been overlooked by some of the critics of the idea of divine action through quantum indeterminisms. For example see Nicholas Saunders, "Does God Cheat at Dice? Divine Action and Quantum Possibilities," *Zygon: Journal of Religion and Science*, 35.4 (September 2000): 517–44, and my response "Divine Action and Quantum Theory," *Zygon*, 35.5 (December 2000): 889–98.

[54] For a more detailed discussion of this objection, see my "Particular Providence and the God of the Gaps," sec. 1.

QUANTUM THEORY AND THE MACROSCOPIC WORLD

George F.R. Ellis

The main part of this essay discusses the nature of physical action and its relation to reductionism and ontology, seen in the light of our present-day understanding of quantum theory. The specific physical topics I consider are the macro-effects attainable by bottom-up actions of micro-effects, and the contrasting feature of top-down action (or whole-part causation). I then turn to the implications for reductionism, and for micro and macro issues of ontology. In the final section, I relate the resulting view to theology, and specifically to the theme of divine action that underlies this conference series.

1 Micro-to-Macro Relations: Bottom-Up Causality

Bottom-up causality is a fundamental feature of the structural hierarchy in the physical world. What happens at each higher level is based on causal functioning at the level below. Hence, what happens at the highest level is based on what happens at the bottom-most level. This is the profound basis for reductionist worldviews.

1.1 The Classical World

At the classical levels of structure (i.e., levels where we can use phenomenological relations that ignore quantum theory), the relation of micro- to macrostructure is in principle deterministic, given the complete structure and relations of the lower level.

Physics and chemistry. The successive levels of order entail chemistry being based on physics, material science on both physics and chemistry, geology on material science, and so on.[1] Given the number of lower-level constituents involved, a detailed description of the lower levels is impractical in most situations, and in conditions close to equilibrium is replaced by a statistical description, with the laws of statistical physics relating the lower-level to the higher-level behaviors.

Biology. The profound discovery of molecular biology is that through the extraordinary nature of biological molecules, a fully mechanistic description applies at the microlevel of living organisms also, including human beings and the human brain.[2] However biological systems are open systems that are far from thermal equilibrium, so statistical physics does not apply (which is why they can apparently violate the second law of thermodynamics). Rather, what happens in such systesms is governed by detailed structural relations and molecular interactions.

1.2 Quantum Basis for "Ordinary" Macrophysics

Underlying classical physics is the extraordinary quantum description of matter at the microlevel, with its apparently irremovable feature of quantum uncertainty.[3]

---

[1] For a more detailed discussion see, George F.R. Ellis, *Before the Beginning: Cosmology Explained* (New York: Bowerdean/Boyers, 1993).

[2] See Neil A. Campbell, *Biology* (Menlo Park, Calif.: Benjamin Cummings, 1990).

[3] Richard W. Robinett, *Quantum Mechanics: Classical Results, Modern Systems, and Visualized Examples* (Oxford: Oxford Univ. Press, 1997); Peter Landshoff, Allan Metherell, and Gareth Rees, *Essential Quantum Mechanics* (Cambridge: Cambridge University Press,

Quantum theory underlies the emission and absorption of light, chemical bonding, etc. In particular, through the Pauli exclusion principle for fermions, together with the principle of complete indistinguishability of elementary particles, quantum theory underlies both the stability of ordinary matter and the periodic table of the elements.[4] Thus, despite the uncertainty principle, quantum physics underlies the stability of ordinary macroscopic causality and the reliability of macrostatistical physics.

The quantum laws have this macro-effect, despite being microphysical laws, precisely because they operate in the same way everywhere and at all times, thereby underlying the various levels of structure and determinism in the physical world.

1.3 Quantum Effects That Are Effective at the Macrolevel

If specifically quantum effects take place at the microlevel, can they have discernible macroscopic results? Can the quantum intrude on the macrolevel?

Amplifiers. There is no problem in principle with individual quantum events (e.g., the arrival of a single photon) having a macro-effect, because electronic photo-amplifiers (widely used in astronomy) are able to produce macro-effects by amplifying individual quantum signals to an essentially arbitrary extent.[5] Similarly, lasers and masers act as amplifiers of quantum events. However, these are in some sense artificial contexts. The important question is whether similar amplifiers occur in the natural world. In particular, can individual quantum events have a macro-effect in living beings?

Two examples will answer both questions affirmatively. The first is the eye. In some species the eye can detect individual photons falling on the retina. The photon is absorbed by a molecule of rhodopsin, eventually resulting in a nervous impulse coming out of the opposite end of the cell with an energy at least a million times that contained in the original photon.[6] The amplification of the incoming signal is due to a molecular cascade of reactions, but with much of the amplification in the initial step, where the single photon-excited rhodopsin passes on the excitation to at least 500 molecules of transducin within one millisecond.

A second example has been presented by Ian Percival, who states that "DNA responds to quantum events, as when mutations are produced by single photons, with consequences that may be macroscopic—leukemia for example."[7] In this case the amplifier is the developmental process by which the information in DNA is read out in the course of the organism's developmental history.[8] A mutation might of course have more beneficial effects than mentioned by Percival (e.g., enhanced cognition). Indeed, mutations caused by cosmic rays may well have played a significant role in evolutionary history.

1999); Richard P. Feynman, *QED: The Strange Theory of Light and Matter* (Princeton, N.J.: Princeton Univ. Press, 1988); Michael E. Peskin and Daniel V. Schroeder, *An Introduction to Quantum Field Theory* (Reading, Mass. Perseus Books, 1999).

[4] For a clear description of this issue, see Alan Durrant, ed., *Quantum Physics of Matter* (Bristol: Institute of Physics, 2000).

[5] Feynman, *QED*.

[6] Mae-Wan Ho, *The Rainbow and the Worm: The Physics of Organisms* (Singapore: World Scientific, 1993), 6.

[7] Ian Percival, "Schrödinger's Quantum Cat," *Nature* 351 (1991): 357.

[8] Lewis Wolpert et al., *Principles of Development* (Oxford: Oxford University Press, 1998).

Coherent implementation of micro-effects. If a micro-effect is implemented simultaneously in many micro-events, a *classical macro-event* can occur because the combined set of microevents, taken together, correspond to such a macro-event. No amplification is then necessary. An example might be emission of photons by excited atoms; if they were all to emit photons at the same time, instead of at random times as would normally happen, a macrovisible light pulse would be emitted at that instant. This can be arranged by suitably adjusting the timing of the microevents in a macrodomain to achieve a macro-event (e.g., the laser).

Essentially quantum effects at the macrolevel. Can quantum effects at the microlevel produce significant but essentially quantum effects at the macrolevel? Yes, as the well-known examples of superconductivity and superfluidity show. Some kinds of quantum effects do not need any amplifier—they spontaneously occur at macroscales. They do however require special conditions—here, low temperature—leading to coordination between states of microparticles. These are examples of cooperative phenomena, where quantum effects undermine the very concepts we typically use to understand bottom-up action. Quantum theory is such that local interactions in specific macroscopic contexts may imply that the usual individual descriptions and behaviors no longer apply, so that the properties of the particles will differ from those they have in isolation. Thus, completely new kinds of behaviors become possible through cooperative effects in a suitable macrocontext (which provides the boundary conditions for the microphysics).

Quantum cooperative effects occur in superconductivity, superfluidity, and the quantum Hall effect. In superconductivity, the electrons—despite their repulsion for each other—form pairs ("Cooper pairs"), which are the basic entities of the superconducting state. This happens by a cooperative process: the negatively charged electrons cause distortions of the lattice of positive ions in which they move, and the real attraction occurs between these distortions.[9] R.B. Laughlin (Nobel Laureate, physics, 1998) discusses the implications in his fascinating Nobel lecture:

> One of my favorite times in the academic year occurs in early spring when I give my class of extremely bright graduate students, who have mastered quantum mechanics but are otherwise unsuspecting and innocent, a take-home exam in which they are asked to deduce superfluidity from first principles. There is no doubt a very special place in hell being reserved for me at this very moment for this mean trick, for the task is impossible. Superfluidity, like the fractional Hall effect, is an emergent phenomenon—a low-energy collective effect of huge numbers of particles that cannot be deduced from the microscopic equations of motion in a rigorous way, and that disappears completely when the system is taken apart. ... The students feel betrayed and hurt by this experience because they have been trained to think in reductionist terms and thus to believe that everything that is not amenable to such thinking is unimportant. But nature is much more heartless than I am, and those students who stay in physics long enough to seriously confront the experimental record eventually come to understand that the reductionist idea is wrong a great deal of the time, and perhaps always... The world is full of things for which one's understanding, i.e. one's ability to predict what will happen in an experiment, is degraded by taking the system apart, including most delightfully the standard model of elementary particles itself.[10]

---

[9] David L. Goodstein, *States of Matter* (New York: Dover, 1985).

[10] R.B. Laughlin, "Fractional Quantisation," *Reviews of Modern Physics* 71 (1999): 863–74. On emergent phenomena in science, see Phil Anderson, "More is Different," *Science* 177 (1972): 377, reprinted in his *A Career in Theoretical Physics* (Singapore: World Scientific, 1994).

The lecture continues to explain how one of the things emergent phenomena can do is create new physics, and particularly the quantum Hall effect (whereby cooperative phenomena between particles with integral charge create quasi-particles with fractional charge).

Quantum entanglement. The underlying fundamental feature is that of quantum entanglement:

> The phenomenon of "entanglement" refers to the fact that the most general quantum description of an ensemble of systems in which each system is composed of two or more subsystems (such as pairs of electrons or photons) does not permit us to assign a *definite quantum state* to each of the individual subsystems... if quantum mechanics gives the correct predictions for experiment and we are not prepared to relax very basic ideas about causality, then *independent of any theoretical interpretation*, the individual particles cannot be conceived of as possessing "properties" in their own right.[11]

This is intimately related to the principle of complementarity, one of the experimental foundations of quantum theory. Its nature is tested by "which-way" experiments,[12] which show that it is not due to interference effects caused by measurement but to entanglement.

A specific example of entanglement is the Bose-Einstein Condensate: "In recent experiments, individual atoms have lost their individual identities for a full ten seconds and behaved as though they were a single superatom, all the particles in the system collapsing into a single quantum state with unusual and novel properties."[13] The experimental realization of two-dimensional and three-dimensional Bose-Einstein condensates, where long-range phase coherence exists and brings quantum behavior to the macroscopic world, can be demonstrated by interferometric techniques.[14] Such correlated electron systems cannot be described in terms of any simple independent-electron picture; rather the electrons behave cooperatively. This undermines the reductionist idea at its very heart: the individual components whose combined properties are to be summed to explain the macrosystem properties do not possess specific properties in their own right!

2 Macro- to-Micro Relations: Top-Down Causality

The complementary feature to bottom-up causality is top-down causality. There are numerous interesting examples of top-down causality in the physical universe when looked at from the macroviewpoint, many of which were discussed in the *Chaos and Complexity* volume of this series. However the idea is central to the theme of the

---

[11] Tony Leggett, "Quantum Theory: Weird and Wonderful," *Physics World* (December 1999): 73–7, 74; see also Chris Isham, *Lectures on Quantum Theory* (Singapore: World Scientific, 1995), 143.

[12] Marlin O. Scully et al., "Quantum Optical Tests of Complementarity," *Nature* 351 (1991): 111; S. Dürr, T. Nonn, and G. Rempe, "The Origin of Quantum-Mechanical Complementarity Probed by a 'Which-Way' Experiment in an Atom Interferometer," *Nature* 395 (1998): 33–7; E. Buks et al., "Dephasing in Electron Interference by a 'Which-Path' Detector," *Nature* 391 (1998): 871–74.

[13] Eric A. Cornell and C.E. Wiseman, "The Bose-Einstein Condensate," *Scientific American* (March 1998): 26. See also Tom Hijmans, "Hydrogen a Quantum Gas at Last," *Physics World* (February 1999): 17; Yvan Castin et al., "Bose Condensate Makes Quantum Leaps and Bounds," *Physics World* (August 1999): 37.

[14] Léon Bloch, T.W. Hansch, and T. Esslinger, "Measurement of the Spatial Coherence of a Trapped Bose Gas at the Phase Transition," *Nature* 403 (2000): 166.

conference series, so I return to it here, particularly in the context of quantum theory. I mention three kinds of examples from the classical world to set the scene (from physics, biology, and society), and then consider two where the issue relates directly to quantum theory.

2.1 Physics and Biology

Top-down causality is ubiquitous in biology and in practical applications of classical physics, but this feature is not normally noted in physics courses because of the emphasis in physics teaching on "isolated systems." Real physical and biological systems are not isolated. Engineers and ecologists are fully aware of this fact—it is central to their disciplines. A few examples will make the point.

Nucleosynthesis. The rates of nuclear interactions depend on the density and temperature of the interaction medium. Hence, their outcomes in the early universe and then in stars and supernovae explosions—all crucial to the later formation of structure in the universe[15]—depend in an important way on top-down action from the environment in which they occur. In each case the microscopic reactions that take place, and thus the elements produced, depend on the environment: in the first case, the density of matter in the universe and its consequent rate of expansion; and in the second, the atmospheric structure of the stars and supernovae, determining which reactions will occur at what rate.

Top-down action is also central to two main themes of molecular biology, even though many texts on the subject[16] emphasize bottom-up (mechanistic) aspects.

Biology: development of DNA codings. The first central theme of evolutionary biology is the development of particular DNA codings (the particular sequence of bases in the DNA) through an evolutionary process that results in adaptation of an organism to its ecological niche.[17] This is a classical case of top-down action from the environment to detailed biological microstructure: through the process of adaptation, the environment (along with other causal factors) fixes the specific DNA coding.[18] There is no way you could ever predict this coding on the basis of biochemistry or microphysics alone.

Biology: reading of DNA codings. The second central theme of molecular biology is the reading of DNA by an organism in the developmental process.[19] This is not a mechanistic process, but a context dependent one—this was the central

[15] See for example, Joseph Silk, *A Short History of the Universe* (New York: Freeman, 1997).

[16] See, for example, Bruce Alberts et al., *Molecular Biology of the Cell* (New York: Garland, 1989).

[17] See Campbell, *Biology*.

[18] Donald T. Campbell: " 'Downward Causation," in Hierarchically Organized Biological Systems," in *Studies in the Philosophy of Biology: Reduction and Related Problems*, F.J. Ayala and T. Dobzhansky, eds. (London: MacMillan, 1974), 179–86; this is discussed in detail in Nancey Murphy, "Supervenience and the Downward Efficacy of the Mental: A Nonreductive Physicalist Account of Human Action," 147–64, Arthur Peacocke, "The Sound of Sheer Silence: How Does God Communicate with Humanity?", 215–48, both in *Neuroscience and the Person: Scientific Perspectives on Divine Action*, Robert J. Russell, Nancey C. Murphy, Theo C. Meyering, and Michael A. Arbib, eds. (Vatican City State/Berkeley, Calif.: Vatican Observatory/Center for Theology and the Natural Sciences, 1999), the volume hereafter referred to as *NAP*.

[19] See Wolpert et al., *Principles of Development*.

insight of Jacob and Monod for the bacterial lac operon in the 1950s. Indeed development is context-dependent all the way down, with what happens before having everything to do with what happens next.[20]

The central process of developmental biology, whereby positional information determines which genes in each cell get switched on and which do not, so determining their developmental fate, is a top-down process from the developing organism to the cell based on the existence of gradients of positional indicators in the body.[21] Without this feature, an organism could not develop in a structured way. Thus, the functioning of the crucial cellular mechanism determining the type of each cell is controlled in an explicitly top-down way. At a more macrolevel, recent research on genes and various hereditary diseases shows that presence in an organism of the gene for a particular disease is typically not sufficient for the disease to occur: the outcome depends on the nature of the gene, on the rest of the genome, and on the environment.

So "mechanistic in principle" works out to "not mechanistic in practice." The macrosituations determine what happens, not specific microfeatures by themselves, which do work mechanistically but in a context of larger meaning that largely determines the outcome. And note particularly that the macro-environment includes the result of conscious decisions (the patient will or will not seek medical treatment for a hereditary condition, for example), so these too are a significant causal factor.

2.2 Human Volition

The last point highlights the fact that consciousness brings a whole new series of effects into the causal network. When a human being has a plan in mind (e.g, a proposal to build a bridge) and this is implemented, then enormous numbers of microparticles (comprising the protons, neutrons, and electrons in the sand, concrete, bricks, etc. that become the bridge) are moved around as a consequence of this plan and in conformity with it. Thus in the real world, the detailed microconfigurations of many objects (which electrons and protons go where) is in fact to a major degree determined by the macroplans that humans have for what will happen, and the way they implement them.[22]

The existence of human volition thus causes a major difference in the causal hierarchy between the branch containing the natural sciences on the one hand, where human action may be ignored, and the branch containing the human sciences on the other, where volition and goal-seeking determine the course of events.[23] Some specific socially important examples of top-down action involving goal-choice are:

- *The Internet* embodies local action in response to information requests, causing electrons to flow in meaningful patterns in a computer's silicon

---

[20] Enrico Coen, *The Art of Genes* (Oxford: Oxford University Press, 1999).

[21] Wolpert et al., *Principles of Development.*

[22] Cf. Karl Popper and John Eccles's discussions of Worlds 1, 2, and 3, and the interactions between these worlds in their *The Self and Its Brain: An Argument for Interactionism* (Berlin: Springer, 1977). This dilemma is concretized in the film "Contact" when the radio astronomers try to decipher a message from outer space but don't know how to interpret the data.

[23] This difference is explicitly represented in the branching causal hierarchy of the sciences presented in Nancey Murphy and George F.R. Ellis, *On the Moral Nature of the Universe* (Fortress Press, Minneapolis, 1995), hereafter *MNU*.

chips and memory, mirroring patterns thousands of miles away, when one reads web pages. This is a structured influence-at-a-distance due to channeled causal propagation and resulting local physical action.

- *Hiroshima*. The dropping of the nuclear bomb at Hiroshima was a dramatic macro-event realized through numerous micro-events (fissions of uranium nuclei) occurring because of a human-based process of planning and implementation of those plans.
- *Global warming*. The effect of human actions on the earth's atmosphere, through the combined effect of human causation moving very large numbers of microparticles (specifically, CFCs) around, thereby affecting the global climate. The macroprocesses at the planetary level cannot be understood without explicitly accounting for human activity.[24]

The Human body and brain. The foundation for all of this is the top-down action in the human body, where the brain controls the functioning of the parts of the body through a hierarchically structured feedback control system, which incorporates the idea of decentralized control to spread the computational and communication load and increase local response capacity.[25] The human body is a highly specific system in that dedicated communication links convey information from specific areas of the brain to specific areas of the body, enabling brain impulses to activate specific muscles (by coordinated control of electrons in myosin filaments in the bundles of myofibrils that constitute skeletal muscles), in order to carry out consciously formulated intentions.[26]

Furthermore, through this process there is top-down action by the mind on the body, and indeed on the mind itself, both in the short term (immediate causation through the structural relations embodied in the brain and body) and in the long term (structural determination through imposition of repetitive patterns). An example of the latter is how repeated stimulation of the same muscles or neurons encourages growth of those muscles and neurons. This is the underlying basis of both athletic training and of learning by rote, but it is much more. Indeed (as emphasized in *MNU*), the nature of our chosen goals and even our morality is in the long term hardwired into our neural circuits, and so realized in the physical microstructure of our brains. Additionally, Western medicine is only now beginning to investigate the important area of the effect of the mind on health.[27]

2.3 Quantum Considerations

Thus top-down effects occur and are central in the real world: the macrolevel influences the microlevels at all stages. This has to be based in quantum theory, for

---

[24] Hans J. Schellnhuber, "Earth System Analysis and the Second Copernican Revolution," *Nature* 402 (2 December 1999): C19–C23.

[25] For illuminating discussion of these features, see Stafford Beer, *Brain of the Firm* (New York: Wiley, 1981).

[26] This is the feature that underlies the discussion of the right-hand branch of the causal hierarchy in *MNU*, where the various aspects of human consciousness act down on the physical environment and are themselves hierarchically structured in a causal fashion in terms of goals and aims, with ethics as the topmost level.

[27] Bill Moyers, *Healing and the Mind* (New York: Doubleday, 1995); Esther Sternberg, *The Balance Within: The Science Connecting Health and Emotions* (New York: W.H. Freeman, 2000).

quantum theory underlies the physical structure at the microlevel. How then is it made explicit? Here I consider a number of ways this effect is embodied in our present-day understanding of quantum theory.

Interaction potentials. First, potentials in the Schrödinger equation, or in the action for the system,[28] represent the summed effects of other particles and forces, and hence are the way that the nature of both simple and complex structures can be implemented (from a particle in a box to a computer or a set of brain connections). These potentials describe the summed interactions between microstates, enabling internal top-down effects. Additionally one may have external potentials imposed in the chosen representation, representing top-down effects from the environment on the system.

Experiments and collapse of the wavefunction. Second, the central additional feature where top-down action takes place is in the quantum measurement process—collapse of the wavefunction to an eigenstate of a chosen measurement system. The experimenter chooses the details of the measurement apparatus (e.g., aligning the axes of polarization measurement equipment), and this decides what set of microstates can result from a measurement process, crucially influencing the possible microstate outcomes of the interactions that happen. Thus the quantum measurement process is partially a top-down action controlled by the observer, determining what set of eigenstates are available to the system during the measurement process. Specific descriptions of quantum theory are based in specific representations and quantities characterizing the physical situation considered. The choice of Hilbert space and the associated operators and functions is made to reflect the experimenter's choice of measurement process and apparatus, thus reflecting this top-down action.

One response is that the measurement process interferes with the system, and bottom-up causation works for the system, as long as it is not being interfered with. But the point here is that it is precisely the fiction of an isolated system that is at fault: such systems do not exist in the real world, and the degree to which there is top-down action from the environment, including the experimenter, is precisely the degree to which the system is not isolated. If you try to construct a "bottom-up" proposal for explaining how the macroscopic measuring apparatus will work through summation of the effects of the microscopic particles that comprise it, this will not work because the process involved in observations due to the measuring macrosystem (collapse of the wavefunction to an eigenstate) is nonunitary—but the Schrödinger or Dirac evolution of the microsystem is unitary.[29] You can't derive the former by summing the latter. Thus, while the effect of Schrödinger evolution of any combination of microsystems is unitary, one cannot reduce the apparatus to a sum of behaviors of its quantum components because its effect is nonunitary.

A further response is that in the natural world, measurement only takes place occasionally, so this is a very rare effect, in physical terms. But one of the key issues then is that we do not have a clear statement of when a "measurement process" takes place in a naturally evolving system with interactions, but without a conscious

[28] See Peskin and Schroeder, *An Introduction to Quantum Field Theory*.

[29] That is, the magnitude of the wavefunction is preserved by the evolution it undergoes (Isham, *Lectures on Quantum Theory*, secs. 6.4.1 and 7.1); and summing any number of unitary actions still results in a unitary action. In contrast, a nonunitary evolution, such as reduction of the state vector associated with a measurement process (ibid., 135), does not preserve this magnitude.

observer present (we do not know when collapse of the wavefunction will take place, and if so along what eigenstates). Certainly it is conceivable that, for example, photons will cause interference fringes without any artificial apparatus being part of the process. It is also clear that plants register the arrival of photon in the process of photosynthesis. But are these cases equivalent to measurement, or not? Whether they are or not, the interception of a photon by a plant leaf alters the evolution of the photon and represents a situation where a macro-effect shapes the possible micro-evolutions.

State preparation. The third feature where top-down action takes place in quantum theory, related to the previous one, is the choice and control of the initial state in an experiment (state preparation). This is also a choice implemented in top-down fashion by the experimenter. Thus the experimental apparatus can be described in (nonquantum) macroterms and we see top-down action of the macroworld on the quantum world both in the measurement process and in the prior preparation process.

Decoherence. The fourth aspect, probably implied by the previous ones but worth mentioning in its own right because of its importance, is decoherence, discussed in depth in other articles in this volume. Here we see the environment acting on the system in a top-down way, effectively removing quantum entanglements: a vector state for a pair of systems becomes a mixed state for one of them, if the degrees of freedom of the other are "traced over." (This is the best you can do if there is some part of the state that cannot be accessed—which is true for the state of the environment that envelopes the system. Indeed, labeling it "the environment" is precisely a statement that this is not the system we are describing in detail.) Consequently, according to Chris Isham, "the notion of an 'isolated system' (which is postulated in formulating the rules of quantum theory) is an idealization that can never really apply: the only truly isolated system is the universe."[30]

The arrow of time. Finally, one cannot tell how a macrosystem will behave in the future on the basis of the laws of physics and the properties of the particles that make up the system alone, because time-reversible microphysics allows two solutions—one the time-reverse of the other—but the macrosystem allows only one of those solutions.[31] The absolute prohibition of one of the allowed microsolutions is mathematically put in by hand to correspond to the real physical situation, where only entropy-increasing solutions in one direction of time occur at the macrolevel; this does not follow from the microphysics. It was the failure to solve this issue that apparently lead to Ludwig Boltzmann's suicide—he brilliantly proved his (macro) H-theorem on the basis of microphysics, and then he realized that this solution worked equally well with both directions of the arrow of time.

Physically, the only known solution to this arrow-of-time problem seems to be that there is top-down action by the universe as a whole, perhaps expressed as boundary conditions at beginning of spacetime,[32] that allows one solution and not the other. One cannot tell which is solution is disallowed by looking at the microlevel alone. Some kind of top-down action must come into play.

---

[30] Isham, *Lectures on Quantum Theory*, 155.

[31] Paul C.W. Davies, *The Physics of Time Asymmetry* (London: Surrey University Press, 1974); Heinz Dieter Zeh, *The Physical Basis of the Direction of Time* (Berlin: Springer, 1992).

[32] See Roger Penrose, *The Emperor's New Mind* (Oxford: Oxford University Press, 1989).

This is related to the quantum measurement issue raised above: collapse of the wavefunction takes place with a preferred direction of time.[33] It is not clear how the direction is determined if we look at the microlevel of the system alone.[34] Thus there is a macrodetermined arrow in the quantum measurement process that cannot be reduced in any known way to the result of microscopic actions. Whether this is sufficient to account for the macro-arrow of time is unclear, but what is clear is that there is already an arrow-of-time problem at the quantum level, and top-down action from the environment (or whole-part causation) is probably the only way to solve it.

3 Reductionism in Light of These Effects

The features discussed above undermine a straightforward reductionist view in several ways. I consider these in turn, and then discuss when reductionism is useful in theory and in practice.

3.1 The Effect of Top-down Action

First, as we have seen, in a hierarchically structured system there is top-down action (or "whole-part causation") from the macro- to the microlevel,[35] as well as bottom-up action from the micro- to the macrolevel. This completely changes the causal relationships between the levels. This is why a simple reductionist view will not work. The idea many have—that reductionism implies that macro-events are completely causally determined by the laws acting at the microlevel—is true but misleading because it is also correct to say that what happens at the microlevel is to a large degree determined by macrostructure[36] and what happens at the macrolevel.

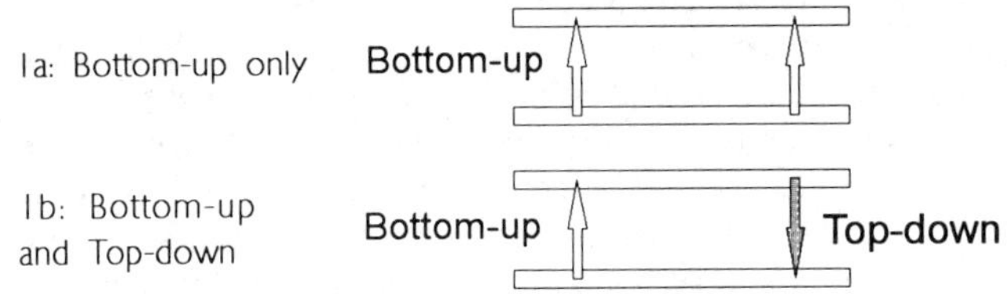

Figure 1. Bottom-up and Top-down action. The fundamental importance of top-down action is that it changes the overall causal relation between upper and lower levels in the hierarchy of structure and organization.

---

[33] Schrödinger or Dirac evolution is time symmetric, and so does not encounter this problem.

[34] See Penrose, *The Emperor's New Mind.*

[35] Peacocke, "The Sound of Sheer Silence."

[36] Understood in terms of constitutive and structural relationships, and boundary conditions; see Robert J. Russell, Nancey C. Murphy, and Arthur Peacocke, eds., *Chaos and Complexity: Scientific Perspectives on Divine Action* (Vatican City State/Berkeley, Calif.: Vatican Observatory/Center for Theology and the Natural Sciences, 1995), hereafter *CAC*; William R. Stoeger, S.J., "The Mind-Brain Problem, the Laws of Nature, and Constitutive Relationships," in *NAP*, 129–46.

This makes a key difference in terms of one's understanding of what is going on. Even when the microlevels are mechanistic and fully deterministic, determinism will be effective within the context of boundary conditions, macroconstraints, and macroinfluences, which largely decide what the actual outcome is. As well as specifying the microphysics, to determine what happens you also need to define the macrostructure and macro-environment, which play a major role in determining what happens at the microlevel and hence at the macrolevel.

Tom Tracy suggests this means that top-down causation consists in a particular way of structuring bottom-up causation.[37] This is correct; precisely this higher-level structuration (physically embodied in the microstructure in terms of hierarchically related feedback loops) gives the whole its higher-level relations and order. Taking the system apart destroys this structure and removes the essential features that make it function. To put it another way, the macrostructure is a high-level view of all the detailed microstructure. To describe the structure as a whole adequately *requires* a macrolevel view and language, for the relevant causal relations are not apparent if one looks solely at the microlevel, and indeed cannot even be described in terms of the language appropriate at that level.

This means one must be very careful in talking about determinism on the basis of bottom-up (mechanistic) action. There are multiple levels of description and meaning described by the phenomenology of each level. The key point is that the higher levels of meaning cannot be *reduced* to a description solely in terms of lower-level concepts. They are compatible with the lower levels, of course; but in a serious they sense determine what actually happens.

Systems thinking. The implication becomes particularly powerful when one takes into account the issue of purposeful action in complex systems. Russell Ackoff[38] comments on the need for systems thinking based on *synthesis* (putting things together), as well as reductionist thinking based on *analysis* (taking things apart), if one wants to adequately understand complex systems. In particular he makes the following interesting point: the machine age is based on the powerful idea of *the machine*, a mechanism that can be used to apply energy to matter in a reliable and predictable way. Analysis "explains" the properties of the machine by analyzing its behavior in terms of the functioning of its component parts (the lower levels of structure). Systems thinking on the other hand "explains" the behavior or properties of the machine by determining its role or function within the higher levels of structure (an automobile exists to provide transport, a computer to process information, scissors to cut material, and so on).

The point is fundamental. The analytic approach ignores the environment and takes the existence of the machine for granted. Questioning how the machine functions enables one to understand its reliable replicable behavior. But this approach completely fails to answer why the machine exists with that specific behavior. Systems analysis in terms of the machine's purpose within the higher-level structure, where it is one of many interacting components, provides this answer—giving another equally valid, and in some ways more profound, explanation of why the machine has the properties it does. This approach determines the rationale, the *raison d'être* of the machine—and a given purpose can usually be fulfilled in a variety of different ways in terms of structure at the microlevel.

---

[37] Private communication.

[38] Russell L. Ackoff, *Ackoff's Best: His Classic Writings in Management* (New York: Wiley and Sons, 1999).

3.2 The Implications of Quantum Theory

The profound implications of the nature of quantum theory is the second main issue and has two major aspects.

The effect of quantum entanglement. Most quantum states are entangled states. This means that instead of thinking of bottom-up action by invariant constituents, one must consider cooperative effects between the constituent components that modify their very nature. Because of quantum entanglement, it is difficult even to talk of individual properties of constituent parts. But if the constituent particles at the microlevel don't even have individual properties, a simplistic reductionist view is undermined.

In practice in many cases, because of decoherence (induced by the top-down effect of the environment) the system may be regarded as composed of individual particles with well-defined properties. But this approximation is valid only because the interference terms are small. In principle the particles have no separate existence. It can be suggested that our worldview should take this seriously, if indeed we take quantum theory seriously.[39]

The effect of quantum uncertainty. Uncertainty at the microlevel undermines reductionism in microsystems, because it means one cannot explain the behavior of the system causally from knowledge of the initial state of its components. Indeed one cannot even attain adequate knowledge of that initial state. The classic example is the two-slit experiment, where one cannot predict the final position of a photon on a screen from one's initial knowledge of the state of the system (where "the system" is the photon, its source, the screen, and the intervening barrier with slits).

Now the response may be that this does not affect reductionism in macrosystems, because well-established statistical effects remove the uncertainty at the macrolevel. But this is not always true; there is a possibility of amplification of the micro-uncertainty to the macrolevel. Indeed that is a central feature of the "Schrödinger's cat" *Gedanken* experiment, but it can occur in much less exotic contexts, for example, in the two-slit photon experiment (with the result amplified to macroscales by a photo-amplifier or other means). Indeed, the standard way the two-slit experiment is carried out is an example of this kind: one cannot determine that the result is uncertain at the microlevel unless one first amplifies that result to a macrostate where it can be observed! The same applies to radioactive decay, where the (macrolevel) Geiger counter acts as the amplifier from the micro- to the macrolevel. The final state of the total system (radioactive atom plus Geiger counter) at any specific time cannot be predicted with finality from a knowledge of its initial state.

3.3 Implications for Reductionism

Taken together, these features mean that simplistic ideas of reductionism ("we are *nothing but* the sum of particles controlled by forces at the microlevel") simply do not hold; they must be replaced by much more sophisticated views integrating the combined bottom-up (microcausation), bottom-bottom (co-operative), and top-down (context-setting) interactions. This kind of integrative view takes seriously the effect of bottom-up microbased causality, but in a wider framework that allows emergent (nonreducible) levels of meaning with real causal functionality.

---

[39] See, for example, David Bohm and B.J. Hiley, "Quantum Theory and the Implicate Order," chap. 15 of *The Undivided Universe* (London: Routledge, 1993).

Reductionism in principle. "Ontological reductionism" is the view that higher-level entities are nothing but the sum of their parts.[40] Expressed differently, as one goes up the hierarchy of structure, no new kinds of metaphysical ingredients need to be added to produce higher-level entities from lower. The metaphysical assumption made is that the parts of an entity or system alone determine the character and behavior of the whole. It is now clear that this formulation misses the crucial ingredient of macro-organization, which is an additional essential aspect of the whole. Thus this idea is not only wrong in the context of quantum theory because of quantum entanglement, when the parts cannot be separated from the whole; it is also wrong at the macrolevel because it does not take seriously macro-organization and top-down action, the basis of higher-level behavior. The key point, in terms of function and meaning, is that the same set of component parts in a different configuration form a different structural entity with different behavior. Therefore, to omit that configuration from one's theoretical description[41] is to omit a key element in the situation. It should be recognized as a metaphysical ingredient of the system.

Reductionism in practice. Additionally, there is a major contrast between reductionism in principle and the realistic limits on reductionism in practice, as is the case when going from quantum physics to chemistry, or even within physics (e.g., explaining the properties of water). In practice we have some broad general principles that we believe apply (e.g., the thermal properties of matter can be explained through the motions of its constitutional particles, and the nature of chemical bonds can be explained through the Schrödinger equation of quantum mechanics in suitable contexts), and a patchwork of specific cases where these can actually be carried out. As a specific example related to the brain, the Hodgkin-Huxley equations that govern nerve impulse dynamics,[42] and so are the basis of electrophysiology, cannot be *reduced* to the laws of physics:

> [The Hodgkin-Huxley equations] can be traced to the laws of physics through chemistry, biochemistry, and the dynamics of membrane ion permeability. On the other hand, the equations are not "ordinary laws of physics" (as Schrödinger pointed out) but "new laws" that emerge at the hierarchical level of the axon to govern the dynamics of nerve impulses. One cannot derive these new laws from physics and chemistry because they depend on the detailed organization of the intrinsic proteins that mediate sodium and potassium current across the membranes and upon the geometric structure of the nerve fibers. Schrödinger's equation is based on the law of energy conservation, while the Hodgkin-Huxley equations don't conserve anything. Time, in Schrödinger's quantum theory, flows with equal ease in either direction, but solutions of the Hodgkin-Huxley equations push forward in one direction with respect to time. There is no parameter in the Hodgkin-Huxley equations that could be predicted or derived from the equations in Schrödinger's theory. None.[43] .

As another example, in the special issue of *Nature* on "Impacts of Foreseeable Science," Leland Hartwell and coworkers suggest that molecular and cell biology

---

[40] For a discussion of the different statements of reductionism, see, for example, Nancey Murphy, "Supervenience and the Nonreducibility of Ethics to Biology," in *Evolutionary and Molecular Biology: Scientific Perspectives on Divine Action*, Robert J. Russell, William R. Stoeger, S.J. and Francisco J. Ayala, eds. (Vatican City State/Berkeley, Calif.: Vatican Observatory/Center for Theology and the Natural Sciences, 1998), 463–90, this volume hereafter *EMB*.

[41] As is implied by the recurrent key phrase, "nothing but."

[42] Alwyn Scott, *Stairway to the Mind* (Berlin: Springer, 1995), 50–4 and 207–10.

[43] Ibid., 52–3.

may have more in common with engineering and computer science than with basic sciences (e.g., the networks that underlie most biological functions come straight from engineering control theory).[44] Furthermore, this kind of understanding has important implications in our understanding of genetics itself[45] and in terms of possible societal implications.[46]

In many cases, then, reductionism, rather than being an actual achievement, is a *guiding paradigm*. Its use is based in a faith that this approach can succeed in principle, even if it cannot often succeed in practice. This faith is supported by those very impressive cases where it can indeed be carried out, together with a general philosophical position as regards the underlying mechanical nature of what is going on at the microlevel. However it usually emphasizes mainly or only the bottom-up interactions taking place, not sufficiently considering the other side: the holistic aspects and associated top-down interactions that have been emphasized here.

Criteria for use of reductionism. The implication of the above is that reductionism is not justified as universally applicable—as a scientific tool one can use at all times. Rather it is identified as a working proposal that is an essential tool for building up our overall understanding, but which will sometimes be applicable and sometimes not. A purely reductionist approach misses the essential ingredient of higher-order structuration, and so is unable to provide a basis for understanding any of the holistic sciences where the essence is in these higher-order relations and where the specific nature of the constituent parts is often largely irrelevant. Thus a competent scientist recognizes the *limits* of reductionism.[47]

The key issue, then, is to decide when a reductionist explanation is appropriate, and when it is not. I suggest that a reductionist explanation is suitable when both

(a) bottom-up causation dominates *and*
(b) the microcomponents maintain their individual properties unaltered,

but not when *either*

(a') top-down causation plays a major role in what happens,[48] *or*
(b') cooperative phenomena between the constituents results in fundamentally changed behavior in those constituents.

The latter will be true in particular if there is a loss of their individual identity so that they no longer have individual properties, as happens in quantum entanglement. The two features (a') and (b') are usually closely related to each other, and are able to lead to new levels of objective emergent nonreducible properties.[49] As Philip Clayton has remarked in conversation, there may be a gradual shading over from

---

[44] Leland H. Hartwell, J.J. Hopfield, S. Leibler, A.W. Murray, "From Molecular to Modular Cell Biology," *Nature* 402 (02 December 1999): C47–C52.

[45] Kenneth M. Weiss, "Is there a Paradigm Shift in Genetics? Lessons from the Studies of Human Diseases," *Molecular Phylogenetics and Evolution* 5 (1996): 259.

[46] Jon Beckwith, "Genes and Human Behaviour: Scientific and Ethical Implications of the Human Genome Project," in *Handbook of Molecular-Genetic Techniques for Brain and Behavioural Research*, W.E. Crusio and R.T. Gerlai, eds. (Amsterdam: Elsevier, 1999).

[47] See, for example, Gregory Bock and Jamie Goode, eds. *The Limits of Reductionism in Biology* (New York: John Wiley and Sons, 1998).

[48] See Arthur Peacocke, *Theology for a Scientific Age: Being and Becoming—Natural and Divine* (London: SCM Press, 1993); idem, "God's Interaction with the World," in *CAC*; idem, "The Sound of Sheer Silence," and references therein.

[49] This is described nicely in Campbell, *Biology*, and Scott, *Stairway to the Mind*.

reductionist to holistic considerations as one moves from particle physics through the biological sciences and the "human sciences" and on to metaphysical questions. It is also important to distinguish between those processes at each level that are governed by deterministic (even if statistically represented) laws, and those that can be viewed as possibly either indeterministic (as in the quantum case) or purposeful, such as human mental states (where "free will" may intervene).[50] In the latter cases, the nature of reductionist arguments we can apply is very different from what is applicable in the former case.

4 Ontology and Quantum Theory

I will not try to enter the major epistemology-versus-ontology argument regarding the wavefunction and its collapse associated with the measurement process, as discussed for example by others at this meeting and by Bohm and Hiley.[51] Rather, I will just touch on two issues that relate to the topic of this volume.

4.1 Quantum Uncertainty

The whole nature of quantum theory is based on the statistical nature of experimental results revealed by careful measurements at microscales. An important question then is whether the measured distribution of results is really statistically random. Might it, for example, be chaotic instead?

Our understanding of random and chaotic processes has improved a great deal since the foundational days of quantum theory. It is just possible—not likely, but possible—that recording the times and positions of arrival of photons in the double-slit experiment might after all reveal a chaotic pattern in the sense of present-day chaos theory,[52] rather than a truly random pattern. That is, if we analyze the pattern of arrival position against time, using our present understanding of chaotic processes, we might find that there is an underlying pattern in this apparently random process. Even though this is unlikely, one may suggest that it is sufficiently important for the foundation of quantum theory that it should be investigated.

An interesting point then arises. In trying to establish whether this is true, we come up against the fundamental limit of probability theory: we can't ever prove the experimental pattern obtained is random![53] We can search for chaotic/fractal signatures in the distribution, and try to detect a pattern in the measured positions and times. If such a pattern is discovered, it shows some hidden mechanism is at work. If none is discovered by any particular search procedure, or series of searches using different procedures, this does not prove there is no pattern, for there is no algorithm that can do this (no number or sequence of numbers can in fact be proved to be completely random). All that we can state is that we have not discovered any pattern so far! Thus the fundamental uncertainty in mathematics and probability theory that is discussed by Chaitin could possibly prevent us from ever *proving* that quantum processes are in fact fully statistically random.

---

[50] See Ackoff, *Ackoff's Best*, for a consideration of the different such cases that can arise.

[51] Bohm and Hiley, *The Undivided Universe*.

[52] Robert L. Devaney, *An Introduction to Chaotic Dynamical Systems* (Reading, Mass.: Addison-Wesley, 1989).

[53] See, for example, Gregory J. Chaitin, "A Hundred Years of Controversy over the Foundations of Mathematics," chap. 1 of his *The Unknowable* (Berlin: Springer, 1999), also available at http/xxx.lanl.gov/Chao-dyn/99090001.

4.2 Quantum Theory Entities and Quantum Origins

Quantum field theory in its full form,[54] applied to the standard model of particle physics (and I would argue that one cannot meaningfully talk about quantum theory in the abstract—it only gains its full meaning in the context of the theory to which it is applied) is immensely complex. Conceptually, it involves *inter alia*:

- Hilbert spaces, operators, commutators, symmetry groups, higher dimensional spaces;
- particles/waves/wavepackets, various kinds of spinors, quantum states/wavefunctions;
- parallel transport/connections/metrics (the importance of connection and parallel transport in quantum theory as well as gauge theories has been experimentally demonstrated[55]);
- the Dirac equation and interaction potentials, Lagrangians and Hamiltonians;
- variational principles that seem to be logically and/or causally prior to all the rest.

Derived (effective) theories, including classical (nonquantum) theories of physics, also have complex abstract structures underlying their use: force laws, interaction potentials, metrics, and so on.

The underlying issue of interest concerns the ontology of all this quantum apparatus, and of higher-level (effective) descriptions. We seem to have two options[56]:

1. These are simply our own mathematical and physical constructs that happen to characterize reasonably accurately the physical nature of physical quantities.
2. They represent a more fundamental reality: Platonic quantities that have the power to control the behavior of physical quantities (and which can be represented accurately by our descriptions of them).

On the first supposition, the "unreasonable power of mathematics" to describe the nature of the particles is a major problem—if matter is endowed with its properties in some way we are unable to specify, but not determined specifically in mathematical terms, and its behavior happens to be accurately described by equations of the kind encountered in present-day mathematical physics,[57] then that is truly weird![58] Why should it be possible that *any* mathematical construct whatever

---

[54] Peskin and Schroeder, *An Introduction to Quantum Field Theory*.

[55] From the "Search and Discovery" section, "Currents in Normal Metal Rings Exhibit the Aharonov-Bohm Effect," *Physics Today* (January 1986): 17; Yoseph Imry and R.A. Webb, "Quantum Interference and the Aharonov-Bohm Effect," *Scientific American* (April 1989): 56.

[56] Bill Stoeger has discussed these issues from a different viewpoint in previous volumes of this series.

[57] For example a Yang-Mills type theory associated with a group structure with SO(3)×SU(2)×U(1) as a subgroup, or a string theory of all forces associated with particular Calabi-Yau spaces.

[58] Bill Stoeger expresses the opinion that any property with which you endow matter will be mathematically expressible in some way. While this may be the case for simple curve fitting

gives an accurate description of this reality, let alone ones of such complexity as in the standard theory of particle physics? Additionally, it is not clear on this basis why all matter has the same properties—why are electrons here identical to those at the other side of the universe? On the second supposition, this is no longer a mystery—the world is indeed constructed on a mathematical basis, and all matter everywhere is identical in its properties. But then the question is, How did this come about? How are these mathematical laws imposed on physical matter? And which of the various alternative forms (Schrödinger, Heisenberg, Feynman, Hamiltonian, Lagrangian) is the "ultimate" one? What is the reason for variational principles of any kind?

Either way, we face the fundamental metaphysical issue: What chooses this set of laws/behaviors and holds it in existence/in operation? And on either view the anthropic issue[59] remains: Why does this specific chosen set of laws that has come into being allow intelligent life to exist? These issues arise specifically in terms of current scientific speculations on the origin of the universe.[60] Most proposals either envisage a creation process starting from some very different previous state, which then itself needs explanation and is based in the validity of the present laws of physics (which are therefore invariant to some major change in the status of the universe, such as a change of spacetime signature from Euclidean to hyperbolic), or else represent what is called "creation from nothing," but in fact envisage some kind of process based on all the apparatus of quantum field theory mentioned above—which is far from nothing! In both cases the laws of physics in some sense pre-exist the origin of the (present phase of) the universe. In the case of ever-existing universes, the same essential issue arises.

The essential point is that it would be helpful to have clarity on the ontology assumed to underlie these creation theories and processes. This is far from the case in most present proposals.

5 Holistic Ontology

Closely related to issues of reductionism are issues of ontology, delimiting what is real, i.e. what actually exists, as compared to what is seen as insubstantial because it is essentially comprised *only* of lower-level objects, and so is without serious ontological status in its own right. In this section I propose a holistic view of ontology, building on the previous proposals by Popper and Eccles[61] of reality as composed of World 1 (the universe of physical entities), World 2 (the psychic reality of an inner life), and World 3 (the world of human culture in the broadest sense of the term), and Roger Penrose's vision of a tripartite reality consisting of a world of

---

and behavior described by differential equations, it does not seem to me to apply to something as complex as the behavior represented by the standard model of particle physics based on gauge theories and the Higgs mechanism, or to the kind of structure represented by M-theory.

[59] See e.g. *MNU*, or George F.R. Ellis, "The Theology of the Anthropic Principle," in *Quantum Cosmology and the Laws of Nature: Scientific Perspectives on Divine Action*, Robert J. Russell, Nancey C. Murphy, and Chris J. Isham, eds. (Vatican City State/Berkeley, Calif.: Vatican Observatory/Center for Theology and the Natural Sciences, 1993), 367–406, and references therein, the volume hereafter *QCLN*.

[60] See *QCLN*. A summary of proposals such as those of Tryon, Hartle, Hawking, Gott, Linde, Turok, and Gasperini is given in George F.R. Ellis, "Before the Beginning: Emerging Questions and Uncertainties," in *Toward a New Millennium in Galaxy Morphology*, D. Block, I. Puerari, A. Stockton, and D. Ferreira, eds. (Dordrecht: Kluwer, 2000).

[61] Popper and Eccles, *The Self and Its Brain*.

conscious perceptions, a physical world, and a Platonic world of mathematical forms.[62] The view I present here takes seriously the issues raised in the previous sections and provides a counter to reductive views of ontology. To prepare the way, I first comment on the reality of the levels in the causal and structural hierarchy, and on the ontological status of human intentions and plans.

5.1 The Higher Causal Levels

A key issue is the relation of hierarchical structuring to ontology. Reductionism is often taken to imply that the higher levels of order are in some sense not real—they are *epi*-phenomena—whereas the lower levels are indeed real. The first point in response is that we don't even know what the lowest levels of structure are (maybe they are quarks, but maybe not), so if taken seriously this reduces us to a state of impotence—we don't know what is really real! But additionally, this absolutist view simply does not take seriously the ontological status of the higher levels. On the view taken in this essay, quarks and electrons are real, as are protons and neutrons, tables and chairs, planets, stars, and galaxies. The fact that each is composed of lower-level entities does not undermine its status as existing in its own right.

The lower levels place remarkably little constraint on what happens at the higher levels, where independent levels of causality and phenomenology come into play. Thus, for example, I can type any text I like on this computer; the deterministic action of the forces between the particles of the machine allow this.[63] The phenomenological causal relations valid at each level in the hierarchy can be studied in their own right. Indeed, we can attain and confirm high representational accuracy and predictive ability for quantities and relations at higher levels, independent of our level of knowledge of interactions at lower levels, allowing us to give well-validated and reliable descriptions at higher levels accurately describing the various levels of emergent nonreducible properties and meanings.

This is why there is an ontological reality of higher levels of the hierarchy, which can be considered as entities existing in their own right and with their own phenomenological laws of behavior. The higher-level behaviors are largely independent of the detailed nature of their component parts at lower levels. An example is digital computers, with their hierarchical logical structure expressed in a hierarchy of computer languages that underlie the top-level user programs.[64] The user does not need to know machine code, and indeed the top-level behavior is independent of which particular hardware and software underlie it at the machine level. The computer has a reality of existence at each level that enables one to meaningfully deal with it as an entity at that level.[65] Another example is that a motor mechanic does not have to study particle physics in order to ply her trade.

The physical reasons allowing this independence of higher-level properties from the nature of lower-level constituents have been discussed by Anderson, Schweber,

---

[62] Penrose, *Shadows of the Mind* (Oxford: Oxford University Press, 1994), sec. 8.7.

[63] Silvan Schweber, "Physics, Community, and the Crisis in Physical Theory," *Physics Today* (November 1993): 34.

[64] See Andrew S. Tanenbaum, *Structured Computer Organisation* (Englewood Cliffs, N.J.: Prentice-Hall, 1990), and George F.R. Ellis, "Ordinary and Extraordinary Divine Action: The Nexus of Interaction," in *CAC*, 359–96.

[65] The conclusive proof of this feature is that one has to pay hard cash to obtain the higher-level benefits!

and Kadanoff.[66] This is a fundamentally important property of physics, underlying our everyday lives and their reality. Its source is the nature of quantum field theory,[67] applied to the microproperties of matter as summarized in the standard model of particle physics; a significant role is played by the behavior of matter characterized by the renormalization group.

5.2 Human Thoughts and Intentions

The second key issue is that I am assigning an ontological reality to human thoughts, intentions, plans, and emotions. As discussed in section 2.2, these demonstrably have causative effects on the course of macro- and hence micro-events. Thus I include them as truly existing macrolevels of causation because our understanding of events dictates much that happens in the causality chain, as do our plans and emotions (cf. the example above of nuclear explosion at Hiroshima).

5.3 A Holistic View of Ontology

The overall question is, What entities should be assigned an ontological reality? The discussion above suggests that there are not only different levels of reality in a hierarchically structured system, all present at the same time in a cohesive way, but that there may be different kinds of reality—different worlds, as suggested by Plato, Popper and Eccles, and Penrose—and that each must be taken seriously.

To approach this, I take as given the reality of the everyday world—tables and chairs, and the people who perceive them—and then assign a reality additionally to each kind of entity that can have a demonstrable causal effect on that everyday reality. Thus, I suggest that an entity can be assigned an ontological reality if it has a presence that results in detectable causal effects in the everyday world. In this case it represents a causal reality that can influence what happens. The problem then is to characterize the various kinds of independent reality that may exist in this sense.

Taking into account the causal efficacy of all the entities discussed above, I suggest as a possible completion of the proposals by Popper and Eccles and by Penrose, that the following four worlds are ontologically real. These are not different causal levels within the same kind of existence; rather, they are quite different kinds of existence, but related to each other through causal links. The challenge is to show first that each is indeed ontologically real, and second that each is sufficiently and clearly different from the others to be considered as separate from them.

5.3.1 World 1—The Physical World of Energy and Matter[68]

This world is hierarchically structured to form lower and higher causal levels, all of whose entities are ontologically real.[69] It is the basic world of matter and interactions

[66] Anderson, "More is Different"; Schweber, "Physics, Community, and the Crisis in Physical Theory"; Leo Kadanoff, *From Order to Chaos: Essays, Critical, Chaotic, and Otherwise* (Singapore: World Scientific, 1993). Regarding this fundamental feature, it would be helpful to have more detailed studies of quantum field theory on the one hand, and of the standard particle/field model—with all its particular symmetry groups, families of particles, and interaction potentials—on the other.

[67] Peskin and Schroeder, *An Introduction to Quantum Field Theory*.

[68] This is Popper and Eccles's World 1, and Penrose's physical world.

[69] Schweber, "Physics, Community, and the Crisis in Physical Theory."

between matter, based at the microlevel on elementary particles and fundamental forces, and providing the ground of physical existence. It is comprised of three major parts:

- WORLD 1A: *inanimate objects* (both naturally occurring and manufactured);
- WORLD 1B: *living things*, apart from humans (amoeba, plants, insects, animals, etc.);
- WORLD 1C: *human beings* (having the unique property of self-consciousness.

All these objects are made of the same physical stuff, but the structure and behavior of inanimate and living things (described respectively by physics and inorganic chemistry, and by biochemistry and biology) are so different that they require separate recognition, particularly when self-consciousness and purposive activity (described by psychology and sociology) occurs. Top-down and bottom-up action take place between these levels in the hierarchy.

5.3.2 WORLD 2—The World of Individual and Communal Consciousness

This world consists of ideas, emotions, and social constructions. Again, all the components of this world are ontologically real (it is clear that they all exist) and are causally effective. It is comprised of three major parts:

- WORLD 2A : human information, thoughts, theories, and ideas;
- WORLD 2B: human goals, intentions, sensations, and emotions;
- WORLD 2C: explicit social constructions.

These worlds are different from the world of material things, and are realized through the human mind and society.[70] WORLD 2A is the world of rationality; WORLD 2B is the world of intention and emotion, and so comprehends nonpropositional knowing[71]; and WORLD 2C is the world of consciously constructed social legislation and convention. Although each is individually and socially constructed in a complex interaction between culture and learning,[72] as discussed above in section 5.2, these are indeed each capable of causally changing what happens in the physical world, and each has an effect on the others.

WORLD 2A is the world of human rationality. It is hierarchically structured with many different components—words, sentences, paragraphs, analogies, metaphors, hypotheses, theories, indeed, the entire bodies of science and literature—and refers both to abstract entities and to specific objects and events. It is necessarily socially constructed on the basis of varying degrees of experimental and observational interaction with WORLD 1, which it then represents with varying degrees of success.

WORLD 2A is represented by symbols,[73] particularly language and mathematics, which are arbitrarily assigned (e.g., in terms of various languages) and which can

[70] They may also be partially realized in the minds of higher mammals other than humans. If so, this does not require a separate ontology, but simply highlights to what extent awareness of that ontology is available to different living species.

[71] This can also include religious experience, including that of a nonpropositional and nonsensory kind.

[72] Peter Berger and Thomas Luckmann, *The Social Construction of Reality* (London: Penguin, 1979).

[73] Terrence Deacon, *The Symbolic Species* (London: Penguin, 1997).

themselves be represented in various ways (sound, on paper, on computer screens, in digital coding, etc.). Thus each concept can be expressed in many different ways, and each is an entity in its own right independent of whichever way it is coded or expressed. Sometimes these concepts correspond closely to entities in the other worlds. However, their claim to ontological reality does not hinge on the reality of the objects or concepts they refer to. This world equally contains concepts of rabbits and fairies, galaxies and UFOs, science and magic, electrons and aether, unicorns and apples—the point being that all of these certainly exist *as concepts*. This statement is neutral about whether these concepts correspond to objects or entities that exist in the real universe (specifically, whether there is or is not some corresponding entity in WORLD 1) or whether the theories in this world are correct (that is, whether they represent WORLD 1 accurately or not). All the ideas and theories in this world are ontologically real in that they are able to cause events and patterns of structures in the physical world.

WORLD 2B is the world of motivation and sensation. It is also ontologically real, for it is clear that its components do indeed exist in themselves. In WORLD 2B, we find that the goals and intentions that cause the intellectual ideas of WORLD2A have physical effects in the real world.

WORLD 2C is the world of language, customs, roles, laws, etc., that shapes and enables human social interaction.[74] It is developed by society historically and through conscious legislative and governmental processes. It gives the background for ordinary life, enabling WORLD 2A and WORLD 2B to function, particularly by determining the means of social communication.

5.3.3 WORLD 3—The World of Aristotelian Possibilities

This world is comprised of the set of all physical possibilities, from which the specific instances of what actually happens in WORLD 1 are drawn. As a world of possibilities, it is ontologically real because of its rigorous prescription of the boundaries of what is possible: it provides the framework within which WORLD 1 exists and operates, and in that sense is causally effective. It is comprised of two major parts:

- WORLD 3A: The world of physical possibilities, delineating possible physical behavior;
- WORLD 3B: The world of biological possibilities, delineating possible biological organization.

These worlds are different from the world of material things, for they provide the background within which that world exists. In a sense they are more real than that world, because of the rigidity of the structure they impose on WORLD 1. There is no element of chance or contingency in them, and they certainly are not socially constructed (although our understanding of them is). They rigidly constrain what can happen in the physical world, and are different from each other because of the great difference between what is possible for life and for inanimate objects.

WORLD 3A delineates all possible physical behavior (and so is a description of all possible motions and physical histories of objects). It describes what can actually occur in a way compatible with the nature of matter and its interactions; only some

[74] For a discussion of such features, see David F. Aberle et al., "On the Functional Prerequisites of a Society," *Ethics* 60 (1950): 100–11, reprinted in Bobbs-Merrill Reprints in the Social Sciences S-1.

of these configurations are realized through the historical evolutionary process in the expanding universe. This world delineates all physically possible actions (different ways particles, planets, footballs, automobiles, aircraft can move, for example). Given these possibilities, what actually happens is determined by initial conditions in the universe, in the case of interactions between inanimate objects, and by the conscious choices made, when living beings exercise volition.

WORLD 3B delineates all possible biological organization[75] (and so is a representation of all possible organisms).[76] This world defines the set of potentialities in biology by giving rigid boundaries to what is achievable in biological processes. It constrains the set of possibilities from which the actual evolutionary process can choose, rigorously delineating the set of organisms that can arise from any possible evolutionary history. This "possibility landscape" for living beings underlies evolutionary theory, for any mutation that attempts to embody a structure that lies outside its boundaries will necessarily fail. Thus even though it is an abstract space in the sense of not being embodied in specific physical form, it strictly determines the boundaries of all possible evolutionary histories. In this sense it is highly effective causally.

Only some of the organisms that can potentially exist are realized in WORLD 1 through the historical evolutionary process; thus only part of this possibility space is explored by evolution on any particular world. When this happens, the information is coded in the hierarchical structure of matter of WORLD 1, and particularly in the genetic coding embodied in DNA, and so is stored via ordered relationships in matter; it is then transformed into various other forms until it is realized in the structure of an animal or plant. In doing so it encodes both a historical evolutionary sequence[77] and structural and functional relationships that emerge in the phenotype and enable its functioning once the genotype is read.[78] In this way, directed feedback systems and the idea of purpose can enter the biological world, which distinguishes the animate from the inanimate world. The structures occurring in the nonbiological world can be complex, but they do not incorporate "purpose" or order in the same sense.[79] Just as WORLD 3A can be thought of as encoded in the laws of physics, WORLD 3B can be thought of as encoded in terms of biological information,[80] which is a core concept in biology, distinguishing the world of biology from the inanimate world.[81]

5.3.4 WORLD 4—The Platonic World of (Abstract) Realities

This world contains realities that are independent of human existence but not embodied in physical form. They can have causal effects in the physical world. This

[75] See sections 3.1 and 3.2 above, and Campbell, *Biology*.

[76] Ian Stewart, *Life's Other Secret* (London: Penguin Books, 1998).

[77] Wen-Hsiung Li and Dan Graur, *Fundamentals of Molecular Evolution* (Sunderland, Mass.: Sinauer, 1991).

[78] Campbell, *Biology*; Wolpert et al., *Principles of Development*; Coen, *The Art of Genes*.

[79] Unless they are designed by intelligent beings precisely in order to store and interpret complex information.

[80] Bernd-Olaf Küppers, *Information and the Origin of Life* (Cambridge: MIT Press, 1990); Clifford A. Pickover, *Visualizing Biological Information* (Singapore: World Scientific, 1995).

[81] Campbell, *Biology*, 8.

world (Platonic realities discovered by human investigation) is perhaps the most contentious. The prime contenders to inhabit such a world are:

- WORLD 4A: Mathematical forms;
- WORLD 4B: Physical laws;
- WORLD 4C: Platonic aesthetic forms.

The existence of WORLD 4A is strongly defended by Penrose,[82] who argues that major parts of mathematics are discovered rather than invented (rational numbers, zero, irrational numbers, and the Mandelbrot set being classic examples). They have an abstract rather than embodied character, this abstract nature being manifested in that the same abstract quantity can be represented and embodied in many symbolic and physical ways. They are not determined by physical experiment, but are independent of the existence and culture of human beings. The same features will be discovered by intelligent beings in the Andromeda galaxy as here, once their mathematical understanding is advanced enough (which is why they are advocated as the basis for inter-stellar communication). This world is partially discovered by humans, and so is represented by our mathematical theories in WORLD 2. That representation is a cultural construct, but the underlying mathematical features it represents are not—indeed like physical laws, they are often unwillingly discovered.[83] This world is causally efficacious in terms of the process of discovery and description (e.g, one can print out the values of irrational numbers or graphic versions of the Mandelbrot Set in a book).

A key question is what (if any) part of logic, probability theory, and physics should be included here. In some as yet unexplained sense, the world of mathematics underlies the world of physics. Many physicists at least implicitly assume that a Platonic WORLD 4B underlies the nature of physical possibilities (WORLD 3A). Another proposal, WORLD 4C, provides a foundation for our sense of beauty. In this essay, I will not pursue those possibilities. It is sufficient for my purpose to note that the existence of a world of mathematical forms, which I strongly support, establishes that this category of world indeed exists and has causal influence.

The major proposal of this section is that all these worlds exist. WORLDS 1–4 are ontologically real and distinct from each other.

5.4 The Relation to Epistemology

What then of epistemology? Given the existence of the various worlds mentioned above, the proposal here is that epistemology is the study of the relation between WORLD 2 and WORLDS 1, 3, and 4. Epistemology attempts to obtain accurate correspondences to quantities in all the worlds by means of entities in WORLD 2A.

This exercise implicitly or explicitly divides the theories and statements of WORLD 2A into: (i) true/accurate representations, (ii) partially true/misleading representations, (iii) false/poor/misleading representations, and (iv) ones where we don't know the situation. These assessments range from statements such as "It is true her hair is red" and "There is no cow in the room" to "Electrons exist" and "Newtonian theory is a very good description of medium-scale physical systems at low speeds and in weak gravitational fields" and "The evidence for UFOs is poor."

---

[82] Penrose, *Shadows of the Mind*.

[83] As, for example, in the cases of zero and of irrational numbers; see Charles Seife, *Zero: The Biography of a Dangerous Idea* (London: Penguin, 2000) for a nice discussion.

This raises interesting issues about the relation between reality and appearance. It is of course known that everyday life gives a different appearance to reality than does microscopic physics—as Eddington pointed out, a table is actually mostly empty space between the atoms that make up its material substance, but in our experience it is a hard object.[84] One can learn much about these relations through solid-state physics and physical chemistry (dealing with the appearance of everyday physical objects), astrophysics (which deals with detection probabilities and appearances of the same astronomical object seen by different detectors, giving very different views of same object at different wavelengths), and relativity theory (giving rigorous rules for the changes in appearances of objects when one changes reference frame). Cosmology also gives useful analogies and arguments about the relation between evidence and theory, particularly because there exist regions beyond the cosmological particle horizon, which we cannot see or examine by any astronomical observation.[85] Similar issues arise in epistemology in much broader contexts.

There is a widespread tendency to equate epistemology and ontology (as in both logical positivism and in extreme relativism, and also in the standard interpretation of quantum theory). This can often be an error, or at least is often unjustified, and a variety of examples can be given where it seriously misleads. This is related to a confusion between WORLD 2 and the other WORLDS discussed here, and it seems to underlie much of what has happened in the so-called "Science Wars" and the Sokal affair.[86] The proposal here strongly asserts the existence of independent domains of reality (WORLDS 1, 3, and 4) that are not socially constructed, and implies that we do not know all about them and indeed cannot expect to ever understand them fully. This ignorance does not undermine their claim to exist in a way that is independent of human knowledge and understanding. The explicit or implicit claim that we can equate epistemology and ontology is in the end simply a case of human hubris.

6 Theological Issues

The theme of this series, "scientific perspectives on divine action," makes the assumption that there is indeed something real to discuss; it asks if the idea of divine action can have a sound physical basis, given our current knowledge of science. Here I adopt both a strong religious position, where God is taken to exist and act, and a strong scientific position, taking seriously our present-day scientific knowledge. Perhaps the most important arguments against God's effective specific action in the world are based on strong ideas of reductionism, underpinned by visions of deterministic bottom-up action in hierarchical systems. These appear to leave no space open for any effective action by God—they seem to restrict one to the deistic view of God simply watching what happens as a result of the inviolate laws of nature that have been invoked and are being kept in place as the foundation of physical existence.

In the previous sections of this essay, I have challenged a strong reductionist view by showing how a more flexible and inclusive approach is warranted and

---

[84] Arthur S. Eddington, *The Nature of the Physical World* (Cambridge: Cambridge University Press, 1928).

[85] George F.R. Ellis, "The Different Nature of Cosmology," *Astronomy and Geophysics* 40 (1999): 4.20–4.23.

[86] See http://www.physics.nyu.edu/faculty/sokal/index.html and http://zakuro.math.tohoku.ac.jp/~kuroki/Sokal/.

underpinned by a broad view of ontology. Additionally, previous essays in this series have argued that quantum uncertainty at the microlevel could underlie the possibility of God acting in the world in a causally effective manner without violating the laws of physics.[87] This has been questioned in various ways. This section reconsiders these issues, after first looking again at ontology in this broader context. As to the specific nature of this action, it is proposed in *MNU* and Paper II that God acts mainly through revelation to individuals, possibly through some "channel of communication" or "causal joint" between God and the physical world, resulting in the authentic religious and spiritual experiences of the faithful. The issue then is whether such a channel can be physically based on quantum uncertainty. I argue here that the answer is "Yes."

6.1 Ontology

Theology has not been included so far in the discussion of ontology. How does what has been said above change if we include it? On a theistic view we must consider an extra kind of world:

6.1.1 World 0—The Foundation

This world constitutes the underlying fundamental reality of God—partially described by theology—and entails the fundamental meanings of the universe and of life.

The key assumption of the theist position is that this world exists.[88] This is highly debatable precisely because it cannot be proven or disproved beyond doubt by scientific, logical, or philosophical argument.[89] The essence of the problem is the transcendent-yet-hidden nature of this metaworld: it has a character totally different from the rest. Indeed, it is not properly describable in terms of the entities they comprehend, but it is nevertheless able to interact with them and influence them in important ways—in particular, by being the ground of their existence.

World 0 provides the ultimate context in which the other worlds are set. Worlds 1–4 are parts or aspects of the "one world" that is the creation of God, depending on World 0 for their existence. On this view, God is *transcendent* and *acts foundationally* through creating Worlds 4, 3 and 1, thus enabling World 2, and is *immanent* and *causally efficacious* in World 2 through revelation and (on the Christian view) incarnation. World 0 does indeed have real physical effectiveness through the channel of communication mentioned above, which directly

[87] See Robert J. Russell, "Quantum Physics in Philosophical and Theological Perspective," in *Physics, Philosophy, and Theology: A Common Quest for Understanding*, Robert J. Russell, William R. Stoeger, S.J. and George V. Coyne, S.J., eds. (Vatican City State: Vatican Observatory Publications, 1988), 343–74; idem, "Special Providence and Genetic Mutation: A New Defense of Theistic Evolution," in *EMB*, 191–224; Nancey C. Murphy, "Divine Action in the Natural Order: Buridan's Ass and Schrödinger's Cat," 325–58, Thomas F. Tracy, "Particular Providence and the God of the Gaps," 289–324, George F.R. Ellis, "Ordinary and Extraordinary Divine Action: The Nexus of Interaction," 359–96, all in *CAC* ("Papers I"); and George F.R. Ellis, "Intimations of Transcendence: Relations of the Mind and God," in *NAP*, 449–74 ("Paper II").

[88] See, for example, George F.R. Ellis, "The Theology of the Anthropic Principle," in *QCLN*, 367–406; *MNU*.

[89] Cf. George F.R. Ellis, "The Thinking Underlying the New 'Scientific' World-Views," in *EMB*, 251–80.

influences World 2. Hence it influences World 1 indirectly through human action (e.g., in the construction of churches and temples, and the spending of resources on missionary and charitable activity). It is also suggested that God is affected by the world (the concept of divine passibility) even while God influences the world. Thus World 0 is influenced by the others.

As Murphy and I argued in *MNU*, a complete set of causal relations must include the foundation of values and ethics. This foundation is implied by the nature of World 0, but it is distinct from that world itself—indeed, it provides the link between World 0 and World 2. The proposal here is that this foundation is to be encompassed by World 5.

6.1.2 World 5—Underlying Purpose

This world contains the set of values and meanings expressing the purpose (*telos*) of God. This world is ontologically real, and it is effective through revelation and discovery.

The key assumption linking morality and theology is that *this world exists in an ontologically real way* (hence the name of our book, "On the Moral Nature of the Universe"), which is of course a highly contentious claim. The suggestion is that while these values embrace justice as well as love, they have a kenotic rather than coercive character; the relations between these aspects is indicated in figure 2. This world influences behavior in World 2 through the revelations we are given of this preferred set of values and meanings. Thus, just as we discern the nature of physical structure and mathematics by a process of investigation and discovery rather than by a process of purely social construction, the same is true for these meanings and associated ethical values, but here the source is ultimately revelation (which may occur in many ways, e.g., being heard in the "still small voice" or being encompassed in the "inner light") and personal experience. They are then causally efficacious, because ethics is the topmost level of the hierarchy of goals (see *MNU*), and so shapes all the underlying goals in important ways, indeed playing a crucial role in the nature of choices made in World 2B, which then determine the effect of human actions on World 1.

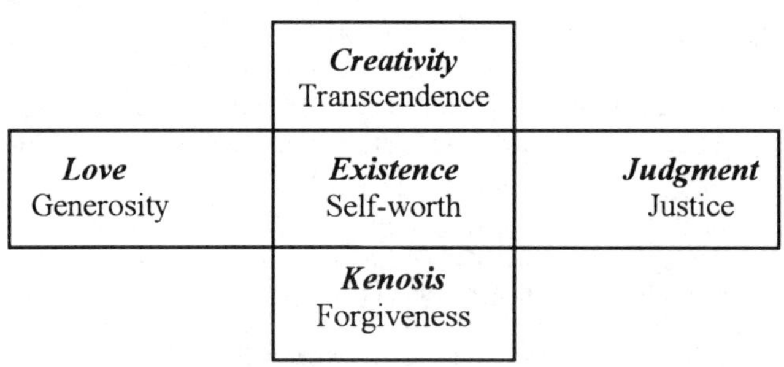

Figure 2. The Big Picture. The existence of God is the foundation of all. Creativity results in the beginning and the end—origins and destiny—of the created order. Love and judgment are two inevitable consequences of the nature of God. Kenosis is the foundation of the profound nature and depth of existence, allowing the tension between love and judgment to be resolved through its transforming nature (expressed symbolically in the cross-shaped nature of the whole). Unity with God (*theosis*) means taking part in all of this. Various intimations of transcendence relate to these various aspects of *theosis*.

This world is distinct from WORLD 0, which is its direct foundation, because that world is of a completely transcendent character and of a different kind than all the rest. WORLD 5 is the basis of the way we can know about WORLD 0, and is of an abstract character similar to WORLD 4 (the Platonic world, found by exploration of logical and mathematical ideas). But it is also different in a very important respect: it is persuasive in nature, engaged by spiritual enquiry and practice[90] rather than by logic and intellectual analysis. Indeed, at its core is a paradoxical nature associated with kenosis and self-sacrifice ("he who would save his life must lose it"), in a profound sense contradicting the world of logical argumentation. However, an understanding of the transformatory nature of this world is available through kenotic experience (e.g., as illustrated in the musical "Les Miserables," based on Victor Hugo's novel).

The overall ontological proposal we have now arrived at is summarized in figure 3 (overleaf). The nature of WORLD 5 determines the subset of WORLD 2B that is compatible with WORLD 0, characterizing the preferred "true" values that God wishes us to embrace freely of our own volition. Given this understanding, we are then in a position to define the proper nature of ethics as the study of the relation between WORLD 5 and WORLD 2. Ethics attempts to accurately relate the preferred values of WORLD 5 to goals and choices in WORLD 2B and WORLD 2C. Thus it explicitly or implicitly characterizes goals and choices in WORLD 2B and WORLD 2C as (i) ethical, (ii) unethical, (iii) neutral, or (iv) undecidable, according to their compatibility with the values and meanings embodied in WORLD 5. Note that this characterization cannot be made without having an ontologically existent WORLD 5, i.e., a set of values and meanings that are preferred in the sense of having a truly normative character, as opposed to purely socially-constructed or socio-biologically evolved sets of values, which arguably do not have such a character.[91] We locate the basis for the existence and character of WORLD 5 in WORLD 0—i.e., the nature of God—and in this way link theology to ethics (see *MNU* for a detailed discussion).

6.2 The Nature of God's Action

The further key issue is theological rather than physical, namely, given the possibility of God's action in the world as envisaged here, how does God use it? The proposal in *MNU* and in other papers in this series is that, first, God's action is *ultimately kenotic* (i.e., of a generous and self-sacrificial nature). This view chooses one particular strand from among those available and identifies it as the one most compatible with a combined scientific, ethical, and theological position that is coherent as a whole and that is present in a profound way in all the major religious traditions. This aspect will not be developed further here, as it has been the subject of much discussion elsewhere. In *MNU* we propose further that the main channel of God's action is through persuasive revelation of this *telos* (purpose) and ethical core of God's nature. Thus we accept the cognitive reality of a preferred set of personal religious experiences and envisage this as the major way God chooses to act in our world.[92] This view takes tested spiritual experience seriously: visions, pre-visions, etc., as witnessed in the vast body of spiritual experience and literature (e.g., Saul on

---

[90] Gordon Wakefield, ed., *A Dictionary of Christian Spirituality* (London: SCM Press, 1983).

[91] See, e.g., Murphy, "Supervenience and the Nonreducibility of Ethics to Biology."

[92] Apart, that is, from the fundamental activities of creating and sustaining the universe, discussed in *QCLN*.

Legend and Guide to Figure 3

FOUNDATION WORLD 0 - *The Underlying Fundamental Reality*
The world of God, partially described by theology, and entailing the fundamental meanings of the universe and of life.

WORLD 1 - *The Physical World of Energy and Matter*
Hierarchically structured, forming lower and higher causal levels whose entities are all ontologically real.

WORLD 2 - *The World of Individual and Communal Consciousness*
The world of ideas, emotions, and social constructions, also ontologically real and causally effective.

WORLD 3 - *The World of Aristotelian Possibilities*
Contains the set of all physical possibilities, from which the specific instances of what actually happens in WORLD 1 are drawn.

WORLD 4 - *The Platonic World of Abstract Realities*
Independent of human existence with causal effects in the physical world, but not embodied in physical form.

WORLD 5 - *The World of Underlying Purpose*
The set of values and meanings expressing the purpose (*telos*) of God, ontologically real and effective through revelation and discovery.

- Each ontological world represents a different kind of reality.
- WORLD 1 provides the substructure for WORLD 2 and is based on WORLDS 3, 4, and 5.
- WORLD 2 *supervenes* on and influences WORLD 1.
- WORLD 3 determines what is possible in WORLD 1.
- WORLD 4, despite its abstract (non-embodied) nature, is causally effective.
- WORLD 0 (on a theological view) is the basis for WORLDS 3, 4, and 5, and hence for WORLD 1, thus enabling the existence of WORLD 2.
- WORLDS 3, 4, and 0 *transcend* WORLD 1.
- WORLD 5 embodies the values inherent in WORLD 0 and causally influences WORLD 2.
- Given the natures of WORLDS 3 and 4, the creation of WORLD 1 involves both necessary relations and contingent features.
- On a nontheological view, WORLDS 0 and 5 will be omitted.

Figure 3. Relations Between the Different Kinds of Worlds

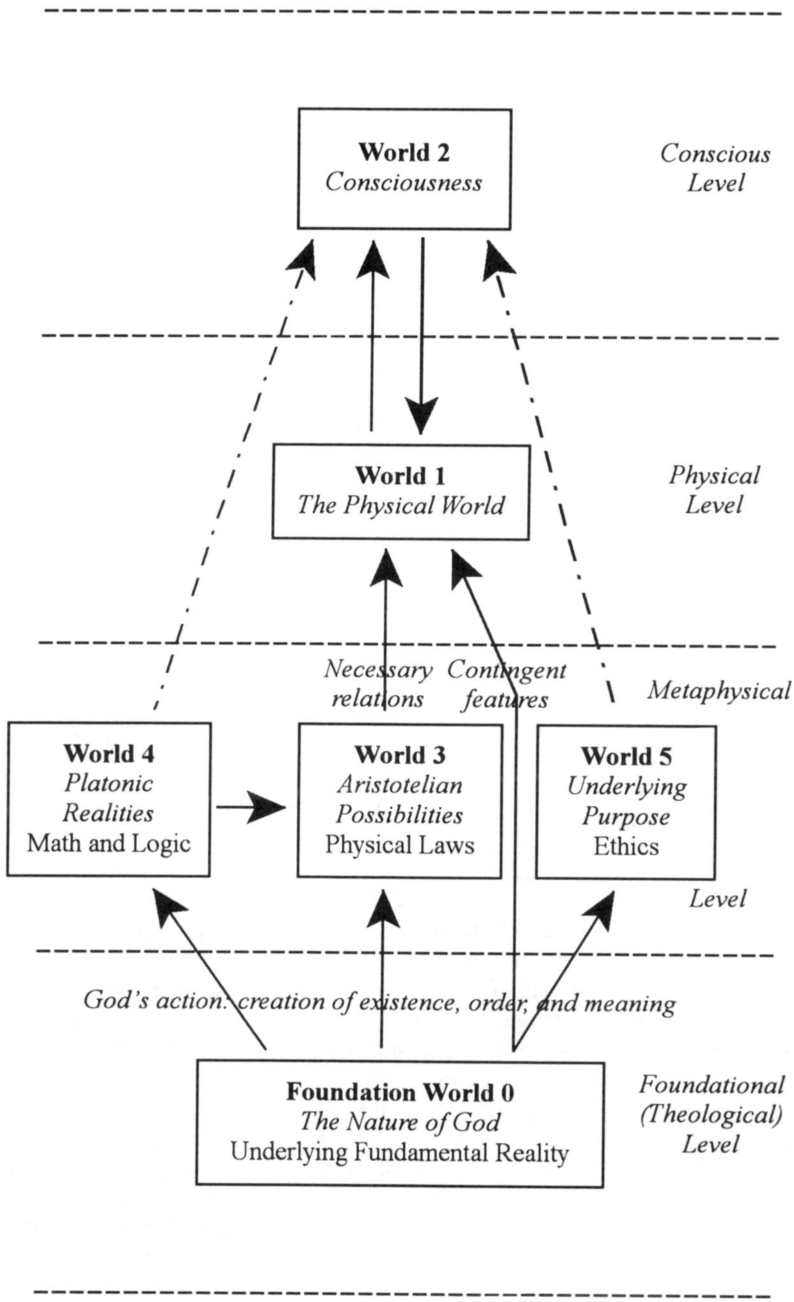

the road to Damascus, Jesus in the Garden of Gethsemane, and the quiet of a Quaker meeting, winnowed by discernment of what is genuine and what is not).[93]

A communication channel. For this position to have serious meaning, God must in some way be able to influence the course of events in the brain without contravening the nature of the physical world. This requires some kind of physically based channel of communication from the underlying transcendent reality to the human mind. These experiences are, therefore, grounded in reality to some degree rather than being purely internally generated with no communication from the transcendent to the finite (i.e., without being related to WORLDS 0 and 5). This channel enables these experiences to be phenomena that are not solely based in the processes of evolutionary psychology, sociobiology, or psychological projection. This view recognizes that each of these causal links takes place, but it argues that each link provides only a partial explanation. Each identifies a strand in the overall pattern of cause and explanation. The channel of communication envisaged here is also part of that pattern, allowing a religious component to these experiences in terms of contact between God and humanity.

6.3 The Physical Base: Quantum Uncertainty?

However, as John Polkinghorne has emphasized, we want this influence to take place within the grain of nature. Divine action must be consistent with physics, not an exception to it (even though this is certainly possible, as Nancey Murphy emphasizes, if one takes seriously a view of God as creator of all). The possibility of divine action within the confines of physics as we know it requires an openness in physical process, allowing a break in the chain of strictly deterministic bottom-up causation, and so enabling God's action to have real causal effect in the physical world. The two obvious possibilities are through chaos theory and quantum uncertainty. Polkinghorne provides support for the first in his series of papers; Russell, Murphy, Tracy, and I provide support for the second in previous papers in this series. I address the second possibility here.

The ontological nature of quantum uncertainty. The first issue in terms of relating quantum uncertainty to divine action is whether quantum uncertainty is ontological and/or epistemological *from the viewpoint of God.* It could be that in physics terms, the process is described by a hidden-variable theory (nonlocal, in accord with the Bell results), in which God or consciousness or both act as hidden variables, and hence are not controllable by the experimenter (cf. Paper II). Whether such a view is viable or not, it seems clear that human views of epistemology and ontology in quantum theory simply cannot be taken to constrain God's knowledge and power, as God is the immanent creator and sustainer of matter.

The proposal that the lack of determinism entailed in quantum uncertainty allows causal efficacy of divine action is commonly criticized on the ground that it implies that action must be statistical (i.e., random) rather than purposeful because quantum outcomes are statistical in nature. But this assumes that quantum uncertainty is ontologically real. To the contrary, Papers I and II argue that the outcome of quantum measurements are fully under God's control, while being apparently random to humans (who lack full knowledge of all the causes entering into their outcomes). In short, quantum epistemology and ontology need not be the same.

---

[93] The process of testing or discernment is discussed in *MNU*, and in John Hick, *The Fifth Dimension: An Exploration of the Spiritual Realm* (Oxford: Oneworld Publications, 1999).

The relation of micro- to macro-events. The second issue relates to quantum bottom-up action and macro-events. If there are changes at the (quantum) microlevel, how can they have significant impact at the macrolevel? Some have suggested that if God's action is at the microlevel, then it can't have a macro-effect. But we have seen in section 1.3 a variety of ways whereby quantum micro-effects can have significant macro-effects. We now revisit them in the present context.

Microbasis for "ordinary" macrophysics. Quantum theory underlies the stability of ordinary macroscopic causality. Given the concept that God is responsible for this quantum action at the microlevel everywhere, this is easily viewed as representing God's action as sustainer of the universe and maintainer of the regularity of physics.[94] This action takes place everywhere and at every time, and hence is a form of realization of the immanence of God. The quantum microlaws have a global macro-effect precisely because they are in operation in the same way everywhere and at all times.

Amplifiers. If specific quantum effects take place—their timing determined by some hidden variable or in some other way so that they are determinate and conveying some signal rather than being controlled by quantum indeterminacy as the orthodoxy has it—can the encoded signal then have macroscopic results? There is no problem in principle, because individual quantum events can have a macro-effect in living beings (cf. the discussion in section 1.3).

Coherent implementation of micro-effects. However, such amplifiers are not really needed for the process in mind. If one has the same broad effect implemented at one instant in many micro-events, this can imply a *classical macro-event* because the combined set of micro-events has been specifically chosen to correspond to such a macro-event. No amplification is then necessary. Such an occurrence would of course be unlikely to result by random processes. Would it then be forbidden by the statistics obeyed by quantum processes? No—for the simple reason that you can't ever violate a statistical law! All that can be said is that an unlikely event has occurred—but this is not forbidden. The long-term statistics would have to add up right, but one would never be able to prove *in any finite time* that they had worked out wrong. The requirement is only that the statistics are satisfied for an infinite number of events—but this can never be attained (a large but finite collection of events is essentially different from an infinite set). So although later events could presumably be organized in such a way as to restore the overall statistics, it is not clear how necessary this is, particularly because measurements of quantum statistics have not so far been carried out in a functioning brain. This might be a useful thing to try to do—if it is possible, which seems unlikely.

6.4 Implications, and Mode of Action

Taken together, these examples show that there is a variety of ways in which suitably actuated quantum effects can have a specific macro-effect and so convey information into the physical world. Certainly for example they might be effective in relation to the brain, as suggested in Papers I and II; indeed, various authors have argued (independent of the issues that are of concern here) that quantum effects might be significant in the functioning of the brain.[95] This possibility strengthens the argument

---

[94] See, for example, *MNU*.

[95] See Nicholas Herbert, *Elemental Mind* (London: Plume Penguin, 1994) and Scott, *Stairway to the Mind*, for summaries of such proposals.

of this section. It is also important to note, as pointed out by Bob Russell, that these are examples of the amplification of quantum effects without calling on chaos theory, and thus without getting entangled in the problem of quantum chaology.[96] Several problems have been raised for the option presented here.

1. Is this a meaningless way of importing essentially (quantum) randomness rather than information into the mind? No. On the present view, as argued above, the events are random as far as physics is concerned, but not as far as God is concerned. This is completely compatible with the known laws of quantum physics.

2. Is this approach futile because quantum effects are unable to cause macro-effects? No. As argued above, micro-events can effect the macroworld quite adequately.

3. Is the quantum influence too fitful and episodic to have significant effect? No. God's action is continuous and immanent, holding and maintaining the laws of physics. The specific action is episodic, and that is quite adequate for the purpose we have in mind. It helps to think of specific examples: sixteen frames per second is fine for a comprehensive vision event, as in Paul on the road to Damascus, and there seems to be no reason that the channel envisaged should not have the required bandwidth (particularly if data compression techniques are used). But this is an extreme case. Much less information is required to convey a quiet conviction that it is right to love your enemy, or that a specific act you have carried out is wrong.

4. Does this view involve a violation of laws of energy conservation? No. Information can be fed in by the coordination of the timing of various energetic events (e.g., timing decays of excited states) with no change in the energy budget of the atoms and molecules concerned. No transfer of energy need be involved.

5. Does it involve God in undignified microtweaking of events? No, no more than when we activate numerous individual cells in our arm to make muscles move in a coordinated way as a result of a decision to move our hand. The detailed implementation is at the microlevel, but the micro-events are coordinated by top-down action so that the micro-events constitute a holistic macro-action, which can be seen by the actor simply as a high-level action. Our mind sets huge numbers of individual neurons into action, resulting in the desired activation, but it does so as the result of an overall macrodecision ("I will move my arm to the right"), not by consciously individually controlling the states of the numerous separate neurons involved. The same would apply in the above envisaged influence of God's thoughts on our brains: it would be envisaged as a top-down action that was enabled by the appropriate set of co-ordinated micro-actions, implemented as a whole by a single conscious decision rather than by conscious enactment of numerous micro-activations.

I suggest that a reasonable case is made here, together with the arguments in *MNU* and Papers I and II, that these effects are sufficient to allow the kind of specific action of God as envisaged in those articles, namely, revelation through mental pre-images and illuminations.

7 Conclusion

This now completes a view of ontology, epistemology, and ethics that is compatible with the worldview expressed in *MNU*. The argument makes substantial claims regarding realism and also clearly separates out ontology and epistemology in a way that may be useful. This view of ontology is summarized in figure 3 above. One can

---

[96] See, for example, Russell's essay in this volume.

characterize WORLDS 1, 3–5 as expressing aspects of "the mind of God." What else might one want to include in that category? One might suggest that WORLD 5 should be expanded to include other aspects corresponding to the various "intimations of transcendence" discussed in Paper II (e.g., aesthetics[97]).

The argument presented here is intended to be of the kind promoted by this series of meetings, namely, of a strictly disciplined kind, logically coherent, and supported at all levels by multiple evidence. It is not claimed to provide the only possible overall causal pattern, but it is certainly claimed to present one self-consistent and logically tight pattern that is fully compatible with present-day science, and which cannot be disproved by scientific argumentation unless the entire edifice of quantum uncertainty is brought into question. It is no more fantastic or speculative than much that presently passes for theoretical physics (e.g., quantum cosmology, M-theory, and brane cosmology), and it is better supported by evidence. The issue is what evidence to take seriously. The view put here takes seriously the world of religious and spiritual experience, and this is the feature that most distinguishes it from mainline scientific argumentation. It takes seriously a broader range of data than is common in scientific argumentation, but the mode of analysis is similar.

Acknowledgment. I thank Ursula Goodenough for useful comments on molecular biology, and participants at the meeting—particularly Philip Clayton, Kirk Wegter-McNelly, Nancey Murphy, and Bob Russell—for helpful comments that have substantially improved the paper.

[97] See for example John W. de Gruchy, *Christianity, Art, and Transformation: Theological Aesthetics in the Struggle for Justice* (Cambridge: Cambridge University Press, 2001).

DIVINE ACTION AND QUANTUM MECHANICS: A FRESH ASSESSMENT

Robert John Russell

1 Introduction

In this essay, I will explore further a thesis about divine action and quantum mechanics whose roots trace back four decades in the field of "theology and science."[1] It has been extensively developed recently by scholars in the decade-long CTNS/Vatican Observatory series of research conferences. The thesis is the following: if quantum mechanics is interpreted philosophically in terms of ontological indeterminism (as found in one form of the Copenhagen interpretation), one can construct a bottom-up, noninterventionist, objective approach[2] to mediated direct divine action in which God's indirect acts of general and special providence at the macroscopic level arise in part, at least, from God's objective direct action at the quantum level both in sustaining the time-development of elementary processes as governed by the Schrödinger equation and in acting with nature to bring about irreversible interactions referred to as "quantum events."

I begin with a few clarifying comments (section 2) before turning to the heart of the essay (sections 3, 4, and 5). Here I first discuss *methodological* issues, including the warrant for a "bottom-up" approach to divine action and the problems of the "multiple interpretability" of quantum mechanics and "historical relativism." Next I turn to two *philosophical* issues: the phenomenological domain of the measurement problem and its relation to the indeterministic form of the Copenhagen interpretation of quantum physics. Then I explore a variety of *theological* issues. Background topics include divine action at the quantum level and general providence, the pervasiveness of divine action, local and global aspects of divine action, and the challenge of special relativity. Central topics include God's action in some or all quantum events and its relation to the problem of human freedom and the challenge of theodicy. In conclusion, I propose that a trinitarian doctrine of God is the most suitable context for locating the "divine action and quantum mechanics" thesis. A

[1] For historical background, see Robert J. Russell, "Special Providence and Genetic Mutation: A New Defense of Theistic Evolution," in *Evolutionary and Molecular Biology: Scientific Perspectives on Divine Action*, R.J. Russell, W.R. Stoeger, and F.J. Ayala, eds. (Vatican City State/Berkeley, Calif.: Vatican Observatory/Center for Theology and the Natural Sciences, 1998), secs. 2.3.1–2, the volume hereafter *EMB*.

[2] For a discussion of such terms as objective and noninterventionist, see Robert J. Russell, "Introduction," in *Chaos and Complexity: Scientific Perspectives on Divine Action*, R.J. Russell, N.C. Murphy, and A. Peacocke, eds. (Vatican City State/Berkeley, Calif.: Vatican Observatory/Center for Theology and the Natural Sciences, 1995), secs. 3.3 and 3.4, esp. figure 1, the volume hereafter *CAC*. For an anthology and careful analysis of the contemporary theological literature on divine action see Owen Thomas, ed., *God's Activity in the World: The Contemporary Problem* (Chico, Calif.: Scholars Press, 1983), hereafter *GAW*, and idem, "Recent Thought on Divine Agency," in *Divine Action*, B. Hebblethwaite and E. Henderson, eds. (Edinburgh: T&T Clark, 1990). For a detailed analysis of the philosophical problems involved, see Keith Ward, *Divine Action* (London: Collins, 1990), and Thomas F. Tracy, ed., *The God Who Acts: Philosophical and Theological Explorations* (University Park, Penn.: Pennsylvania State University, 1994).

brief appendix lays out directions for future research on the philosophical implications of quantum mechanics and their relevance for divine action, including a proposed architecture of philosophical issues, an exploration of implications of Bell's theorem, and a comparison of nonlocality and (in)determinism in Bohm's formulation and the Copenhagen interpretation.

2 Clarifications

The general position of noninterventionist, objective, special divine action actually includes several distinct approaches: (i) agential models of God's interaction with the world; (ii) agential models in combination with embodiment models of the God/world relation; (iii) agential models deployed through complex metaphysical systems, such as process philosophy and neo-Thomism. This essay will focus on the first approach, which, in turn, includes three versions distinguished primarily by their focus on inter- or intra-level causality: top-down causality, whole-part constraints, and bottom-up causality.[3] Though this essay will focus on bottom-up causality, like most scholars I believe that a combination of all three will eventually be needed for an adequate account of objective, noninterventionist divine action.

In the bottom-up approach, God is thought of as acting at a lower level of complexity in nature to influence the processes and properties at a higher level. To qualify as a noninterventionist approach, the lower level must be interpretable philosophically as ontologically indeterministic. A number of scholars[4] have focused

---

[3] For a discussion of how a bottom-up approach relates to possible top-down approaches, as well as why a bottom-up approach is essential in the context of the early evolution of life, see Russell, "Introduction," in *CAC*, sec. 4.3.

[4] Karl Heim, *The Transformation of the Scientific World* (London: SCM Press, 1953); Eric L. Mascall, *Christian Theology and Natural Science: Some Questions in Their Relations* (New York: Longmans, Green & Co., 1956); William G. Pollard, *Chance and Providence: God's Action in a World Governed by Scientific Law* (London: Faber and Faber, 1958); Mary Hesse, "On the Alleged Incompatibility Between Christianity and Science," in *Man and Nature*, Hugh Montefiore, ed. (London: Collins, 1975); Donald M. MacKay, *Chance and Providence* (Oxford: Oxford University Press, 1978); Nancey Murphy, "Divine Action in the Natural Order: Buridan's Ass and Schrödinger's Cat," 325–58, Thomas F. Tracy, "Particular Providence and the God of the Gaps," 289–324, and George F. Ellis, "Ordinary and Extraordinary Divine Action: The Nexus of Interaction," 359–96, all three in *CAC*; Ian G. Barbour, "Five Models of God and Evolution," *EMB*, 419–42; see as far back as idem, *Issues in Science and Religion* (New York: Harper & Row, 1971); Mark W. Worthing, *God, Creation, and Contemporary Physics* (Minneapolis: Fortress Press, 1996), esp. 130–46; Christopher F. Mooney, *Theology and Scientific Knowledge: Changing Models of God's Presence in the World* (Notre Dame: Notre Dame Press, 1996), esp. chap. 3, 108–10; Philip Clayton, *God and Contemporary Science* (Edinburgh: Edinburgh University Press, 1997), esp. chap. 7, 8. Some scholars have raised objections to the approach taken by these scholars. See, for example, Arthur Peacocke, "God's Interaction with the World: The Implications of Deterministic 'Chaos' and of Interconnected and Interdependent Reality," in *CAC*, 279–81. For an interesting recent response to Peacocke in terms of quantum indeterminacy, see John J. Davis, "Quantum Indeterminacy and the Omniscience of God," *Science and Christian Belief* 9.2 (October 1997): 129–44. and Peacocke's reply in the same volume. See also John C. Polkinghorne, "The Metaphysics of Divine Action," in *CAC*, esp. 152–3, articles in Niels H. Gregersen et al., eds., *Studies in Science & Theology 1996: Yearbook of the European Society for the Study of Science and Theology*, vols. 3 and 4, *The Concept of Nature in Science & Theology, Parts I and II* (Geneva: Labor et Fides, 1997), articles in *Science and Christian Belief* 7.2 (October 1995), and George Murphy, "Does the Trinity Play Dice?" *Zygon* 51.1 (March 1999).

on quantum mechanics because it deals with the lowest levels in nature (i.e., fundamental particles and physical interactions) and because it can be given such an interpretation. Their work will serve as sources for the current essay. First, however, I need to stress what the approach adopted in this essay does *not* claim.

1. This approach does *not* "explain how God acts" or even constitute an argument *that* God acts.[5] Instead it assumes that warrants for the belief in divine action come from extended theological arguments whose sources lie elsewhere (including scripture, tradition, experience, and reason).

2. It does *not* constitute either an epistemic or an ontological "god of the gaps" argument.[6] An *epistemic* gaps argument is based on what we don't know. It invokes God to explain things that we don't yet understand but that science will eventually explain. Our approach, instead, is based on what we do know about nature, assuming that quantum physics is the correct theory and that it can be interpreted philosophically as telling us that nature is ontologically indeterministic. In this approach, what we know is that nature provides the necessary but not the sufficient causes for quantum events to occur.

An *ontological* gaps argument assumes that natural processes are ontologically deterministic. Nature lacks what are called "causal gaps"[7] or breaks in the order of event causation. If nature itself lacks such causal gaps, God must act in special events to create these gaps. Such an account of particular divine action is clearly interventionist: in order to act in nature, God must intervene in these processes by suspending them and violating the laws that describe them. But this approach is theologically problematic because it pits God's special acts against God's regular action, the latter of which is seen to be the underlying cause of nature's regularities.

Instead, our approach is *noninterventionist*: God has created the universe *ex nihilo* such that some natural processes at the quantum level are insufficiently determined by prior natural events. One could say that nature is "naturally" indeterministic. Thus God does not suspend natural causality but creates and maintains it as ontologically indeterministic. God does not violate the laws of quantum physics but acts in accordance with them. In essence, God creates the universe such that quantum events occur without sufficient natural causes and acts within these natural processes and together with natural causes to bring them about.

The theological warrants for a noninterventionist account of divine action include the following: objective special providence is achieved without contradicting general providence (since God's particular acts, being noninterventionist, do not violate or suspend God's routine acts as represented in the "laws of nature"); God as the transcendent creator *ex nihilo* of the universe as a whole is the immanent on-going creator of each part (*creatio continua*); God's intentions are disclosed "in what we

---

[5] See Russell, "Special Providence and Genetic Mutation," sec. 4.1.

[6] Michael Ruse, *Can a Darwinian Be a Christian? The Relationship Between Science and Religion* (Cambridge: Cambridge University Press, 2001), 86–8, 91–2, provides a thoughtful and often conciliatory approach to the relations between Darwinism and theism. Unfortunately, though, he reiterates the charge that the appeal to quantum mechanics is an epistemic form of the "gaps" argument without discussing previous responses by Tracy, Ellis, Murphy, and myself. He adds to it the claim that it raises the problem of theodicy. I think the latter is a valid point, but again, it is one that I have discussed in "Special Providence and Genetic Mutation," sec. 5.2, and that I treat in some detail below.

[7] Here I am again following Tracy's usage in his "Particular Providence," sec. 1.1.

know, not in what we don't know," as Dietrich Bonhoeffer urged[8]; noninterventionist objective special divine action offers a robust response to atheistic challenges to the intelligibility and credibility of Christian faith, since the presence of "chance" in nature does not imply an absent God and a "pointless" world but an ever-present God acting with purpose in the world.

3. It does *not* reduce God to a natural cause, nor does God's direct[9] action at the quantum level give rise to phenomena that cannot be explained by science. It affirms that science is characterized by methodological naturalism, and thus it abstains from viewing "God" as an explanation within science.[10] Instead, God's direct action at the quantum level is hidden in principle from science, supporting the integrity of science and yet allowing science to be integrated fruitfully into constructive theology where "God" as an explanation of natural events is appropriately and fully developed.

4. It does *not* propose that God alters the wavefunction between measurements, makes measurements on a given system, or alters the probabilities of obtaining a particular result.[11] Instead, God together with nature (i.e., as mediated divine action[12]) determines what happens during a "quantum event." This claim represents

---

[8] Dietrich Bonhoeffer, *Letters and Papers from Prison*, enlarged ed. (London: SCM Press, 1972/1979), 311. See Tracy, "Particular Providence," 289.

[9] God may be thought of as acting directly at the quantum level (more precisely, the effects of God's direct action may occur at the quantum level). The events we attribute to God at the macroscopic level are their indirect result. A direct, or basic, act is one for which there is no prior act (such as willing my arm to move), and one which may initiate a sequence of acts resulting in an indirect act (such as my arm moving). Thus divine acts of general and special providence at the ordinary, classical level are mediated and indirect divine acts that arise from God's direct acts mediated in, through, and by quantum processes. Such providential acts can equally be seen as a form of God's ongoing, continuous creative action. See Tracy, "Particular Providence and the God of the Gaps," in *CAC*, 295–6.

[10] This approach thus differs from that of "Intelligent Design" since it does not introduce concepts such as agency or designer into scientific theory. Instead it argues that when quantum physics is introduced into theology through the lens of philosophy, it offers a new theological approach to noninterventionist divine action.

[11] Nicholas T. Saunders, "Does God Cheat at Dice? Divine Action and Quantum Possibilities," *Zygon* 35.4 (September 2000): 517–44, offers a helpful overview of the kinds of interpretations of quantum physics and of the theological notions of providence and divine action. He then delineates four ways of relating divine action and quantum mechanics. The first three are the ones I have mentioned here: that God alters the wavefunction between measurements, makes measurements on a given system, or alters the probabilities of obtaining a particular result. They do not seem to describe the actual positions of *any* of the principal scholars in theology and science, nor does Saunders claim that they do.

I agree with Saunders that I and several others probably fit into his fourth approach: as Sanders' puts it, "God ignores the probabilities predicted by orthodox quantum mechanics and simply controls the outcomes of particular measurements." (I would rather say that God acts with nature to bring about the outcomes of particular measurements consistent with the probabilities given before the event occurs.) Saunders' acknowledges that he does not find any specific problems with this approach, except that it requires us to work within a particular philosophical position. I agree with him, but I think that this is unavoidable. I have discussed this problem extensively in previous publications and return to it below.

[12] One can think of God as acting either in, through, and together with the processes of nature (mediated) or as acting unilaterally (unmediated). In the latter case, often called "occasionalism," all events in the world occur solely through God's action. Occasionalism denies that there are natural causes in the world and undercuts the importance of science in discovering and in representing them mathematically. As Murphy stresses, any adequate

a particular philosophical interpretation of quantum mechanics usually referred to as "the Copenhagen interpretation."[13] A variety of scientists have supported ontological indeterminism, including such contemporaries as Chris Isham, Paul Davies, and Ian Barbour.[14] This alone, of course, is not a warrant for adopting indeterminism, only a recommendation.

Clearly this interpretation involves a number of complex issues, including such "external" problems as historical relativity and multiple interpretability, and such "internal" problems as the meaning of measurement, quantum event, quantum indeterminism, and more generally the problem of a realist interpretation of quantum mechanics, including the referential status of the wavefunction, etc. I will seek to respond to these issues in detail below, but I should say from the outset that one generic problem is unavoidable: we must adopt one or another philosophical interpretation whenever we incorporate the results of science (or any other field of knowledge) into a wider intellectual context, particularly into constructive theology. The key is to hold one's philosophical interpretation explicitly, tentatively, and hypothetically as a lens through which to ask questions about the relation between science and theology, not foundationally as the basis of one's theological position (as for example in natural theology or physico-theology).

5. It does *not* limit the relation between quantum mechanics and divine action to special providence. Instead it views the domain of quantum mechanics as giving rise to the general features of the ordinary macroscopic world (i.e., general providence/continuous creation) and to particular events within it (i.e., special providence).

Quantum processes underlie and give rise to the general features of the world of ordinary experience and Newtonian physics.[15] These processes fall into two classes. First are the processes that produce macroscopic properties such as the impenetrability of matter (and thus the extension of matter in space), the chemical properties of the elements (including color and valency), and the electrical and thermodynamic properties of solids (such as conductivity and specific heat). Fermi-Dirac (FD) statistics describe these processes and explain why they lead to the associated macroscopic properties. Particles that obey FD statistics, such as electrons and protons, are called fermions. Second are the processes that "glue the everyday world together," i.e., that produce the electroweak, strong, and gravitational interactions, and that create such macroscopic "quantum" phenomena as superfluidity and

---

account of divine action must avoid both occasionalism and deism (in which God's action is restricted to a single event, the beginning of the world); Murphy, "Divine Action in the Natural Order," 332.

[13] Again, what is crucial here is that the inclusion of a philosophical interpretation is not an option; the only option is which interpretation is to be chosen.

[14] Chris J. Isham, *Lectures on Quantum Theory: Mathematical and Structural Foundations* (London: Imperial College Press, 1995), 131–2; Paul Davies, *Quantum Mechanics* (London: Routledge & Kegan Paul, 1984), 4; Ian G. Barbour, *Religion in an Age of Science* (San Francisco: Harper & Row, 1990), 123.

[15] For earlier detailed discussion see Robert John Russell, "Quantum Physics in Philosophical and Theological Perspective," in *Physics, Philosophy, and Theology: A Common Quest for Understanding*, R.J. Russell, W.R. Stoeger, and G.V. Coyne, eds. (Vatican City State: Vatican Observatory Publications, 1988), hereafter "Quantum Physics"; idem, "Special Providence and Genetic Mutation." How macroscopic phenomena first arose out of the quantum processes of the very early universe remains a profound problem. Here I simply take it for granted that we can describe both our ordinary experience using classical science and our subatomic data using quantum physics, and look to their relation.

superconductivity. Here the statistics are Bose-Einstein (BE), and the particles, such as photons or gravitons, are bosons.[16]

The mathematical equations that represent FD and BE statistics are radically different in the quantum realm of low energies and temperatures, but as we move to "room temperature," both approach the Boltzmannian equation that characterizes classical statistics (i.e., a Gaussian "bell" curve).[17] This fact leads to another striking aspect of the relation between quantum and classical physics. If we look at statistics from an *epistemological* perspective, "classical chance" is grounded mathematically in and arises smoothly in the appropriate limit from quantum statistics. But if we look at statistics from an *ontological* perspective, the result is far more complex. Recall that Boltzmannian statistics originated in classical physics and the context of ontological determinism.[18] On the other hand, FD and BE statistics arise within a quantum mechanical framework suggestive of ontological indeterminism. So if we are interested in *ontology* and start with Boltzmannian statistics, we are led in opposite directions: to *determinism* if we stay within the framework of the classical world in which it originated, and to *indeterminism* if we move to the quantum world and derive Boltzmannian statistics from FD and BE statistics. How strange it is that the classical, everyday world, where Boltzmannian statistics point to causal determinism, is actually the product of a quantum world whose FD and BE statistics point instead to ontological indeterminism!

From a theological perspective, God's noninterventionist action at the quantum level[19] gives rise to the creation of the general features of the classical world

---

[16] Technically, superfluidity and superconductivity involve both FD and BE statistics, as Carl York pointed out (private communications). FD and BE statistics are intimately connected to the indistinguishability of fundamental particles ("all electrons are identical") and their spin: ψ is anti-symmetrized for fermions (which carry odd spin) and symmetrized for bosons (which carry even spin). Indistinguishability and spin, in turn, are strictly quantum features, and yet they too can be seen as giving rise to the ordinary features of the classical world. The space-like correlations in these statistics are also intimately related to the problem of nonlocality in quantum physics, as Bell's theorem reveals (discussed below). A full discussion of spin-statistics requires a relativistic treatment of quantum physics, such as given by Dirac. Thus, in a strict sense, it lies outside the confines of nonrelativistic quantum physics, although quantum statistics can be warranted at least in part on the basis of indistinguishability.

[17] FD statistics, $1/(e^{E/kT} + 1)$, and BE statistics, $1/(e^{E/kT} - 1)$, both approach Boltzmann statistics, namely $1/e^{E/kT}$, at energies $E >> kT$. Here E is the energy of the system, k is Boltzmann's constant and T is the equilibrium temperature of the system. At low energies, BE statistics still resemble the classical form, but FD statistics are strikingly different. See for example figures 11:1–3 and table 11:1 in Robert Eisberg and Robert Resnick, *Quantum Physics of Atoms, Molecules, Solids, Nuclei, and Particles* (New York: John Wiley & Sons, 1974), chap. 11.

[18] Here, bulk properties of solids, liquids, and gases were derived mathematically from a statistical treatment of the deterministic interactions between their component parts (e.g., the kinetic theory of gases).

[19] God may act at other levels in nature should they, too, be open to an indeterministic interpretation. This would apply most clearly in the domain of neurophysiology and thus involve an analysis of the neuro- and cognitive sciences. See Robert J. Russell, Nancey Murphy, Theo Meyering, and Michael A. Arbib, eds., *Neuroscience and the Person: Scientific Perspectives on Divine Action* (Vatican City State/Berkeley, Calif.: Vatican Observatory/Center for Theology and the Natural Sciences, 1999), hereafter *NAP*.

described above, as well as to their sustenance and physical development in time, or what we would routinely call general providence (or continuous creation).[20]

Quantum processes also underlie and give rise to specific effects in the macroscopic world in several ways.[21] One way is through those phenomena, such as superfluidity and superconductivity, which, though found in the ordinary classical world, are really "bulk" quantum states—what George Ellis in this volume calls "essentially quantum effects at the macro level." Another, and quite different, way is through specific quantum processes, which, when amplified correctly, result in particular classical effects in the classical world. It is the latter that will be the focus of this essay and will be thought of in terms of special providence. Obvious examples range from such jury-rigged situations as "Schrödinger's cat" to such routine measurement devices as a Geiger counter. But the production of specific effects in the macroscopic level from quantum processes includes a whole range of phenomena such as the animal eye responding to a single photon, mental states resulting from quantum events at neural junctions,[22] or the phenotypic expression of a single genetic mutation in an organism (resulting, for example, in sickle-cell anemia or cancer). Indeed one may argue that the evolution of life on earth over the past 3.8 billion years depends in part on such "biological amplifiers" as the genotype-phenotype relation, which expresses the effects of quantum mechanics within genetic mutations at the macroscopic level of individual organisms and populations.[23] Moreover, the amplification of microscopic to macroscopic states in most of these processes does not rely on chaos theory. Therefore, contrary to the claim by some scholars, we need *not* deal with the unresolved problem of "quantum chaology" in this approach to divine action.[24]

Thus God's action at the quantum level can be seen as bringing about, in a non-interventionist mode, both the general features of the world we describe in terms of

[20] George Ellis makes this same point nicely in this volume, sec. 2.1; note his references as well. See also Russell, "Quantum Physics," 344–6; Murphy, "Divine Action in the Natural Order," sec. 4.3; Russell, "Special Providence and Genetic Mutation," sec. 2.3.2.

[21] See Russell, "Quantum Physics." It is widely asserted that individual quantum events always "average out" at the macroscopic level, thus making quantum mechanics irrelevant to special providence and human free will. Instead, the "Schrödinger cat" argument provides an elegant way to combine both general and special providence on the same quantum "template."

[22] Ellis actually discusses two possibilities: (i) coherent firings in large arrays of neurons leading to a holistic response in a region of the brain (here "amplification" is an almost inappropriate term), and (ii) localized firings in microtubules that are amplified to macroscopic effect, following the suggestions of Roger Penrose; see George F.R. Ellis, "Intimations of Transcendence: Relations of the Mind and God," in *NAP*, 472; idem, "Ordinary and Extraordinary Divine Action," in *CAC*, 369–71.

[23] See Ellis, in this volume, sec. 2; Barbour, "Five Models of God and Evolution," in *EMB*, 426. For an extended discussion of quantum mechanics, evolutionary biology, and divine action see Russell, "Special Providence and Genetic Mutation."

[24] Polkinghorne, "Metaphysics of Divine Action," section 4.1. Also see Polkinghorne's contribution to this volume, secs. 4 and 5. See also Fred Sanders, *The Image of the Immanent Trinity: Implications of Rahner's Rule for a Theological Interpretation of Scripture* (Berkeley: GTU dissertation, unpublished, 2000), 540. Although quantum chaos is *not* a problem for the present approach relating divine action and quantum physics, it *is* a serious problem when one tries to relate chaos theory, at least in its present state, to divine action, particularly when an appeal is made to quantum physics to provide those variations in initial conditions of specifically chaotic systems that give rise to the appearance of "openness" in deterministic, closed systems.

general providence (or continuous creation) and those specific events in the world to which *special providence* refers.

3 Methodological Issues

1. *Is a "bottom-up" approach to divine action warranted, and does it exclude other approaches?* We should *not* see the present focus as a *general* limitation or restriction of divine action to "bottom-up" causality alone.[25] Instead, I see the present argument as located within a much broader context, namely the theology of divine action in personal experience and human history, because that is primarily where we, as persons of faith, encounter the living God. For this, we clearly need to consider a variety of models, including both "top-down," "whole-part," and "bottom-up" causes and constraints, and their roles within both embodiment and non-embodiment models of agency, with particular emphasis on the "mind-body" problem and human agency.[26] Moreover, I believe we will eventually need to work out the detailed relations between these models by integrating them into a consistent and coherent, adequate and applicable metaphysical framework.

The question here, though, is why and how God might be thought of as acting within nature via a form of bottom-up causality. Granting that God is the creator of the universe *per se*, maintaining the efficacy of nature, whose regularities, which we call the laws of nature, manifest God's faithfulness and rational intelligibility as creator, and granting also that these laws have just the right statistical ingredients to allow for the production of "order out of chaos" as part of God's creative actions, and granting that in some situations, such as our personal encounter through faith with God, it is highly appropriate to introduce top-down language about God's action, can we nonetheless adequately understand God's action within the physical, astrophysical, molecular, and evolutionary processes out of which we arose as expressing God's intention in ways that go beyond that of maintaining the existence of these processes and allowing their built-in "potentialities" to work themselves out over time? And can such an understanding of God's action be rendered in an intelligible way if we restrict ourselves to top-down causality or to whole-part constraint *alone*?

I believe it cannot. Top-down causality is helpful when considering the action of conscious and self-conscious creatures that are genuinely open to God's action and that have at least some capacity to respond to it. But it is hard to see what constitutes the "top" through which God acts in a top-down way when no conscious, let alone self-conscious, creatures capable of mind/brain interactions have yet evolved. Remember, we are trying to understand God's action in the universe over its full twelve to fifteen billion year history, including the production of first- and second-generation stars, planetary systems, and eventually the evolution of organisms at least on Earth over a period of nearly four billion years, ranging from the simplest primitive forms to the present vastly rich profusion of life. Moreover, if God acts at

[25] As Barbour notes, most authors who explore this approach also insist on eventually combining these approaches; Barbour, "Five Models of God and Evolution," 432–3.

[26] I agree with Murphy's 1995 assessment that Arthur Peacocke has given "the most compelling account to date of the role of top-down causation in accounting for God's continuing action." Her reference was to Peacocke, *Theology for a Scientific Age*, to which could now be added a variety of his articles, including "Biological Evolution—A Positive Theological Appraisal," in *EMB*, 357–76, and "The Sound of Sheer Silence: How Does God Communicate with Humanity?" in *NAP*, 215–48. See Murphy, "Divine Action in the Natural Order," 326, fn. 3.

the "top" level of complexity at a given stage in evolutionary history, that level of complexity must be ontologically open, that is, it must be described by laws that can be interpreted in terms of ontological indeterminism. Yet, until the evolution of organisms capable of even primitive mentality, the "top" levels would presumably have been within the domain of the "classical" sciences and the ontological determinism of Newtonian physics. On the top-down approach special divine action would thus be unintelligible *without* intervention from the epochs of early galactic, stellar, and planetary formation on up through those early stages of evolutionary biology prior and leading to the development of a central nervous system. But if we omit this early period from our discussion of special providence, then we once again risk a radical limitation on special divine action: God's special action can only occur after a sufficient degree of biological complexity has been achieved, but it cannot be effective within the processes by which that degree of complexity is achieved. For both these reasons, then, the top-down strategy seems stymied.[27]

Perhaps we should try whole-part constraint arguments instead. The challenge here is to find phenomena in nature that display holistic characteristics and that point to ontological indeterminism. The ecological web is often cited as a candidate, due to its inherent complexity and seemingly endless openness to external factors, but in my opinion it fails to be a candidate for noninterventionist divine action because of the underlying determinism of the processes involved, no matter how complex or inter-related they might be.

Thus on critical reflection, and contrary to the hopes of most previous attempts at theistic evolution, it seems unlikely that top-down or whole-part approaches are of much value for interpreting physical processes and biological evolution at the pre-cognitive and even pre-animate era in terms of special divine action. Unless one returns to the quantum level, where holism and indeterminism are displayed everywhere and at all times since "$t = 0$," I see little hope that God's action within the early stages of physical, astrophysical, and biological phenomena can be described in noninterventionist ways using either whole-part constraint or top-down causal arguments.

2. *The historical relativism and multiple interpretability of quantum mechanics.* The next two problems are also methodological. First, why should we take quantum physics seriously if it will one day be replaced by a new physical theory? Second, how can we take quantum physics seriously in discussing a theology of divine action given the fact that quantum mechanics is subject to a variety of equally valid, and radically distinct, philosophical interpretations?[28]

In response to the first problem, one option would be to disregard all theories that are at the frontier of science, including quantum physics, sticking instead with proven theories such as classical physics. If we did so, we would be on surer grounds for drawing conclusions about the world, since we know precisely where the limits of applicability lie for such theories. For example, we know precisely in which

---

[27] Murphy outlines similar problems with a strictly top-down approach to divine action in Murphy, "Divine Action in the Natural Order," 4.1.

[28] Actually this is a concrete example of the multiple interpretability and historical relativity that inevitably surround any scientific theory. How these factors affect the philosophical and theological discussions of a scientific theory is a crucial methodological issue lying at the heart of *any* conversation about "theology and science." A decision regarding it is required of every scholar in the field. I will try to describe mine here, though all too briefly. See also Russell, "Special Providence and Genetic Mutation," sec. 4.2; idem, "Quantum Physics."

domains classical physics applies for all practical purposes, namely in the limits of Planck's constant $h \rightarrow 0$ and the speed of light $c \rightarrow$ infinity.[29]

I don't agree with this overly cautious approach for two reasons. First, classical physics is in principle false. As a useful theory for practical needs, like engineering or planetary exploration, it is excellent. But as a fundamental theory of nature, its explanation of the world is wrong. As Charles Misner has remarked, the theories that we know are "proven" are the ones that have been the most clearly falsified! Second, it is within this classical view of nature as a closed causal system that the theology of previous centuries has operated—and much of contemporary theology still does! Many of the atheistic challenges to divine action have ignored the quantum mechanical aspects of nature and presupposed classical science and a mechanistic, deterministic metaphysics, as Ellis has pointed out.[30] Thus their arguments, too, are fundamentally flawed. So sticking only with proven theories is out.

As is well known, quantum mechanics can be given a variety of philosophical interpretations.[31] The Copenhagen interpretation is, arguably, the most widely held view by physicists and philosophers of science. According to Jim Cushing, it essentially involves "complementarity (e.g., wave-particle duality), inherent indeterminism at the most fundamental level of quantum phenomena, and the impossibility of an event-by-event causal representation in a continuous spacetime background."[32] Although rooted in the work of Niels Bohr, the term "Copenhagen interpretation" includes several distinct versions. Bohr himself stressed the epistemic limitations on what we can know about quantum processes. Compared with their effortless union in classical physics, spacetime description and causal explanation become complementary (necessary but mutually exclusive) aspects of a quantum account of microscopic processes.[33] Bohr also believed that quantum formalism applies to individual systems, compared with Einstein's statistical view in which the

---

[29] But note Berry's careful discussion of this issue in this volume.

[30] George F.R. Ellis, "The Thinking Underlying the New 'Scientific' World-Views," in *EMB*, 251–80.

[31] In 1966, Ian Barbour provided what is still one of the most helpful surveys of these interpretations. Barbour, *Issues in Science and Religion*, chap. 10, sec. III. See also Barbour, *Religion in an Age of Science*, 101–4. For a more recent and accessible account see Nicholas Herbert, *Quantum Reality: Beyond the New Physics* (Garden City, N.Y.: Anchor Press; Doubleday, 1985). For a technical survey of the philosophical problems in quantum physics see Jammer, *The Philosophy of Quantum Mechanics*; Michael Redhead, *Incompleteness, Nonlocality, and Realism: A Prolegomenon to the Philosophy of Quantum Mechanics* (Oxford: Clarendon Press, 1987); James T. Cushing and Ernan McMullin, eds., *Philosophical Consequences of Quantum Theory: Reflections on Bell's Theorem* (Notre Dame: University of Notre Dame Press, 1989); Abner Shimony, "Conceptual Foundations of Quantum Mechanics," in *The New Physics*, Paul Davies, ed. (Cambridge: Cambridge University Press, 1989); James T. Cushing, *Quantum Mechanics: Historical Contingency and the Copenhagen Hegemony* (Chicago: University of Chicago Press, 1994); Isham, *Lectures on Quantum Theory*.

[32] Cushing, *Quantum Mechanics*, 24.

[33] In his famous 1927 Como lecture Bohr argued that "the spacetime coordination and the claim of causality, the union of which characterizes the classical theories, [are] complementary but exclusive features of the description, symbolizing the idealization of observation and definition respectively." For a convenient source and translation, see Jammer, *The Philosophy of Quantum Mechanics*, 86–94. See also Cushing, *Quantum Mechanics*, 28.

formalism applies to ensembles only.[34] Heisenberg both supported the completeness of quantum mechanics and developed his own realist, indeterministic version of the Copenhagen interpretation in which the measurement process actualizes potential characteristics of the quantum system. His interpretation suggests that the unpredictability that arises during measurement has an ontological basis and is not simply epistemological.[35] Ian Barbour cites Henry Margenau who writes, "the uncertainty does not reside in the imperfection in our measurements, nor in man's ability to know; it has its cause in nature herself." As Barbour puts it, "if this interpretation is correct, indeterminacy is an ontological reality."[36] In sum, Cushing concludes that, "On the Copenhagen interpretation of quantum mechanics, physical processes are arguably, at the most fundamental level, both inherently indeterministic and nonlocal. The ontology of classical physics is dead."[37]

Other interpretations of quantum mechanics include: ontological determinism (the neo-realism of Einstein/incompleteness and Bohm/"hidden variables"); many worlds (Everett); quantum logic (Gribb; Finkelstein); consistent histories (Clarke, Griffiths, Omnès, Gell-Mann, Hartle); and consciousness creates reality (von Neumann, Wigner, Stapp). With this in mind, some have argued that we modify the basic equations of quantum mechanics (e.g., Shimony's philosophically motivated exploration of stochastic modifications of the Schrödinger equation).[38] Since their discovery in the 1960s, Bell's theorems have underscored the nonlocal and

---

[34] See Cushing, *Quantum Mechanics*, for a discussion of Leslie Ballentine's arguments about Bohr versus Einstein. Cushing views Stapp's interpretation as close to Ballentine's statistical approach.

[35] Werner Heisenberg, *Physics and Philosophy: The Revolution in Modern Science* (New York: Harper, 1958); idem, *Physics and Beyond* (New York: Harper & Row, 1971). Heisenberg apparently had a "two truths" view of the relation between science and religion, with religion as a set of ethical principles. See for example idem, *Across the Frontiers*, Peter Heath, trans. (New York: Harper & Row, 1974/1971), chap. XVI. He also argued that "the extension of scientific methods of thought far beyond their legitimate limits of application led to the much deplored division" between science and religion; idem, *Philosophic Problems of Nuclear Science* (Greenwich, Conn.: Fawcett Publications, 1952), chap. 1.

[36] Henry Margenau, "Advantages and Disadvantages of Various Interpretations of the Quantum Theory," *Physics Today* 7 (1954), quoted in Barbour, *Issues in Science and Religion*, 303–4.

[37] Cushing, *Quantum Mechanics*, 32.

[38] What is particularly interesting here is that Shimony not only argues for one philosophical interpretation against its competitors, but that he allows his philosophical commitments (i.e., to realism) to drive his scientific research program in new directions that seek to revise current physics; Shimony, "Search for a Worldview which can Accommodate Our Knowledge of Microphysics," in *Philosophical Consequences of Quantum Theory*, Cushing and McMullin, eds., 25–37, esp. 34. His interest in a modified version of quantum mechanics provides an excellent example of how one's philosophical and theological commitments can play a positive influence in the construction of new and empirically successful scientific theories. In essence, the creative mutual interaction between theology, philosophy, and science can include not only the critical analysis and incorporation of scientific results in constructive theology, but also the positive role played by theological and philosophical commitments in the construction of new theories in science (i.e., the "context of discovery"). Obviously, for such theories to count as scientific, they must be delimited by the assumptions of methodological naturalism and prove their worth empirically. See Robert J. Russell, "Theology and Science: Current Issues and Future Directions," 2000, Part 3-E, available on the Internet at www.ctns.org.

particularly the nonseparable character of quantum phenomena, making each of the earlier interpretations more problematic.[39] How then are we to decide which interpretation or modification is right and reliable for a discussion of divine action, and if we cannot decide, what might be a reasonable way to proceed?

My response is fourfold. First, why single out quantum mechanics? Every scientific theory is open to competing metaphysical interpretations; indeed, metaphysics is *always* underdetermined by science, although some theories, like classical physics, seem strongly to favor one interpretation (e.g., determinism) over others. So this concern about quantum mechanics applies, in principle, to *any* metaphysical interpretation of *any* scientific theory. Indeed, the warrant for choosing a specific metaphysical interpretation of any scientific theory is an issue not only for theists but equally for naturalists or atheists.

Second, none of these interpretations returns us to an entirely classical view of the world; to one extent or another, all of them require a reconstruction of our philosophy of nature. This might seem obvious, but it actually addresses what is a subtle problem in the literature. Bohm's interpretation, being deterministic and describing nature in such classical terms as particles, forces, and trajectories, can seem like a less problematic option than Bohr's epistemology, with its wave-particle complementarity, or Everett's many-worlds ontology. But in fact Bohm's advantages are bought at a heavy price: the determinism suggested by Bohm is not strictly classical, but highly nonlocal. Bohm's view is also nonmechanical, involving the quantum potential and instantaneous action-at-a-distance. (We shall return to the metaphysical problems raised by Bohm's approach in some detail below.) Thus even if we adopted Bohm's approach we would not simply fall back into the safe haven of classical metaphysics (if indeed it ever were so, or we ever wanted to!); instead we would inherit yet another set of thorny issues that I will label "Bohmian determinism." Indeed, this fact can actually be used to our advantage: a careful comparison of Bohmian and Copenhagen views, as suggested below, might help us understand just what is meant on both sides by (in)determinism.

Third, my approach is best seen as a form of constructive theology with a focus on nature (what Barbour calls a "theology of nature"), *not* a form of natural theology, let alone physico-theology. Hence a change in science or its philosophical interpretation would challenge the constructive proposal at hand, but not the overall viability of a theology of divine action in nature, whose primary warrant and sources lie elsewhere in scripture, tradition, reason and experience.

Finally, I think we should welcome the specificity of this approach and follow it as far as it can take us. By illuminating the concrete implications of a noninterventionist approach to objective special divine action in light of a particular interpretation of quantum physics, the strengths as well as the limitations of the approach are revealed, which in turn should lead to further insight and new areas of research.

3. *The approach taken here*. With these responses in mind, my approach will be an explicitly "what if" strategy: I will engage in the theological conversation with quantum mechanics by choosing one particular philosophical interpretation (ontological indeterminism within the general Copenhagen interpretation), stating clearly that this choice is being made, stressing that it may one day prove no longer tenable (presumably for scientific reasons—but philosophical or even theological reasons could also play a role in either initially choosing or later changing

---

[39] See for example Redhead, *Incompleteness, Nonlocality, and Realism*; Cushing and McMullin, eds., *Philosophical Consequences of Quantum Theory*.

interpretations[40]), and proceeding to explore the philosophical and theological implications of this interpretation in full awareness of the tentativeness of the project—but engaged in it nevertheless.

My choice of the Copenhagen interpretation means that I will need to respond to a number of key issues that arise within this interpretation. The most important issues will be the "measurement problem" and the associated "collapse of the wave equation," as well as the meaning of a "quantum event." All of these are involved in the claim of ontological indeterminism with its presupposition that quantum mechanics can be given a (critical) realist interpretation. I will then need to work out the implications of these issues for our understanding of divine action and embed it in a broader theological context. This process will occupy most of the remaining portions of this essay.

4 Philosophical Issues—The Measurement Problem within the Copenhagen Interpretation of Quantum Mechanics

We turn now to a key issue in the Copenhagen interpretation of quantum mechanics, the measurement problem. There are, of course, various formulations of this problem, each raising complex issues for a realist understanding of quantum mechanics. According to Chris Isham, the way we understand the measurement problem depends on our interpretation of the formalism and, in particular, on what one means by the reduction of the state vector. The measurement problem, in turn, is part of a "quaternity of problems" all posed to the realist (but avoided by instrumentalists and pragmatists): (i) the meaning of probability; (ii) the role of measurement; (iii) the reduction of the state vector; and (iv) quantum entanglement. Although their classical analogues allow for a clear resolution from a realist perspective, Isham shows that the quantum versions do not.[41] For Jeremy Butterfield, the measurement problem is important because it illuminates and underscores the problem of quantum indefiniteness from a realist perspective. If, as realists claim, quantum physics applies to everything physical, the indefiniteness of the microrealm should be endemic in the macrorealm—it should be transmitted to the macrorealm, but apparently is not. Indeed, indefiniteness should manifest itself in macrostates that blatantly contradict our ordinary experience of definite states.[42]

For the limited purposes of this essay, I want to distinguish between two issues regarding the measurement problem from a critical realist perspective: (i) its phenomenological domain, i.e., what sorts of physical processes should be called "measurements"? and (ii) its relation to ontological indeterminism. When discussing the mathematical structure of the wavefunction and its implications for divine action below, I will stress again the challenge posed to a realist interpretation.[43]

---

[40] See again Shimony, "Conceptual Foundations of Quantum Mechanics," 34.

[41] Isham, *Lectures on Quantum Theory*, chap. 8.

[42] Butterfield, in this volume, describes four strategies to solve the measurement problem: modify the Schrödinger equation or ascribe additional (though not "hidden") variables, and pursue each assuming that the macrorealm is either definite or not definite.

[43] An excellent example of the challenge that quantum mechanics poses to realism is given by the wavefunction ψ. On the one hand, ψ can be thought of as a mathematical function defined on a multidimensional configuration space; for n particles, configuration space is $3n$-dimensional. Thus to represent the quantum state of two particles in three dimensional physical space requires a six-dimensional configuration space. From this perspective, a realist (versus, say, a Platonic) interpretation of ψ is problematic at best. (Abstraction increases as

1. *The phenomenological domain of the measurement problem.* We begin with a well-known distinction that arises in the Copenhagen interpretation between (i) the *time development* of the wavefunction ψ of a quantum system, as governed by the deterministic Schrödinger equation, and (ii) the *irreversible interaction* between the quantum system and other systems. *Ex hypothesi*, these systems must be of such size and complexity that their interaction with the quantum system is, at least in practice, irreversible, i.e., the Schrödinger equation does not apply. Irreversible interactions are routinely called "measurements," but they are not limited to interactions with the ordinary world around us; instead, they include phenomena ranging from what we can call, for want of better terms, "micro-macro," "micro-meso," and "micro-micro" interactions.[44]

Micro-macro involves interactions between elementary particles and "classical measuring devices," such as the response of a Geiger counter to an alpha particle, but it also includes any irreversible interaction between an elementary particle and an ordinary object, such as the absorption of a photon by an animal retina or an electron by a TV screen. Clearly micro-macro interactions entail a vast range of *natural* phenomena from the physical and biological sciences, as well as those involving *human artifacts* (e.g., Geiger counters). As stated above, the evolution of life depends on such biological amplifiers as genotype-phenotype-population arrangements. But, contrary to the views of some scholars,[45] I claim that the domain of the measurement problem is far more extensive than this, for it also involves irreversible micro-meso and micro-micro phenomena.[46]

Micro-meso includes all those interactions between elementary particles and (sub-) microscopic objects with enough degrees of freedom to make the interaction irreversible (at least in practice). Examples include the capture of an electron by a dust particle in interstellar space, the decay of atoms in solids (such as radioactivity), the interaction between bound and free particles (such as the absorption or emission of a photon by an atomic electron in a crystal solid), and the making or breaking of atomic and molecular bonds (such as hydrogen bonding during genetic mutations of DNA). All of these, too, constitute a measurement since they are irreversible, even though their scale is "micro-meso."

---

one moves from configuration space to Hilbert space.) On the other hand, elementary texts on quantum mechanics routinely treat ψ as a physical wave in ordinary three-dimensional space, and not without precedent: de Broglie favored a physicalist interpretation of quantum "waves," while Schrödinger (and later Bohm) recognized their imbedding in configuration space. For an excellent discussion and references, see Cushing, *Quantum Mechanics*. For the difference between de Broglie and Schrödinger, see Cushing's comments, 105 (and fn. 72) and 120. Cushing tells us (124) that Schrödinger began with a realist interpretation of the wavefunction, but quickly ran into the problems posed by its configuration space context. For the gloss on Bohm, see 149.

44 Since we are working within the Copenhagen interpretation, we have not invoked consciousness in accounting for the measurement process. Thus references to "macro" might involve laboratory instruments, but not conscious observers *per se*. However, see Butterfield, in this volume, for complex ways of including consciousness in the analysis of measurement.

45 "Measurement involves an intervention by our everyday world into the quantum world" (Polkinghorne, *The Quantum World*, 60). Also, see him in this volume, secs. 1 and 4.

46 Although phenomena such as superfluidity and superconductivity are not specifically what I mean by "micro-macro" interactions, they do, as Ellis in this volume points out, represent essentially quantum effects at the macro level.

Micro-micro interactions would normally be considered reversible and governed by the Schrödinger equation, and thus would not constitute "measurements." Examples include proton-proton scattering in free space and pair-production and annihilation in the vacuum. However, if such interactions occur within a complex environment they could well be irreversible and thus constitute "measurements." Proton-proton scattering in the presence of heavy nuclei would be an example.[47]

In summary, the term 'measurement' should *not* be restricted to micro-macro interactions, let alone to those "macro" interactions that involve laboratory experiments. Instead, the term 'measurement' should include *all irreversible interactions* in nature from micro-micro to micro-macro. What is crucial, then, to making an interaction a "measurement" is not that it involve something "macro" but that it is irreversible.

2. *The measurement problem as the basis for the indeterministic interpretation of quantum physics*. The measurement problem can now be stated (but, alas, not solved!) easily: How are we to understand measurements by using quantum physics if measurements cannot be described by applying the Schrödinger equation to them and if we are not to alter quantum physics?[48] Within the Copenhagen interpretation, the response is stark: the measurement problem is not really a "problem to be solved," but a *synonym* for those processes not governed by the Schrödinger equation. Since causes are represented by the Schrödinger equation (as formal cause) and the potential V contained in that equation (as the efficient cause), the inapplicability of the Schrödinger equation to a measurement is the basis for the philosophical claim of ontological indeterminism. Since the outcome of a measurement is not describable in terms of the Schrödinger equation, we can infer that there are necessary (e.g., material) causes but not sufficient (in particular, efficient) causes to bring about the measurement.

We can also see why the phrase "the collapse of the wavefunction" is used to describe "what happens" during a measurement. The wavefunction ψ, which had evolved deterministically in time under the influence of the classical potential V and according to the Schrödinger equation, changes discontinuously from a superposition of states to a specific state. This is also a convenient place to offer a more precise definition of the term 'quantum event' than one customarily finds in the literature. I propose that we restrict our usage of the term to what we are calling "measurements," that is, those interactions that are irreversible regardless of whether they are micro-macro, micro-meso, or micro-micro interactions. Conversely, the time-

---

[47] Note here the crucial role of irreversibility in defining "measurement." In order to distinguish a measurement from the ordinary time-development of a quantum system as governed by the Schrödinger equation, we must refer to irreversibility. But this term is usually borrowed from thermodynamics, which reflects yet another profound problem at the heart of quantum physics: thermodynamics is, arguably, not a fundamental theory, whereas quantum physics is. Why, therefore, would irreversibility play such a fundamental role in quantum physics? For the sake of this essay, I will use "irreversible" as though its meaning were self-evident, although this is overtly not the case. For a complex discussion, see Michael Berry's essay in this volume.

[48] As is well known, one attempt to address this problem was to assume two separate ontologies: classical and quantum. The Schrödinger equation governed the latter, but not the former. Thus when classical objects were seen as interacting with quantum processes, a measurement—in both restricted (laboratory) and general (micro-macro) senses—occurred. The problem is that if we insist that classical objects are made of quantum processes, the basis for the ontological distinction breaks down and the measurement problem remains.

development of the wavefunction between measurements is *not* to be thought of as a series of quantum events.[49]

In this approach, then, the measurement problem and ontological indeterminism are two sides of the same coin: the measurement problem is that aspect of quantum physics to which ontological indeterminism is specifically addressed. For the purposes of this essay, we will stay within the Copenhagen interpretation. This allows us to say that for quantum events or measurements to occur, nature provides the necessary but not the sufficient causal conditions, or what Barbour calls a "weak form of causality."[50] I emphasize the deeply unresolved status of the measurement problem, but I hope that by using it in this specific way we can proceed to explore the case for divine action and quantum physics.[51]

To summarize, within (at least one variety of) the Copenhagen interpretation, ontological indeterminism, the measurement problem, the collapse of the wavefunction, and the meaning of quantum event all merge into one conceptuality: a quantum event is an irreversible interaction (at all scales in physics from micro-micro to micro-macro), in which the Schrödinger equation ceases to govern the time-evolution of the wavefunction ψ describing both the system and that with which it irreversibly interacts. Instantaneously ψ collapses from a superposition of states to one state. The fact that the resulting state is unpredictable in advance, i.e., that it cannot be explained by a deterministic law, is the basis for the philosophical interpretation that such an event is ontologically indeterministic. In short, we find both the determinism described by the Schrödinger equation between quantum events and the indeterminism characterizing quantum events. In the following I shall refer to "ontological indeterminism'" in the strict sense as referring to quantum events.

5 Theological Issues

A variety of theological issues now emerge in the general relation between divine action and the Copenhagen interpretation of quantum physics as we look more closely at the thesis we are exploring here. I will separate them into background issues and crucial issues.

5.1 Background Issues

1. *Divine action at the quantum level and general providence*. God creates *ex nihilo* and sustains the existence of quantum systems as they undergo time-evolution (governed by the Schrödinger equation) and as they undergo irreversible interactions (quantum events, measurements) with other micro- and macro-systems whose existence God also sustains. The time evolution of quantum systems applies to isolated systems, such as elementary particles traveling through relatively empty intergalactic space, or to the very early universe. It also applies to elementary particles bound together, as atoms and molecules undergo time evolution in conformity with the Schrödinger equation. Quantum events arise when micro-systems interact irreversibly with each other or with more complex, molecular or macro-

---

[49] I will return to this point in my critique of process philosophy /theology below.

[50] Barbour, *Issues in Science and Religion*, 304; note his reference to Northrop.

[51] In a similar way Ellis acknowledges the unsettled issues surrounding measurement but proceeds to discuss quantum physics and divine causality; see Ellis, "Ordinary and Extraordinary Divine Action," 369.

scopic systems. (Here I am not considering those irreversible interactions that lead indirectly to significant changes in the world, and are thus interpreted in terms of special providence.)

The point here is that during both time evolution and irreversible interactions, particles and systems retain their FD or BE properties,[52] and these properties account for the classical properties of bulk matter that we experience as the ordinary world of nature and describe in terms of the classical laws of nature and classical statistics (i.e., epistemic chance). It is to this world of ordinary experience that we attribute God's general providence (or continuous creation), namely the ongoing creation and sustenance of the general features of the classical world together with the emergence and evolution of physical, chemical, and biological[53] novelty in nature. Thus what we routinely take as general providence arises *indirectly* from God's *direct* action of sustaining in existence quantum systems and their properties during both their time-evolution and their irreversible interactions. In short, God (indirectly) creates macroscopic structures and interactions, as well as classical chance, as a result of quantum processes and statistics.

In previous writings, I pointed to a watershed accomplishment in theology and science when, in the 1970s, Arthur Peacocke[54] shifted the discussion of chance from a conflict model, "law versus chance," as urged by atheists such as Jacques Monod (unfortunately, a formulation all too often accepted by Christians who reject evolution) to an integrative framework, "law and chance." As a result of this shift, Christians could claim that God acts through both law and chance to create physical, chemical, and biological novelty in nature. Still, the meaning of chance in this context may not be adequate for a genuine sense of God's noninterventionist action. Instead, I suggest that we now face a second and even more fundamental—and promising—shift in our discussion of "law and chance" in light of quantum physics: a shift from the meaning of chance in classical physics and biology (i.e., chance in the Boltzmannian sense of our epistemic ignorance of underlying causal processes, which is not helpful for the agenda of noninterventionist divine action) to the meaning of chance in quantum physics (i.e., chance in the Copenhagen version of ontological indeterminism, which is open to noninterventionist divine action, as well as chance represented by FD and BE statistics and their relation to order at the classical level). Rather than saying that God directly creates by turning chaos into new and novel forms of order,[55] we could say from a quantum perspective that God indirectly creates order and novelty in the classical realm by directly creating a quantum mechanical universe with its combination of quantum events and FD/BE statistics and by acting as continuous creator in time within the indeterminism of quantum events. God is thus truly the God of both order and novelty.

---

[52] A fuller warrant for including the discussion of FD and BE statistics would require relativistic quantum mechanics, and this lies beyond the scope of this essay; here we simply have introduced it in relation to the symmetry properties of the wavefunction.

[53] When applied to the realm of molecular and evolutionary biology, the relations are further complicated, since the micro-macro processes are involved in all genotypic-phenotypic relations, including those that have little effect on a species, as well as those that, accumulated over time, lead to species differentiation and in turn to what might be called general and special providence. See Russell, "Special Providence and Genetic Mutation."

[54] See in particular Arthur Peacocke, *Creation and the World of Science* (Oxford: Clarendon Press, 1979) and his many subsequent publications.

[55] Ilya Prigogine's "order out of chaos" program, adapted so creatively by Peacocke.

2. *Divine action and special providence within the domain of measurement: is "pervasive" a more helpful term than "ubiquitous" or "episodic?"* We have argued that God's direct act of sustaining quantum systems in existence—both during their time-evolution and during the occurrence of quantum events—results indirectly in those features of the world that we attribute theologically to God's general providence. But the more important claim of this essay is that quantum events, i.e., all irreversible interactions in nature from micro-micro to micro-macro, constitute the domain in which God's direct, noninterventionist action can lead indirectly to special events at the macroscopic level, events that we can interpret theologically in terms of special divine action or special providence.

John Polkinghorne and others[56] have been concerned that this approach would lead to an "episodic" account of divine action for at least three reasons: (i) the concept of measurement is limited to processes that involve the quantum and classical levels (or what I have called irreversible micro-macro interactions); (ii) such interactions only occur from time to time, and (iii) they relate quantum mechanics to chaos theory and thus raise the technical problems associated with quantum chaos. I would share their caution about this approach, too, if these concerns were persuasive. However, as I hope I have shown here, the concept of measurement is not defined by the levels involved (i.e., micro-macro, micro-micro) but by irreversibility. Quantum chaos is not necessarily (or even typically) involved in such irreversible interactions.

What then about the "episodic" nature of such interactions? In fact, such interactions can occur at any time and place in the universe where the deterministic time-development of the quantum phenomena governed by the Schrödinger equation is disrupted by an irreversible interaction (measurement), as is evident from the examples given in the previous discussion. Previously I have used the term 'ubiquitous'[57] to suggest this comprehensive characteristic, since the term 'episodic' sounds far too occasional. But I am persuaded that both terms unduly emphasize distinct aspects of what is in reality a single complex situation. A term is needed that suggests that noninterventionist divine action can be related to the sudden disruptive aspect of quantum processes that can occur anywhere, but not to the continuous time development of the system governed by the Schrödinger equation. An appropriate term for such divine action might be 'pervasive', and I shall use this term in future writings. With this understanding in place, I hope that concerns about this approach being episodic can be put to rest.

---

[56] See Polkinghorne, "Metaphysics of Divine Action," 152–3, and secs. 4, 5 of Polkinghorne's contribution to this volume where he uses the term "episodic" to describe the limitations of this approach. Sanders apparently agrees with Polkinghorne's claim that "measurements are relatively infrequent events, and thus any theory of divine action linked to them is likely to be highly episodic in nature." Sanders, *The Image of the Immanent Trinity*, 541, 532–3. I have previously responded to this claim in some detail. See Russell, "Special Providence and Genetic Mutation," particularly sec. 3.2.

[57] See my previous response in "Special Providence and Genetic Mutation," 211–2. There I did *not* mean "ubiquitous" in the sense that both (i) the time-evolution of a quantum system and (ii) its irreversible interaction with other systems are the domain of noninterventionist direct divine action and, in turn, of indirect special providence in the macroscopic world. But surely this was evident since it was the indeterminism implied by quantum physics that allowed us to think of noninterventionist direct divine action in the first place, and indeterminism obviously does not apply to the time-evolution of quantum systems governed by the deterministic Schrödinger equation.

3. *Is divine action local or global?* Before proceeding, we should inspect an implicit assumption, namely that God's action in relation to ψ should be thought of as an unambiguously "local" action. Instead I will propose two claims.[58]

First, the mathematical features of the wavefunction ψ used in elementary quantum mechanics, and the parametric role of both space and time variables in defining ψ, suggest that God's action in relation to ψ occurs globally in space and time.[59] To see this we start with the general[60] wavefunction ψ such that $\psi = \psi(x, t)$. In principle, ψ is defined[61] for $-\infty \leq x \leq +\infty$ and for $-\infty \leq t \leq +\infty$, with x and t both serving formally as parameters of ψ. We can view this in at least three ways: (i) we can stipulate the spatial shape of the wavefunction everywhere along the x-axis at a given moment of time; (ii) we can describe the spatial shape everywhere along the x-axis as it changes in time; (iii) we can specify its amplitude (height) at a particular point in space as it changes in time. Now (ii) is probably the closest we come, in very rough terms, to the classical conception of a particle with a well-defined location in space at a moment in time, such that we can write $x = x(t)$.[62] Thus, right from the outset, an important aspect of the nonlocality of the quantum conception of matter is built in. Our conception of divine action in relation to ψ must reflect this view. We must take care not to presuppose an unambiguous locality to God's action when it is conceived in relation to ψ. We may think about divine action as "localized" by thinking of it in relation to the region in space where ψ is relatively large, somewhat in the way we refer to the "location" of the particle represented by ψ, as long as we keep in mind the fact that this is a rough way of speaking and do not fall tacitly into the classical conception of matter—or divine action.[63]

Second, the concept of God as acting to bring about a quantum event (i.e., the collapse of the wavefunction) is as much a global as a local event, regardless of whether this event leads indirectly to an instance of special providence. Consider a simple physical process: a particle is emitted at time t_0 and propagates freely through

[58] However, these claims presuppose a realist interpretation of quantum mechanics in general, and of ψ as referring, even if only partially, to the physical world. But a variety of profound problems are associated with any such realist interpretation of ψ, not the least of which is that ψ is typically formulated in an abstract space called "configuration space," mentioned above in fn. 43. Such challenges to realism should be borne clearly in mind in the following discussion.

[59] Again, this tends to presuppose a "physical space" approach instead of "configuration space" and this would be highly problematic when considering a quantum state composed of more than one system. However, see Cushing, *Quantum Mechanics*, fn. 33, 251–2.

[60] For simplicity, we will work strictly in configuration space, although a momentum-space formulation is certainly an option, too. Again, for simplicity, we restrict the discussion to one spatial dimension, x.

[61] Of course, to be physically admissible, ψ must be normalized properly and thus be square integrable.

[62] In essence, the classical ontology is of a fully localized material object whose properties include its place in space, and this place can change in time, allowing an "$x = x(t)$" conceptuality. In the Copenhagen interpretation of quantum mechanics, however, we picture ψ as defined everywhere in space and in time.

[63] Of course there are qualifications here. Consider, for example, a wavefunction bounded by an infinite square well of length L (such that $\psi = 0$ when $0 \leq x \leq L$). Although wavefunctions of this type are useful for various practical purposes, infinite square wells do not exist in nature. In principle, the ubiquity of ψ always holds, and thus the caution about presupposing a classical assumption of "locality" in conceptualizing special divine action.

space until it is detected at time t_1, let's say one hour later. The motion of the particle between t_0 and t_1 is governed by the Schrödinger equation, and its wavefunction ψ is a uniformly expanding sphere centered on the source. (To be more precise, the particle is described by a wavepacket whose maximum value, ψ_{max}, describes a uniformly expanding sphere, but one that is everywhere nonzero.) Now, at t_1 the particle is detected and its wavefunction collapses instantaneously and unpredictably to a state representing the particle at the detector. We may make the additional assumption that its detection has significant consequences in the world, which we interpret theologically in terms of special providence, but this is irrelevant to the present issue.[64] What then can this suggest about the relation between God's action at the quantum level and the collapse of the wavefunction?

First we should keep in mind the previous point: God is active everywhere in space and time in relation to ψ as it extends throughout space and evolves in time. Indeed, one might say that the "general action" of God is God's action in maintaining the regular time development of ψ as described by the Schrödinger equation, much as we understand God's general providence as maintaining the world in its bulk, macroscopic configurations. Still for convenience let us think in terms of the peak in ψ (ψ_{max}) as it expands spherically, since for all practical purposes this represents a spherical wavepacket about to "collapse."

Now, at the moment of collapse, ψ changes discontinuously from a light-hour sphere, ψ_s, to a fully localized wavepacket ψ_x. Thus the irreversible interaction or quantum event involving the particle and the detector is represented here by the juxtaposition of, and discontinuous transition between, the global ψ_s and the local ψ_x that co-characterize and co-constitute what we mean by the collapse of the wavefunction. If we are to think of God's action in relation to this event, then it, too, must have both a global and a local character: God acts *globally* on ψ_s to bring about the "collapse" by causing a *local* transition from a nonzero to a zero amplitude everywhere on a sphere one light-hour in radius except at the location of the detector. Finally, if we then assume that the detection of the particle leads to a macroscopic event that we interpret as an act of special providence, then the concept of special providence, which refers to significant local macroscopic events in history and nature, comes about by God's action at the quantum level globally *and* locally.[65]

4. *Divine action, quantum physics, and the challenge of special relativity*. So far we have discussed several general issues related to divine action and quantum physics. Before turning to more detailed issues, we should note that this discussion has tacitly assumed the classical view of space and time found in Newtonian-Galilean physics. Special relativity (c. 1905) poses important issues for quantum physics and thus for our discussion of divine action.[66] It would be good to mention these briefly before turning to more detailed issues. Indeed, we shall see that some of the reasons given for not pursuing divine action in terms of quantum physics stem from the problems with special relativity and not from the issues that we will later consider. I will discuss scientific issues first, and then theological issues raised by them.

[64] Bear in mind, though, that it is an example of mediated and indirect divine action.

[65] At the same time, God's action in regard to both ψ_s and ψ_x is fully global in the general sense that both wavefunctions, in principle at least, extend infinitely in both space and time.

[66] I will not extend this essay to include relativistic quantum mechanics, the union of quantum physics and special relativity, and its heir, quantum field theory.

From a scientific perspective, the Copenhagen interpretation in particular is challenged by special relativity in several ways. First, special relativity undercuts the classical assumptions of a global present and a universally unique rate of time's flow. Both the Schrödinger equation and the measurement problem presuppose these assumptions. Thus, in light of special relativity, it becomes crucial to ask how we are to pick out the physically correct surface of simultaneity on which the Schrödinger equation governs ψ and on which ψ collapses, as Jeremy Butterfield and Raymond Chiao stress in this volume. Second, special relativity can be given alternative ontological interpretations—much as alternative interpretations pervade quantum physics—namely, the "block universe" and "flowing time" views.[67] Which of these ontologies are we to adopt in a relativistic reformulation of the Copenhagen interpretation? These are serious problems for quantum physics. On the other hand, however, it is crucial to note that quantum mechanics is consistent with special relativity in a crucial way: violations of Bell's inequalities need not violate relativity's "first signal principle" (i.e., instantaneous causal action-at-a-distance). This is a subtle point, since space-like correlations do exist and their presence undercuts local realism, as we shall discuss briefly below.[68] Additional insight is also shed on the relation between present and future by the "temporal nonlocality" that Raymond Chiao describes in this volume. Here once-related events in the present and the future display a Bell-like correlation, which undercuts the classical relation between present and future.

Theologically, special relativity challenges the problem of "time and eternity" that lies behind what I have proposed about divine action and quantum physics. How, for example, does God know what action to take in the present to bring about events of special providence in the future in light of special relativity? There are actually a variety of nested problems and issues here. Two will suffice for the present discussion.

The first is the "block universe" versus "flowing-time" interpretation mentioned above. Chris Isham represents one widely held view: the "block universe" perspective in which the future (and the past) are as real as the present. We may not know what the future holds, but from the perspective of eternity, God's knowledge of the future is perfect. But can God—or can we—act to change things in the present, and thus the future, in this scenario, and does quantum indeterminism make a

[67] See for example Chris J. Isham and John C. Polkinghorne, "The Debate Over the Block Universe," in *Quantum Cosmology and the Laws of Nature: Scientific Perspectives on Divine Action*, R.J. Russell, N.C. Murphy and, C.J. Isham (Vatican City State/Berkeley, Calif.: Vatican Observatory/Center for Theology and the Natural Sciences, 1993), 134–44.; Robert J. Russell, "Time in Eternity," *Dialog* 39.1 (Spring 2000): 46–55. Thus even if special relativity is given a "world-line flowing time" interpretation, one should be careful about referring to God's action in terms of the "world-as-a-whole" and "the future," as well as divine action in a specific event. A closely related problem exists for all theologies—trinitarian, dipolar, panentheist, and process—claiming that God experiences the world as a whole in a moment of time.

[68] These issues are extraordinarily subtle. Cushing claims that Bohm gives us a preferred frame for instantaneous action, and thus allows for "true becoming"—which may sound strange, since it is also a completely deterministic theory in which what becomes is fully predetermined. He has also argued that Bohm's approach allows for action-at-a-distance but without remote signaling either, and that it offers a unique solution to the problem of simultaneity in special relativity. Michael Redhead, however, claims that Bohm's approach is inconsistent with a stronger requirement, the "philosophically grounded invariance principle." See their essays in this volume.

difference to our answer here? John Polkinghorne, like many others, rejects the "block universe," with its apparent contradiction of our experience of time and free will, and opts instead for a "flowing-time perspective" in which the future has no ontological status and thus cannot be known by us or God. Here God's providential involvement in a genuinely open world is more like the "master chess player" who may not know the outcome of a specific game in advance but who is certain to win. But again, how do we make physical sense out of the "present" or uniformly flowing time in light of special relativity?

I think both of these options are valuable but problematic. Hence, as I have indicated previously,[69] I am attempting to construct a third alternative that draws on the strengths of the previous scenarios. I call it an "*event/world-line* flowing time" interpretation of special relativity.[70] The project includes a relationally-based ontology of events in which the status of "present," "future" and "past" is attributed to relations between events rather than to the events themselves. It then uses this ontology to explore the conception of time and eternity as developed in trinitarian doctrines of God. I believe this move will alleviate some of the problems raised by the "block universe" versus "flowing-time" debate. In any case, though, one can always argue that God does not foresee the future in the sense of seeing the future from the present, but rather by seeing the future in its own state as present.[71]

Second, Arthur Peacocke has argued that, given ontological indeterminism, even God can have only limited, probabilistic knowledge of the future outcome of quantum processes. Thus, if God knows the future by predicting it from present knowledge, even God can only have a probabilistic knowledge of the future.[72] My response is that the ontological indeterminism of quantum processes does not stop God from bringing about a particular outcome because, as I have just indicated, God sees, not foresees, the future. God brings about the future not by predicting it from the present, as we do, but by knowing the future in its own future present.[73]

5.2 Crucial Issues

We are now ready to move directly to the key questions in the debate on divine action and (nonrelativistic) quantum physics. My central thesis is that God acts in quantum events to bring about, or actualize, one of several potential outcomes; the collapse of the wavefunction occurs because of divine and natural causality. But does God act in every quantum event or only in some? And what are the theological implications for human freedom and the problem of evil in nature? To respond to these questions, it will be helpful to focus carefully on the responses given by Murphy, Ellis, and Tracy as they have explored these and other crucial issues.

---

[69] Russell, "Special Providence and Genetic Mutation," 221.

[70] The challenge to "divine purpose" is more complex still in the context of "biological chance," i.e., the uncorrelated inter-relation between mutations at the level of molecular biology and change at the level of environment and population (á la Monod).

[71] Thus I would not agree with Sanders' claim; see Sanders, *The Image of the Immanent Trinity*, 535.

[72] Peacocke, "God's Interaction with the World," 279–81.

[73] I hope eventually to formulate my response in a way that is consistent with special relativity and the irreducibility of flowing time and free will.

1. *Does God act providentially (general and/or special) in all, or only in some, quantum events?* Nancey Murphy[74] has given what I consider one of the most important accounts we have of divine action in relation to quantum physics. Her arguments have been pursued in helpful ways by George Ellis,[75] as we shall see below. Murphy starts with the claim that God acts intentionally in *all* quantum events. She begins by providing two theological criteria for an acceptable theory of divine action: it should enable us to distinguish between events that are special acts of God and those that are not, and it should leave room for "extraordinary acts of God."[76] These criteria are needed if we are to allow for divine revelation through natural and historical events, to account for the practice of petitionary prayer, and to respond to the problem of evil (theodicy), with their associated entailments about human agency, natural goodness, and the regularity and autonomy of natural processes. Moreover, because Murphy's approach depicts God's action as mediated (God acting together with nature), it avoids making God the sole determiner of the processes of nature (occasionalism). Because it is a "bottom-up" approach to divine causation, God can effect the behavior of macroscopic objects without intervening in the everyday world. By viewing God as an indirect participant in every macro-level event, God is kept from becoming a "competitor with processes that on other occasions are sufficient in and of themselves to bring about a given effect."[77]

Murphy points to the close relation between her work and that of William G. Pollard. Unlike Pollard, though, Murphy claims that her approach does not portray God as unilaterally determining, and thus dominating, all events in the world, nor does it undercut human freedom. Instead it limits bottom-up divine action by allowing for top-down causation and it stresses God's respect for the integrity and rights of creatures. In doing so, Murphy sees her approach as steering a path between two extremes: making God responsible for all the randomness, purposelessness, and evil in the world, or undercutting any possibility for divine action within the course of nature and history.[78]

Tom Tracy[79] has also developed an elegant account of divine action in light of quantum physics. According to Tracy, a theory of noninterventionist divine action requires a world that is both "open and ordered, smoothly integrating chance and law." Quantum physics provides this: the probabilistic distribution of quantum events gives rise to ordered, deterministic structures at the macro-level, yet ontological openness remains because quantum events are not "uniquely specified by antecedent conditions." God's action remains hidden in nature. Tracy then asks, is it more helpful to think of God as acting in all quantum events, as Murphy does, or in only some of them? In order for Murphy's argument to work, he contends that

---

[74] Murphy, "Divine Action in the Natural Order."

[75] Ellis, "Ordinary and Extraordinary Divine Action"; idem, in this volume.

[76] Murphy prefers this term instead of "miracle." See Murphy, "Divine Action in the Natural Order," 331. In private correspondence, Murphy indicates that she now thinks that Jesus' resurrection should be placed in an entirely separate category from other "miracles," since it can't be the result merely of God's guiding quantum events.

[77] Ibid., 343.

[78] Ibid., 355–6.

[79] Tracy, "Particular Providence." Tracy clearly indicates that his thought on this issue is not settled. He is instead exploring a particular option to test its strengths and weaknesses—a research approach that I find highly congenial.

she must provide a developed account of top-down causation. But because the effects of wholes on parts are mediated by the bottom-up interactions of the parts, it remains unclear how freedom can appear as a top-down effect within a system of deterministic bottom-up causal relationships. Accordingly, Tracy explores the alternative idea that God both creates a world with ontological indeterminism and chooses whether or not to act in a given event in light of its impact on the course of nature and history.[80]

Let me first say that I find Murphy's approach helpful for several reasons. The idea of God acting in all quantum events supports the theological claim that God does more than sustain the existence of all events and processes; in fact, God sustains, governs, and cooperates with all that nature does. This idea offers us a subtle but compelling way to interpret God's action as leading to both general and special providence. I think this point is so crucial that I will repeat my previous argument here: Schrödinger's cat makes it clear that God's action at the quantum level results in two quite different kinds of macroscopic effects. It produces the *ordinary* world of the cat and Geiger counter (the ordinary physics of solid matter and Ohm's law, the routine biology of metabolism, etc.), which we describe as general providence. But it also results in *specific differences* in the ordinary world—the cat living instead of dying—when God acts in one way instead of another in a specific quantum event. For example, God acts with nature so that the particle is emitted now and not later, or it is emitted in the $+x$ direction rather than $-x$, etc. Which way God acts determines (indirectly) a specific result in the ordinary world. Thus we may attribute special providence to the cat being spared from death and granted life in the crucial moment. In fact, it is precisely the nature of the measurement problem, namely the collapse of the wavefunction from a superposition of states to a single state, that might allow us to combine Murphy's pervasiveness of divine causality with Tracy's concern for the event to be objectively special: God acts in this event as in all events (God's action is never "more" or "less" but the same, equally causative). Still in this occasion, with two states superposed before the event, God will chose one state in particular and not the other, the one destined to promote life, thus conveying God's intentionality in this particular event. We can thus interpret this particular event, in which the cat lives instead of dying, in terms of objective special providence without restricting God's action to that event, and yet still maintain the objectively revelatory character of that particular event.

The chief virtue of Tracy's option is that it provides a more intuitive connection between the idea of God's occasional action at the quantum level and God's special providence in the everyday world. Still, it seems less clear how God's general providence could be based on God's occasional action at the quantum level. Murphy's approach, unlike Tracy's, conforms with the principle of sufficient reason, which I find a highly attractive philosophical advantage—although I agree with Tracy that, at least in principle, God need not create a world in which the principle of sufficient reason holds.

In sum, Murphy's approach (and possibly Tracy's too) delivers just what is needed for noninterventionist objective, special providence. It involves *objective* special providence, for the actual fact is that the cat lives when it might have died; it is objective special *providence* since it truly conveys God's intentions through the event of the cat living; and it is *special* providence because it is that event that we use to refer to God's providence against the assumed backdrop of the general

[80] Ibid., 321–2.

situation itself: the cat purring, the sun shining, the apparatus functioning routinely, and so on. Most importantly, it is *noninterventionist* objective special providence because it is an act of objective special providence that God achieves without violating or suspending the ongoing processes of nature and the laws that describe them. So in short, God causes all the processes of the ordinary world (general providence), but a few of them genuinely convey special meaning because the choices God makes in causing them, and not the other options available to God, bring them about. I am not persuaded, however, that either Murphy's or Tracy's approach deals adequately with the problems of human freedom and theodicy. In the following two sections, I will sketch an approach I have been developing as a third option that attempts to combine the advantages of their views.

2. *Quantum physics, divine action, and the problem of human freedom*. The problem of free will, as formulated in the modern period, is the following: how are we able to act freely in the world if, as in the classical science picture, deterministic laws govern us somatically? Actually the problem only arises on an *incompatibilist/ libertarian* account of free will (which I adopt here). Many scholars have seen quantum indeterminism as a way out of the impasse: perhaps the human mind, through some form of "top-down" causality (e.g., mind/brain causality), can objectively influence the movements of the body, making the enactment of free choices possible. Ian Barbour notes that as early as the 1920s physicists Arthur Eddington and Arthur Compton sought to relate quantum indeterminism to volition.[81] This idea is pursued in this volume by George Ellis, who argues that the mind is necessary to collapse the wavefunction and to give a complete account of natural events, which quantum physics by itself cannot supply. This, however, raises a concern I have pointed out previously: how do we allow God's action to determine the quantum events that occur in my body and still allow for my own mind/brain to determine them? I will call this the problem of "somatic overdetermination."[82]

Before turning to it, though, I want to focus on the sub-problem of free will and quantum indeterminism. It is important to note here that Murphy does not see quantum indeterminism as essential to human freedom. She does appeal to the self-limitation of God in respecting the "natural rights" of creatures and of thus creating a dependable environment necessary to human agency. However, she argues that top-down causation does *not* depend on quantum indeterminacy at the bottom level. She cites Don Campbell's example to show how top-down causation could work even if all biological processes were deterministic.[83] I am not convinced by her response; in my view, the somatic enactment of incompatibilist human freedom requires lower-level indeterminism, and thus when we add the possibility of divine action we return to the problem of somatic overdetermination.

Tracy, too, is concerned with the issue of free will, asking how freedom can "appear as a top-down effect within a system of deterministic bottom-up causal

---

[81] Barbour, *Issues in Science and Religion*, 133, 305–14, particularly 308; Arthur Eddington, *The Nature of the Physical World* (Cambridge: Cambridge Univ. Press, 1928); Arthur Compton, *The Freedom of Man* (New Haven, Conn.: Yale Univ. Press, 1935).

[82] Russell, "Special Providence and Genetic Mutation," 215, point 2.

[83] See Nancey Murphy's careful discussion in her "Supervenience and the Downward Efficacy of the Mental: A Nonreductive Physicalist Account of Human Action," in *NAP*, esp. 154–7. If Murphy adopts a compatibilist view then it would be clearer why she doesn't need quantum indeterminism.

relationships."[84] It was precisely this concern that led him to explore the alternative option regarding divine action. Unfortunately, Tracy does not provide a detailed response there, either. Ellis, too, has stressed the problem of free will and quantum indeterminism to the extent of "inverting it" in a beautiful way: starting from his assumption of divine kenosis and the intention of God to create a universe where moral action is possible, Ellis argues that there must be openness in physical laws, so that morality and special divine action are possible. Thus, just as Murphy and others insist that the macroscopic world must be regular for moral agency to function, Ellis demands there be causal gaps, using Tracy's term, at the microscopic level for it to be enacted.[85]

But this takes us back to the larger problem: somatic overdetermination. My suggestion is to start with the scenario that God acts in all quantum events in the universe until the rise of life and consciousness anywhere.[86] God then increasingly refrains from determining outcomes, leaving room for top-down causality in conscious, and preeminently in self-conscious, creatures. This would be a version of the standard "solution" to the problem of free will, namely God's voluntary or metaphysically necessary self-limitation,[87] but seen now as a temporal development of the limitations, from minimum to maximum.

3. *Quantum physics, divine action, and the challenge of theodicy.* The problem of theodicy is a perennial issue for theism: if God is purely good and if God can really act in history, why doesn't God minimize the evil done by humanity (i.e., "moral evil")? When we expand the scope of divine action to include the evolutionary history of life on earth, the question becomes: Why doesn't God act to minimize suffering, disease, death of individual organisms, and extinction of species (i.e., "natural evil")?[88]

Of course, theodicy has been discussed extensively in the "theology and science" literature,[89] where its subtle connection to the problem of human freedom has

---

[84] Tracy, "Particular Providence," 316–9.

[85] Ellis, "Ordinary and Extraordinary Divine Action," 393.

[86] See Russell, "Special Providence and Genetic Mutation," secs. 3.3, 4. This approach might also shed light on the profoundly hard problem of the origins of sin in an evolutionary perspective.

[87] In this sense, my approach is compatible with either a neo-orthodox or a process view of divine self-limitation. I wish to note, however, that Ted Peters rejects the use of "divine limitation" in general as a "zero-sum" view of freedom. Instead he argues for a "both-and" view theologically. In future work I wish to consider the issue of quantum physics, divine action, and human freedom from the perspective that Peters offers.

[88] It is one of the most powerful arguments used by atheists in their rejection of attempts to accommodate Christianity and Darwinian evolution. See for example Richard Dawkins, *A River Out of Eden* (New York: Basic Books, 1995). In fact, the argument goes back to Darwin's own writings. For the pertinent reference to Darwin's letter to Asa Gray, May 22, 1860, see Ruse, *Can a Darwinian Be a Christian?*, 130. It is noteworthy that, even while suggesting some creative ways in which Christianity and Darwinism might find a bit of common ground (or at least some appreciation for their respective positions), Ruse underscores the fundamental problem for that common ground raised by pain and suffering in the natural world; ibid., 91–2. Ruse refers specifically to the thesis being explored here, but he does not discuss the response to the problem of theodicy in this reference, although he, too, suggests that a theology that stresses the suffering of God might be relevant to Darwinian evolution; ibid., 134, and Russell, "Special Providence and Genetic Mutation," sec. 5.2.

[89] Barbour, *Religion in an Age of Science*, chap. 8, pt. 4; Denis Edwards, "Original Sin

frequently been stressed. Arthur Peacocke provided an elegant example of this connection as far back as 1979, when he wrote: "[I]t seems hard to avoid the paradox that 'natural evil' is a necessary prerequisite for the emergence of free, self-conscious beings."[90] But theodicy becomes a particularly intense issue in light of the present thesis regarding a noninterventionist approach to objective, special divine action. In 1995, for example, George Ellis put the problem eloquently: "[T]here has to be a cast-iron reason why a merciful and loving God does not alleviate a lot more of the suffering in the world, if he/she has indeed the power to do so."[91] Does the approach of either Murphy or Tracy in relating divine action and quantum physics provide such a reason?

In response to the challenge of theodicy, Murphy calls on her notion of God's respect for the integrity or "natural rights" of all creatures. Being noncoercive, God's action is consistent with human freedom and thus addresses, in part, the issue of theodicy as "moral evil." But what of theodicy as "natural evil"? I am not entirely clear how Murphy would respond here. She makes a passing reference to the "free-process" defense proposed by Polkinghorne in analogy with the traditional "free-will" response.[92] Nevertheless, it raises several concerns. One is that it may be irreducibly tied to other concepts, such as top-down causality, which cannot fit, even analogously, at the much less complex domain of physics and early biology. Another is that, while it accounts for why God does not interfere in cases of natural evil where God's interference would undermine the conditions for the possibility of human freedom (i.e., regularity/predictability), it may not be able to account for why God does not interfere in those cases where human freedom is unaffected, including the vast sweep of pre-human (and pre-sentient?) evolution.[93]

---

and Saving Grace in Evolutionary Context," in *EMB*, 377–92; David Ray Griffin, *God, Power, and Evil: A Process Theodicy* (Philadelphia: Westminster Press, 1976); Gary Emberger, "Theological and Scientific Explanations for the Origin and Purpose of Natural Evil," *Perspectives on Science and Christian Faith* 46.3 (September 1994): 150–8; John F. Haught, "Evolution, Tragedy, and Hope," in *Science & Theology: The New Consonance*, Ted Peters, ed. (Boulder, Colorado: Westview Press, 1998); Philip J. Hefner, *The Human Factor: Evolution, Culture, and Religion* (Minneapolis: Fortress Press, 1993), 271; Nancey Murphy and George F. Ellis, *On the Moral Nature of the Universe: Theology, Cosmology, and Ethics* (Minneapolis, Minn.: Fortress Press, 1996), sec. 4.1; Ruth Page, *God and the Web of Creation* (London: SCM Press, 1996), esp. 91–105; Peacocke, *Theology for a Scientific Age*, chap. 8, sec. 2e; Polkinghorne, *The Faith of a Physicist*, esp. 81–7, 169; Robert J. Russell, "Entropy and Evil," *Zygon* 19.4 (December 1984): 449–68; Worthing, *God, Creation, and Contemporary Physics*, 146–56. A frequent source for these ideas is John Hick, *Evil and the God of Love*, rev. ed. (San Francisco: Harper & Row, 1966).

90 Peacocke, *Creation and the World of Science*, 166.

91 Ellis, "Ordinary and Extraordinary Divine Action," 360.

92 Note that her reference does not occur specifically in the context of theodicy. Murphy, "Divine Action in the Natural Order," 342. See Polkinghorne, *Science and Providence*. I have worked along similar lines in developing Polkinghorne's approach in term of thermodynamics. Robert J. Russell, "The Thermodynamics of 'Natural Evil'," *CTNS Bulletin* 10.2 (Spring 1990): 20–5.

93 For a helpful discussion, see Daniel Howard-Snyder, "God, Evil, and Suffering," in *Reason for the Hope Within*, M.J. Murray, ed. (Grand Rapids, Mich: Eerdmans, 1999), esp. 96–8. and his references to Peter van Inwagen, William Rowe, and, interestingly, Quentin Smith. His conclusion should give us pause: "My sense is that we have no idea how God would be justified in permitting the isolated suffering of nonhuman animals at Nature's hand." For a classic version of the challenge of theodicy involving animal pain, see John Stuart Mill,

Since her 1995 essay on quantum physics, Murphy has worked with Ellis to develop a detailed theodicy in their work on the "moral universe."[94] There they explicitly reject the Augustinian response to theodicy, arguing instead for an Anabaptist approach grounded in a kenotic view of God's action that takes natural evil seriously, utilizes Murphy's work on quantum physics and divine action, and moves to the suffering of Christ on the cross. Clearly, Murphy and Ellis offer a promising approach to the challenge of theodicy.

Tracy, as we saw, explored the alternative view of divine action, citing the problem of theodicy encountered by Murphy's approach as a reason for his choice. But does Tracy's option help us here? It is not clear to me how restricting God's action really helps matters: why does God not act in those events, or refrain from acting in others, if this would alleviate suffering, etc.? Tracy has also discussed the impossibility of assessing the extent of suffering compared to the goals met by these processes.[95] I find this helpful in showing the difficulty of such an assessment, and the naïveté with which such difficulty is normally overlooked, but the search for an acceptable response to theodicy must move beyond the philosophical framework of this approach to a fully-developed theology of redemption. I believe it is here that we will find something like the "cast-iron reasons" that Ellis so rightly demands—reasons that will have the form of the cross.

4. *Embedding "divine action and quantum physics" in a broader theological framework.* In essence, the question now is how to locate our work on divine action and quantum physics in the context of a fully developed and robust systematic theology. At this point, a number of promising options are available. With Murphy and Ellis, Barbour, Peacocke, Polkinghorne, Edwards, Peters and many others in the "theology and science" conversation, I believe we must look to a kenotic theology that respects human freedom and focuses on the passibility and suffering of God: through the cross and the atonement of Christ, God redeems the world, suffering with and taking on the pain and death of all creatures. We could explore the route Murphy and Ellis have taken, or pursue the "theologies of nature" articulated by Peacocke and Polkinghorne, or explore the directions taken by other scholars in "theology and science." However, I am still persuaded by Barbour's argument some thirty years ago that "an elaborated metaphysics is needed if we want to relate rather than simply juxtapose divine causation, natural causation, and free human causation."[96] Owen Thomas has recently underscored the lasting centrality of this problem, asserting that the most promising options are the metaphysical systems of neo-Thomism and Whitehead[97]; I would add to these the metaphysical framework of Wolfhart Pannenberg and other theologians exploring the doctrine of the Trinity.

It would be natural to explore divine action and quantum physics from the perspective of process theology. Ground breaking research in "theology and science"

Three Essays on Religion (London: Longmans, Green, Reader & Dyer, 1875).

[94] Murphy and Ellis, *On the Moral Nature of the Universe*, chap. 10, sec. 4. See my response in Robert J. Russell, "The Theological Consequences of the Thermodynamics of a Moral Universe: An Appreciative Critique and Extension of the Murphy/Ellis Project," *CTNS Bulletin* 19.4 (Fall 1998): 19–24.

[95] Tom Tracy, "Evolution, Divine Action, and the Problem of Evil," in *EMB*, sec. 3. Also see the extensive discussion in Howard-Snyder, "God, Evil, and Suffering," sec. 6, of what he calls "the argument from amount."

[96] Barbour, *Issues in Science and Religion*, 430.

[97] Thomas, "Recent Thought on Divine Agency," 35–50.

has already come from a variety of scholars who work in differing ways within the broad outlines of process theology, including Ian Barbour, Charles Birch, John Cobb, Jr., David Griffin, and John Haught. These scholars draw on a crucial aspect of Whiteheadian metaphysics: namely, that reality consists of "actual occasions" that perish as they come to be, an idea highly reminiscent of "quantum events." Such actual occasions experience the causal efficacy of the past by prehension, are characterized by inherent novelty, and respond freely to God's inviting, subjective lure. Process theology views God as active in all levels of nature, stressing God's respect of human free will and God's kenotic and redemptive suffering with all creatures.[98]

The similarity between "actual occasions" and "quantum events" may not be entirely surprising. One of the advantages attributed by process scholars to Whitehead's philosophical system is its compatibility with science.[99] Whitehead himself claimed to offer a conceptual framework suited to science in general and quantum mechanics in particular.[100] But, as Abner Shimony has pointed out, Whitehead may have been reflecting on very early stages in the development of quantum mechanics when he constructed his "philosophy of organism" in the mid-1920s, and not on quantum mechanics as we now know and use it.[101] Moreover, important differences appear to exist between Whitehead's philosophy and quantum mechanics. After a detailed comparison, Shimony has concluded that "the discrepancies... between Whiteheadian physics and current microphysics constitute strong disconfirmation of Whitehead's philosophy as a whole."[102] One discrepancy is particularly relevant here: from a Whiteheadian perspective, the temporal atomicity of actual occasions underlies and gives rise to what we take to be enduring objects, but from a quantum perspective, such atomicities are "quantum events"

[98] Process scholars argue that the inclusion of God's subjective lure to evoke a response from creatures offers a creative new approach to noninterventionist divine action at various levels of organization and complexity in nature. See Barbour, *Religion in an Age of Science*, 232–4; John F. Haught, "Darwin's Gift to Theology," 402–5, Charles Birch, "Neo-Darwinism, Self-Organization, and Divine Action in Evolution," secs. 4, 8, both in *EMB*. The problem here is that one has to explain how divine agency is effective in the domains of chemistry, biology, and early evolutionary life, if the result of a succession of actual occasions is described classically by deterministic laws and epistemic (not ontological) chance. Even with the metaphysical richness of the subjective lure, I believe we need quantum mechanics to offer the indeterministic framework in which actual occasions can "make a difference"—and then we have to face the apparent discrepancies between process philosophy and quantum mechanics discussed immediately below. For the related problem of sentience, top-down causation, and consistency with science, see Barbour, *Religion in an Age of Science*, 224–7. I also have theological reservations about the way process theologians treat such crucial issues as the bodily resurrection of Jesus and the eschatological perspective of a new heaven and earth, and in turn creation *ex nihilo*. These reservations would remain even if the issues to be discussed between Whitehead and quantum physics were settled.

[99] For a careful and balanced assessment of this compatibility, see Barbour, *Religion in an Age of Science*, chap. 8, esp. pt. 3.

[100] Alfred North Whitehead, *Science and the Modern World* (New York: The Free Press, 1925), chap. 8; idem, *Process and Reality*, corrected ed., David Ray Griffin and Donald W. Sherburne, eds. (New York: The Free Press, 1978), 94–5, 238–9, 254.

[101] According to Shimony, "Quantum Physics and the Philosophy of Whitehead," in *Search for a Naturalistic World View: Volume II, Natural Science and Metaphysics* (Cambridge: Cambridge University Press, 1993/1965), chap. 19, esp. 291–2, Whitehead never refers to the new quantum theory in the exposition of his system.

[102] Ibid., parts II and III, and 303.

between which quantum systems undergo a continuous and deterministic time-development governed by the Schrödinger equation.

The story, though, is far from over. In his attempt to reformulate quantum physics, Shimony has introduced a stochastic term that addresses precisely this discrepancy, making his proposal closer to Whitehead's view of indeterminism (where chance pervades each actual occasion and hence the trajectory of an isolated particle) than it is to the indeterminism of current quantum physics (where it is focused strictly on quantum events). Shimony also suggests that Whitehead's concept of the concrescence of an actual occasion may contribute to a clearer understanding of the collapse of the wavefunction.[103] Other scholars too, including Henry Folse, Jr., Charles Hartshorne, William Jones, and Henry Stapp, have provided careful responses to the problematic relation between quantum physics and Whiteheadian philosophy.[104] Whether these suggestions and concerns will prove fruitful is an open and intriguing question, particularly as it suggests, once again, the creative role philosophy can play in the construction of new scientific theories (see footnotes 38 and 58).

Rather than look to process theology, I propose we locate the problem of divine action and quantum physics in an explicitly trinitarian doctrine of God. In *The Crucified God*, which I take to be a landmark in twentieth-century Protestant theology, Jürgen Moltmann pointedly argues that only a move from a "weakly Christianized monotheism" to a fully articulated trinitarianism can respond to the theological problem of the cross.[105]

The challenge for this approach, however, is that this understanding of the cross is linked theologically to Christian eschatology, including the bodily resurrection of Jesus, the general resurrection in the parousia, and the transformation of this universe into the new creation to come. Although Moltmann sees this, it is given a central place in the proleptic trinitarian theology of Wolfhart Pannenberg. It is only through the theology of reconciliation that the challenge of theodicy can be met, and reconciliation means both the end and the transfiguration of the world. "Only in the light of the eschatological consummation may [the verdict 'very good'] be said of our world as it is in all its confusion and pain."[106]

---

[103] Ibid., 309. Shimony proposes a hybrid between the most radical elements in quantum theory and the philosophy of organism, but in my view the input is almost entirely from quantum physics after the fact and not *a priori* from process metaphysics (chap. 19, esp. 303–4). Shimony also points to Whitehead's treatment of an *n*-particle system as being at odds with a quantum treatment and leading to "revolutionary philosophical implications" (300–2).

[104] Henry P. Stapp, "Quantum Mechanics, Local Causality, and Process Philosophy," *Process Studies* 7.4 (Winter 1977): 173–82; Charles Hartshorne, "Bell's Theorem and Stapp's Revised View of Space-Time," *Process Studies* 7.4 (Winter 1977): 183–91; William B. Jones, "Bell's Theorem, H.P. Stapp, and Process Theism," *Process Studies* 8.1 (Spring 1978): 250–61; Henry J. Folse, Jr., "Complementarity, Bell's Theorem, and the Framework of Process Metaphysics," *Process Studies* 11.4 (Winter 1981): 259–73. See also the two recent issues of *Process Studies*, vols. 26.3–4 (1997), guest edited by Timothy Eastman and devoted to the question of the relation between process thought and physics.

[105] Jürgen Moltmann, *The Crucified God: The Cross of Christ as the Foundation and Criticism of Christian Theology* (Minneapolis: Fortress Press, 1993), 236. At the same time, he claims that only a theology of the cross can extricate us from the perpetual warfare over the problem of evil between theism, which is "tantamount to idolatry," and its "brother" atheism. Ibid., 250, 221.

[106] Wolfhart Pannenberg, *Systematic Theology*, 3 vols., G. W. Bromiley, trans. (Grand

But we now find ourselves at "ground zero" of what is arguably the most powerful challenge to Christian theology in its encounter with science: how are we to understand eschatology in light of physics, biology, and Big Bang cosmology? I do not think that noninterventionist divine action will be of significant help with these issues. The resurrection of Jesus involves "more than a miracle," namely, the eschatological transformation of the fundamental conditions of nature, and not an extraordinary event within an unchanged natural backdrop, as described by this essay on special providence through noninterventionist divine action. I am currently beginning a major research project aimed at these issues.

I do, however, expect quantum physics to play some role in the overall approach to this vast problem, particularly through the way Pannenberg reformulates the concept of divine action in both creation (and thus providence) and redemption in terms of the Spirit of God. He has suggested that we use the concept of field in modern physics in order to talk about the Spirit and divine action.[107] Pannenberg's promising suggestion invites a number of responses, the principal one here being that his understanding of field comes from the context of classical field theory, as seen in both Faraday's and Einstein's work. When we move to the context of quantum physics and then to quantum field theory, a number of dramatic new features occur, as we have seen already. John Polkinghorne has underscored several of these in his critique of Pannenberg's use of the concept of field: superposition, nonlocality, and entanglement,[108] to which I would add the relation between determinism in Bohm's work and ontological indeterminism within the Copenhagen interpretation, the unification of such classically separate concepts as "matter" and "interaction" through the nonclassical nature of quantum statistics, and the concept of the "filled" quantum vacuum and its suggestion of a "meonic" view of spontaneous creation and annihilation.[109] Hopefully these discussions, in turn, will contribute at least indirectly to the central issue of eschatology and scientific cosmology, towards which our focus on divine action and quantum physics has slowly but inexorably led.

Acknowledgment. I wish to thank Nancey Murphy, John Polkinghorne, and Kirk Wegter-McNelly for their helpful comments on this essay, and all the participants for a most enjoyable conference.

Rapids, Mich.: Eerdmans, 1998), 3:chap. 15, sec. 5, 645. See also Pannenberg's comments on Barth's response to eighteenth-century theodicies.

[107] Ibid., 1:382ff; idem, "The Doctrine of Creation and Modern Science," in *Cosmos as Creation: Theology and Science in Consonance*, Ted Peters, ed. (Nashville: Abingdon Press, 1989), esp. 162–7; idem, *Toward a Theology of Nature: Essays on Science and Faith*, Ted Peters, ed. (Louisville, Ky.: Westminster/John Knox Press, 1993), esp. chaps. 5, 6, 7.

[108] See for example John Polkinghorne, "Pannenberg's Engagement with the Natural Sciences," *Zygon* 34.1 (March 1999): 151–8.

[109] Ernest Simmons has developed this approach in relation to divine kenosis. See his recent article, "Toward a Kenotic Pneumatology: Quantum Field Theory and the Theology of the Cross," *CTNS Bulletin* 19.2 (Spring 1999): 11–6.

Appendix—Directions for Future Research on Quantum Physics, Its Philosophical Implications, and Their Relevance for Divine Action

Are there ways to move out of the Copenhagen interpretation and get a broader perspective on the problem facing *any* philosophical interpretation of quantum physics, a perspective worth pursuing for its theological relevance? Here I will tentatively suggest three areas that seem worth pursuing. I propose we attempt to find a way to gain insights from each of the leading interpretations in a broader program of research, acknowledging that these interpretations and insights often conflict with each other, yet seeking ways to bring them into a larger picture so that each can contribute to the ongoing interaction with theology.

A1 Architecture of Philosophical Issues

My first suggestion is to sort out which features are general enough to be found in most or perhaps all interpretations. Superposition and nonlocality are likely candidates. Raymond Chiao in this volume distinguishes between quantum nonlocality as displayed in the Aharonov-Bohm effect, the Tunnel effect, and the Einstein-Podolsky-Rosen effect. Each stems from the superposition principle (i.e., quantum interference), but the first two involve single-particle interference, while the third involves an "entangled state" between two particles. Jim Cushing uses locality and separability interchangeably in his discussion of Jarrett locality, Jarrett completeness, and Howard's factorizability.[110] He suggests in this volume that either objective reality or locality must be given up. Cushing's option is for nonlocality, and he stresses the distinction between separability and locality, taking "relational holism" seriously.

Other issues and features seem to arise in closer association with individual interpretations of quantum physics. For example, the measurement problem and ontological indeterminism are integral to the standard Copenhagen interpretation, but there are a variety of attempts to resolve the measurement problem. Some work within the perspective of the Copenhagen interpretation broadly conceived, including those who propose to modify the Schrödinger equation, either through the introduction of nonlinear terms or through the inclusion of stochastic factors, and those attempt to understand consciousness (the observer's mind) as bringing about the collapse of the wavefunction. Others seek to resolve the measurement problem by interpreting quantum physics in such a way that it simply does not arise. The most notable of these are Bohm's quantum potential interpretation and the branching of reality in the many-worlds and many-minds approaches discussed by Butterfield in this volume.

This suggests that we can begin to lay out what I will call the "architecture of philosophical problems" in quantum physics. A first sketch might be as follows:

A. generic features:
- i. superposition (interference)
- ii. nonlocality, including:
  - a. single-particle nonlocality (the Aharonov-Bohm effect; the Tunnel effect)
  - b. multi-particle (entanglement) nonlocality (the Einstein-Podolsky-Rosen effect)
- iii. nonseparability/relational holism

B. interpretation-specific features:
- i. Copenhagen: the measurement problem
  - a. accept the measurement problem and ontological indeterminism
  - b. overcome the measurement problem by:
    - 1. modifying the Schrödinger equation with
      - i. nonlinear terms
      - ii. stochastic terms
    - 2. von Neumann: introducing consciousness
- ii. Bohm: the quantum potential/nonclassical determinism
- iii. Everett: many worlds, many minds

---

[110] Cushing, *Quantum Mechanics*, 56–60; here he also discusses nonlocality in Bohm's theory.

The task will then be to see how the interpretation-specific features give particular expression to the generic features as we study the relation between each interpretation of quantum physics, philosophy, and finally, theology.

A2 Implications of Bell's Theorem Independent of Quantum Theory

A second strategy is to unpack the implications of the actual data from quantum processes without getting entangled (!) in quantum formalism and its competing interpretations. One source is the data showing the violation of Bell's theorem, which may give us more direct access to these more general features of the quantum world, but without requiring us to get at them directly through the lens of quantum mechanics. These data will have to be accounted for by any future theory that replaces quantum physics. Thus any insights they give us regarding divine action will be less vulnerable to the problems of multiple interpretability and historical relativism.

For the purposes of the interaction with theology, this view of Bell's theorem suggests that we needn't wait for the philosophical controversies to be "settled" regarding quantum physics before engaging with it. We might use the "leverage" of Bell's results to pursue the conversation and allow some of the philosophical uncertainties to play themselves out on their own.[111]

A3 Comparison of the Meaning of Nonlocality and (In)determinism in Bohm's Formulation and the Copenhagen Interpretation

A final suggestion for further work is to compare the approach to quantum physics by Bohm and Bohr in order to uncover a clearer understanding of the similarities and differences in the meanings they give to such key terms as (in)determinism and nonlocality and the resulting hints about quantum ontology. There are several reasons for such a comparison. First, the mathematical route from the Schrödinger structure of Copenhagen to the semi-classical context of Bohm is so straightforward that one can almost view them as formally equivalent, though the ontologies differ remarkably. Thus to compare quantum (in)determinism and nonlocality to classical determinism and locality, we will first move from Schrödinger to Bohm (who is close to Newton), and then from Newton to as close to Schrodinger as possible. Second, comparison helps us avoid the tacit assumption that Bohmian determinism is more like the classical worldview than is Copenhagen indeterminism. Clearly Bohm does limit the fundamental role of indeterminism in the Copenhagen approach by offering a deterministic alternative. However, as Jim Cushing and others[112] have stressed, Bohmian determinism is highly nonclassical in several important ways, making an explicit comparison with Newtonian determinism crucial. Third, the meaning of Bohmian nonlocality differs from that of Copenhagen. Thus a comparison of Bohm and Bohr should help clarify just what the Bohmian "deterministic alternative" really involves, what one means by Copenhagen indeterminism, and how nonlocality come to play in both approaches. Let us begin with the mathematical route from the Copenhagen formulation to that of Bohm, and compare the results with classical mechanics.

1. From Schrödinger to Bohm. As is well known,[113] we can start with the Schrödinger equation

---

[111] An instructive example is the Mermin machine, which shows how quantum data challenge local realism without explicitly invoking quantum mechanics; N. David Mermin, "Is the Moon There When Nobody Looks?", *Physics Today*, 38 (April 1985): 38. See also idem, "Can You Help Your Team Tonight by Watching on TV? More Experimental Metaphysics from Einstein, Podolsky, and Rosen," in *Philosophical Consequences of Quantum Theory*, Cushing and McMullin, eds.

[112] Cushing, *Quantum Mechanics*; see also George Greenstein and Arthur G. Zajonc, eds., *The Quantum Challenge: Modern Research on the Foundations of Quantum Mechanics* (Boston: Jones and Bartlett Publishers, 1997); John Polkinghorne, in this volume.

[113] See for example Cushing, *Quantum Mechanics*, Appendix 1.1, 60–63.

$$-\left(\frac{\hbar^2}{2m}\right)\nabla^2\psi + V\psi = i\hbar\frac{\partial\psi}{\partial t} \tag{57}$$

and show that an additional term, which Bohm called the quantum potential U

$$U = -\left(\frac{\hbar^2}{2m}\right)\left(\frac{\nabla^2 R}{R}\right), \tag{58}$$

is added to Newton's law to give a modified form of classical mechanics:

$$\frac{d\boldsymbol{p}}{dt} = -\nabla(V + U). \tag{59}$$

This move requires us to change ontologies from Copenhagen to Newton, with the crucial addition of a de Broglie-like pilot wave ψ that governs the particle's motion. Bohm considers a particle of mass m following a well-defined trajectory with position $\boldsymbol{x}$ and momentum $\boldsymbol{p} = m\boldsymbol{v}$. Here $\boldsymbol{x}$ and $\boldsymbol{p}$ are the "hidden variables" in Bohm's account, and our knowledge of them is statistical in the classical sense: the probability $P(\boldsymbol{x}, t)$ of finding the particle at $\boldsymbol{x}$ and t is given by $P = |\psi|^2$. We assume that P is conserved. For the purposes of calculation, it is convenient to represent ψ as $Re^{iS/\hbar}$ where $R(\boldsymbol{x}, t)$ and $S(\boldsymbol{x}, t)$ are real functions. In a crucial move, Bohm defines the momentum $\boldsymbol{p}$ in terms of the partial phase S through the "guidance condition" $\boldsymbol{p} = \nabla S$.

2. *From Newton almost to Schrödinger.* We could also reverse the process and see how close we can get to the Schrödinger equation using classical mechanics as our starting point.

Thus, if we start with Newton's second law:

$$\frac{d\boldsymbol{p}}{dt} = -\nabla V \tag{4}$$

and follow Bohm in setting $\psi = Re^{iS/\hbar}$, $\boldsymbol{p} = m\boldsymbol{v} = \nabla S$, $P = |\psi|^2$, and in assuming that probability P is conserved, we will obtain:

$$-\left(\frac{\hbar^2}{2m}\right)\left[\left(\frac{i}{\hbar}\right)\left(R\nabla^2 S + 2\nabla R\cdot\nabla S\right) - \left(\frac{R}{\hbar^2}\right)\left(\nabla S\right)^2\right] + VR = -R\frac{\partial S}{\partial t} + i\hbar\frac{\partial R}{\partial t}. \tag{5}$$

This is a *truncated* version of the Schrödinger equation. When written in terms of R and S, the full Schrödinger equation takes the following form:

$$-\left(\frac{\hbar^2}{2m}\right)\left[\nabla^2 R + \left(\frac{i}{\hbar}\right)\left(R\nabla^2 S + 2\nabla R\cdot\nabla S\right) - \left(\frac{R}{\hbar^2}\right)\left(\nabla S\right)^2\right] + VR = -R\frac{\partial S}{\partial t} + i\hbar\frac{\partial R}{\partial t}. \tag{6}$$

We can summarize our results as follows:

- Schrödinger equation (1) → Bohm's modified classical mechanics (3)
- Standard classical mechanics (4) → truncated Schrödinger equation (5)

In one sense this result is completely obvious: if we know that the Schrödinger equation leads to the addition of the quantum potential U, then leaving it out of the Newtonian picture means it will be "subtracted" from the Schrödinger picture (6) leaving us equation (5). To emphasize this point, we can rewrite (5) as

$$-\left(\frac{\hbar^2}{2m}\right)\nabla^2\psi + V\psi - \left[-\left(\frac{\hbar^2}{2m}\right)\left(\frac{\nabla^2 R}{R}\right)\psi\right] = i\hbar\frac{\partial\psi}{\partial t}. \tag{7}$$

This is clearly the Schrödinger equation minus the quantum potential U.

In another sense, the result is intriguing, for it explicitly shows how the sources of the nonlocal and nonmechanical features associated strictly with the quantum potential U in the context of Bohm's interpretation carry over and are placed within the context of the Schrödinger equation when one moves to the Copenhagen interpretation. In particular, the quantum potential U, which acts as a *separate* factor in Bohm's "$U + V$" picture, is built into the $\nabla^2\psi$ term in the Schrödinger picture. In essence, if the Newtonian picture were correct, we could get a Schrödinger-like equation and still have classical physics, but the equation would not be a complete wave equation, since the $\nabla^2\psi$ term would be incomplete: it is missing the crucial term, $\nabla^2 R$.

I propose we view this result in the following way. The Bohmian formulation, with its delineation between and linear addition of $V + U$, allows us to separate out quantum (nonlocal and nonmechanical) aspects from the classical (local and mechanical) aspects of the governing equation $d\mathbf{p}/dt = -\nabla(V + U)$; all of the uniquely quantum aspects of this equation are carried in one term, ∇U. The Schrödinger formulation, on the other hand, seamlessly combines the term U with the rest of the mathematical "machinery" available from the classical picture to produce one term, $\nabla^2\psi$. In this sense all of the nonlocal and nonmechanical aspects of U are hidden in and mingled with the classical aspects to yield the term $\nabla^2\psi$.

This allows us to make a further point. One could ask how many of the "quantum" features of the Copenhagen picture are carried by the wavefunction ψ and how many by the Schrödinger wave equation. The answer, regarding ψ, is straightforward: features such as superposition, entanglement, quantum statistics, etc. We know this answer immediately because we explicitly and intentionally build them into the wavefunction. But which quantum features does that leave out? Now, from a comparison with the Bohmian picture we can conclude that the Schrödinger equation carries all those quantum aspects that we attribute to the quantum potential U; moreover, they are carried precisely within the $\nabla^2\psi$ term. This is all the more interesting since the usual motivation for the Schrödinger equation in physics textbooks, and particularly for the $\nabla^2\psi$ term, is so straightforward.[114] Thus it is all the more surprising to see how much of the overall quantum picture arises from these seemingly minimal assumptions.

In summary, then, the Bohmian formulation is highly nonclassical, involving nonlocal and nonmechanical features simply not found in the Newtonian picture. Bohm does not offer a return to classical determinism in comparison with the quantum indeterminism of Bohr. Instead both Copenhagen quantum indeterminism and Bohmian quantum determinism are highly nonclassical. The use of either view in a discussion of divine action thus requires a thorough rethinking of the conversation compared to its traditional context.

Finally, we may analyze the significance of the quantum potential's contribution to the "nonclassical" aspects of Bohm's formulation. Here I will follow the illuminating discussion by Greenstein and Zajonc.[115]

Consider the double-slit experiment from Bohm's perspective. The trajectory of each particle is influenced both by the slit through which it passes (note that it passes through only *one* slit!) and by the quantum potential U. The quantum potential, in turn, depends on the "pilot wave" ψ, which is conditioned by the entire experimental arrangement, including the fact that there are *two* slits. U has broad plateaus cut by "deep valleys... where U changes quickly, leading to a strong quantum force [which] guides the particles into the interference maxima and away from the minima."[116] Now, close either slit and the wavefunction—and thus the quantum potential—changes instantaneously, causing a force that alters the particles motion. But the nonlocality of U is even more complex that this.

The quantum potential does not fall off with distance, because U depends on R, which appears in the numerator and denominator. In this sense, the quantum potential U brings the

---

[114] For example, in light of the Davisson-Germer experiment, we represent the state function of a particle as a de Broglie wavefunction and see that its second derivation $\boldsymbol{r}$ varies as its first derivation in t.

[115] Greenstein and Zajonc, *The Quantum Challenge*, chap. 6.

[116] Ibid., 145 and figures 6–11, 6–12.

influence of the whole system to bear on each part with an intensity and immediacy that we do not see with the classical potential V, even though the influence of either U or V can come from arbitrary distances.

Consider also a many-particle problem: Here ψ is a function of the coordinates of all n particles $\psi(x_1, x_2, \ldots, x_n, t)$. The force on the ith particle is a function of the gradient of the total potential $V + U$ at the particle's coordinates, x_i, making the problem seem like ordinary mechanics. But the force on each particle due to U depends on the position of *all* the particles in the system through the factor R because $U = -\left(h^2/2mR\right)\left(\nabla_1^2 + \nabla_2^2 + \ldots + \nabla_n^2\right)R$. Thus it depends on the coordinates of all of the particles, both through the ∇^2 terms and through the factor $R = R(x_1, x_2, \ldots, x_n)$, and not just on the coordinates of the particle at x_i. As Cushing stresses, "the many-body quantum potential entangles the motion of the various particles."[117] In essence, the force is a function of a *local* gradient on a *nonlocal* potential U as well as on a *local* potential V. It thus combines both classical and nonclassical features in producing the net acceleration of each individual particle.

Finally, quantum nonlocality is highly *nonmechanical* in the sense that the quantum potential U depends not only on the positions of the other particles, but also on their wavefunctions and thus on the state of the entire system. As Greenstein and Zajonc note, Bohm's interpretation "... goes beyond simple nonlocality, and calls upon us to see the world as an undivided whole. Even in a mechanical world of parts, the interactions between the parts could, in principle, be nonlocal but still mechanical. Not so in the quantum universe."[118]

---

[117] An important exception arises with independent systems in which the wavefunction factors out and the quantum potential reduces to a linear sum of terms for each system. See Cushing, *Quantum Mechanics*, 62–3.

[118] Greenstein and Zajonc, *The Quantum Challenge*, 148. In a helpful example, Greenstein and Zajonc show how even in Bohm's case the motion of electrons in an atom is not mechanical in the way the motion of the planets is.

Glossary

An asterisk (*) preceding a word indicates a relevant entry under that (or a closely similar) heading.

Aharonov-Bohm effect. A quantum *interference effect (involving a *phase shift) arising from a particle encircling a region of spatially confined magnetic flux, despite the fact that the particle does not penetrate the flux region.

Anti-correlations. Consistently opposite behaviors of two particles observed in *Einstein-Podolsky-Rosen experiments.

Bell's theorem and inequality. A realistic local theory—that is, one in which all observables at all times have an observer-independent existence ("objective reality"), and for which there are no instantaneous influences or connections between spatially separated regions ("locality")—leads to certain mathematical constraints on the possible results of *measurement applied to a two-particle system (the Bell inequality). Since this equality appears to be violated experimentally, any empirically adequate theory (more specifically, *quantum theory) must violate either objective reality or locality, or both.

Bohmian quantum mechanics. A version of *quantum theory in which *probabilities arise from unavoidable ignorance of certain relevant factors ("hidden variables"). The theory is nonlocal, completely deterministic, and empirically indistinguishable from standard *quantum mechanics. Its hidden variables are the positions of the particles, which follow trajectories at all times.

Causality. The principle of causality states that each event is the effect of a specific cause. Some philosophers understood the cause to be something after which the corresponding effect follows necessarily. This seemed in agreement with *classical physics, in which the initial conditions of a given process uniquely determine the process itself. In the light of *quantum mechanics, in which processes unfold indeterministically, the principle of causality has to be generalized to incorporate the situation in which a given cause may be followed by many alternative effects, each of them with a certain *probability.

Chance. Common usage is based on intuition rather than on scientific definition. We say that an event happens by chance if it is unpredictable, or if its occurrence is improbable, or if it is difficult to establish its cause. The closest scientific term is "random event." In the calculus of *probability an event is called random if one can ascribe to it a number (lying between 0 and 1) representing the *probability of its occurrence.

Chaos. In dynamics, a type of instability in which initially close trajectories rapidly diverge, thereby frustrating long-term predictability because of great sensitivity to the fine detail of initial circumstances.

Classical Logic. *See* quantum logic.

Classical physics. Physics of a Newtonian type, in which the properties of *realism, *causality, *determinism, and *locality hold.

Collapse of the wavefunction. According to the standard version of *quantum theory, a state that is a *superposition of several different states, each corresponding to a different result of measuring a particular observable, on the actual *measurement of that observable will collapse into the single state corresponding to the value uniquely obtained on that occasion. This collapse is not a consequence of the *Schrödinger equation, but must be imposed on the theory as an extra condition. *See also* wavefunction.

Complementarity. An idea, much emphasized by Niels Bohr, that *quantum theory involves mutually exclusive modes of description, each complete in itself and complementary to the other mode, as in wave/particle duality, or in the possibility of framing accounts either in terms of particle positions or in terms of particle momenta.

Consistent histories. An interpretation of *quantum theory based on sets of histories (time-sequences of propositions derived quantum theoretically), where the set is governed by a mathematical condition (consistency or *decoherence) that ensures that the *probabilities of different histories are compatible with a classical account.

Copenhagen interpretation. A family of interpretations of *quantum theory, originally deriving from ideas due to Niels Bohr, which emphasizes quantum indeterminacy and the significance of *complementarity, and which lays stress on the role of classical apparatus in defining the nature of *measurement and in determining its actual outcome.

Correlations. *See* anti-correlations.

Counterfactual reasoning. Reasoning about states of affairs in an alternative possible world in which events take a course different from that actually occurring in this world.

De Broglie wave. The wave associated with a particle, such as an electron. The wavelength of the particle is inversely related to its momentum.

Decoherence. In wave physics (such as *quantum mechanics), the destruction of *interference brought about by uncontrolled external disturbances (such as that due to some form of background radiation). In *quantum theory, this environmental effect can lead to the rapid generation of almost-classical patterns of *probability.

Density matrix. The representation of a *quantum state as a matrix (a square array of numbers). Any *quantum state can be represented in this way but, since a pure *quantum state can also be represented as a vector, the density matrix is mostly associated with representing a mixture of states, that is to say, representations of situations in which both quantum and classical *probabilities are present.

Determinism, physical. The physical state of the world at any given time fixes its state at all subsequent times.

Einstein-Podolsky-Rosen (EPR) effect. An effect in which two particles emitted from a common source exhibit correlations upon coincidence detection. *Quantum theory predicts that these correlations can be greater than would be expected on the basis of local reality. *See also* Bell's theorem.

Emergence. A level is said to emerge from a lower level when the entities at the higher level cannot be fully understood in terms of concepts referring solely to the lower level. In this sense, chemistry may be said to emerge from physics.

Entanglement, quantum. The phenomenon whereby the *quantum state of a composite system cannot be decomposed into a single product of states referring to individual constituents. The latter are, therefore, inextricably entangled with each other (as in the *Einstein-Podolsky-Rosen effect).

EPR effect. *See* Einstein-Podolsky-Rosen effect.

Gauge invariance. A symmetry principle in quantum physics that states that the phase of the *wavefunction can be changed arbitrarily without inducing any observable physical effect. The stronger principle of *local gauge invariance* requires that these variations in phase can be different at different points. This requirement necessitates the introduction of some kind of compensating vector potential, as in the case of electromagnetism.

***Gedanken* experiment.** A thought experiment that tests the applicability of basic concepts and theories.

Heisenberg's uncertainty principle. *See* uncertainty principle.

Hidden variables theory. *See* Bell's theorem and inequality, Bohmian quantum mechanics.

Holism. The properties of a complex system cannot be understood purely in terms of the properties of its component parts.

Indeterminism. A given past is consistent with a variety of possible futures.

Indistinguishability. In *classical physics, identical particles (such as two electrons) are nevertheless distinguishable because their trajectories can be followed throughout their motion, thereby maintaining the definition of their separate identities. In *quantum theory there are no observable trajectories. In consequence quantum identical particles are also indistinguishable. This lead to *quantum statistics.

Interference. A phenomenon in which the combined effects of superposed waves reinforce or cancel each, according to whether the waves are in phase (crest coinciding with crest) or out of phase (crest coinciding with trough). This leads to alternating zones of maximal and minimal intensity, as in the double slit experiment.

Invariance. *See* gauge invariance, Lorentz invariance.

Local gauge invariance. *See* gauge invariance.

Locality. *See* Bell's theorem and inequality.

Logic. *See* quantum logic.

Lorentz invariance. In special relativity, the laws of physics take the same form when viewed from different inertial reference frames. The corresponding kinematic variables are related by transformations discovered by Lorentz.

Measurement. The experimental determination of the value of an observable quantity.

Measurement problem. In *classical physics, *measurement is unproblematically the observation of what is actually the case. It will always yield the same result for the same variable measured in the same state. In quantum physics by contrast, *measurement may yield a variety of possible outcomes, differing from each other on different occasions of *measurement applied to the same state. Only the *probability of obtaining a particular result can be calculated in the theory. The essence of the measurement problem is to understand why a *specific* result obtains on a specific occasion.

Nonlocality vs. nonseparability. Nonlocality: physical effects can be transmitted from one location to another faster than the speed of light. Nonseparability: the component parts of a composite system do not have individual properties independent of the properties of the whole system.

Phase. *See* wavefunction.

Pilot-wave interpretation. Pioneered by *de Broglie and *Bohm, this interpretation of *quantum theory proposes that some physical quantity is "preferred," in that for all quantum systems it has a definite value at all times. This value evolves deterministically, in accordance with a "guidance equation," rather like a pilot guiding a ship.

Probability. The probability of an event A belonging to a set of events E is a number p, lying between 0 and 1, ascribed to A and satisfying the condition that the sum of all such numbers ascribed to all the events belonging to E adds up to 1. According to the frequency interpretation of probability, p will be the ratio of the number of times A occurs to the total number of events. *See also* chance.

Quantum logic. Classical logic depends upon the principle of the excluded middle: there is no possibility intermediate between "A" and "not-A." In *quantum theory, however, the *superposition principle permits the mixing of states with "A" and "not-A," producing just such a middle term undreamed of by Aristotle. Consequently, *quantum theory calls for the employment of a different kind of logic.

Quantum mechanics. The fundamental physical laws that rule the microworld of molecules and smaller, where matter is described by waves that interfere and resonate, and precise simultaneous *measurement of position and velocity is impossible.

Quantum state. As in any physical theory, the state of a system in *quantum theory is defined by the theory's specification of the system's properties. The main contrast with states in *classical physics is that in *quantum theory even the most complete specification is probabilistic: for each physical quantity, the state specifies *probability distributions for the quantity's various possible values. If the state is specified as precisely as possible, it is called a *pure state*; otherwise it is called a *mixed state*. The properties of the latter will involve a combination of classical and quantum *probabilities. *See also* density matrix.

Quantum statistics. Statistics describes the behavior of collections (ensembles) of particles. Because in *quantum theory identical particles are *indistinguishable, interchanging two identical particles in the description of a quantum system must lead to no change of physical properties. It turns out that this implies that the *wavefunction of the total system must either be symmetrical under interchange of identical particles (an option called *Bose-Einstein statistics*) or antisymmetrical (*Fermi-Dirac statistics*). Particles obeying Fermi-Dirac statistics (fermions) satisfy the *exclusion principle*: there cannot be more than one particle in each state. Particles obeying Bose-Einstein statistics (bosons), on the other hand, display a tendency to favor being in the same state (Bose condensation).

Quantum theory. *See* quantum mechanics.

Realism. *See* Bell's theorem and inequality.

Schrödinger's cat. A hypothetical experiment described by Schrödinger in which a cat is enclosed in a box containing a lethal device triggered by a radioactive atom which has a 50% chance of decaying in the observation period. Applying *quantum theory to the entire system results in the apparently absurd conclusion that when the box is opened the cat is found in a *superposition of the state in which it is alive and the state in which it is dead.

Schrödinger's equation. A differential equation that determines the (smooth) variation with time of the *wavefunction.

Semi-classical. The borderland between *quantum mechanics and its classical (Newtonian) approximation. Mathematical complexities bedevil and enrich the limiting process in which one set of laws reduces to the other.

Singularity. A place where a mathematical quantity or geometric structure is ill-defined, such as the focus of a lens (where the intensity of the light rays is infinite), or time at the north pole.

Spin. An intrinsic angular momentum associated with particles in *quantum theory.

Statistics. *See* quantum statistics.

Stern-Gerlach apparatus. An apparatus whose inhomogeneous magnetic field deflects an atomic beam in directions that depend on the *spin state of the atoms.

Superposition principle. In *classical physics a particle must be in one of the mutually exclusive states "here" or "not-here." *Quantum theory permits the mixing (superposition) of states that classically are immiscible, allowing for instance states that are a combination of "here" and "not-here."

Tunnel effect. A barrier penetration effect in which a quantum particle traverses a classically forbidden barrier, for example passing through a region that classically it would not have sufficient energy to enter.

Uncertainty principle. In *quantum theory, variables can be grouped in pairs (position/momentum, time/energy) corresponding to the fact that both members of the pair cannot simultaneously be measured with unlimited accuracy. For example, the more certainly a particle's position is known, the more uncertain must be knowledge of its momentum.

Wavefunction. A particular mathematical representation of a *quantum state. The wavefunction (conventionally denoted by ψ) is frequently expressed as a function of position (x). It takes values that are complex numbers (involving i, the square root of -1). A positive quantity called the square of the modulus of the wavefunction then gives the *probability of finding the particle at x. Another quantity, relating to the balance between the real and imaginary parts of ψ, is the *phase of the wavefunction. (If ψ is written in the form $re^{i\theta}$, the *probability and *phase are r^2 and θ, respectively.) *See also* Schrödinger's equation.

CONTRIBUTORS AND CONFERENCE PARTICIPANTS

Michael Berry, Royal Society Research Professor, Department of Physics, University of Bristol, England.

Jeremy Butterfield, Fellow, All Souls College, Oxford University, England.

Raymond Chiao, Professor of Physics, University of California, Berkeley, California, USA.

Chris Clarke, Visiting Professor, Faculty of Mathematical Studies, University of Southampton, England.

Philip Clayton, Professor and Chair of the Philosophy Department, California State University (Sonoma), USA.

George V. Coyne, S.J., Director, Vatican Observatory, Vatican City State, Italy.

James T. Cushing, Professor of Physics and of Philosophy, University of Notre Dame, Notre Dame, Indiana, USA.

George F.R. Ellis, Professor of Applied Mathematics, University of Cape Town, Rondebosch, South Africa.

Michael Heller, Professor of Philosophy, Pontifical Academy of Theology, Cracow, Poland.

John Lucas, Fellow, Merton College, Oxford University, England.

Ernan McMullin, O'Hara Chair Emeritus of Philosophy, and Director Emeritus, Program of History and Philosophy of Science, University of Notre Dame, Notre Dame, Indiana, USA.

Nancey Murphy, Professor of Christian Philosophy, Fuller Theological Seminary, Pasadena, California, USA.

Ted Peters, Professor of Systematic Theology, Pacific Lutheran Theological Seminary and the Graduate Theological Union, and Program Director of the CTNS Science and Religion Course Program, Berkeley, California, USA.

John Polkinghorne, Past President and now Fellow, Queens' College, Cambridge, and Canon Theologian of Liverpool, England.

Michael L.G. Redhead, Centennial Professor of Philosophy, The London School of Economics and Political Science, Centre for Philosophy of Natural and Social Science, London, England.

Robert John Russell, Professor of Theology and Science in Residence, Graduate Theological Union, and Founder and Director of the Center for Theology and the Natural Sciences, Berkeley, California, USA.

Abner Shimony, Emeritus Professor of Physics and Philosophy, Boston University, Boston, Massachsuetts, USA.

William R. Stoeger, S.J., Staff Astrophysicist, Vatican Observatory Research Group, and Adjunct Associate Professor of Astronomy, Steward Observatory, The University of Arizona, Tucson, Arizona, USA.

Owen C. Thomas, Professor of Theology Emeritus, Episcopal Divinity School, Cambridge, Massachusetts, USA, currently adjunct professor at the Graduate Theological Union, Berkeley, California, USA.

Thomas F. Tracy, Phillips Professor of Religion, Department of Philosophy and Religion, Bates College, Lewiston, Maine, USA.

Kirk Wegter-McNelly, doctoral candidate, Graduate Theological Union, Berkeley, California, USA.

Wesley J. Wildman, Associate Professor of Theology and Ethics, Boston University, Boston, Massachusetts, USA.

Carl York, retired Professor of Physics at UCLA and Chair of the Board of Directors, Center for Theology and the Natural Sciences, Berkeley, California, USA.

Name Index

Subject Index

VATICAN PRESS